PETROGRAPHY

An Introduction to the Study of Rocks in Thin Sections

Second Edition

Howel Williams
Late of the University of California, Berkeley

Francis J. Turner
University of California, Berkeley

Charles M. Gilbert
University of California, Berkeley

W. H. Freeman and Company
San Francisco

Project Editor: Judith Wilson

Copy Editor: Amy Einsohn

Designer: Gary Head

Production Coordinator: Bill Murdock

Illustration Coordinator: Richard Quiñones

Artist: Georg Klatt

Compositor: Graphic Typesetting Service

Printer and Binder: The Maple-Vail Book Manufacturing Group

Library of Congress Cataloging in Publication Data

Williams, Howel, 1898-
  Petrography : an introduction to the study of rocks in thin sections.

  Includes bibliographies and index.
  1. Petrology. 2. Thin sections (Geology) I. Turner, Francis J.   II. Gilbert, Charles M., 1910- .
  III. Title.
  QE434.W73 1982   552   82-5072
  ISBN 0-7167-1376-4   AACR2

Printed in the United States of America

1 2 3 4 5 6 7 8 9 0   MP   0 8 9 8 7 6 5 4 3 2

Contents

PART TWO

SEDIMENTARY ROCKS

PART THREE

METAMORPHIC ROCKS

Detailed Contents

PART TWO

SEDIMENTARY ROCKS

PART THREE

METAMORPHIC ROCKS

Preface

Our aims in this book are much the same as they were in the first edition: We are concerned primarily with the description of common rocks as they appear in thin section beneath the petrographic microscope. We have assumed that our readers have already learned how to identify rocks within broad limits by use of a hand lens; indeed, it is most desirable that observers have available the hand specimens from which the thin sections have been cut. Readers must already be acquainted with the principles of optical mineralogy, for this is the very basis of petrographic technique. We devote considerable, but, we hope, not undue attention to two unifying themes, each treated at an elementary level, both of which contribute geological significance to purely descriptive study. First, the relevant aspects of the modes of origin of igneous, sedimentary, and metamorphic rocks are treated in a manner consistent with modern concepts of petrogenesis.* This topic requires an elementary knowledge of chemistry and, preferably, an inkling of the nature of the thermodynamic assessment of phase equilibria and the possible interactions between the mineral components of rocks, the concrete evidence of which exists in textural detail observed through the microscope. As an aid to thinking in thermodynamic terms, we offer Appendix A, written in the simplest possible terms and illustrated with plots of specific data

---

*These concepts are elaborated in standard petrology texts.

for albite and for water. The second thematic topic is the field occurrence and global distribution of principal classes of common rocks, with special emphasis on the context of tectonics and time.

With proper selection of rock types, *Petrography* provides the basis for a one-semester introductory course involving six hours of laboratory work and such explanatory lectures as meet the requirements of any specific class. This book will also furnish the descriptive background for more advanced courses in any of the three main branches of petrology. Of course, we hope that it will continue its usefulness as a reference for professional work in most fields of geology.

We completely omit discussion of some of the most fruitful techniques practiced today in the laboratory investigation of rocks, for example, microscopic oil-immersion, x rays, electron-microprobe, and electron-microphotography. The main difference between these complementary methods is the scale of observation. However, their goals are the same: understanding the mineralogical make-up of rocks, the particle-to-particle textural relations of component phases on every scale (in both space and time), and the evaluation of mineral and rock chemistry within the limits of each approach.

No new rock names have been coined, though a few, pertaining to rocks that have recently assumed new significance in geology, have been added to the list appearing in the first edition. On the contrary, we have striven to limit the number of terms, especially in igneous petrology, a subject area in which terminology appears to have proliferated needlessly. Alternate terms that we consider superfluous have been relegated to footnotes marked:°. Each of these terms has a page reference in the index; in this way, students will always be able to find information on a particular rock, despite vagaries of terminology in the technical literature.

Some rock types, of course, are much more abundant on a global scale and some are of greater genetic significance than others; among igneous rocks, basalts and granites are notable examples. However, description of rare but significant rocks may demand at least as much text as is devoted to more voluminous or more common rocks. Chapters treating, respectively, siliceous plutonic rocks (including granites) and silica-deficient feldspathoidal plutonic rocks (including nepheline syenites) are of comparable length. Greenschists, widespread in both space and time, and blueschists, whose distribution in both respects is much more limited, receive comparable discussion. In brief, neither the abundance nor the geological significance of a rock is to be measured by the pages devoted to its description.

Many students and teachers would prefer to see rock types neatly arranged in tables and charts, since such graphic devices aid memory and tend to satisfy the desire for regularity and order in nature (the very impulse that leads scientists to stipulate natural laws). A few such aids are included here; but we feel that "pigeonhole classifications," however satisfying to the order-minded pedagogue or student, may convey a false impression of exactitude where none exists. Our aim is to stress the gradational relationships among individual rocks within any natural or artificially designated group of rocks. This topic is elaborated especially in the chapters on igneous petrography.

The thirty years since the first edition have seen spectacular advances in all branches of petrology and in relevant peripheral aspects of geology: theoretical and experimental geochemistry, chemical composition and spatial relationships of phases identified and analyzed on the submicroscopic scale, geochronology, and plate tectonics. Of course, the composition and texture of rocks have not changed within this period, but they are seen in a different and more revealing perspective than they were three decades ago. In response to such changes, we found it necessary to extensively revise the content and arrangement of the original text, while retaining the approach and general coverage. Chapters 1, 2, 9, 14, and 16 where much attention is devoted to petrogenesis and the tectonic backgrounds of common rocks, have been completely revised; many parts were entirely rewritten. More attention is now paid to certain rocks whose geological significance has become much more fully understood during these last thirty years, including: basalts, carbonatites, limestones, the three lithologic ingredients of ophiolites (serpentinites, spilites, cherts), blueschists, rocks affected regionally by incipient metamorphism (clay sediments, graywackes, basalts), and a group of rocks that appears to have been transported virtually unchanged from the upper mantle. Tectonic environments of specific rock groups receive somewhat greater emphasis than previously.

This revised edition, like its predecessor, has not been easy to write, and as before, we hope that it will not be difficult to read. The preparation of any introductory text presents particular difficulties, especially with respect to classification, selection of material, mode of presentation, and relevant references. Few instructors will agree on the naming and grouping of rocks, the order in which they should be taken up, or the relative emphasis to be placed on each type or class. Each instructor will be guided by his or her own predilections, the demands of the course, and the role of petrography courses in any individual geological curriculum. References are confined to a few easily accessible, classic, comprehensive works and a few specialized works (written in English) that are helpful in illustrating or elaborating specific details relevant to the petrography of some particular type or group of rocks.

Principal responsibility for writing and revising this edition was divided as follows: Part One, igneous rocks, Turner and Gilbert; Part Two, sedimentary rocks, Gilbert; Part Three, metamorphic rocks, and Appendix A, Turner. To avoid confusing the reader with regard to cross-references, the usage and petrogenic approach adopted in Parts One and Three are consistent, as far as possible, with recent petrological texts by Turner.

All petrographic microdrawings, except two (Figure 15-8 by C. O. Hutton and Figure 14-4 by R. S. Creely), are by our late coauthor Howel Williams and, except for Figure 4-1, have been retained from the previous edition. About half of them depict rocks from well-known localities within the United States. The remaining half refer to classic provinces elsewhere (especially those designated as type localites by the European founders of petrography) or to areas of special interest to one or the other of the three original authors.

We cannot adequately acknowledge the encouragement we have received over the years from so many colleagues and former students who have provided material and support in the preparation of both editions of *Petrography*. It is particularly appropriate, however, to acknowledge our special indebtedness to A. V. Cox, R. V. Dietrich, W. G. Ernst, and R. M. Gates for critical assistance in the preparation of the present edition; and to Amy Einsohn and Judith Wilson for skill and tolerance in copy editing.

Finally we, the two surviving authors, wish to pay tribute to the two men who conceived the original project, spiritedly persuaded us into coauthorship, and brought the complete project to fruition as the original edition: Howel Williams, geologist, and William H. Freeman, publisher. It was with the active encouragement of the former that this revision was undertaken. Accordingly, and in affectionate memory of a stimulating, critical, and generously forbearing colleague and friend, we dedicate this work to the late Howel Williams.

Charles M. Gilbert
Francis J. Turner
March 1982

PART ONE
IGNEOUS ROCKS

1

Magma and Igneous Crystallization

NATURE AND SOURCES OF MAGMAS

Igneous rocks are formed by cooling and solidification of *magma*—hot mobile rock matter made up wholly or in appreciable part of a liquid phase having the composition of a silicate melt. The chemistry of this melt, by analogy with chemical analyses of igneous rocks, is usually expressed by weight percentages of a dozen principal component oxides—SiO_2, Al_2O_3, FeO, MgO, CaO, Na_2O, K_2O, and others—and parts per million of a host of trace elements (Mn, Ni, Sr, Ba, Rb, U, S, Cl, F, etc.). There is an important group of minor constituents, totaling no more than a small percentage by weight, that because of their capacity to escape from the crystallizing magma can be termed (following the usage of S. J. Shand and others) the *fugitive* components—commonly loosely referred to as "volatiles"—of the magma. Chief among these are H_2O and CO_2. The actual chemical species that can be identified in the melt phase are of a different nature. They include metallic ions (Fe^{2+}, Fe^{3+}, Mg^{2+}, Na^+ . . .), $(OH)^-$, $(HCO_3)^-$, and Cl^- held within a discontinuous, fluctuating matrix of variously linked Si, Al, and O atoms analogous to embryonic and imperfect crystalline structures. At the time of emplacement within the crust or eruption on the surface, magmas commonly are to some degree heterogeneous; swimming in the melt phase are suspended crystals and even bubbles of gas generated as the melt, on release from pressure, becomes oversaturated in its fugitive components.

Now there is no doubt that most volcanic rocks are produced from magmas that were once largely liquid. But the immediate parentage of many deep-seated (plutonic) igneous bodies at the time of their emplacement is more problematical. It is doubtful, for example, whether some large bodies of granitic rock were actually emplaced in an entirely or even largely liquid condition. Opinions on this question have changed over the years with prevailing fashion of geological thinking. Thirty years ago many geologists argued that most of the great granitic batholiths of mobile belts and of the ancient crust are metasomatic in origin, the result of "granitization" of preexisting rocks. This was the era of the stimulating "granite controversy" sparked especially by the writings of H. H. Read. Other geologists have advocated models of selective melting of sediments in the lower levels of thick geosynclinal prisms and of underlying older siliceous crustal rocks. Others again place the immediate sources of granitic magmas at still deeper levels, in the upper mantle itself. But whatever their origins batholithic "granites," because of their intrusive relations and their mobility (presumably conferred by the presence of a pervasive molten phase) must be classed as igneous rocks; and in this book, as in most others today, they are so treated.

Within limits imposed by universal preponderance of SiO_2 among their oxide components, the chemical and mineralogical variety among igneous rocks is indeed wide. Yet there are two contrasting groups of rocks, each rather narrowly limited in composition, that together compose by far the greater part of the igneous component of the exposed crust. These have been referred to in general terms as granodioritic (or granitic) and basaltic: the first represented mainly by siliceous intracrustal (plutonic) rocks, the second principally by surface flows (volcanic rocks) that are much lower in silica. From this well-established generalization grew the long-prevalent concept of two corresponding worldwide types of repeatedly generated primary magma from which most other kinds of magma were presumed to be generated by various means. Of the overwhelming preponderance of these two broad types there can be no doubt; but from one province to another each type encompasses a significant compositional spectrum. Moreover there is no obvious criterion (though some based on trace-element and isotopic chemistry have been proposed) of primary status in any specific instance. From time to time the existence of primary magmas of other compositions has been proposed. Today, with rapidly increasing knowledge of rock chemistry—particularly as regards trace elements and variation in isotopic compositions of Sr, U, Pb, O, and other key indicators of magmatic sources—it is found impossible to relate all types of igneous rocks to just a few (let alone two) global primary magmas. True, there is good reason to believe that within specific magmatic provinces, limited both geographically and in time, certain types of magma have been derived directly from one or more parental types. These magmas generally predominate over their derivatives in any province; and more often than not their compositions are roughly basaltic or broadly granodioritic.

In any case the *ultimate* sources of all magmas intrusive into the crust or poured out on the surface lie in the mantle. The crust itself from the earliest times onward has been formed of materials of such an origin, and these have been repeatedly reworked in successive cycles of weathering, erosion, sedimentation, and metamorphism to produce the lithologically heterogeneous crust as we see it today. With the passage of time magmas have been regenerated again and again during fresh cycles of complete or partial melting, not only of mantle rocks but probably also of rocks of mixed origins in the deeper crust. Even during the life span of a single major volcano—perhaps covering 1 or 2 million years—fresh batches of magma seem to have been generated by episodic melting at deep levels beneath the site of eruption. Against this background the term *primary magma* is disappearing from current usage and will not be employed in this book.

The lithologic character of rocks comprising the upper mantle clearly is of prime interest to students of igneous phenomena. A multitude of seismic and gravitational data collectively require the existence of a relatively light solid crust separated by a sharp interface—the Mohorovicic discontinuity (or Moho)—from underlying denser, essentially solid rocks composing the upper or outer mantle. The depth of the Moho is about 6 km to 8 km below the ocean floors, and between 50 km and 70 km beneath the surfaces of continental cratons. The average density of rocks in the upper mantle, viewed on the large scale of seismic and gravitational observation, is 3.3 to 3.4 g cm^{-3}. Global patterns of large-scale volcanism also place limitations on the choice of possible mantle rocks; for these must certainly be capable of generating large volumes of basaltic magma (in the broad sense of the term). It is also possible, as some geochemists aver from isotopic and minor-element patterns in igneous rocks, that other and more siliceous magmas have been repeatedly generated in the upper mantle. But such propositions are based on geochemical models, necessarily of a somewhat ephemeral nature, and these are beyond the scope of petrography.

Among known igneous and metamorphic rocks sporadically exposed at the earth's surface, only a few types (all insignificant in global bulk) qualify as candidates for possible mantle materials: *peridotites* (magnesian olivine-rich rocks), especially two-pyroxene varieties (lherzolites) containing garnet; *kimberlites,* magnesian olivine-rich rocks with abundant biotite; and *eclogites,* rocks of basaltic bulk composition composed principally of the sodic diopsidic pyroxene omphacite and magnesian garnet. The petrography of each of these groups of rock will be described under the appropriate heading in Chapters 8 and 20.

Further discussion of the petrography of the upper-mantle rocks is postponed until the reader has become more familiar with rocks such as those just mentioned that seem to qualify as samples directly derived from that source (pp. 251–258).

MAGMATIC EVOLUTION

Topics of prime interest to the petrologist—and hence in the background of petrography—concern the evolution of derivative magmas from common parental magmas, and the regular patterns that are displayed in the mineralogy and chemistry of resulting lineages of associated igneous rocks. Three general mechanisms are thought to be responsible for evolutionary processes reflected in common rock associations: differentiation of homogeneous magma; assimilation, partial or complete, of solid rock by liquid magma; and mixing of magmas of separate identity. These we shall shortly consider individually.

First, however, it must be remembered that magmas, generated in the first place by partial melting of mantle or deep crustal rocks, are unlikely ever to become superheated much above the temperature of incipient crystallization. Even allowing for the effects of reduced pressure during ascent into the upper levels of the crust, most magmas are likely to be in equilibrium with crystalline phases during their history of final emplacement and subsequent cooling. So magmatic evolution depends on the normal sequence of crystal-liquid equilibria that can be expected in a silicate melt cooling to ultimate solidification. This topic is eminently suitable for laboratory investigation—starting with systems studied at atmospheric pressure. So, in order to understand some of the most familiar petrographic features of common igneous rocks in the light of experimentally established patterns of crystallization behavior and their evolutionary implications, the reader must make a brief and superficial excursion into the experimental field.

Crystal-Melt Relations in Simple
Experimentally Investigated Silicate Systems

The Experimental Approach

Almost from the moment of its inception in the first decade of this century, geochemists in the Geophysical Laboratory of the Carnegie Institution (Washington, D.C.) embarked on what was to become a most extensive experimental program devoted to crystal-melt equilibria as they develop during cooling of silicate melts of selected composition. The influence of the total output on concepts of igneous petrogenesis, with an emphasis on a particular potential mechanism of magmatic differentiation—fractional crystallization—has been immense. And the project is still being actively and successfully pursued. Many distinguished workers have made important contributions to the project; but it will always be associated with N. L. Bowen, who first synthesized the experimental data with petrographic and field data to provide a comprehensive model of igneous petrogenesis. No student of igneous rocks can afford to ignore this work; for through Bowen's exposition (1928), amplified and modified in the light of later experiments, the full significance of many of the most familiar petrographic and chemical characteristics of igneous rocks now stand revealed. This fact and the general impact of the experimental approach the reader can readily appreciate by referring to a few very simple well-documented examples.

The general procedure is to investigate the nature of crystal-melt equilibria, covering a limited range of composition set by two or three chemical components, as observed at some constant pressure—in the first instance at $P = 1$ atmosphere (1.013 bar). Geochemists use a phase-composition-temperature diagram to bring together the data accruing from many experiments, each of which was conducted on a mixture (initially a melt) of fixed composition. Comprehensive systems of two components (e.g., $NaAlSi_3O_8$–SiO_2 or $NaAlSi_3O_8$–$CaAl_2Si_2O_8$) are conventionally represented by plots of temperature against composition. For three-component systems (such as $CaAl_2Si_2O_8$–Mg_2SiO_4–SiO_2) it is customary to use a triangular compositional plot contoured for temperature. In this elementary introduction we shall treat only systems of two components; for if one allows for lowering of temperature and modification of detail, the types of crystallization behavior that they illustrate also characterize more complex multicomponent systems more closely analogous to magmas.

Theoretical Constraints

The configuration of any phase-equilibrium diagram is dictated by thermal properties* of the crystalline phases in question, their capacity for mutual solid solution, and constraints imposed by the phase rule. Application of the latter to crystal-melt equilibria investigated at some constant pressure is attended by simplifying stipulations:

1. In addition to compositional variables pertinent to solid solutions and the melt phase the only remaining "unknown" that must be specified to define the state of the system is temperature T (pressure P having already been specified as 1.013 bar).
2. Where the system consists of a single phase—for example the melt above the crystallization curve, or homogeneous crystalline plagioclase below the solidus in the system $CaAl_2Si_2O_8$–$NaAlSi_3O_8$—the number of components c likewise is 1 (since by definition c cannot exceed ϕ, the number of coexisting phases).
3. In consequence of stipulation 1 the "variance" referred to in the following discussion is smaller by 1 than the variance ω given by the phase rule

$$\omega = c + 2 - \phi$$

   which applies to equilibrium systems subject to variation in both T and P.
4. The phase rule applies to any heterogeneous system of specified chemical composition in a state of internal equilibrium. The familiar phase diagrams of petrology, however, depict equilibrium within a convenient range of composition (a "system") limited by the components selected. Any specific composition within this range defines a system subject to the constraints of the phase rule.

Eutectic Crystallization

The simplest pattern of behavior is eutectic crystallization of phases whose capacity for mutual solid solution is negligible. Examples are

---

*Numerical values of these were largely unknown in Bowen's day. In fact, some of these properties (e.g., the melting entropy of albite) Bowen computed from the configuration of the experimentally determined results half a century before they were determined calorimetrically.

albite-cristobalite, diopside-anorthite, and the ternary eutectic diopside-anorthite-forsterite. Consider now the system $CaMgSi_2O_6$–$CaAl_2Si_2O_8$. The temperature-composition diagram (Figure 1-1A)—like all others constructed for similar purposes (e.g., Figures 1-2, 1-3)—is divided into three sections, in order of decreasing temperature:

1. A field of homogeneous liquid above the *liquidus*, the curve of commencing crystallization (*UEV*).
2. A sector comprising two or more fields each representing coexisting liquid and crystalline phase.
3. A field of crystalline phases below the *solidus* curve (*LM*).

The composition and temperature of any liquid that can coexist with a crystalline phase must fall on the liquidus curve.

Any system of specific composition (e.g., X or Y) in a constant-pressure diagram is univariant. Above the liquidus its state is completely defined by fixing the only remaining variable, T. The same may be said for a system of coexisting crystalline phases below the solidus; for the composition of the system determines the relative proportions of the two end-member phases, and T is the only remaining variable to be specified. Within each field of coexisting liquid and crystals there are two unknowns, T and liquid composition; but if the value of one is arbitrarily designated (e.g., $T = T_2$) that of the other (composition Z_l) automatically follows. Of special importance is the eutectic point E. It is unique within the comprehensive system, for here three phases, forsterite, anorthite, and liquid coexist at constant temperature T_e—the lowest temperature at which any liquid within the range of Figure 1-1A is stable—in a state of invariant equilibrium. To specify coexistence of all three phases automatically defines the temperature and the composition of the liquid.

Observe now what happens when liquid X_i is cooled in a system of composition X to complete solidification. The liquid follows the cooling path X_iX_l unchanged. At X_l (temperature T_1) crystalline anorthite begins to separate and continues to do so as liquid composition changes down the cooling path X_lE. At the eutectic point E temperature and liquid composition both remain constant with simultaneous separation of crystalline anorthite and diopside until the liquid is used up. The result is a mixture of anorthite (70%) and diopside (30%) that cools without further change. Any liquid such as Y_i, on the $CaMgSi_2O_6$ side of the eutectic, follows a path (Y_lE) with crystallization of diopside, to end up with simultaneous crystallization of the two solid phases at E. All paths of

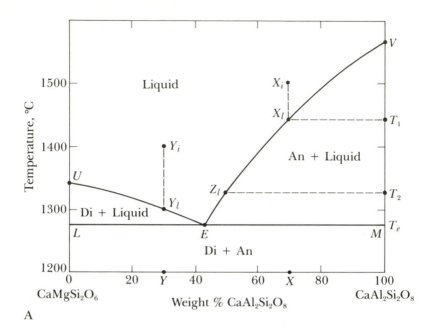

Figure 1-1. Crystal-Melt Equilibria at Atmospheric Pressure in Three Two-Component Systems Showing Eutectic Crystallization. See text for description.

equilibrium crystallization lead ultimately to the eutectic point. Note how the presence of a second component $CaMgSi_2O_6$ greatly lowers the temperature (by an amount determined by the thermodynamic properties of the two end members) at which crystalline anorthite and melt can coexist in equilibrium.

Suppose that at Z_l somewhere along the path X_lE, all previously crystallized anorthite were separated (as an early crystal fraction) from the remaining melt. The latter, following the unchanged path Z_lE would ultimately yield a crystalline fraction (late differentiate) Z with equal proportions of $CaMgSi_2O_6$ and $CaAl_2Si_2O_8$ thus notably enriched in diopside compared with the parent liquid X_i from which it was derived. No degree of fractionation of liquid in contact with crystalline anorthite can enrich it in $CaMgSi_2O_6$ beyond the eutectic composition (57% by weight). Nor can liquids in contact with crystalline diopside ever become enriched in $CaAl_2 Si_2O_8$ above 43% by weight.

Figure 1-1B illustrates eutectic crystallization of albite and pure iron olivine (fayalite) in the system $NaAlSi_3O_8$–Fe_2SiO_4. Residues of extreme fractionation of basaltic magmas lead toward an analogous eutectic, which, it should be noted, is greatly enriched in albite relative to fayalite.

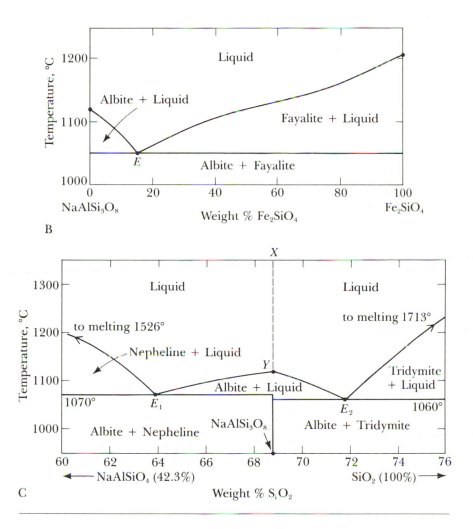

The petrologically significant part of the system NaAlSiO₄–SiO₂ is shown in Figure 1-1C. It is divided into two independent subsystems at the composition of the intermediate phase NaAlSi₃O₈ (albite). All liquids to the left of the line XY lead ultimately to the albite-nepheline eutectic E_1. All liquids to the right of XY, in contrast, must finish crystallizing at the albite-tridymite eutectic point E_2. Final purely liquid fractions may approximate either E_1 or E_2; but no liquid at any stage can cross the division XY between the two subsystems—a property that places a severe constraint on petrogenesis of feldspathoidal basalts at pressures within the crust.

The position of the eutectic point in any system is determined by the thermal properties of the participating phases: the melting temperature

T_m(in degrees Kelvin) and the increase in entropy ΔS_m accompanying the crystal → liquid transition. The eutectic tends to be enriched in the low-melting member (in Figure 1-1A, diopside); but this effect can be offset by a low value of ΔS_m, which signifies minimal disruption of crystal structure in the course of the crystal → melt transition. Note that the value of ΔS_m depends on ΔH_m, the "latent heat" of melting, and is an inverse function of T_m. (At the melting point $\Delta S_m = \Delta H_m/T_m$; cf. Appendix A.)

Among rock-forming minerals the SiO_2 polymorphs and to a lesser degree the alkaline aluminosilicates ($KAlSi_3O_8$, $NaAlSiO_4$, etc.) have low melting entropy and hence tend to build up to fairly high concentrations in late liquid fractions approaching eutectic composition. In the system $CaAl_2Si_2O_8$–SiO_2, where cristobalite melts at 1713°C and anorthite at 1553°C, the two end members have about equal molar concentrations in the eutectic liquid (1368°C). In the system $NaAlSi_3O_8$–Fe_2SiO_4 the difference in melting points of the end members is less than 100 degrees Celsius (albite, 1120°C; fayalite, 1205°C). Yet the eutectic point stands at almost 85% $NaAlSi_3O_8$.

Incongruent Melting

The system Mg_2SiO_4–SiO_2 (Figure 1-2) is complicated by the existence of a crystalline phase of intermediate composition—pyroxene $MgSiO_3$ (SiO_2, 59.8%)—which is stable along a limited sector of the liquidus (*RE* in Figure 1-2). This pyroxene P on being heated to melting gives rise, at 1557°C, to two complementary phases: a somewhat more siliceous liquid R (SiO_2 60.8%) and crystalline forsterite Mg_2SiO_4 (SiO_2 42.7%). The point R at which two crystalline phases and a liquid coexist must be invariant. Melting thus goes on to completion at 1557°C, the ultimate product being approximately 94% liquid, 6% forsterite. The liquid is now free to move up the liquidus, changing composition by continuous reaction with forsterite, which becomes eliminated in the process. As the last of the forsterite disappears the liquid, P_l, now at 1577°C, has attained the composition $MgSiO_3$ and remains unchanged in composition as temperature rises (along $P_l → P_i$).

Incongruent melting, a rather common phenomenon among igneous minerals, introduces an element of flexibility into possible crystallization paths. Consider a liquid X_i (composition $X = SiO_2$ 58%) cooling from 1620°C. At X_l (1604°C) forsterite separates, and continues to do so as the liquid follows the path X_lR. At the invariant point R reaction between

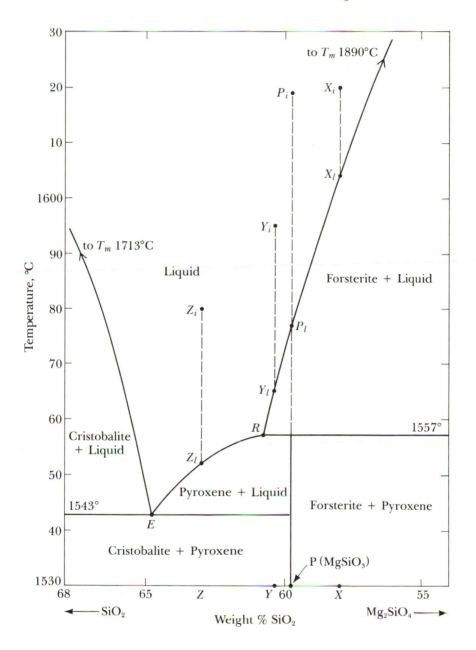

Figure 1-2. Crystal-Melt Equilibria at Atmospheric Pressure in the Two-Component System Mg₂SiO₄–SiO₂, with an Incongruently Melting Intermediate Phase MgSiO₃. See text for description.

liquid and forsterite sets in, to yield pyroxene as an additional phase. The process continues at constant temperature until the liquid R ultimately becomes eliminated; and the end product is pyroxene plus forsterite. Cooling of liquid P_i follows a comparable path which is the reverse of the melting path of $MgSiO_3$: Although forsterite is the first crystal phase to form, the end product is purely crystalline pyroxene $P.$

Liquids on the limited sector P_lR have special petrogenic significance; for though they contain a slight excess of SiO_2 above the pyroxene composition, they first precipitate the much less siliceous phase forsterite. Thus Y_i ($SiO_2 = 60.4\%$) follows the cooling path $Y_i \rightarrow Y_l$, and at the latter point forsterite begins to crystallize. This continues along the liquidus from Y_l to R. Reaction between liquid and the small amount of crystalline forsterite with which it is then associated now sets in. Pyroxene crystallizes copiously as forsterite is gradually incorporated into the melt (to keep the liquid composition constant at R). When the last forsterite disappears the end product is abundant pyroxene plus small quantities of liquid R. Cooling now resumes as pyroxene continues to crystallize and the liquid follows the path $R \rightarrow E$. At E, a second invariant point of three-phase equilibrium, pyroxene and cristobalite crystallize as a eutectic mixture until solidification is complete.

The petrogenic significance of incongruent melting is that it permits a liquid in the forsterite field to transgress the $MgSiO_3$ compositional boundary while still precipitating crystals of forsterite. X_l, P_l, Y_l, and Z_l all represent liquids which by appropriate crystal fractionation could ultimately evolve from a much less siliceous forsterite-generating parent liquid. Correspondingly varied solid products of the evolutionary line would include

pyroxene + forsterite (from X_l)
pyroxene alone (from P_l)
pyroxene + cristobalite (from Y_l or Z_l)

The extreme limit of fractionation is given by E (93% pyroxene, 7% cristobalite).

Crystallization in Solid-Solution Series

The system $NaAlSi_3O_8$–$CaAl_2Si_2O_8$, obviously of crucial significance in patterns of magmatic crystallization, illustrates crystallization paths leading to a chemically continuous series of homogeneous solid solutions (high-temperature plagioclase). Liquidus and solidus curves (Figure 1-3A) are both smoothly continuous and meet at the respective melting points of the pure end members albite and anorthite. They enclose an undivided field of coexisting plagioclase crystals and liquid whose compositions (since such two-phase systems must be univariant at constant pressure) are automatically fixed at any arbitrarily selected temperature: X_s and X_l at 1500°C; Y_s and Y_l at 1400°C; Z_s and Z_l at 1300°C.

Liquid X_i cools unchanged until it impinges on the liquidus at X_l; plagioclase X_s now begins to crystallize and changes composition continuously, remaining homogeneous the while, from X_s to x_s as the liquid changes continuously from X_l to x. The last of the liquid is eliminated as the solid phase reaches the homogeneous composition x_s. Similarly an initial liquid follows the path $Y_i \rightarrow Y_l \rightarrow y$; and plagioclase crystallizing first, Y_s, is progressively made over by reaction with the liquid to its ultimate composition y_s.

The possibilities of fractional crystallization within the limits of the system are infinite. By appropriate fractionation on the $X_i \rightarrow X_l \rightarrow x$ line of liquid descent (with removal of all crystals at point Y_l) the small amount of residual liquid would reach composition y, and its crystalline product would be plagioclase y_s.

The olivine series (Mg_2SiO_4–Fe_2SiO_4) also exemplifies continuous solid solution (Figure 1-3B). The respective melting extremes are Mg_2SiO_4 1890°C and Fe_2SiO_4 1205°C. Liquid residues consequently become greatly enriched in Fe with respect to Mg.

Application to Magmatic Crystallization

The preceding discussion concerns crystallization of the simplest of silicate melts (anhydrous two-component systems) under extreme conditions—atmospheric pressure and temperatures unrealistically high compared with magmatic temperatures. To apply the conclusions so reached (and all of Bowen's early work was in a similar category) to magmatic crystallization requires modifications in accordance with general thermodynamic prediction and the results of experiments conducted at high pressures in the presence of a water-vapor phase.

First, magmas are systems of many components. To introduce even a third or a fourth component into the simple systems depicted in Figures 1-1 to 1-3 results in drastic lowering of minimal crystallization temperatures.

Second, pressures of a few kilobars may be enough to eliminate or

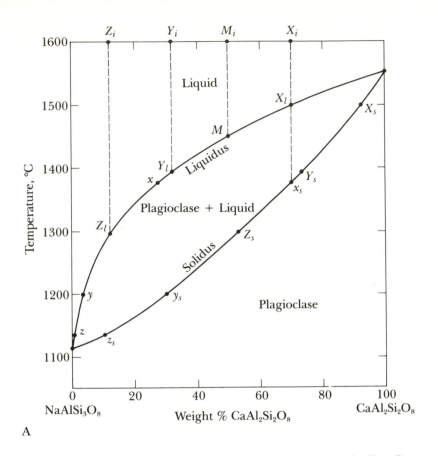

Figure 1-3. Crystal-Melt Equilibria at Atmospheric Pressure in Two-Component Systems Showing Complete Solid Solution Between End Members.

A. Plagioclase system.

B. Olivine system.

even reverse the phenomena of incongruent melting. Orthoclase melts incongruently at atmospheric pressure (to yield leucite and a silica-excess liquid) but no longer does so at pressures above 2 or 3 kb. At upper-mantle pressures ($P > 10$ kb) even magnesian pyroxene may melt congruently.

Third, increased pressure alone increases the melting points of typical igneous minerals; but when water enters the system and water pressure is maintained equal to the load pressure P, the lowering effect can be drastic indeed. Thus an initial melt Ab_{50} An_{50} in the anhydrous two-component system (M_i in Figure 1-3A) begins to crystallize (yielding pla-

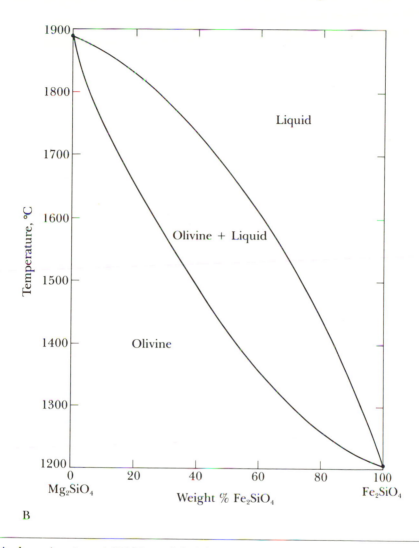

B

gioclase An$_{84}$) at 1450°C and finishes, if equilibrium is maintained, at about 1290°C. In the presence of excess water at $P = 5$ kb the onset of crystallization is delayed to about 1320°C (with An$_{75}$ as the first crystalline phase), and the final liquid is not eliminated until the temperature has fallen to about 1180°C.

However, both thermodynamic prediction and high-pressure experiments demonstrate that, making due allowance for such modifications, we can expect to find in magmatic crystallization the same *kinds* and *diversity* of behavior that have been so clearly elucidated by experiments conducted at low pressure.

Differentiation

This term covers any process whereby an initially homogeneous magma gives rise ultimately to contrasted fractions—fluid or crystalline—of different compositions. Certain physically possible mechanisms, once the subject of wide debate, have been rejected by most petrologists in the light of laboratory experiment, or at the most relegated to minor roles. These include differential *ionic migration* within a completely molten mass of magma under gradients of temperature or pressure; division of a cooling initially homogeneous melt into two separate liquid fractions—a model of *liquid immiscibility* that even today has its adherents;* gaseous transfer of some fugitive chemical species that could become preferentially concentrated in rising gas bubbles. But far more important than any of the three processes just mentioned is the *fractionation of magma resulting from crystallization*.

Certain minerals in igneous rocks are normally associated because they crystallize at about the same temperature. Olivine and labradorite are typical associates; so are quartz and fayalite; and so are orthoclase and oligoclase. Some pairs of minerals, however, are seldom seen together; among these are olivine and albite, and muscovite and labradorite. These relationships imply fractional crystallization of cooling magma along patterns suggested by experimental work. As crystallization proceeds, there is always a tendency for equilibrium to be maintained between the solid and liquid phases. To maintain this equilibrium as the temperature drops, early crystals may react with the liquid and change in composition. The reaction may be progressive, so that a continuous series of homogeneous solid solutions is produced. In the case of the plagioclase feldspars, for instance, the first-formed crystals are those rich in calcium; as reaction continues and the temperature falls, the crystals become progressively more sodic (cf. Figure 1-3A). Changes of this kind, in the classic terminology of Bowen, constitute *continuous reaction series*. Certain ferromagnesian minerals, however, as cooling and reaction continue, are transformed at definite temperatures into other minerals, of different crystal structure—a process in some cases the reverse of incongruent melting. Olivine, for example, may be transformed into hypersthene (cf. Figure 1-2), or augite into hornblende. Such abrupt changes characterize *discontinuous reaction series*.

---

*Unmixing of a liquid sulfide from a liquid silicate fraction would certainly seem a likely explanation for the existence of blebs of pyritic sulfide scattered through some rocks of basaltic composition.

Bowen illustrated igneous reaction series by means of the now familiar diagram based on combined petrographic and experimental data and reproduced here in modified form:

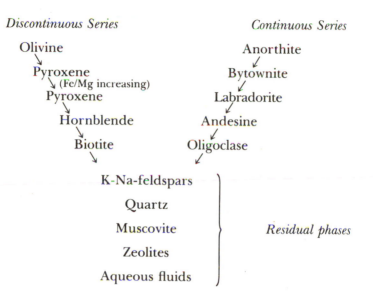

Note how the last minerals to appear are those, as predicted by thermodynamic reasoning, with low melting entropy ΔS_m (quartz and the alkaline aluminosilicates).

Early, high-temperature members of both series generally crystallize together. That is why rocks like gabbro, which contain olivine and magnesian pyroxenes, usually contain calcic feldspars. Low-temperature minerals likewise tend to go together, so that mica, alkali feldspars, and quartz are closely associated in such rocks as granite. For the same reason certain minerals tend to be antipathetic; thus quartz and bytownite, and orthoclase and labradorite, are seldom seen side by side.

When reaction between crystals and liquid goes to completion, the minerals of the final rock are obviously not those that formed first but precisely the opposite ones. But if reaction is incomplete, because of too rapid cooling or for other reasons, early members of both reaction series may persist as relics in the final rock. That is why one observes zoned feldspars and crystals of one ferromagnesian mineral enveloped by shells of another.

Now when we consider the two reaction series listed above, certain variations must be borne in mind. Each ferromagnesian member of the discontinuous series has isomorphous varieties, those rich in magnesium preceding those rich in iron in the normal order of formation. The magnesian olivine, forsterite, usually forms at the same time as calcic plagioclase, while crystallization of iron-rich olivine, fayalite, is generally delayed until quartz and orthoclase are being precipitated. Other factors modify the normal sequence. The time when amphiboles and micas appear is affected by the ever-changing concentration of H_2O and F in the magma. Much depends also on differences in degree of fractionation in the two reaction series, for this influences the ratios between the various elements in the residual liquid. For example, the ratio of alkalies to silica plus alumina in the residual liquid is important in determining whether or not the rock will contain feldspathoids, such as nepheline, or iron-soda silicates, such as aegirine and riebeckite.

Fractionation of magma by crystallization is accentuated by the tendency of heavy minerals to sink and of light ones to float. *Crystal settling* and *crystal flotation* are therefore important in producing variation among igneous rocks. Concentration of olivine and calcic feldspars near the bottoms of thick sills and flows testifies to the settling of these early-formed crystals, just as abundance of leucite crystals among the first ejecta of some eruptions of Vesuvius testifies to upward flotation of that light mineral. An original basaltic magma in a volcanic reservoir may become crudely stratified by this process so that the kind of lava discharged depends partly on the volume extruded and partly on the level of the reservoir tapped by the eruptive conduit. Basic plutonic sheets 5km or more thick habitually display gravity-controlled layering, with early dense members of the two reaction series concentrated downward. *Gas streaming* might possibly contribute to differentiation, for bubbles conceivably may attach themselves to crystals and buoy them upward, or perhaps they may cause interstitial liquids to rise. Obviously effective is *filter-press action*, whereby the last residual liquids in a crystal mush are squeezed from one place to another in the magma chamber, or migrate into cracks cutting material that has already solidified. Such action results from continued upsurge of fresh magma, from opening of tension cracks in the walls of the reservoir, and from concurrent compression due to earth movements.

Whatever the mechanism of fractional crystallization, the early crystalline fractions comprise anhydrous minerals—plagioclase, olivine, pyroxenes—whose removal leaves the later melt fractions progressively enriched not only in water but in other fugitive components such as CO_2, F, and Cl (to which the term *volatile* has long been loosely applied). The ultimate fate of these components depends on the composition of the magma and on the environment of final crystallization—plutonic (high pressure) or volcanic (low pressure). Water is highly soluble in most silicate melts at temperatures of 1000°C or so under pressures of a few kilobars. However, with slow cooling under plutonic conditions it begins to enter some of the common silicate phases, notably amphiboles and micas—minerals whose appearance is generally favored by a relatively high content of Na_2O and K_2O in the magma. Fluorine also is readily accommodated in phlogopitic biotites and in some amphiboles; phosphorus tends to be withdrawn from the melt as a special phase, apatite, which also accommodates fluorine or chlorine.

Alternative possible paths of crystallization in the later stages when fugitive components play an increasingly important role are too varied to warrant any general terminology such as was once the accepted fashion. But one or two of these older terms nevertheless remain useful. We may speak in a rather loose way of a *pegmatitic* stage (favored by slow crystallization under high pressure) when the fugitive content of the silicate melt rises to boiling point, and thereafter crystalline phases (e.g., alkali feldspar, quartz, micas, apatite), a residual silicate melt, and a gas phase coexist until the melt is exhausted. For some time after complete consolidation, plutonic rocks and thickly piled surface flows are liable to be affected, usually to a minor degree, by gas (or in the ultimate stage by hot aqueous solutions) streaming upward and outward from a still hotter source deeper down. Terms such as *hydrothermal* and *pneumatolytic* (respectively signifying effects supposedly due to hot water and to gas) have been applied to mineralogical effects imprinted at this stage—partial conversion of feldspars to finely divided aggregates of clay minerals, chloritization of biotite, veining of olivine crystals with a serpentine mineral, growth of boron-rich minerals like tourmaline or of fluorine-bearing phases such as topaz and some micas. In this book we adopt the comprehensive term *deuteric* to cover all the great variety of incipient phase changes imprinted on the hot consolidated rock in the later part of its cooling history.

Assimilation

Evolution of magma may also be influenced by reaction with wall rocks. If magma is superheated above the temperature at which crystallization can begin, it may become contaminated by melting of the reservoir walls; it is not probable, however, that this often occurs, especially where plutonic magmas are involved. But consider a magma that has already begun to crystallize. Say that it has a granitic composition, that crystals of hornblende and oligoclase are already separating from the liquid, and that the wall rocks are gabbros composed essentially of augite and labradorite. Now labradorite is an earlier member of the continuous reaction series than oligoclase, and augite is an earlier member of the discontinuous series than hornblende. The magma cannot dissolve either the labradorite or the augite, for it is already effectively supersaturated with respect to them. What happens instead is a complex reaction whereby these foreign minerals are made over into hornblende and oligoclase, minerals that are in equilibrium with the liquid at that particular time. Consider next a hotter magma from which magnesian olivine is crystallizing, and say that the wall rocks again include crystals of augite, a later member of the discontinuous reaction series. Under these conditions the augite could become dissolved in the magma; and, to supply the requisite heat of fusion, an equivalent amount of olivine would separate from the liquid. What kind of assimilation takes place thus will depend on what minerals constitute the wall rocks and on what minerals are crystallizing from the adjacent magma. In any event the magma is contaminated, and the rocks ultimately formed from it are of hybrid origin. Such hybrid rocks are particularly common near the margins of large plutonic bodies. Some diorites, for instance, originate in this way by reaction of granitic magma with walls of gabbro or limestone.

Mingling of Magmas

Hybrid rocks, particularly volcanic and shallow intrusive rocks, may also be produced by mixing of partly crystallized magmas. Suggestive evidence comes from the close association within individual lava flows of plagioclase phenocrysts of widely different composition, many of which seem to be out of equilibrium with the groundmass. Examples occur in the San Juan volcanic province of Colorado, where basalts, andesites, and rhyolites were erupted in quick succession from the same or adjacent vents; other examples have been noted in volcanic rocks in Japan and California, where rhyolites containing zoned plagioclase phenocrysts with calcic cores are closely associated with basalts in which the zoned feldspars have sodic cores. Certain andesitic lavas of Hakone volcano, Japan, exhibit reverse zoning of both plagioclase and pyroxene phenocrysts: The rims of the plagioclase are more calcic than the cores, and the rims of pyroxene contain more magnesium and less iron than the cores. Individual thin sections of andesite from the Pleistocene volcanoes of California commonly display an assortment of early-formed plagioclase crystals (phenocrysts) ranging from calcic (bytownite) to sodic (oligoclase) in mean composition; some are zoned; some riddled with glass inclusions; clearly several crops of crystals with different thermal histories are represented. Some welded tuffs in Costa Rica consist chiefly of small particles of glass whose notably different refractive indices denote a wide range in liquid composition. Comparable rocks in Mexico show mixtures of large glassy fragments, some dark, some light in color, and again clearly of contrasted composition. And the pyroclastic flow deposits of the Valley of Ten Thousand Smokes, Alaska, contain a mixture of simultaneously erupted white rhyolitic pumice and dark andesitic scoria, both products of explosively effervescing fresh magmas. Mixing of contrasted magmas in subvolcanic reservoirs, hybridization even on a microscopic scale, and simultaneous explosive ejection of contrasting magmatic materials together constitute what seems to be a common volcanic phenomenon.

IGNEOUS ROCK ASSOCIATIONS AND THE CONCEPT OF MAGMATIC LINEAGES

Petrographic Provinces

The concept of kindred rock types crystallizing from magmatic lineages that stem from some common parental magma is more than a century old. It was developed especially by Alfred Harker in the first decade of this century; and since his time it has played an important role in the development of ideas in igneous petrogenesis (petrology), leaving its mark on petrography as treated in texts and explored in the laboratory.

The starting point is to recognize the existence of *petrographic provinces*. Any broad region within which igneous rocks are approximately the same age (within a few million years, but commonly less) and share general chemical and mineralogical features in common (though varying notably among themselves in other characteristics) constitutes such a province. Recognition of its existence is based on field mapping, petrographic examination, and chemical analysis of representative rock types. Well-known examples are the volcanic provinces of the Hawaiian Islands, (Pliocene to Recent); the western isles of Scotland (Hebridean province, early Tertiary); the islands of the mid-Atlantic Ridge (Pliocene to Recent); and the Roman province of Italy (Holocene), which includes the highly potassic lavas of the Somma-Vesuvius volcano. Two contrasting types of plutonic provinces are the Sierra Nevada of California (dominantly granodioritic and later Mesozoic in age) and the colossal Precambrian Bushveld Complex of South Africa (gabbros, with associated pyroxenites, peridotites and granitic members).

The step from observed contemporary field association to inferred common parentage and mutual genetic connection of varied rocks within a specific province necessarily involves speculation. All such speculation must be within constraints imposed by thermodynamic argument and experimental data. Some of the bold concepts proposed by our predecessors, in the days when the nature of these constraints was not generally understood and experimental data were few, nevertheless have stood up well to the tests of modern theory and experiment. Other generalizations have had to be abandoned. Today most petrologists would accept this proposition: that much of the mineralogical and chemical variety displayed within any province conforms to a pattern that has been shaped by magmatic evolution along paths stemming from a common parent magma or source material; and that fractional crystallization or melting has played the dominant role in the process.

Chemical Variation Diagrams

A convenient and primarily descriptive device to highlight aspects of chemical patterns in specific provinces is some form of variation diagram. Each type of diagram is based on data drawn from chemical analyses of representative rocks within a single province or even a smaller unit such as a composite volcano or pluton. Every diagram is thus unique, though some may be sufficiently similar to suggest parallel evolutionary processes that tend to recur at different times and places.

The most comprehensive type of variation diagram, widely used since the first decade of the century, is a rectangular plot of metallic oxides against the common datum SiO_2, with all values expressed as weight percentages taken directly from chemical analyses. This type of diagram is illustrated in Figure 1-4, which represents the volcanic rocks (basalts, andesites, dacites, and rhyolites) of the Crater Lake volcano, Oregon as described in 1942 by H. Williams. Note how alkaline oxides increase and oxides of Ca, Fe, and Mg all decrease with increasing SiO_2—a general feature highly typical of many petrographic provinces.

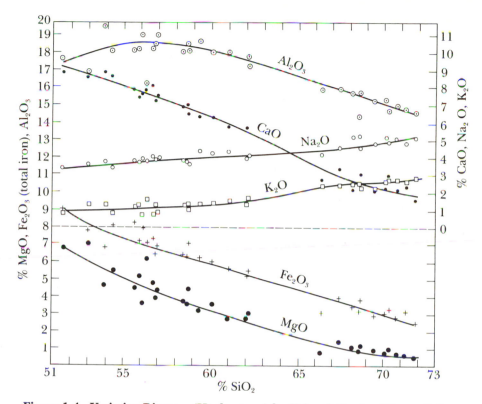

Figure 1-4. Variation Diagram (Harker type) for Volcanic Rocks of Crater Lake, Cascades Volcanic Province, Northwestern United States. (After Williams, 1942.)

Other types of diagrams in current use are triangular plots of three selected components expressed either as oxides (weight percentages) or as elements (atomic percentages). In every case the data, extracted from chemical analyses, are recalculated to a total of 100 to permit construction of the triangular plot. Most generally used is the *AFM** diagram whose three components, recalculated to total 100, are expressed in two alternative forms:

Weight percentages of $A = (Na_2O + K_2O)$; $F = (FeO + Fe_2O_3 + MnO)$;
$$M = MgO,$$

or

Atomic percentages of $A = Na + K$, $F = Fe + Mn$, $M = Mg$.

---

*Not to be confused with the *AFM* projection of metamorphic petrology.

Examples drawn from well-documented provinces are illustrated in Figure 1-5. The value of three-component plots lies in their ability to demonstrate subtleties of variation not obvious in the more comprehensive rectangular diagram (Figure 1-4). Note, for example, how in many

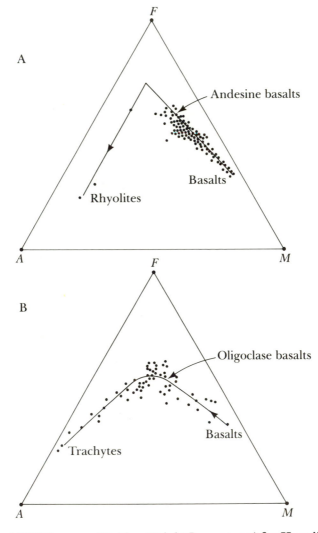

Figure 1-5. AFM Diagrams (Oxides, Weight Percentages) for Hawaiian Lavas. (Modified from G. A. MacDonald and T. Katsura, *Journal of Petrology, vol. 5*, 1964, p. 109).

A. Tholeiitic basalts and differentiates.

B. Alkaline basalts and differentiates.

basalt-dominated provinces (e.g., the Hawaiian Islands) the most marked chemical trend, as SiO_2 rises from $\sim 47\%$ to $\sim 52\%$, is an increase in the ratio F/M (from 0.4 to 0.7 in Figure 1-5A,B). Over the same interval A/ (F + M) also rises from 0.05 to 0.15 in series A; and from 0.04 to 0.2 in series B.

Chemical Variation in Magma Series

A variation diagram expresses chemical characteristics of the rocks on which it is based. Yet these same characteristics in some way must reflect a pattern of serial variation within corresponding magmas from which the rocks themselves have crystallized. But what is the relation between rocks and magmas? How truly does the chemistry of one mirror that of the other? In this book we cannot pursue these important questions in depth. Nevertheless, if only to sharpen vigilance of petrographic obser-vation, it is appropriate to note several possible relationships, each of them seemingly valid in particular circumstances.

1. Uniformly fine-grained or glassy volcanic rocks may be assumed to be rapidly quenched volcanic liquids. In such cases—and only in these as Bowen pointed out in 1928—one may legitimately equate the chem-ical pattern of a rock series with that of an equivalent magma series. The variation diagram based on rock analyses truly characterizes a magmatic lineage—in Bowen's words a "liquid line of descent."
2. In massive stratified basic sheets (some of them 10 km or more thick), which comprise the most spectacular plutonic deployment of basaltic magma in stable cratons, successive layers are largely made up of crystals that settled out at successive stages, partly under gravity con-trol, from liquids of quite different composition. Rock compositions here are in no way equivalent to magma compositions. Cumulate lay-ers of olivine and pyroxene, others of calcic plagioclase and pyroxene, others of almost pure plagioclase represent successive crystalline frac-tions separated from a cooling body of basic magma which itself evolved simultaneously along a somewhat parallel line. To reconstruct the pattern of chemical variation in the *liquid* involves measurements and estimates of relative volumes of the crystalline fractions and assumptions regarding the initial volume and composition of the par-ent body of magma. This is a formidable, perhaps nearly impossible, task. The outcome is at best only one possible model of magmatic

evolution. The best known and most carefully constructed model of this kind is that of L. R. Wager and his devoted co-workers—the famous Skaergaard trend (Figure 1-6A) representing a model of evolution based on an exhaustively studied layered basic intrusion of Eocene age in eastern Greenland.

3. Texture alone prohibits direct correlation between rock analyses and magma chemistry in two broad classes of rock. The first includes volcanic rocks with *porphyritic* texture; for in these, two generations of crystalline phases are clearly distinguishable—large well-formed crystals of an early generation enclosed in a fine-grained or glassy matrix that represents a subsequently quenched liquid residue. The other group includes all the granite-granodiorite clan, and many other common types of plutonic rock as well. The texture is *granitoid*, uniformly coarse in grain. There are few if any certain textural clues to distinguish between possible early cumulates and crystalline products of later liquid residues, if indeed such ever existed as separate entities. Yet analyses of representative rocks in many extensive plutonic provinces (e.g., Figure 1-6B) yield oxide plots that approximate smooth curves just as neatly as in variation diagrams constructed for uniformly fine-grained and glassy volcanic series. Do curves such as that of Figure 1-6B represent real liquid lines of descent? Or is the smoothness of the curve merely illusory? There are as yet no certain answers to such questions.

Alkalinity and Silica-Saturation in Magma Series

Clearly it is unsafe in most cases to equate the chemical analysis of a rock specimen with the composition of some homogeneous parent melt. But

---

Figure 1-6.

A. AFM diagrams for the computed liquid trend of fractionation, Skaergaard pluton. Atom percent. O is parental magma (chilled marginal gabbro); X is liquid composition when the lowest exposed rocks began to form (70% solidification); Z is a late granophyre (other granophyres shown as open circles). Numbers in parentheses show percentage solidified. (Modified from Carmichael, Turner, and Verhoogen, 1974.)

B. Variation diagrams for volcanic rocks of Yellowstone volcanic province (full lines) and plutonic rocks of the Lower California batholith (broken lines). (After E. S. Larsen, *Geological Society of America Memoir 29*, 1948.)

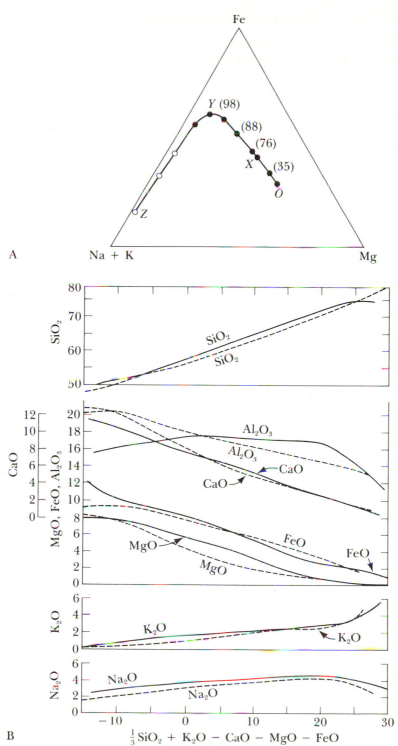

Fe

Y (98)

(88)

(76)

X

(35)

O

Z

A

Na + K   Mg

80

70

SiO₂

60

50

SiO₂

SiO₂

12

10

8

6

CaO

4

2

0

20

18

16

14

12

10

MgO, FeO, Al₂O₃

8

6

4

2

0

Al₂O₃

Al₂O₃

CaO

CaO

MgO

MgO

FeO

FeO

6

4

K₂O

2

0

K₂O

K₂O

6

4

Na₂O

2

0

Na₂O

Na₂O

−10   0   10   20   30

B   ⅓ SiO₂ + K₂O − CaO − MgO − FeO

it is still legitimate to categorize general features of the chemistry of a magma series in terms of selected aspects of the chemical pattern of a variation diagram based on a series of representative rock specimens. Two qualities, whose significance in this connection was recognized from the outset by Harker, concern what may be called alkalinity and silica-saturation.

Harker distinguished (and the distinction still remains valid) two extensive classes or rocks (and so of magmas) based on alkalinity:

Rock analyses of *alkaline series* consistently show high ($Na_2O + K_2O$) with respect both to CaO and to SiO_2. Even in basaltic members with SiO_2 less than about 50%, the combined weight percentage ($Na_2O + K_2O$) equals or exceeds that of CaO. In consequence the groundmass of such a rock is likely to show small amounts of K-feldspar, biotite, or nepheline. In highly alkaline series, in fact, nepheline is a major constit-uent. Quartz is unknown, except as a minor constituent in siliceous mem-bers (trachytes) of some more weakly alkaline series.

In *calcalkaline series*, in contrast, ($Na_2O + K_2O$) equals or exceeds CaO only at the silicic end of the series (SiO_2 typically > 60%), as exemplified in the Crater Lake province, Figure 1-4. Feldspathoids are unknown; but quartz is widely distributed and is an essential constituent of the more siliceous rocks (e.g., dacites, rhyolites, granodiorites). Potassic minerals, K-feldspar, and biotite likewise appear only at the siliceous end of the series.

Not all rock series fit neatly into one or other of these categories. Should one, for example, categorize the most extensive of all volcanic rocks—the basalts of the ocean floor—as calcalkaline simply because they show none of the traits that are generally associated with the notion of alkalinity?

The criterion of silica saturation was developed with reference to individual rock composition. But it can usefully be extended to a rock or magma series that individually embraces a narrow range of chemical composition. Ocean-floor basalts, continental flood basalts, and the mas-sive basaltic piles of some oceanic islands such as the Hawaiian group and Iceland comprise rock series of this kind. So we shall confine our attention to the silica-saturation status of basaltic rocks.

Certain minerals have been shown experimentally to be incompatible with siliceous melts;* and in nature these are never associated with quartz

---

*This situation can be explored profitably from the thermodynamic standpoint—making use of the concept of silica *activity* (a parameter related to ΔG_r for reactions involving quartz).

of the same magmatic generation. Feldspathoid minerals obviously fall into this category. So also does magnesian olivine—except for a narrow range of liquid composition analogous to the segment *PR* in Figure 1-2. Feldspathoids and magnesian olivines themselves may therefore be designated as *undersaturated* in SiO_2 by contrast with corresponding *silica-saturated* phases (alkali feldspars and magnesian pyroxenes). On mineralogical grounds alone it is possible then to distinguish two corresponding classes of basaltic rocks:

> Undersaturated basalts contain abundant olivine and, in one group, nepheline or other feldspathoid as well. Corresponding magma series fall within a broad category designated *alkali-olivine-basaltic type.*
> Saturated basalts completely lack feldspathoids and contain only minor olivine, which is confined to an early generation (phenocrysts) that crystallized from liquids with slight excess of SiO_2 analogous to those on the segment *PR* in Figure 1-2. Corresponding magma series fall within a broad *tholeiitic type.*

Certain characteristics of the individual chemical analysis will place any rock unequivocally in the undersaturated or the saturated class. The definitive criteria appear in a strictly programmed computed expression of chemical composition (in terms of simplified phases analogous to minerals in composition) termed the *norm* (see Chapter 2): presence of normative olivine or nepheline in the one group, presence of quartz in the other.

In many provinces plentiful basaltic members, erupted during a major volcanic cycle fall entirely within one or the other class. Accordingly the magma series in question may be designated as undersaturated or saturated with respect to SiO_2.

MAGMA CHEMISTRY AND TECTONIC PATTERN:
A HINT OF FUNDAMENTAL RELATIONSHIPS

In his early quest for global magmatic patterns Harker provisionally postulated two sets of relationships, the one geographically, the other tectonically oriented. The first, a proposed twofold global division of Tertiary and Recent volcanism into an Atlantic (alkaline) and a Pacific (calcalkaline) megaprovince, proved, in the light of later and increasingly diversified information, to be greatly oversimplified. It has since been abandoned. The second proposal was unexpectedly fruitful and achieves singular significance as a preview of certain aspects of the present tendency to think along tectonic lines. Briefly, Harker conceived his alkaline (Atlantic) type of magma series as related to regional rifting (as along the mid-Atlantic Ridge and the Rifts of eastern Africa). The calcalkaline series he correlated with regional folding as exemplified by current activity along the circum-Pacific mobile belts. It was not long before H. E. Gregory recognized that the fundamental control was tectonic, geographical correlation being of secondary significance. Thus was established, in the years before World War I, the existence of some kind of fundamental relation between chemical pattern in magma series and tectonic pattern with its various implications regarding depth regimes and source materials of magma genesis.

In recent years, with vastly increased information in such fields as geochronology, rock chemistry, and regional petrography (especially within the ocean basins), and under the influence of developing plate-tectonic theory, Harker's simple concept has been modified and greatly elaborated. To conclude this chapter we note some of the recent generalizations that relate magma types to tectonic pattern (see Carmichael, Turner, and Verhoogen, 1974, pp. 662–664).

1. By far the most voluminous magmas to reach the earth's surface are tholeiitic. Generated in the upper mantle, they rise to the surface along zones of major rifting. Their most notable products are ocean-floor tholeiitic basalts (with a very low K_2O/Na_2O ratio) spreading outward from sites of eruption along the mid-Atlantic rift and similar linear sources in other oceans. On the continental surfaces, too, there are immense accumulations of tholeiitic flows whose eruption accompanied regional fracture of the continental crust in various episodes of continental fragmentation and drift. Such are the Mesozoic basalts and associated diabase sheets of southern Africa (Karroo province) and, across the Atlantic gulf, central and southern Brazil (Paraná province), both erupted during separate early episodes in the breakup of the Gondwana land continent.

2. Along other sites of oceanic or continental rifting, and during later stages of volcanism at sites previously characterized by outpouring of tholeiitic lavas, erupted magmas are consistently alkaline. Here alkali-olivine basalts and even more alkaline basic lavas with conspicuous feldspathoid (e.g., basanite and nephelinite) appear to represent the parent magmas, again welling up from sources in the upper mantle. More siliceous alkaline lavas (trachytes, phonolites) are thought to

represent late differentiates. Classic examples are displayed in the extensive continental rift system stretching south from the Red Sea to the Rift Valley region of East Africa, and again in the discontinuous line of recently active volcanic islands that extends along the full length of the mid-Atlantic Ridge.

3. By far the greatest plutonic development of basic magma is seen in great gravity-stratified gabbro-dominated sheets, some of them 10 km in thickness, that since the earliest times have invaded the crust of stable continental cratons. Again the parent magmas appear to fall in the general tholeiitic composition range. Classic examples include the Precambrian Stillwater (Montana) and Bushveld (South Africa) complexes and the Eocene Skaergaard intrusion of Greenland.

4. Calcalkaline lavas ranging from tholeiitic basalt through andesite to dacite and rhyolite are characteristic volcanic products of active island arcs of the circum-Pacific belt. The more siliceous members are characteristic of segments where arcs impinge on borderlands of continental crust. On seismic evidence the sources of these island-arc volcanics are thought to be located on Benioff zones. Exactly comparable calcalkaline magma series are found, however, well within continental borders with no hint of any relation to Benioff zones. Such are the Holocene Yellowstone and Eocene San Juan provinces of North America, both situated 1500 km inland from the Pacific margin of the continent. Calcalkaline lavas may appear in yet other continental settings. For example, they are associated with lavas of both tholeiitic and alkali-olivine-basalt parentage in the Eocene Hebridean province of western Scotland (here volcanism was apparently synchronous with fragmentation of this sector of the ancient North Atlantic craton). Source materials of calcalkaline magmas probably include both mantle and deep crustal rocks.

5. Calcalkaline plutonic (granodioritic) complexes make up the most conspicuous igneous element in the continental crust. They tend to be located along or near mobile belts, and the calcalkaline plutonic episode commonly correlates with the later stages of orogeny. But such correlation is not universal, and some belts of younger granite plutonism (as in the Cenozoic batholiths of the Rockies) are located far from recognizable sites of Benioff zones.

6. In contrast with calcalkaline and silica-saturated basic magmas as these appear respectively in granodioritic and stratified sheet complexes, plutonic manifestations of alkaline magmas are insignificant in volume. They have considerable petrographic interest, however, in view

of their mineralogical diversity. Alkaline plutons appear to favor a continental environment of relative tranquillity. At the basic end of the series are differentiated sheets such as that of Shonken Sag, Montana, in the western United States. The more siliceous rocks include feldspathoidal syenites whose best known development is seen in numerous plugs and other minor intrusions of later Mesozoic age cutting the Precambrian cratons of southern Africa and Brazil. Some of these intrusions overlap, others postdate episodes of basaltic outpourings in the same broad regions.

7. Spilites are highly sodic albite basalts that are widespread in eugeosynclines. Once thought to be products of a special type of basic magma correspondingly rich in sodium, they are now recognized as basalts of normal chemistry (mostly tholeiitic) that have become enriched in sodium during the course of low-grade regional metamorphism (see pp. 112–115).

Every one of these generalizations is open to future modification and present exception. Within the course of a few million years successive volcanic cycles on the same general site may show a marked change in magmatic pattern—perhaps indicating change in depth and chemical nature of source materials. Monotonous repetition of *identical* magma types in identical tectonic situations recurrent in geographical locale and in time is hardly to be expected. Already some magma types seem to have a unique place in time: Andesine anorthosites of the Proterozoic and highly magnesian ultramafic flows (komatiites) of the earlier Precambrian come to mind in this connection. Igneous phenomena like those connected with most other major geological processes indeed show perceptible pattern. But against the background of regularity this pattern provides, departure and nonconformity will become obvious to the critical petrographer as he relates his material to chronological and structural aspects of its field environment. At this point we must leave the preliminary contemplation of magmas and magma series and turn our attention to a more immediate objective, the microscopic study of magmatic products—the igneous rocks themselves.

GENERAL REFERENCE

Throughout the chapters on igneous rocks repeated reference is made to three general and readily accessible works:

A. Johannsen, *A Descriptive Petrography of the Igneous Rock*, 4 vols. Chicago: University of Chicago Press, 1931–1939.
Remains the standard, comprehensive account of igneous petrography, supplemented by chemical (analytic and normative) data.

N. L. Bowen, *The Evolution of the Igneous Rocks*. Princeton: Princeton University Press, 1928; reprint ed., New York: Dover, 1956.
The foundation stone of modern experimentally based igneous petrology.

I. S. E. Carmichael, F. J. Turner, and J. Verhoogen, *Igneous Petrology.* New York: McGraw-Hill, 1974.
An updated account of the genesis and distribution of igneous rocks, supplemented by modern chemical data (analytical, normative, trace-element, and isotopic). The authors present expanded statements on most of the petrogenic topics we briefly refer to, and our work is consistent with theirs.

2

Characteristics and Classification of Igneous Rocks

The distinctive features of igneous rocks in general and of individual rock types (granite, basalt, etc.) in particular relate to field occurrence, bulk chemistry, component minerals, and texture. The last two of these are of principal concern to the petrographer, although one also must bear in mind possible field and chemical implications of what one observes in the laboratory. In this book, therefore, in contrast with more comprehensive works that cover the whole field of petrology, the emphasis is primarily on mineralogy and texture.

MINERALOGICAL COMPOSITION OF IGNEOUS ROCKS

In common igneous rocks the principal chemical components (SiO_2, Al_2O_3, CaO, MgO, etc.) are largely or completely accommodated in three- to six-phase combinations of a few widely distributed minerals: quartz, feldspars, pyroxenes, amphiboles, micas, iron-titanium oxides, olivines, and feldspathoids (the two last-named in silica-deficient rocks). An additional phase in some rapidly chilled volcanic rocks is residual glass. The relative proportions of these *essential* constituents determine the two most readily perceived properties of any rock—its color and its density. Accordingly, igneous minerals are customarily grouped in two loosely defined color categories: *felsic*, light in color and of relatively low density (G mostly 2.5 to 2.7) and *mafic*, darker in color and significantly more dense (G mostly 3 to 3.6).

With the exception of quartz, most essential minerals of igneous rocks are highly variable in composition, reflecting complex and widespread solid-solution. Further complexities are introduced by temperature-sensitive polymorphism. With such features in mind we now review in outline some petrographically significant features of mineral chemistry.

Felsic Minerals

Quartz, SiO_2

Quartz is characteristic of all the more siliceous types of igneous rocks ($SiO_2 > 66\%$). Below 870°C at atmospheric pressure it is the stable form of silica.

Feldspars

Feldspars are the most abundant and widely distributed group of igneous minerals, whose ideal component end members are $KAlSi_3O_8$, $NaAlSi_3O_8$, and $CaAl_2Si_2O_8$. There are two solid-solution series in each of which two of the three components play the dominant role. Neglecting in each case the third component these series approximate $KAlSi_3O_8$–$NaAlSi_3O_8$ (the *alkali-feldspar* series) and $NaAlSi_3O_8$–$CaAl_2Si_2O_8$ (the *plagioclase* series). The crystallographic character, and hence the optical properties, of each series are complicated by polymorphism connected with degree of ordering of Al^{3+} and Si^{4+} ions in tetrahedral lattice sites. Five chemical-crystallographic series of feldspars, the first four of which are alkali feldspars (Figure 2-1), can be distinguished beneath the microscope:

Alkali feldspars

Sanidine-anorthoclase series, (K, Na) $AlSi_3O_8$–(Na, K) $AlSi_3O_8$: disordered high-temperature alkali feldspars, monoclinic at the potassic, triclinic toward the sodic end.

Orthoclase, (K, Na) $AlSi_3O_8$: metastable partially ordered, monoclinic potassic feldspar crystallizing in the medium range of igneous temperatures.

Microcline, (K, Na) $AlSi_3O_8$: ordered triclinic potassic feldspar stable at relatively low temperatures.

Perthites, (K, Na) $AlSi_3O_8$: potassic feldspars—microcline or orthoclase—enclosing finely intergrown albite in most cases formed by exsolution. Microcline-perthite is the most stable microscopically recognizable

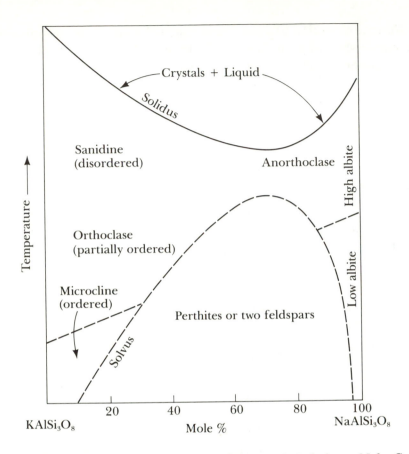

Figure 2-1. Nomenclature of Stable Alkali Feldspars in Relation to Molar Composition and Temperatures Between 400°C and Solidus Temperatures. All phase boundaries are diagrammatic. Solidus is for volcanic and subvolcanic pressures.

form of alkali feldspar in the lower range of igneous temperatures. (Most optically homogeneous high-temperature alkali feldspars of volcanic rocks are now known to have exsolved to discrete potassic and sodic end members intergrown on a scale so small that their separate identities can be recognized only by x rays or the electron microscope. Mineralogists working on this very small scale term them *cryptoperthites*.)

Plagioclase series

Triclinic feldspars encompassing the continuous range albite, $NaAlSi_3O_8$, to anorthite, $CaAl_2Si_2O_8$. It is customary to divide the series arbitrarily on the basis of optically defined criteria as follows:

Albite	An_0 to An_{10}
Oligoclase	An_{10} to An_{30}
Andesine	An_{30} to An_{50}
Labradorite	An_{50} to An_{70}
Bytownite	An_{70} to An_{90}
Anorthite	An_{90} to An_{100}

Most igneous plagioclase falls within the range An_{20} to An_{80}. Complications arising from polymorphism (order-disorder phenomena) and cryptoperthitic exsolution, though of mineralogical interest as clues to cooling histories, play no part in classic petrography. Plagioclase with visible intergrown inclusions of K-feldspar is called *antiperthite*.

Feldspathoids

Feldspathoids include a number of structurally unrelated sodium and potassium aluminosilicates in which the atomic ratio $(Na + K)/Si$ exceeds that (1:3) of alkali feldspars. They are confined to silica-deficient aluminous rocks, and typically are associated with alkali feldspars. The commonest feldspathoid species are:

Nepheline	$Na_3KAl_4Si_4O_{16}$
Sodalite	$Na_8Al_6Si_6O_{24}Cl_9$
Leucite	$KAlSi_2O_6$

Mafic Minerals

Olivines, $(Mg, Fe)_2 SiO_4$

Magnesian olivines (rich in forsterite, Fo) are widespread among igneous rocks with a relatively low content of SiO_2—notably in basaltic lavas and their deep-seated chemical equivalents. Olivines approaching Fe_2SiO_4 (fayalite) are relatively rare; but, unlike their more magnesian counterparts, they are stable in the presence of excess SiO_2 and so may appear in quartz-bearing rocks with high FeO/MgO.

Pyroxenes

Pyroxenes are chain silicates in which the principal participating ions are Ca^{2+}, Mg^{2+}, Fe^{2+}, and in one series $(Na^+ Fe^{3+})$; aluminum and titanium play significant but subordinate roles in some species. There are three main solid-solution series (see Figure 2-2):

1. *Orthopyroxenes*, (Mg, Fe) SiO_3, orthorhombic—the enstatite-bronzite-hypersthene-ferrosilite series (En–Fs). Common hypersthenes contain less than 50% $FeSiO_3$ (the ferrosilite molecule, Fs).
2. *Augite-pigeonite series*, monoclinic Ca Mg Fe pyroxenes with $CaMgSi_2O_6$ (diopside), $CaFeSi_2O_6$ (hedenbergite), $MgSiO_3$ (En) and $FeSiO_3$ (Fs) as ideal end members. Augites are significantly aluminous and some, characterized by a violet tint in thin section, contain TiO_2 as well. At igneous temperatures there is an immiscibility break between the more calcic and the more magnesian members, *augites* and *pigeonites* respectively. In some volcanic rocks the gap is bridged by a metastable quench product, *subcalcic augite*. The only certain optical criteria for distinguishing the three varieties is optic axial angle (2V, in all cases positive; axial plane, except in pigeonite, parallel to {010}): 2V in augites ranges from 60° in types approximately Ca(Mg, Fe) Si_2O_6 to 40° in less calcic types;* in subcalcic augites 2V is consistently low, 10°–30° being typical; pigeonites typically are uniaxial or have low 2V with the axial plane normal to (010).
3. *Sodic pyroxenes*, deep green monoclinic pyroxenes ranging from *aegirines* (almost pure acmite, $NaFe^{3+}Si_2O_6$) to acmite-diopside-hedenbergite solid solutions. They characterize igneous rocks with a content of sodium too high, compared with Al_2O_3, to be accommodated completely in the alkali feldspars or feldspathoids [in both of which (Na + K)/Al = 1].

Amphiboles

The common igneous amphiboles are *hornblendes*—chemically complex monoclinic double-chain silicates that approximate $NaCa_2(Mg,Fe)_4$ $Al(Al_2Si_6O_{22})$ $(OH)_2$ with limited substitution of K for Na, of Fe^{3+} for Al, and of F for (OH). Members of the deep-brown kaersutite series may contain as much as 5% to 10% TiO_2. In blue alkaline hornblendes ($NaFe^{2+}$) may substitute to a limited degree for ($CaFe^{2+}$). Deep blue *riebeckite*, $Na_2Fe_3^{2+}Fe_2^{3+}$ (S_8O_{22}) $(OH)_2$, occurs in some sodic volcanic rocks. *Arfvedsonite*, deep blue in color, is another alkaline hornblende found mainly in sodic plutonic rocks.

---

*In highly titaniferous varieties 2V may be much lower; but these are distinguished from subcalcic augites and pigeonites by their distinctive violet color.

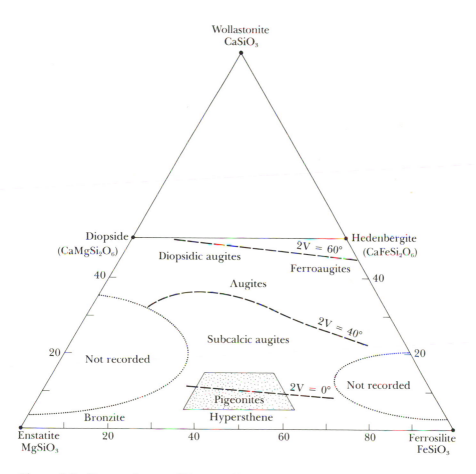

Figure 2-2. Nomenclature of Igneous Pyroxenes Defined in Simplified Terms in the Three-Component System CaSiO₃–MgSiO₃–FeSiO₃. Approximate loci for three values of 2V in monoclinic members are shown as dashed lines. The field of pigeonite is stippled. (Data from W. A. Deer, R. A. Howie, and J. Zussman, *Rock-Forming Minerals,* vol. 2. New York: Wiley, 1963.)

Biotites

Most igneous micas are biotites, $K(Fe,Mg)_3(AlSi_3O_{10})(OH)_2$, in which Fe^{3+} may substitute for Mg with compensating substitution of Al for Si.

Iron-Titanium Oxides

The common opaque minerals of igneous rocks fall into two isomorphous series with iron and titanium oxides as end members; they are almost ubiquitous, but are usually in minor quantity. The α-series (*ilmenites*) is rhombohedral—actually ilmenite-hematite solid solutions $FeTiO_3$–Fe_2O_3 with up to 25% Fe_2O_3. The β-series (*magnetites*) is a complete isometric magnetite-ulvospinel solid-solution series, Fe_3O_4–Fe_2TiO_4. Exsolution and oxidation during cooling commonly lead in both series to development of internal intergrowths of iron-titanium oxide combinations. The TiO_2 polymorphs rutile and anatase are much rarer.

Minor (Accessory) Minerals

Most igneous rocks contain small amounts of minor (accessory) minerals into which enter elements that are excluded from or fail to be completely accommodated in the principal phases: phosphorus and fluorine in *apatite* $Ca_5(PO_4)_3(OH,F)$; zirconium in *zircon*, $ZrSiO_4$; excess titanium and minor thorium in *sphene*, $CaTiSiO_5$; chromium in a *chromiferous spinel*, $(Fe, Mg)O(Cr,Al)_2O_3$; excess aluminum of some granites in *muscovite*, $KAl_2(AlSi_3O_{10})(OH)_2$; sulfur in various sulfides, notably pyrite, FeS_2; carbon in *calcite*, $CaCO_3$, or *siderite*, $FeCO_3$; titanium and rare-earth elements in perovskite, $(Ca, Na, Fe^{2+}, Ce)(Ti,Nb)O_3$. Petrographic identification of these throws light on the evaluation of trace-element patterns recorded by geochemists in igneous rock series.

TEXTURES OF IGNEOUS ROCKS

Texture and Structure

It is difficult to draw a clear distinction between the terms *structure* and *texture*. In general, however, *structure* refers to large-scale features recognizable in the field, such as banding, lineation, and jointing. *Texture* refers to degree of crystallinity and grain size and to the fabric or geometrical relationships between the constituents of a rock. These textural features cast much light on the conditions under which igneous rocks consolidate from their parent magmas, for they are controlled by the rate and order of crystallization; and these in turn depend on the initial temperature, composition, content of fugitive components, and viscosity of the magma and the pressure under which it solidifies.

Texture as the Petrographic Record of Cooling History

Even during ascent from the depths many magmas have already started to crystallize. The solidification process, continuing through the period of magma emplacement, is mainly effected and carried on to completion within the individual lava flow, dike, sill, or pluton. Continuous solidification involves nucleation and growth of crystals of successively generated mineral phases at the expense of an ever-diminishing quantity of melt in an environment of progressive cooling. It ceases when the melt phase is exhausted or when falling temperature reduces the rate of effective crystallization to zero.

The span of time available for the operative processes is limited by *rates* of cooling. These depend upon a number of factors, by far the most influential being the minimum dimension of the cooling body—thickness of a flow or sill, diameter of a major pluton—and to a lesser degree a blanketing effect controlled by depth of emplacement. Other things being equal, thin surface flows cool much more rapidly than massive deep-seated plutons. Some idea of the wide latitude in possible time span can be gained from quantitative cooling models developed by J. C. Jaeger. He postulates a sheet of basaltic magma, thickness D, at liquidus temperature (i.e., on the point of crystallizing) instantaneously injected into cool country rock and subsequently crystallizing without internal convection. The computed time span for complete crystallization at the center of the sill—over a liquidus-to-solidus interval of 200°–300°C—is found to be directly proportional to D^2. Values computed, on reasonable assumptions regarding the thermal properties of magma, igneous, and country rock, for sills respectively 1, 10, 100, and 2000 m thick are 5 days, 15 months, 1,300 years, and 50,000 years. Corresponding average cooling rates range from about 50 degrees per day to 5 degrees per 1000 years. Similar computations show that the margins of thick sheets would become completely solid to a thickness of 2 m within a year, increasing to 20 m in 100 years.

Understandably, in view of this wide latitude of possible cooling regime, there is corresponding diversity among commonly recorded igneous textures. But there is much of discernible regular pattern as well, and herein valuable material is provided for descriptive petrography. Obviously, to grasp the significance of textural data we must first look briefly at the nature and controls of the crystallization process as it operates in magmas cooling at widely differing rates in different geological environments.

Physical Controls of Textural Evolution

Evolution of texture during solidification of magma is governed by two principal sets of physical controls. First are externally imposed variables: temperature T, time t, and to a more limited degree pressure P (which we largely neglect in the following discussion). Second are the thermodynamic properties of participating phases,* whose separate identities in any given case depend on magma composition and to some degree on conditions of consolidation. Most important of these are functions deriving from *entropy* which itself has a special and subtle relation to temperature.

Melting Entropy of a Crystalline Phase

The reciprocal phenomena of melting and crystallization in a one-component system are accompanied by sharp changes in the degree of internal order (related to geometric arrangement and motions of constituent particles). An index of the change is the quantity of heat absorbed during melting or its equivalent emitted during crystallization. Closely related to heat of melting ΔH_m (cal mole $^{-1}$) *at equilibrium temperature* T_E is *entropy of melting*, ΔS_m (cal mole $^{-1}$ degr $^{-1}$), defined as

$$\Delta S_m = \frac{\Delta H_m}{T_E}$$

(with temperature being expressed in kelvins). The value of the concept of entropy is that, as just defined, it is a numerical expression of the

---

*Readers unfamiliar with thermodynamic reasoning are referred to Appendix A.

degree of internal *disorder* in the system.* At T_E melting and crystallization are accompanied by entropy changes of identical magnitude but opposite sign—positive for melting (increasing disorder), negative for crystallization.

The numerical value of ΔS_m at any given pressure is a unique thermodynamic function for every crystalline phase. Expressed for mutual comparison in calories per atom (of the gram-formula weight), values range from ~ 0.3 in crystalline SiO_2 to ~ 1.9–2 in pyroxenes, up to 2.6 in iron-titanium oxides. Entropy of melting is intimately related to all aspects of melting-crystallization behavior, including respective rates of nucleation and growth of crystals.[†]

Free Energy of Crystallization

As temperature falls below T_E, the quantity of heat ΔH_c emitted during crystallization from the now metastable melt changes continuously, as also does the entropy of crystallization ΔS_c. At the same time the value of $T\Delta S_c$ becomes increasingly greater than ΔH_c. The difference

$$\Delta G_c = \Delta H_c - T\Delta S_c,$$

negative in sign, is by definition the *free energy* change within the system resulting from the phase transformation—now a spontaneous and irreversible process. ΔG_c (sign negative) is a measure of the motivating force that makes the crystallization process thermodynamically possible.

---

*This, our first reference to entropy, emphasizes one of its two main aspects, namely the degree of disorder within a system considered on what physicists call the *microscopic scale*— that of individual component particles. But the *definition* of entropy is framed in terms of *macroscopic* data, the results of laboratory measurement and experiment on actual systems— each, of course, consisting of a host of particles. (In this usage even chemical analyses of tiny samples conducted with an electron microprobe fall in the macroscopic category.) The standard definition of entropy is embodied in an empirical equation

$$S_T = \int_0^T C_P \ln T$$

where S_T is molar entropy of the system at temperature T (in kelvins) and some specified constant pressure; and C_P is the heat capacity (a macroscopic property) of the system at that same pressure. This relation is further elaborated and illustrated in Appendix A.

[†]The cited numerical values cannot be extrapolated directly to crystallization of a phase from a melt (such as a magma) of more complex composition. But they indicate in a general way how crystallization rates can vary among simultaneously generated minerals of different composition.

Whether crystallization actually is effective and remains so as ΔG_c increases with falling temperature depends on kinetic factors; for opposing the crystallization process are forces that increase as the temperature falls. Under some circumstances a disordered phase (undercooled liquid or glass) may survive unchanged at temperatures below T_E (even at room temperature). This phase, however, is metastable in that ΔG_c (the criterion of stability) has a high negative value.

Undercooling Factor

The degree of undercooling of a melt in which crystallization is in progress at temperature T_c is expressed in later discussion as

$$\Delta T = T_E - T_c$$

where T_E is the liquidus temperature of the melt at its present composition. Note that except for single-component systems, melts of eutectic composition, or other invariant points on the phase diagram, the liquidus temperature falls steadily as crystallization proceeds. Thus in the later cooling history of a magma, ΔT has a value considerably less than the total temperature drop since crystals first began to form (at some point higher on the liquidus).

Kinetic Controls of Igneous Texture

Process Rates in Relation to Temperature

The theory of rates of chemical processes is based on energy properties of particles—atoms and ions—rather than on the thermodynamic properties of macroscopic phases (crystals, fluids) as measured in calorimetric experiments. A general equation relates the rate k (units per second) of any chemical process to temperature in kelvins:

$$k = Ae^{-E/RT}$$

where A is a "frequency" factor expressing the number of particles participating in the process in unit time; e is the base of natural logarithms, 2.71828; E is activation energy necessary to stimulate a particle into participation in the process; R is the gas constant, 1.9872 cal mole^{-1} degr^{-1}.

Significant with regard to present discussion is the general implication that chemical rates increase exponentially with T, that is, $\ln k \propto T$.

Viscosity of Magmas

Viscosity η of a liquid is conventionally expressed in terms of the reciprocal of the rate of internal strain (flow) under specified conditions. So for the general relation cited above we substitute $\ln \eta \propto 1/T$.

At given T and P the viscosity of a melt depends on chemical composition and directly related internal structure on the atomic scale. High viscosity is typical of siliceous and feldspathic melts in which there is extensive linkage of (SiO_4) and (AlO_4) tetrahedrons. Values are greatly reduced by the presence of certain negative ions, notably $(OH)^-$, so that in residual magmatic liquids the effects of falling temperature and increasing concentration of SiO_2 and alkali feldspar may be offset by simultaneous buildup of H_2O and other fugitive components.

The viscosity of a rhyolitic glass (metastable highly undercooled melt) containing 6.2% water at 2 kb confining pressure was found to increase tenfold with a drop in temperature from 800°C to 700°C. A comparable effect[1] accompanies reduction of the water content from 6.2 to 4.3 %.

Crystallization Rates

Of prime petrographic interest are magmatic crystallization rates as these are reflected in igneous textures. Most relevant experiments and theoretical discussions treat simple one-component systems—for example, SiO_2 or $Na_2Si_2O_5$—in which a single crystalline phase separates continuously from a melt of the same unchanging composition. But the data and generalizations that emerge can be applied qualitatively to magmas—multicomponent systems in which the composition of the melt phase (except in special circumstances) changes continuously as one or more crystalline phases separate from a diminishing volume of melt.

Crystallization of each mineral grain is a two-stage process: *nucleation* of an embryonic crystallite from the melt is followed by *growth* of the nucleus so formed to present crystal dimensions. Nucleation and growth about different individual centers operate simultaneously within a given sample of melt. Each therefore is subject to the same conditions of cooling; but their separate responses to identical physical conditions are fundamentally different when viewed at the atomic-ionic level (see Carmichael, Turner, and Verhoogen, 1974, pp. 147–169). The two corresponding rate patterns, therefore, though in general respects rather similar, differ significantly in details; and in these differences lies the explanation for much of the known variety among common igneous textures.

Growth of a crystal from a stable nucleus is readily visualized in terms of physical properties of the participating phases, as directly measured in the laboratory and recorded in standard tables. The motivating force is the free energy change ΔG_c (sign negative) that accompanies transformation of an undercooled melt to a more stable crystalline phase. The magnitude of ΔG_c increases with the degree of undercooling ΔT below the liquidus temperature T_m of equilibrium crystallization; so that a convenient index of motivation is ΔT. Opposed to the influence of ΔG_c is the effect of viscosity η which reduces the capacity of ions to diffuse through the melt to the face of the growing crystal. The magnitude of viscosity, it will be remembered, increases exponentially as the temperature falls, that is, with increasing ΔT. The rate of growth at any specified value of ΔT is the net product of the two mutually opposed influences. Experimentally based plots of growth rate against ΔT (Figure 2-3) yield curves that at first rise steeply, then peak as negative influence of viscosity starts to overtake the positive influence of ΔG_c, and then fall more gently toward an effective zero where further growth can no longer be detected. Growth thus virtually ceases in a melt undercooled to a few hundred degrees below T_m.

Comparison of different systems shows that, other things being equal, a high value of ΔS_m favors rapid crystal growth.

Nucleation rates cannot be treated in such simple terms. Because the surface area of a very small particle is relatively large compared with the volume, a significant contribution to the free energy of the crystalline nucleus is made by its surface energy. For any given value of undercooling ΔT as computed for large crystals, the magnitude of ΔG_c (sign negative) is correspondingly reduced. The smaller the nucleus and the lower the value of ΔT, the more effective becomes the influence of the surface-energy contribution. The net result is a complex situation, changing continuously with increasing ΔT and with falling T, that is beyond the scope of present discussion. We simply note that experiment, backed by theoretical treatment of simple systems at the atomic level, brings out some generalizations that can probably be applied, exercising due caution, to nucleation in magmas:

1. Some degree of undercooling is a necessary prerequisite for spontaneous nucleation of a crystalline phase.

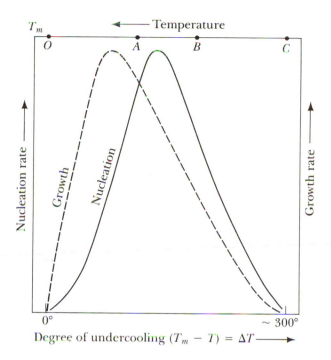

Figure 2-3. Diagrammatic Representation of Respective Rates of Nucleation and Crystal Growth in One-Component Systems, in Relation to Melting Temperature, T_m, and Degree of Undercooling, ΔT, of the liquid phase. See text for details. (Cf. data for $Na_2Si_2O_5$ as depicted by Carmichael, Turner, and Verhoogen, 1974, p. 165).

2. The net rate of nucleation for a specified degree of undercooling again expresses the resultant of two opposing influences. The motivating force is a complex free-energy function involving surface energy, ΔS_m and ΔT, and the opposing force is an activation-energy term analogous to viscosity and increasing exponentially with falling temperature. As with crystal growth the nucleation rate at first rises sharply, peaks, and then falls to effective zero as ΔT increases to an ultimate value of a few hundred degrees (Figure 2-3).

3. Scanty experimental evidence suggests that the nucleation peak is located at a ΔT value perhaps 30 to 50 degrees beyond the growth peak.

4. Among different crystalline phases there is an inverse correlation, other things being equal, between nucleation rate and the value of melting entropy, ΔS_m—the reverse of the situation for growth rates.

Texture in Relation to Rates of Cooling and Crystallization

One-Component Systems

Figure 2-3 illustrates the changing relative roles of nucleation and crystal growth during progressive undercooling of a melt in a one-component system (crystals and melt identical in composition). Crystallization ceases, in any individual case, at some value of ΔT where the melt phase is exhausted or where the effective nucleation rate becomes zero. At just what point this happens depends on the rate of cooling of the crystallizing melt. To illustrate the possible variety in rate patterns and so in the texture of the final product, we note three mutually gradational patterns of undercooling, all of them possible in a melt of given composition but cooling at markedly different rates.

1. Slow cooling: degree of undercooling limited to some span within OA. Throughout the crystallization process crystals grow relatively rapidly about sparsely nucleating embryos. Resultant crystalline aggregates are uniformly coarse grained.
2. Cooling at moderate rates: total span of undercooling somewhere between OA and OB. Early rapid growth about few nuclei during cooling along OA is followed by a phase of rapid nucleation and slower growth between A and B. Resultant crystalline aggregates consist of crystals of two generations: large early crystals set in a matrix of smaller grains formed during the later stage of cooling. If cooling is slightly more rapid a residue of melt may survive beyond C, the point of zero nucleation, to yield a small amount of interstitial glass.
3. Very rapid cooling: the bulk of the melt is carried beyond C without significant nucleation. The product is glass* possibly enclosing scattered nuclei and even small crystals that formed in the earlier stages of cooling.

Magmatic Systems

Crystallization of multicomponent systems such as magmas will show the same general relation of undercooling span to rate of crystallization as that just described (Figure 2-3). But there are three complicating influ-

---

*In this book the term *glass* is used in the strictly petrographic sense to refer to microscopically isotropic, highly viscous, rapidly undercooled (quenched) products of silicate melts; it is used without regard to thermal properties or degree of internal ordering as revealed by x rays.

ences: (1) The total span of temperature along the liquidus surface from the beginning to the end of crystallization may be several hundred degrees—even in systems of only two or three components (e.g., plagioclase or diopside-plagioclase). (2) Despite the tendency toward uniform overall kinetic pattern neither nucleation nor growth curves are identical for two phases crystallizing simultaneously from the same melt. (3) Kinetics of crystallization can be strongly influenced by compositional changes (especially a build-up of fugitive components) in the melt fraction. Possible effects of the above influences will be considered briefly in turn.

Figure 2-4 illustrates four mutually gradational possibilities for undercooling with progressive crystallization in a two-component system, the plagioclase series.

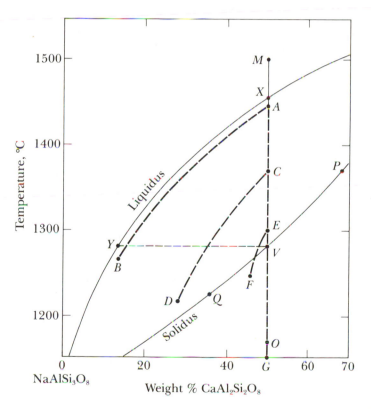

Figure 2-4. Hypothetical Paths of Changing Composition of Crystallizing Melts of the Same Initial Composition, *M*, with Falling Temperature at Atmospheric Pressure in the Anhydrous System NaAlSi$_3$O$_8$−CaAl$_2$Si$_2$O$_8$. Cooling rates: *XAB, very slow; XCD,* moderate; *XEF,* rapid; *XVG,* very rapid.

1. Slow cooling, with crystallization beginning at A, after which melt composition and temperature follow the course AB. Over a total temperature drop of 150 degrees or so, undercooling ΔT never exceeds a few degrees. This is a close approximation to equilibrium cooling (along XY), and the end product is a mass of large unzoned crystals of the same composition V.
2. Moderately rapid cooling, with the beginning of effective crystallization delayed to C, after which the melt composition changes along CD. Cooling is too rapid to permit free ionic diffusion into and from the core of the growing crystal—a required condition to maintain equilibrium (and uniformity of crystal composition) at every stage. The end product is an aggregate of compositionally zoned crystals with the albite content increasing from a central core to the rim (P to Q on the solidus curve).
3. Rapid cooling of the melt along MEF. Crystallization is limited to some sector EF partly below the solidus curve, where undercooling is now high enough for nucleation to outstrip growth. The end product is a mass of small grains of nearly uniform composition V (that of the parent liquid M).
4. Very rapid cooling could carry a large residual fraction of melt to G, beyond O the limit of effective nucleation. The end product would be a mixture of crystals and glass of approximately the same composition, V.

Consider now two crystalline phases separating simultaneously from the same melt—for example, pyroxene and plagioclase crystallizing from basaltic melt. Superposed on the effect illustrated in Figure 2-3 is that of difference between the respective entropies of melting ΔS_m of the two mineral phases. In a slowly cooling melt (OA in Figure 2-3) nucleation of pyroxene is retarded and growth accelerated (because of higher ΔS_m) compared with rates of the same processes for plagioclase (with lower ΔS_m). Herein lies a possible explanation of a common texture (ophitic, p. 61) in products of slow and moderately slow cooling of basaltic magma: Large crystals of augite enclose much more numerous smaller crystals of plagioclase.

Finally, kinetic properties, notably viscosity, of the diminishing melt fraction may change profoundly with progressive solidification. Viscosity, it will be remembered, is the main contributor to opposition forces that slow down nucleation and growth rates with falling temperature. Increasing viscosity due to build-up of SiO_2 in the residual melts can slow crystallization to a standstill while a substantial fraction of melt still survives as glass. Reduced viscosity accompanying increase in water concentration can have an exactly opposite effect, permitting crystallization to reach completion (within the span *OA* of Figure 2-3) while nucleation is still slow compared with growth rates (as in pegmatitic rocks).

GENERAL TEXTURAL CHARACTERISTICS

Petrography, like any other branch of science, requires special terms to describe the phenomena peculiar to its field. For the modern petrographer, however, history has bequeathed an elaborate textural nomenclature encumbered by superfluous technical words. To satisfy the requirements for this book, we have arbitrarily selected what appear to us the most useful terms and have rejected others, including some equally valid synonyms. Some terms that we consider superfluous for our present purpose have been relegated to single footnotes marked ° or °°, each with a page reference in the Index.

Degree of Crystallinity and Grain Size

Some igneous rocks, such as granite, are composed wholly of crystals and are therefore classed as *holocrystalline*. Others, such as obsidian, consist entirely of glass; still others, including many lavas and shallow intrusive rocks, contain both glass and crystals.° Minute crystals, usually of tabular or prismatic habit, are called *microlites* provided they are birefringent; if they are even smaller, spherical, rodlike and hairlike isotropic forms, they are called *crystallites* (Figure 2-5).

Consider next the *grain size*. In this respect igneous rocks vary greatly; some are so fine grained that individual crystals cannot be distinguished separately even with a hand lens, while others contain crystals centimeters or even tens of centimeters in length. If most of the constituents are so small as not to be visible to the unaided eye, the rock is called an *aphanite* and its texture *aphanitic*. Some aphanitic rocks prove on microscopic examination to contain glass; others are completely crystalline. If the individual component crystals in such rocks can be distinguished with

---

°*Holohyaline* and *hypocrystalline,* respectively.

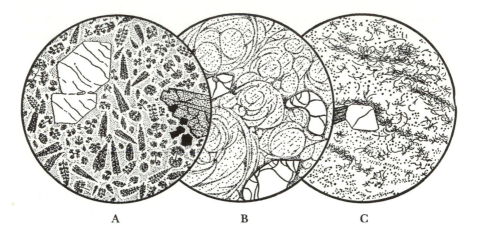

A   B   C

Figure 2-5. Rhyolitic Pitchstones with Microlites and Crystallites

A. Isle of Arran, Scotland. Diam. 1 mm. Phenocrysts of quartz, augite, and magnetite in a glassy matrix crowded with arborescent microlites of green hornblende, around which the glass is clear.

B. Meissen, Saxony. Diam. 2 mm. Phenocrysts of quartz with corroded outlines and conchoidal fractures, in a matrix of glass showing perlitic cracks. Trains of spherical crystallites emphasize the fluidal banding.

C. Turtle Mountains, California. Diam. 1 mm. Hornblende and sanidine phenocrysts lie in a matrix of glass rich in spherical and hairlike crystallites.

a microscope, the texture may be called *microcrystalline;* if they are too minute to be distinguished with a microscope even under high magnification, the texture is *cryptocrystalline.** Most igneous rocks are composed largely of crystals big enough to be identified with the aid of a hand lens alone.° Where their grain size is more or less uniform, the grain is said to be *fine* if the average diameter of individual grains is 1 mm or less; *medium* if between 1 mm and 5 mm; *coarse* if between 5 mm and 3 cm; *very coarse* if more than 3 cm. These values, of course, are merely guides to uniform usage.

The degree of crystallinity and the grain size of any igneous rock depend mainly on its cooling history during the period of solidification, but partly too on the chemical composition of the magma as this affects its viscosity. Rate of cooling and depth-controlled pressure are all-impor-

---

*Some materials that are optically isotropic prove to be cryptocrystalline when examined by x-ray diffraction.
°Phaneritic texture.

tant; for on the former depends the degree of undercooling below liquidus temperatures throughout the cooling and solidification process,
while high pressures tend to prevent release of the fugitive components
of the crystallizing magma, thereby maintaining its viscosity at minimal
levels.

Holocrystalline rocks of medium and coarser grain size are typical of
massive plutonic intrusions. They have crystallized slowly from magmas
intruded perhaps at 800°C–1000°C and then cooled slowly (maintaining
much of their original content of fugitive components) because of their
size, shape, or depth of cover. During the final stages of consolidation,
residual liquids tend to become greatly enriched in fugitive components
(H_2O, CO_2, F, and others). All these factors combine to produce a regime
(with emphasis on minimal undercooling and decreasing viscosity) that
favors rapid crystal growth in a liquid medium within which nucleation
rates continue to be relatively slow. In the very last stages of crystallization
the ultimate liquid residues, drawn into cracks and fissures in the now
largely crystalline body, may have become so enriched in fugitive components that actual boiling ensues. Migration of ions essential to crystal
growth is now greatly facilitated, and individual crystals may develop to
dimensions of several or even tens of centimeters—in some cases even
several meters. Under such conditions extremely coarse-grained rocks
known as *pegmatites* develop. Expectably these contain more than their
share of fugitive components now fixed in such minerals as micas, tourmaline, apatite, fluorite, and others.

Finely crystalline, aphanitic, and glassy textures reflect rapid cooling
conducive to a high degree of undercooling and increasing viscosity.
Solidification is therefore accomplished mainly under a regime favoring
rapid nucleation while crystal growth rates are steadily declining. A fine-
grained crystalline aggregate results. If cooling is rapid enough even the
nucleation rate reaches an effective zero, while much of the liquid (now
extremely viscous) remains. The final product is then glass. These general conditions are commonly met in extrusive magmas or within small
intrusive bodies emplaced at shallow depth. In both situations, prevailing
low pressures facilitate loss of fugitive components from the liquid
magma, thus further increasing viscosities and compounding the effects
due to rapid cooling in the volcanic environment. Because magmas rich
in SiO_2 or containing a high concentration of complex ionic groups with
alkalies and aluminum are more viscous than less siliceous magmas rich
in such ions as Ca^{2+}, Mg^{2+}, and Fe^{2+}, abundant glass is more often
present in such lavas as rhyolite, dacite, trachyte and phonolite than in

basaltic rocks. Even basalt, however, may chill rapidly enough to produce glass. Glass of any kind is metastable at ordinary temperatures and therefore, given sufficient time, tends to crystallize or *devitrify*, and it is also readily altered if affected by circulating solutions or if subjected to increased temperatures as a result of deep burial. For this reason ancient terrestrial glasses are extremely rare.

The careful student will beware of supposing that coarse-grained igneous rocks are necessarily of deep-seated origin, or that fine-grained ones are necessarily volcanic in the field sense of the term. The margins of some plutonic bodies, though never glassy, may be aphanitic. Examples are known of shallow dike rocks in which glassy margins, produced by sudden chilling of magma against cold walls, enclose very coarse-grained material that crystallized from slow-cooling magma rich in H_2O, CO_2, or F. In fact, the central part of a thick basaltic lava flow may be just as coarse grained as many plutonic rocks, even though its crust and bottom consist chiefly of glass.

Igneous Fabrics and Order of Crystallization

Our next concern is with the *fabric* of rocks—that is, with the shape and mutual relationships between grains, or, in other words, with the forms of grains and the manner of their articulation. Grains completely bounded by their own rational crystal faces are called *euhedral*; those only partly bounded by crystal faces are called *subhedral*; and those devoid of crystal faces, and often of irregular shape, are referred to as *anhedral*.° Grains may also be characterized by particular habits, such as columnar, acicular, fibrous, tabular, prismatic, equant, and platy.

Geometric-crystallographic boundary relations between contiguous grains of different minerals as seen in thin sections of igneous rocks tend to conform to regular patterns. Quartz in plutonic rocks almost always occurs as aggregates of anhedral grains. Sphene, zircon, and apatite habitually take the form of small isolated euhedral crystals. Biotite crystals abutting against feldspar or quartz in granitic rocks tend to develop sharp boundaries parallel to {001}. These and numerous comparable generalizations raise intriguing questions regarding the possible influ-

---

°Widely used terms synonymous with *euhedral, subhedral,* and *anhedral* are *idiomorphic, hypidiomorphic,* and *allotriomorphic,* respectively.

ence of such diversified controls as properties inherent to the minerals in question, kinetics of crystal nucleation and growth, and chronological sequences of crystallization from parent magmas.

A century ago Rosenbusch proposed a general "order of crystallization" of igneous minerals that, as the "Rosenbusch rule," became incorporated into classic petrographic lore. This "rule" was based on general petrographic criteria, none of them without value, but all of them now known to be far from infallible. The three principal criteria are worth noting since each can be applied (with due caution) to particular textures, especially to those of volcanic rocks:

1. When grains of one mineral habitually enclose those of another, the second is of earlier origin.* However, small late crystals also may be enclosed within an earlier host, a relationship that is clearly due in many instances to quite different controls: exsolution from a once homogeneous solid solution (e.g., exsolved lamellae of orthopyroxene enclosed in augites of plutonic rocks); late crystallization of small droplets of melt occluded in rapidly growing early crystals. Finally, diabases very commonly display a texture (ophitic, see p. 61) in which small sharply defined plagioclase crystals are enclosed in much coarser host grains of augite; yet the two are now believed to have crystallized simultaneously but at different rates of nucleation (possibly stemming from differences in entropy of melting).

2. Where two sets of crystals differ conspicuously in size, the larger comprise an earlier, the smaller a later generation. This interpretation is consistent with the generalized kinetic cycle of slow followed by rapid nucleation. The criterion of relative size can be applied with confidence to volcanic rocks containing two distinct size fractions of a single mineral, for example, plagioclase in andesites. In plutonic rocks, other factors, especially postmagmatic responses to circulating pore fluids and to changing conditions of stress during slow cooling, may be responsible for ultimate differences in grain size among the constituents.

3. Early-formed minerals growing freely in an enclosing melt assume euhedral outlines. Crystals of later phases, in contrast, grow in partial mutual contact as the volume of melt diminishes and tend to build

---

*Note, however, that a thin section gives only a two-dimensional picture of an actual three-dimensional situation, and so may give a false impression of spatial relations between contiguous crystals having deeply embayed or otherwise irregular outlines.

aggregates of subhedral and anhedral grains. Both propositions are true enough. But euhedral outlines are no guarantee of early origin. Solution theory suggests that accessory constituents like apatite and sphene, both typically euhedral, can crystallize only when the content of key minor elements (in the above cases phosphorus and titanium) in residual fluids has reached some threshold concentration. Such minerals are likely to be not the first but the last in the crystallization sequence. Again during the later stages when crystals of two different phases are growing in mutual contact, surface energy (itself depending on crystal structures) determines which phase shall exert its own crystal outlines against its neighbors. By analogy with a well-established characteristic of metamorphic textures (p. 439) the capacity for euhedralism should tend to diminish in the order ortho-, chain, sheet, framework silicates. Minerals such as zircon, garnet, and sphene should, on this assumption, rank high, while feldspars and quartz should be placed at the bottom of the scale.

The Rosenbusch rule, once a principal tenet of textural interpretation, is now of little more than historical interest. But for these very reasons the modern student of petrography should be aware of its provisions. According to Rosenbusch the general sequence of crystallization from igneous melts is as follows:

1. Minor constituents such as zircon, apatite, sphene.
2. Mafic silicates in the order: olivine, pyroxenes, amphiboles, micas (ortho-, chain, sheet silicates).
3. Feldspars in the order calcic to sodic plagioclase, alkali feldspars; quartz.

We now realize that such a diversified phenomenon as crystallization of magma defies expression in such simple terms, and that fabric characteristics once attributed solely to generalized sequence in time can develop under the influence of other and more subtle controls. The nearest approach to a satisfactory chronological generalization perhaps is the notion of parallel simultaneously operating reaction series suggested by Bowen (see p. 19). Yet even this is but a greatly simplified picture of magmatic crystallization, with numerous exceptions and obvious possibilities for elaboration in specific instances. While calcic plagioclase, for instance, usually precedes sodic plagioclase in sequence of formation, both oscillatory and reverse zoning may be found in individual crystals, and not uncommonly the late-crystallizing grains of plagioclase in the groundmass of a rock are more calcic than some of the large, early-formed feldspars. Mingling of partly crystallized magmas and assimilation of calcic wall rocks are two of several possible explanations for such anomalies. It should be remembered, moreover, in considering the members of the discontinuous reaction series, that each has isomorphous varieties, and that those rich in magnesium usually antedate those rich in iron. The iron-rich olivine fayalite, for example, generally forms at a very late stage, often with quartz or tridymite, whereas the magnesian mica phlogopite usually forms at an early stage in basic rocks, along with olivine and highly calcic feldspars.

Specific Textures Defined

We may now continue discussion of fabrics by defining the main textural terms applied to igneous rocks. If most of the minerals in a holocrystalline rock are approximately equidimensional or equant and of more or less uniform size, the texture is said to be *granular* or *equigranular*. Pegmatites generally have extremely coarse granular texture, most plutonic rocks are coarse or medium granular, and many dike rocks and lavas are finely granular or microgranular. The commonest granular texture, by far, is one in which some constituents are euhedral, some subhedral, and the rest anhedral. Such a texture is called *subhedral granular* (Figure 2-6A), but because it is well exemplified by most granites, it is commonly referred to as *granitic texture*. If almost all of the constituents are anhedral, the texture is called *anhedral granular* (Figure 2-6C), a texture typically displayed in the felsic dike rocks called aplite, where it is referred to as *aplitic texture*, and in many medium- and coarse-grained ultrabasic and basic plutonic rocks.

Many igneous rocks contain relatively large crystals in a distinctly finer-grained or glassy matrix, or *groundmass*. Such notably inequigranular rocks are called *porphyritic*, the large crystals being termed *phenocrysts* (Figure 2-6B). Phenocrysts, always large with respect to the groundmass crystals, may be so small that a microscope is needed to detect them; they are then termed microphenocrysts and the texture is *microporphyritic*. Rocks in which the phenocrysts are gathered in distinct clusters are called *glomeroporphyritic* (see Chapter 3, Figure 3-1B). If phenocrysts lie in a

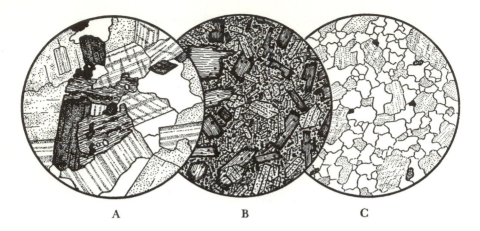

A   B   C

Figure 2-6. Igneous Textures

A. Subhedral granular texture in granodiorite. Diam. 3 mm. Benton Range, Mono County, California. Euhedral and subhedral crystals of green hornblende and brown biotite, the latter containing inclusions of apatite and secondary sphene. Subhedral crystals of plagioclase, and more poorly formed crystals of partially altered orthoclase (stippled), with clear, anhedral, interstitial patches of quartz.

B. Porphyritic texture in mica lamprophyre. Diam. 2 mm. Boundary Butte, Navajo Reservation, Utah. Euhedral prisms of diopside and flakes of zoned biotite, in a matrix of altered sanidine microlites, opaque oxides, and calcite.

C. Anhedral granular texture in granite aplite. Diam. 3 mm. Near Wellington, Nevada. Interlocking anhedral grains of quartz, microcline, orthoclase, and albite, with accessory hornblende and magnetite.

matrix composed of glass, the texture is called *vitrophyric*; and if aphanitic or glassy rocks are completely free of phenocrysts, they are termed *aphyric* or *nonporphyritic*.*

Porphyritic textures originate in many ways. Magma may begin to crystallize slowly at depth, where large crystals can develop about sparse nuclei, and then, before complete solidification, it may be injected between cold wall rocks or be erupted at the surface, so that the residual liquid crystallizes rapidly to form a fine-grained groundmass or congeals almost at once to produce a glassy matrix. There is no justification, however, for invoking change in the environment of a crystallizing magma

---

*The root *phyr-* or *phyre*, which appears in many igneous rock names, is merely a letter combination abstracted from *porphyry*, itself a name from antiquity denoting a particular rock purple in color and having this structure (the Greek word *porphyros* means "purple").

to explain all porphyritic textures. Some are the result of continuous crystallization while cooling slowly under uniform pressure: At first the phenocrysts grow freely about sparse nuclei until undercooling reaches the critical stage for rapid nucleation, when the remaining liquid yields a crop of small crystals that, if undercooling is rapid enough, may be set in a small residue of glass. Again there are granites with huge euhedral crystals of K-feldspar or of oligoclase set in a normal granitic ground-mass; and at least in some cases these "megaphenocrysts" (more properly called *porphyroblasts,* see p. 439) are of late metasomatic origin.

In some siliceous igneous rocks, particularly granites, granite peg-matites, and granophyres, quartz is intergrown with alkali feldspar, either through simultaneous crystallization from a eutectic mixture or through replacement of one mineral by the other. In these intergrowths the quartz is commonly cuneiform, resembling runic inscriptions on a background of feldspar. The resultant texture is therefore termed *graphic* (Figure 2-8A). Less often, other pairs of minerals are thus intergrown. Somewhat similar to the graphic intergrowth is the *myrmekitic* texture, which is characterized by minute wormlike or fingerlike bodies of quartz enclosed in sodic plagioclase, usually oligoclase. It results from replace-ment of the marginal parts of potassic feldspars, especially at contacts with plagioclase, owing to late magmatic or post-consolidation reactions.

In some rocks, especially gabbros, diabases, and basalts, laths of pla-gioclase may lie in a matrix of coarse subhedral augite or pigeonite, so that in thin sections the feldspar laths, whose average length does not exceed the diameters of the pyroxene grains, appear to be largely or entirely enclosed in pyroxene. Such intergrowths are called *ophitic* (Fig-ure 2-7B). If the average length of the plagioclase laths exceeds that of the pyroxene grains, and the latter only partly enclose a number of the former, the texture is *subophitic* (Figure 2-7C). At one time ophitic texture was thought to reflect a chronological crystallization sequence: plagio-clase followed by pyroxene. Bowen (1928) convincingly argued for an alternative model of simultaneous crystallization of both phases, and this view soon gained general acceptance. The size anomaly can be explained in terms of crystallization kinetics controlled by marked difference between the respective melting entropies, ΔS_m, of feldspar and pyroxene.

Rocks in which numerous grains of various minerals in random ori-entation are completely enclosed within large, optically continuous crys-tals of different composition are said to have a *poikilitic* texture (Figure 2-7A). Such rocks may have a mottled luster. In many acid plutonic rocks, plates of K-feldspar enclose abundant laths of plagioclase, and in some

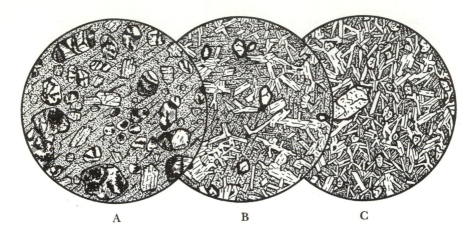

A   B   C

Figure 2-7. Igneous Textures

A. Poikilitic texture in hornblende peridotite, Odenwald, Germany. Diam. 3 mm. A single crystal of hornblende encloses rounded granules of serpentinized olivine and subhedral prisms of fresh diopside.

B. Ophitic texture in basalt, Kauai, Hawaiian Islands. Diam. 3 mm. Large plates of pigeonite partly enclosing laths of labradorite, and granules of olivine marginally altered to iddingsite.

C. Subophitic texture in basalt, Medicine Lake, California. Diam. 2 mm. Crystals of augite partly enveloping some of the feldspars and partly interstitial between them. One phenocryst and abundant small granules of olivine.

ultrabasic rocks, like the peridotite of Figure 2-7A, plates of hornblende surround ovoid granules of olivine and pyroxene. In most cases poikilitic texture probably does express a crystallization sequence, the host phase being the latest to finish crystallizing. This topic is further developed in connection with the crystallization cycle in massive layered basic intrusions (see p. 119).

Coronas (reaction rims) envelop some minerals in igneous rocks. Olivine, the earliest member of the discontinuous reaction series, may be surrounded by later members, such as pyroxene or amphibole, owing to incomplete reaction of the olivine with the melt in which it was temporarily suspended. Or coronas may be formed by postmagmatic reactions between adjacent minerals, induced either by deuteric solutions or by still later low-grade metamorphism. Many of these secondary coronas, sometimes distinguished as *kelyphitic rims*, are marked by concentric shells with a radial fibrous texture. Rims of this kind are especially common in basic and ultrabasic rocks (Figure 2-8B,C). In some troctolites, for exam-

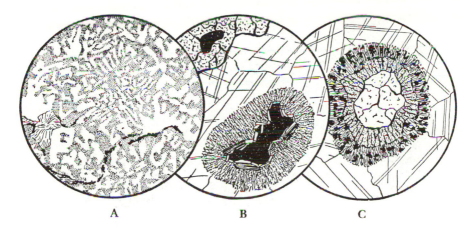

Figure 2-8. Igneous Textures

A. Micrographic texture in granophyre, Rosskopf, Vosges, Germany. Diam. 2 mm. Cuneiform intergrowth of quartz and altered orthoclase. In lower part of section are granules of magnetite and flakes of hematite and lithium mica.

B. Kelyphitic rims around green spinel in troctolite, Quebec. Diam. 2 mm. In upper part of section, green spinel is included in pyrope garnet; in lower part, the spinel is enveloped by a rim of anthophyllite and pale phlogopite, surrounded in turn by a radiating fibrous intergrowth of tremolite and actinolite. These rims result from reaction between the spinel and the labradorite that makes up the rest of the section.

C. Kelyphitic rim around olivine in gabbro, Quebec. Diam. 2 mm. The olivine is enclosed by a shell of hypersthene, around which is a second shell composed of actinolite and green spinel. The rest of the section consists of labradorite.

ple, kernels of olivine are enveloped by jackets of orthopyroxene or amphibole or both, which are wrapped, in turn, by shells of green amphibole and spinel. These rims are produced by reaction between olivine and calcic feldspar. Similar concentric fibrous rims may be observed around grains of pyrope garnet in some ultrabasic rocks.

In many lavas and "hypabyssal" rocks, especially basalts and diabases, the angular interstices between the feldspars are occupied by ferromagnesian granules, usually olivine, pyroxene, or iron-titanium oxides, of random orientation. The resultant texture is called *intergranular* (Figure 2-9A). Or the interstices may be filled with glass, cryptocrystalline material, or nongranular deuteric and secondary minerals, such as serpentine, smectite, chlorite, calcite, zeolites, and sodalite; the texture is then called *intersertal* (Figure 2-9B). And just as the intergranular texture grades into the subophitic when pyroxenes begin to mold themselves

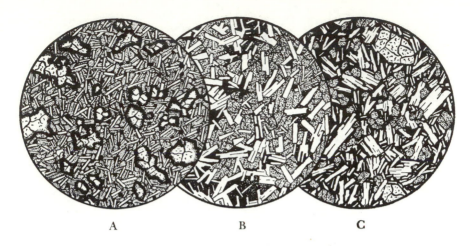

A   B   C

Figure 2-9. Igneous Textures

A. Intergranular texture in picrite basalt, Kilauea, Hawaii. Diam. 2.5 mm. Corroded phenocrysts of olivine rimmed with magnetite and hematite in an intergranular matrix composed of laths of labrodorite and interstitial grains of augite and pigeonite.

B. Intersertal texture in tholeiitic diabase, Northumberland, England. Diam. 2 mm. Augite and labradorite occur in ophitic intergrowth; between them are irregular pools of dark-brown glass.

C. Hyaloophitic texture in basalt, Pedregal, Mexico. Diam. 2 mm. Olivine, green diopsidic augite, and laths of labradorite lie in a matrix of dark, iron-rich glass.

around the ends of the feldspars, so does the intersertal texture merge into the *hyaloophitic* as the glass and nongranular minerals begin to envelop the feldspars. Intersertal texture also grades into the so-called *hyalopilitic* texture, typical of many lavas, in which glass occupies minute interspaces between microlites of feldspar in haphazard orientation (Figure 2-10C).

The holocrystalline matrix of some dense rocks consists of tightly appressed microlites, generally of feldspar, interwoven in irregular fashion. Such a groundmass is called *felty*. If, as in many andesites and trachytes, the crowded microlites of feldspar are disposed in a subparallel manner as a result of flow and their interstices are occupied by micro- or cryptocrystalline material, the texture is called *pilotaxitic* (Figure 2-10B) or *trachytic* (Figure 2-10A).

Expanding gases in lavas and shallow intrusions often form cavities or

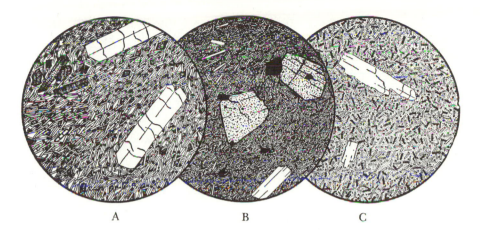

A   B   C

Figure 2-10. Igneous Textures

A. Trachytic texture in trachyte, Castello d'Ischia, Italy. Diam. 2 mm. Phenocrysts of sanidine and of golden-yellow, oxidized aegirine-augite, in a fluidal groundmass of subparallel sanidine laths with intergranular aegirine-augite, aegirite, and iron oxides, plus accessory apatite and sphene. Many triangular and polygonal spaces between the sanidine laths are occupied in intersertal fashion by analcite or sodalite.

B. Pilotaxitic texture in hypersthene andesite, Mount Rainier, Washington. Diam. 2 mm. Phenocrysts of hypersthene and labradorite, in a groundmass of andesine microlites with interstitial cryptocrystalline material and specks of augite and iron oxides. The fluidal banding is much less pronounced than in rocks of trachytic texture.

C. Hyalopilitic texture in pyroxene dacite, Weiselberg, northern Germany. Diam. 2 mm. Phenocrysts of labradorite, together with microlites of andesine-oligoclase and slender prisms of pigeonite of random orientation, in a matrix of clear brown glass.

vesicles. Usually these vesicles are spherical or ovoid, or have arcuate outlines, but many are highly irregular. They may subsequently be filled with deuteric or secondary minerals, such as opal, chalcedony, chlorite, calcite, and zeolites, to form *amygdules.* Features of this kind are never found in plutonic rocks. These rocks, however, may contain irregular cavities, often a centimeter or two across, into which large subhedral and euhedral crystals project. These are referred to as *drusy* or *miarolitic* cavities. In some lavas, between laths of feldspar of random orientation, there are minute, angular cavities shaped more or less like glassy areas in Figure 2-9B. The feldspars bordering such cavities are of the same size and composition as in the areas between the cavities, whereas the crystals projecting into miarolitic cavities are generally larger than those of the surrounding rock and often of different composition.

In siliceous lavas, particularly those rich in glass, radial aggregates of acicular and fibrous minerals are common. These are known as *spherulites* (see Chapter 5, Figure 5-12B). Some are actually spheroidal; others, owing to mutual interference during growth, are polygonal; others are elongated; still others have quite irregular outlines. A few of these spherulites are composed entirely of feldspar, but most of them are intergrowths of cristobalite or tridymite with alkali feldspar. Generally these radial aggregates result from devitrification of glass at relatively high temperatures during the cooling episode after the lavas ceased to flow (see pp. 186–187). Spherulitic texture is exceptional among basic rocks. It may be seen, however, in some basalts and diabases, especially in flows marked by pillow structures and in the glassy margins of dikes and sills. These radial or sheaflike bodies in basic rocks are called *varioles* (see Chapter 3, Figure 3-5C), and the resulting texture is called *variolitic*. Most varioles consist of divergent plagioclase fibers, with or without interstitial glass, or of plagioclase fibers intergrown with granules of pyroxene, olivine, or iron oxides. Another radial texture may be seen in certain medium- and fine-grained dike rocks that contain irregular interlocking laths of alkali feldspar, arranged in crudely divergent groups.

Finally there are various textures best described as *clastic* on account of the fractured appearance of the constituents. Among these textures are the *pyroclastic*, typical of the fragmental products of volcanoes, discussed in Chapter 9. Some intrusive magmas continue to move even after they are almost wholly crystallized, so that many of their crystals become granulated and rounded by rubbing together during differential flow. Excellent examples of such *protoclastic* or *autoclastic* texture may be noted near the margins of large granitic and granodioritic intrusions, and also in anorthosites. Crushing and fragmentation of crystals may also result from post-consolidation movements. The resultant textures are called *cataclastic*; if granulation and shearing of the crystals are extreme, the texture is called *mylonitic*. It may be difficult or even impossible to distinguish protoclastic from cataclastic texture, but the former, owing to crystallization of residual liquids after magmatic flow came to an end, is sometimes characterized by veins and interstitial patches of unstrained and uncrushed material between the granulated constituents.

To special textures commonly developed by crystal settling and accumulation, particularly in the lower levels of basic intrusions, the general term *cumulate texture* has been applied (see pp. 119–120).

NOMENCLATURE AND CLASSIFICATION

No formal rules govern the nomenclature of igneous rocks. Some names are of ancient origin, usually a heritage from miners' jargon; a few are derived from the names of the component minerals of particular rocks; most are based on the names of their type localities. Altogether more than a thousand names have been coined, though it is doubtful whether even the specialist has need for as many as a hundred. Happily there is less tendency nowadays to propose new names for unusual rocks and for minor variants of common ones. It would be better if many variants had been designated simply by affixing mineral prefixes to well-established names. As with textural terms the nomenclature of rock types has proliferated unnecessarily, and drastic selection of rock names is required to keep the working vocabulary to a minimum consistent with everyday needs.

Confusion has been increased by different shades of connotation attached even to widely used rock names. This situation cannot be completely avoided, for one rock may require different names as the basis of classification varies to suit the needs of alternative approaches, for example, petrographic versus chemical. A basalt in one scheme becomes an andesite in another, and a diorite becomes a gabbro, or vice versa, depending on whether distinctions are based on the nature of the feldspars, or on abundance of dark minerals, or on silica percentage. It may seem especially unfortunate that the commonest igneous rocks are usually those most vaguely defined. We should remember, however, that our confusion is only in part a result of conflicting classifications; it results no less from the fact that all rock types grade into one another. "Pigeonhole classifications," with names comfortably enclosed between parallel lines, are useful as aids to memory and for purposes of comparison, but they give a false impression of accuracy and are quite misleading when they suggest clear-cut distinctions where none exist (as, for example, at the basalt-andesite transition). The student, once he has gained experience, will realize that only an artificial classification can be rigid, and that looseness of definition is not necessarily an evil. He must be content when assigning a name to a rock to define his basis of classification, knowing that others may prefer a different name because they favor some other basis. The study of rocks, it has been well said, should start in the field, continue with the microscope, and finish with the crucible. A rock may be given one name on the ground of field occurrence and from hand-lens examination, only to require another when it is studied in thin section, and perhaps a third when it is chemically analyzed. These are difficulties that cannot be avoided. A single scheme for classifying rocks that is both logical and practical does not seem attainable. Different schemes have different objects in view. One may give chief weight to manner of occurrence; another may emphasize coarseness and texture; a third may consider mineral content of prime importance; and a fourth may be founded primarily on chemical composition. Most schemes have merit, but none can fully combine the advantages of all. The following notes attempt to summarize some alternative approaches currently adopted in describing and classifying igneous rocks to suit the needs of different petrologic problems and situations. Fuller discussions will be found in some of the standard works listed at the end of Chapter 1.

Volcanic and Plutonic Rocks

Field mapping and attendant scrutiny of rock samples reveal the geometric form, extent, and the general lithology of an igneous body. Seen thus, two broad categories of igneous rocks are readily distinguished— one comprising surface flows and the other major, cross-cutting, presumably deep-seated intrusions. Not all igneous bodies, however, precisely fit one or the other of these groups. Rosenbusch recognized this and added a third and in some respects intermediate category to accommodate rocks of minor intrusions. He advocated a primary threefold division: effusive rocks, dike rocks, and deep-seated rocks. Today many geologists follow the same plan, but the three divisions are now more generally designated *volcanic, hypabyssal,* and *plutonic.* Many hypabyssal rocks are intimately associated, and indeed chemically and petrographically identical with volcanic rocks of thick lava flows. One of the most spectacular deployments of basaltic magma takes the form of massive dikes and sills of great extent invading sediments of the upper continental crusts. They are composed of coarse-grained diabases which, however, are indistinguishable from rocks of individual flows of truly volcanic piles. Less commonly, hypabyssal rocks may have closer affinities with plutonic counterparts especially where they cross-cut or are direct offshoots from major plutons of the same or closely similar age.

Clearly, on the basis of petrographic evidence alone, the hypabyssal group has no individual identity; so from the viewpoint of the petrographer, Rosenbusch's threefold division becomes impracticable. Nevertheless, even without knowledge of field relations, one may distinguish most plutonic rocks from volcanic ones, and vice versa, solely on the basis of texture and mineral composition. Monomineralic rocks such as dunite, anorthosite, and pyroxenite are unknown among lavas. Minerals stable at high temperatures and low pressures, such as sanidine and leucite, and metastable phases such as tridymite and glass, are seldom seen in plutonic rocks, whereas some other minerals, like muscovite, tourmaline, cancrinite, and microcline, which form at lower temperatures or higher pressures, are rarely found in volcanic rocks. And because slow cooling facilitates exsolution (unmixing), as in perthite, the minerals in which its effects are optically recognizable are far more common among plutonic rocks. Besides, some hydroxyl-bearing minerals that are stable under plutonic conditions tend to break down in a volcanic environment of high temperature and low pressure. Few volcanic rocks, therefore, except quickly cooled, glass-rich types, contain unaltered biotite or hornblende; in holocrystalline lavas, formed by slow cooling, these minerals are generally replaced by oxyhornblende, pyroxene, and granular iron-titanium oxides. The quartz phenocrysts in volcanic rocks often contain inclusions of glass in euhedral cavities, but such inclusions are extremely rare in the quartz of plutonic rocks. Moreover, the optical properties of plagioclase in plutonic rocks commonly differ from those in extrusive rocks, so that it has become possible as a result of recent studies to distinguish a high-temperature volcanic series of feldspars (disordered) from a low-temperature plutonic series (relatively ordered).

Textural differences are even more striking. Most plutonic rocks are holocrystalline, medium- or coarse-grained, and nonporphyritic; and their main constituents occur in anhedral or subhedral grains. Many volcanic rocks are porphyritic, with euhedral phenocrysts in a fine-grained or partially glassy groundmass; others are aphanitic or composed largely of glass. In many volcanic rocks groundmass minerals, especially feldspars, tend to euhedral outlines.

For these reasons, we prefer the twofold division recognizing only a volcanic and a plutonic class. The distinctive criteria are entirely petrographic, with strong emphasis on texture. Most rocks that could be classed in the field as hypabyssal fall within the volcanic category as petrographically defined. Others do not: diabases, lamprophyres, and kimberlites come to mind. But the petrographer has no obligation to force them into either the plutonic or the volcanic category; nor need he collect such disparate types into a third category. Either course would distort reality. In this book such rocks are simply treated separately from the two categories that do successfully accommodate the vast majority of igneous rocks.

Chemical and Quasi-Chemical Classifications

Objectives and Limitations

In order to discuss problems relating to the nature, origin, and evolution of magmas, to compare igneous rock series, and to interpret chemical analyses themselves, it is desirable to have chemical classifications of igneous rocks. Obviously, these tell nothing of rock textures and only hint at mineral content. Thus they ignore the cooling and crystallization history, for magmas of generally similar (though not necessarily identical) chemical composition may yield rocks of many different textures and markedly different mineral content—witness obsidian, porphyritic rhyolite, and granite.

To what degree, moreover, does the composition of an igneous rock correspond with that of a magma? As demonstrated particularly by Bowen, the degree to which bulk rock analyses represent compositions of liquid magmas can be established only through scrutiny of textures. An aphanitic or purely glassy rock generally can be assumed to represent a chilled liquid of exactly the same chemical composition (apart from deuteric alteration and hydration; see pp. 27–28). The same applies to the uniformly fine groundmass of a porphyritic rock. But with respect to medium- and coarse-grained holocrystalline rocks of major plutons and all porphyritic volcanic rocks, such assumptions are at least doubtful and in some cases certainly erroneous. Phenocrysts tend to become locally concentrated by sinking or rising under gravity; the resultant porphyritic rock may thus represent a concentrated crop of early crystals entrapped in a later liquid fraction. Many plutonic rocks, especially those of the gabbro family, consist essentially of crops of crystals that have accumulated on the floors of extensive magma chambers, thus becoming separated and sealed off from later liquid fractions of the parent melt. A sequence of cumulate fractions, each of which is a different plutonic rock, cannot be regarded as a liquid line of descent, for none of the successive liquid fractions, nor even the undifferentiated parent magma, have chemically equivalent counterparts among the plutonic rocks to which they gave birth.

Bulk chemical analyses contain a wealth of useful information and indeed are indispensable to the geochemist and petrologist. But a rock classification based solely on chemical composition, without reference to texture and mineralogy, has somewhat limited applicability, and for petrographic purposes is completely inadequate.

Significance of SiO_2

All igneous rocks are composed largely of silicates, with or without quartz, and their predominant chemical component is SiO_2. Yet the weight percentage of SiO_2 in common rocks ranges from about 40% to 75%. Content of SiO_2 is thus an obvious parameter for purposes of classification, and a fourfold grouping of igneous rocks on this basis has long had a place in the standard vocabulary of geology. The four groups are designated *acid*, *intermediate*, *basic*, and *ultrabasic*. Rocks with more than 66% SiO_2 are called acid, those with 52% to 66% intermediate, those with 45% to 52% basic, and those with less than 45% ultrabasic. This terminology perpetuates the archaic notion of SiO_2 as a component of "silicic acids," the metallic oxide components of mineral silicates once being regarded by contrast as "bases."

In the absence of chemical analyses, SiO_2 content of the rock as a whole can be approximately inferred from the assemblage of minerals present in the rock. By geologists in general and by most petrographers, the terms acid, intermediate, basic, and ultrabasic are loosely used in a quasi-chemical, qualitative sense. Pertinent values of SiO_2 percentage for the principal mineral constituents of igneous rocks serve as a useful guide.

Felsic minerals
$\begin{cases} \text{quartz, 100; alkali feldspars, } 64-66; \text{ oligoclase, 62;} \\ \text{andesine, } 59-60; \text{ labradorite, } 52-53; \text{ bytownite, 47;} \\ \text{leucite, } {\sim}54; \text{ nepheline, } {\sim}40-44; \text{ kalsilite, 39.} \end{cases}$

Mafic minerals
$\begin{cases} \text{magnesian and diopsidic pyroxenes, } 50-55; \text{ augites,} \\ 47-51; \text{ titaniferous augites, } 46-47; \text{ hornblendes,} \\ 42-50, \text{ biotites, } 35-38; \text{ opaque oxides, 0.} \end{cases}$

Rhyolite and granite contain, on an average, about 72% SiO_2; hence they are regarded as acid. Intermediate types include syenite (average

59%), diorite (average 57%), and monzonite (average 55%). Basic types are represented by gabbro and basalt (average 48%), and peridotite (average 41%) is a typical ultrabasic rock. Acid rocks are usually richer in alkalies and poorer in calcium, iron, and magnesium than basic ones and are therefore generally paler in color owing to a smaller content of mafic minerals.

Percentage of silica bears little relationship to percentage of quartz in a rock. Of two rocks containing the same amount of silica, one may be devoid of quartz, while the other may carry as much as 35% by volume; and two rocks having the same content of quartz may differ in silica content by as much as 15%. In brief, when silica percentage is used as a basis of classification, it brings together many mineralogically dissimilar rocks. Clearly some more subtle, but related, parameter than silica percentage alone is responsible for the varied mineralogical patterns displayed by common igneous rocks. The controlling factor, the *activity* of silica in the magma, will be considered more fully on pages 77–78; it can be thought of as the "effective concentration" of silica and depends not on silica content alone but also on pressure, temperature, and *relative* concentrations of *all* other components in the magma. Nevertheless, the old and well-established terms *acid, intermediate, basic,* and *ultrabasic,* as defined by silica percentage, still serve a useful purpose for general qualitative description, as in this book.

Silica content is especially valuable in distinguishing between rocks wholly or mainly composed of glass, for it can readily be determined by measurement of refractive indices, as shown in Figure 2-11. It may also be used to classify fine-grained rocks, even those that are holocrystalline. Specimens from a petrographic province are first selected to represent the full range of rock types and are then analyzed chemically. Finely crushed samples of the analyzed rocks are fused to glass beads, and the indices of refraction of the beads are measured. The indices are then plotted against the silica percentages to provide a characteristic curve for the petrographic province. When that has been done, other rocks from the same province may be crushed and fused so that their silica content may be found without recourse to analyses.*

---

*Today petrographers and geochemists alike must be alert to select the simplest and quickest of available techniques to obtain chemical data adequate to their purposes. The glass-bead technique referred to here is a case in point. In any given situation, will it provide the necessary data more efficiently than more sophisticated modern methods of microanalysis?

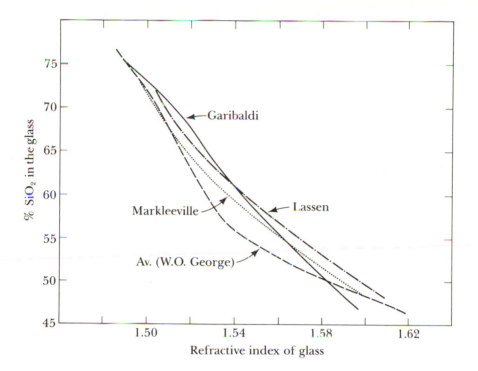

Figure 2-11. Relation Between Silica Content and Refractive Indices of Volcanic Glasses. The curve made by W. O. George is for natural glasses; the other three curves are for artificial glasses produced by fusion of volcanic rocks belonging to the calcalkaline igneous series. The Markleeville curve is based on Tertiary lavas of the Sierra Nevada, California; the Lassen and Garibaldi curves are based on Pleistocene and younger lavas of the Cascade Range of northern California and British Columbia respectively.

Normative Classification

An elaborate chemically based scheme of precise quantitative classification and complementary unambiguous terminology was proposed early in the century by four leading American petrologists, W. Cross, J. P. Iddings, L. V. Pirsson, and H. S. Washington. Known as the *normative* or *C.I.P.W. classification*, it utilizes the complete chemical analysis as its sole basis.[2]

Following a precisely stipulated procedure (nowadays carried out by a simple computer program), the analysis of any rock sample is recalculated water-free to a standard form, the *norm*, expressed as weight percentages of simple "molecules" analogous in composition to selected ideal minerals (or their component end members). Compositions of most igneous rocks can be stated completely in terms of some combination of thirteen normative components:*

Quartz	Q	SiO_2
Corundum	C	Al_2O_3
orthoclase	*or*	$KAlSi_3O_8$
albite	*ab*	$NaAlSi_3O_8$
anorthite	*an*	$CaAl_2Si_2O_8$
nephelene	*ne*	$NaAlSiO_4$
leucite	*lc*	$KAlSi_2O_6$
diopside	*di*	$\begin{cases} CaMgSi_2O_6 \\ CaFeSi_2O_6 \end{cases}$
hypersthene	*hy*	$\begin{cases} MgSiO_3 \\ FeSiO_3 \end{cases}$
olivine	*ol*	$\begin{cases} Mg_2SiO_4 \\ Fe_2SiO_4 \end{cases}$
magnetite	*mg*	Fe_3O_4
ilmenite	*il*	$FeTiO_3$
apatite	*ap*	$Ca_5(PO_4)_3$

Certain combinations of normative components—such as $(Q + ne)$, $(Q + ol)$, and $(ne + hy)$—are unstable with respect to chemically equivalent stable combinations. Thus in the equation

$$NaAlSiO_4 + 4MgSiO_3 \rightarrow NaAlSi_3O_8 + 2Mg_2SiO_4$$

$$\quad ne \qquad\qquad hy \qquad\qquad ab \qquad\qquad ol$$

the right-hand pair is stable. Unstable pairs are seldom found in igneous rocks, and then only when the texture indicates failure to maintain equilibrium during the crystallization process. One such example is fine-grained quartz or tridymite in the groundmass of basalt containing corroded phenocrysts of magnesian olivine. The program for computing the norm was designed to eliminate combinations inferred from general petrographic experience to be unstable—an inference since borne out by thermodynamic data.

---

*In some cases, additional components may be needed, e.g., Fe_2O_3, $CaSiO_3$, or Na_2SiO_3.

On the basis of the norm, every analysis is given an approximate symbol and assigned to a corresponding classificatory pigeonhole. The terminology developed to this end is too cumbersome and, in many respects, too artificially contrived for practical use as a medium for petrographic classification. Nevertheless, the scheme as a whole is ideal for codifying and storing data on the bulk chemistry of igneous rocks. Moreover, the norm itself has proved to be invaluable as a condensed expression of petrologically significant aspects of the total major-element chemistry of any igneous rock. For example, normative quartz Q, as contrasted with nepheline *ne* or olivine *ol*, has obvious genetic implications. More subtle, but highly significant nevertheless, are the relative proportions of *di*, *hy*, and *ol* in norms of basaltic rocks (see pp. 96–97).

Petrography is directly concerned not with the *norm* but with the actual mineral composition, the *mode*. The two are of course related, but they are by no means identical. Modal hornblende appears in the norm as some combination of *an*, *di*, *hy*, and *ne*; biotite in quartz-bearing rocks as *or*, *hy*, and *C*; muscovite as *or* and *C*. Clearly then, the normative classification is unsuited to petrography. But the norm, taken alone, is a most useful adjunct that brings out the significance of certain rather subtle but petrographically recognizable aspects of igneous mineralogy—such as the composition of pyroxenes in basalts (see pp. 97; 100); and it is indispensable for allocating glassy and very fine-grained rocks to the same categories as their chemically equivalent crystalline counterparts in any petrographic classification.

Shand's Classification and the Silica-Saturation Concept

An ingenious and detailed quasi-chemical classification of igneous rocks was developed by S. J. Shand,[3] utilizing those chemical parameters that most influence the distinctive mineralogical characteristics of common individual rock types. Certain minerals, never or only exceptionally associated with quartz (or tridymite) in igneous rocks, Shand designated as *undersaturated* with respect to SiO_2. They include common magnesian olivines, all feldspathoids, and minor or rare minerals such as melanite, perovskite, melilite, corundum, and spinel. By contrast, minerals commonly found associated with quartz, Shand called *saturated*, even though some of them contain no SiO_2: all feldspars, pyroxenes, amphiboles, and micas, the common iron-titanium oxides, and numerous accessory minerals (tourmaline, almandine, sphene, zircon, apatite, and others). Underlying the saturation concept is the assumption (now borne out by laboratory experiments and thermodynamic data) that saturated minerals are stable, unsaturated minerals unstable, in contact with melts that simultaneously precipitate quartz (or tridymite); in this respect, departure from equilibrium is negligible. The exception referred to earlier in this paragraph arises where phenocrysts of unsaturated minerals (usually corroded in outline) are in obvious disequilibrium with a groundmass containing quartz or siliceous glass.

On this basis, Shand separated igneous rocks into three *mineralogically* defined chemical classes, the third of which he subdivided further:

1. *Oversaturated rocks*, containing saturated minerals and quartz (or tridymite). All acid rocks are oversaturated, but not all oversaturated rocks are acid; some are intermediate, and a few even basic.
2. *Saturated rocks*, containing only saturated minerals.
3. *Undersaturated rocks*, containing undersaturated and in most cases also saturated minerals.
   a. *Nonfeldspathoidal* division in which feldspathoids are absent; the common unsaturated phase is magnesian olivine.
   b. *Feldspathoidal* division, containing one or more feldspathoids (nepheline the commonest) with or without other undersaturated minerals. Magnesian pyroxenes are never present.

Saturation in Al_2O_3, second in general abundance to SiO_2, provides a second chemical criterion, also expressed in mineralogical terms for subdividing each of the above groups into four subgroups. Excess or deficiency of Al_2O_3 is measured with reference to the ratio $Al_2O_3 / (K_2O + Na_2O + CaO) = 1$, as exemplified by all feldspars and feldspathoids. This parameter, therefore, is reflected in the nature of the mafic minerals. Accordingly, Shand proposed four groups:

1. *Peraluminous* rocks: $Al_2O_3/(K_2O + Na_2O + CaO) > 1$. Excess Al_2O_3 is accommodated in micas and certain minor constituents (corundum, tourmaline, topaz, Fe–Mn garnet). Most peraluminous rocks belong to the acid plutonic class; some of these contain cordierite or sillimanite.
2. *Metaluminous* rocks: $Al_2O_3 / (K_2O + Na_2O) > 1 > Al_2O_3 / (K_2O + Na_2O + CaO)$. Hornblende and aluminous augites are typical of metaluminous rocks; less commonly melilite or such combinations as biotite and olivine.

3. *Subaluminous* rocks: $Al_2O_3 / (K_2O + Na_2O) \approx 1$. Mafic minerals are all nonaluminous—olivine, orthopyroxene, diopside.
4. *Peralkaline* rocks: $Al_2O_3 / (K_2O + Na_2O) < 1$. The typical mafic minerals are sodic pyroxenes (aegirine, aegirine-augite) and amphiboles (riebeckite, arfvedsonite); and, in some others, the place of Al_2O_3 is taken by Fe_2O_3, ZrO_2, or TiO_2. Such rocks form in the final stages of fractional crystallization of sodic magmas.

In detail, Shand's classification seems perhaps too cumbersome and too "artificial" for classroom use or for general adoption among "hard rock" geologists. Yet it has subtle genetic implications, and it underscores certain elements of mineralogical pattern that most clearly bring out aspects of magma firmly established in general use since they conveniently relate petrography to geochemistry. Frequent references are made, for example, to "oversaturated" and "undersaturated" (in SiO_2) and to "peraluminous" versus "peralkaline" rocks and magmas.

There are obvious analogies between Shand's concept of silica saturation and normative composition. But the parallel is not exact. Most oversaturated rocks are quartz-normative; but so too are some saturated rocks, for example, in the andesite family. Olivine basalts typically are olivine-normative. But some, especially those in which modal olivine is confined to corroded phenocrysts, have small amounts of Q in the norm. Here modal departure from equilibrium is automatically eliminated from the norm by instructions written into the program.

Significance of SiO_2 Activity

In all chemically based classifications, a principal role is assigned to content (weight percentage) of SiO_2. Yet two rocks of identical silica content (say 53% or 57%) may differ strikingly with respect to silica saturation: in one quartz, in the other olivine or nepheline, may appear in both mode and norm. The problem has been explored, with promising results, by I. S. E. Carmichael (e.g., Carmichael, Turner, and Verhoogen, 1974, pp. 50–59), who points out that the controlling thermodynamic factor is the *activity*, $a_{SiO_2}^{melt}$, of SiO_2 in the melt. Under stated conditions, activity a is a function of molar concentration X and can be written in the above case.

$$a_{SiO_2}^{melt} = \gamma X_{SiO_2}^{melt}$$

where γ is an activity coefficient that depends on P, T and relative concentrations of all other components in the melt. Activity can be regarded loosely as an "effective concentration" that expresses the capacity of the melt to yield SiO_2 to crystalline phases with which it is in equilibrium. The role of other components such as Al_2O_3 now becomes clear; for their presence in a silicate melt permits the building up of imperfect compound silica-retentive "molecular" or ionic groups within the melt itself, thus lowering the value of $a_{SiO_2}^{melt}$.

Under given physical conditions, a saturated phase such as $MgSiO_3$ can crystallize from a melt only when $a_{SiO_2}^{melt}$ exceeds some limiting value. Below this level a corresponding undersaturated phase, Mg_2SiO_4, may crystallize instead at any given temperature. This limiting value of $a_{SiO_2}^{melt}$ is different for each of a succession of undersaturated phases crystallizing from the same melt. Petrography gives significant though only qualitative information in this connection. We have already seen that while in some basalts alkali feldspar (saturated) and magnesian olivine (undersaturated) coexist, hypersthene and feldspathoids are incompatible. Evidently at temperatures of magmatic crystallization the respective stability fields of hypersthene-melt and nepheline-melt do not overlap; and the upper limit of $a_{SiO_2}^{melt}$ permitting separation of magnesian olivine from the magma must be significantly higher than that at which crystallization of nepheline occurs.

Petrographic and experimental data also lead to broad conclusions regarding activity of Na_2O in magmas. The plagioclase sequence bytownite to oligoclase expresses in a qualitative fashion a progressive increase in $a_{Na_2O}^{melt}$. In successive melt fractions derived from basaltic magmas, *concentrations* of both SiO_2 and Na_2O increase simultaneously. But while in tholeiitic series *activities* of both components in the melt likewise increase, in series generated from alkaline basaltic magmas $a_{SiO_2}^{melt}$ maintains a more or less even level with increasing $a_{Na_2O}^{melt}$ (see Carmichael, Turner, and Verhoogen, 1974, p. 58).

In the SiO_2 activity of magmas, $a_{SiO_2}^{melt}$, Carmichael envisages a potential numerical parameter that one day might be used in classifications of igneous, and more particularly volcanic, rocks. But because $a_{SiO_2}^{melt}$ depends on so many variables, it will always be difficult to assign to it numerical values except between the broadest limits. Values have indeed been computed for some very simple transformations (over a range of temperature) that bear on the control that $a_{SiO_2}^{melt}$ exerts upon the mineralogy of volcanic rocks (Carmichael, Turner, and Verhoogen, 1974, pp. 50–59). These include:

1. SiO_2 glass $\rightarrow$ quartz
2. Forsterite + SiO_2 glass $\rightarrow$ Enstatite
3. Nepheline + SiO_2 glass $\rightarrow$ Albite
4. Perovskite + SiO_2 glass $\rightarrow$ Sphene

In all cases, increase in $a_{SiO_2}^{glass}$ favors the transformation in the direction of the arrow.

The problem is how to use this information with respect to the occurrence of perovskite, nepheline, magnesian olivine, and quartz (or tridymite) as stable phases in volcanic rocks. All that can now be said is that from magmas of the same SiO_2 content, as the value of $a_{SiO_2}^{magma}$ increases first perovskite, then nepheline, then magnesian olivine will be eliminated and ultimately quartz may crystallize. Quartz, olivine, and nepheline are three critical indices of the degree of SiO_2 saturation employed by Shand. Perovskite may be used indirectly as an index for disappearance of plagioclase, for these two phases do not coexist (as do nepheline and perovskite) in volcanic rocks.

Qualitative relationships between SiO_2 content, SiO_2 saturation, and $a_{SiO_2}^{magma}$ are indicated in Figure 2-12 which illustrates *in a purely schematic way* the relative positions of the respective stability fields of quartz, magnesian olivine, nepheline, and perovskite with respect to SiO_2 activity of magmas represented by some of the principal kinds of volcanic rocks. The $a_{SiO_2}^{magma}$ scale (on the vertical coordinates) is both nonquantitative and flexible with reference to both coordinates; the horizontal scale (SiO_2 content of rock) is precise. Approximate locations of named volcanic rocks are indicated, but with no attempt to define their compositional limits. Finally, a few broken lines have been drawn to indicate possible magmatic trends (lineages) as postulated by various authors (see Carmichael, Turner, and Verhoogen, 1974, p. 58).

Petrographic Classifications

Strictly petrographic classifications should be based entirely on textural and mineralogical characteristics recognizable in hand specimens and beneath the microscope. Several classifications essentially of this kind are currently in general use. All utilize the same main criteria. They differ, however, regarding the relative emphasis placed on individual characteristics, on the degree of refinement and precision with which classes and subdivisions are defined, and on the extent to which chemical data

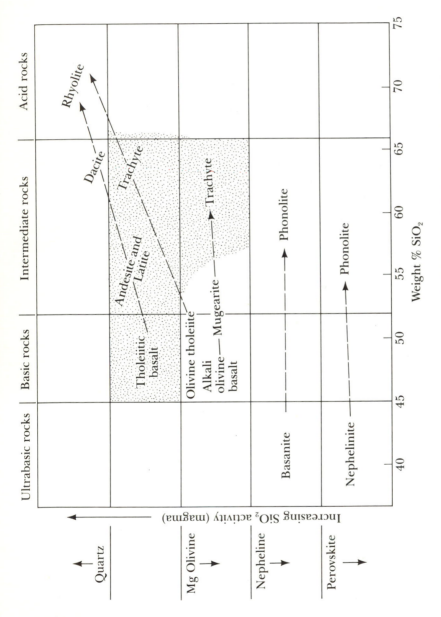

Figure 2-12. Diagrammatic Illustration (Vertical Scale Variable) of Relation Between SiO₂ Content (Weight Percentage), SiO₂ Saturation, and SiO₂ Activity in Magmas Represented by a Few Typical Volcanic Rocks. Location of rocks named is approximate only. Silica-saturated types occur within the stippled area. Blank field at upper right represents oversaturated rocks; that below and to the left represents undersaturated types. Possible magmatic trends are drawn as straight lines; curvature and slopes of these lines have no significance in this diagram since the vertical scale is flexible. See text for details.

(especially normative composition) are used to supplement the data of petrography. Since the aim of this book is to describe rocks in the simplest terms rather than to discuss classifications of rocks, no attempt will be made to set out and appraise individual systems. So, in this section we first review the criteria, and then, after a brief digression on the problem of precision in classification, outline the plan here adopted to present systematically the descriptive matter of igneous petrography.

Textural Criteria

From the first, grain size has been used as a primary basis for dividing igneous rocks, and with good reason: it is a measurable property, and it depends mainly on cooling history and thus has broad genetic significance. We have already seen how grain size, supplemented by other textural detail (porphyritic, ophitic, granitic, etc.) can be used to divide all igneous rocks into two groups with contrasted cooling histories—*volcanic* and *plutonic*. Since this nomenclature carries implications of field occurrence (a matter strictly beyond our present scope), some petrographers prefer to distinguish between *fine-*, *medium-*, and *coarse-grained* types, with the intermediate limits drawn at 1 mm and 5 mm. Or in accordance with the twofold subdivision of volcanic and plutonic rocks, a twofold distinction can be made between *finer-* and *coarser-grained* rocks, the boundary grain size being approximately 1 mm. Taken alone, however, without recourse to field occurrence, such purely textural distinctions inevitably result in some degree of ambiguity and inconsistency. Thus, many ophitic diabases (volcanic rocks) belong in the coarser-grained class. And chilled margins of deep-seated truly plutonic intrusions may be fine-grained enough to fall within the volcanic category as defined purely by texture.

Color Index

The most obvious characteristic of a rock specimen is its color, ranging from almost white (in some granites) to nearly black (in basalts). Color of holocrystalline rocks in a general way reflects the proportion of mafic to felsic minerals, and provides an immediate rough index of mineral composition; with few exceptions, dark rocks are rich in pyroxenes, amphiboles, olivine, and less commonly, biotite; light rocks are rich in feldspars (with or without quartz or feldspathoids). For most purposes, it is sufficient to classify rocks on a color basis as light- or dark-colored. For greater precision, a color index, *CI*, is used, expressing the simple ratio of dark- to light-colored constituents as estimated by microscopic measurement (or merely by rapid inspection, see Appendix B) or alternatively as computed from the norm. Here, we use a simple scheme of classification by color index, dark-colored, CI >40; light-colored, CI <40.*

Modal Mineralogy

The mineralogical criteria employed today are much the same as Rosenbusch used a century ago. Feldspars are prominent constituents of all igneous rocks except ultramafic and ultrabasic types; and since these minerals show a wide and regular pattern of compositional range, they are ideal to use as a primary basis of classification. Quartz is more restricted in its occurrence, as are its undersaturated antitheses, feldspathoids and olivine. Presence or absence of any of the three has significant and thus genetic implications and so provides another useful criterion for classification. Individually the common mafic minerals pyroxene, hornblende, and biotite are used at a lower level to define corresponding varieties of more comprehensive rock types; collectively they define the color index.

Significance of Feldspars

For rocks carrying significant alkali feldspar, most classifications stress the relative proportions of alkali feldspar to plagioclase. The alkali feldspar of unaltered volcanic rocks is either sanidine (monoclinic) or anorthoclase (triclinic).[†] Alkali feldspars of plutonic rocks include microcline and orthoclase, both typically perthitic. For classificatory purposes perthite is treated as a single entity, ignoring the fact that its albite

---

*Standard works such as those of A. Johannsen (1931–1939) and S. J. Shand (1943) use other limits to define four color classes. Shand, for example, uses a special color terminology; *leucocratic, CI* <30; *mesocratic, CI* 30–60; *melanocratic, CI* 60–90; *hypermelanic CI* >90. The first and third of these terms are rather widely used as synonyms of "light-" and "dark-colored"; or the prefixes *leuco-* and *melano-* may simply be attached to accepted names of rock types—e.g., leucogabbro, melanodiorite.

[†]The two form a continuous solid-solution series, anorthoclase comprising the more sodic members of the series (molar $NaAlSi_3O_8$ >65%). The two phases do not coexist in igneous rocks; records to the contrary are based on failure to recognize anorthoclase as such in sections that lack lamellar twinning.

component—a product of late exsolution—is strictly a plagioclase. Likewise, primary albite in certain sodic granitic rocks is treated as alkali feldspar.*

Many rocks of intermediate and basic composition lack alkali feldspar but contain abundant plagioclase. Grouping of these is based primarily on composition (optically determined) of plagioclase. The significance of this criterion must be obvious to anyone even casually familiar with patterns of crystal-melt behavior as experimentally determined in relevant systems (e.g., Figure 1-3). Nonetheless, from microscopic measurements (now reinforced by microprobe data) it has long been obvious that the composition of plagioclase in many igneous rocks is far from uniform. Concentric compositional zoning of individual crystals, and in porphyritic rocks marked differences among phenocrysts themselves as well as between phenocrysts and groundmass crystals, are commonplace. Clearly only approximate estimates of average plagioclase composition can have any value for rock classification. It would be time-consuming, and indeed futile, to attempt any more refined estimates on the basis of microscopic measurements alone. The normative ratio $an/(an+ab)$, if available, is a more reliable index.

Significance of Quartz

In all classifications a major emphasis is placed on the presence or absence of quartz. Oversaturated rocks with more than about 5% modal quartz are grouped according to relative proportions of quartz, alkali feldspar, and plagioclase (excluding albite). Clearly it is undesirable to separate rocks with only minor quartz from otherwise mineralogically identical rocks containing none. In such cases, the word "quartz" is used adjectivally to designate the oversaturated condition: e.g., quartz syenite, quartz gabbro. Ambiguity of course arises regarding the upper percentage limit of "minor" quartz. Should this be 5%, 10%, or (as in the IUGS classification) as high as 20%? It matters little provided the limit adopted for any specific purpose is clearly defined.

---

*Volcanic rocks in eugeosynclines commonly contain plentiful albite as the sole feldspar (e.g., spilites, keratophyres). But today many petrologists consider this to be a product of postmagmatic metasomatism of original igneous feldspar, involving introduction or large-scale redistribution of Na_2O. In this book, rocks of this kind are treated as metamorphic, so that modal albite plays no role in the classification of unaltered volcanic rocks.

Significance of Feldspathoids

The presence of a feldspathoid unmistakably marks a low level of silica saturation (i.e., of silica activity $a_{SiO_2}^{magma}$ in the parent magma). All classifications distinguish, therefore, between feldspathoidal and nonfeldspathoidal rocks—with the provision that, as with quartz, the presence of minor feldspathoid can be designated by a qualifying phrase such as nepheline-bearing trachyte (or syenite).* Feldspathoidal rocks can be further divided according to whether the feldspathoid is or is not accompanied by plagioclase. Absence of plagioclase in feldspathoidal rocks indicates the lowest level of $a_{SiO_2}^{melt}$ attained in basic magmas.

Significance of Mafic Minerals

Presence or absence of magnesian olivine is widely used to distinguish undersaturated from saturated basic rocks. Less significant but useful to separate varieties of the same broad rock type are pyroxene, hornblende, and biotite. Where two or more are used to define a given variety, they are written in order of increasing abundance. Thus, hypersthene-augite andesite contains more augite than hypersthene.

Problem of Precision and Refinement

Regularity, order, and predictability in natural phenomena are basic premises to scientific investigation and discovery. But equally obvious is universal departure from order in broadly ordered natural systems—as exemplified in the modern approach to crystallography. And closely allied to the notion of disorder is the transitional nature of boundaries between sets of natural, gradational phenomena—for example, between the respective structures of silica melts and crystalline SiO_2, or of sanidine and microcline. Analogous conditions in the realm of petrography pose questions of great significance regarding rock classification. What degree of classificatory refinement is attainable or even desirable, and how sharply should interdivisional boundaries be drawn in framing a classification of igneous rocks? And to what degree—regardless of precision, refinement, or artificially imposed symmetry—should a classification reflect natural field associations and genetic, especially evolutionary, characteristics of individual rock types and groups? Answers to these and

---

*The term *nepheline syenite* has long been preempted to cover rocks in which nepheline and alkali feldspar are the two major felsic constituents.

related questions will never be unanimous, for they depend on the purpose for which a classification is designed and they also reflect the personal attitudes of its architects. To illustrate the nature of the general problem we examine briefly some aspects of two well-known classifications, each based on essentially the same mineralogical criteria: the one brand-new, as developed by A. L. Streckeisen and henceforth designated *IUGS classification* to signify subsequent approval by the International Union of Geological Sciences; the other first put forward by Rosenbusch and still flourishing (with minor modifications), time-honored and seasoned by a century's use.

The IUGS classification starts with two propositions, neither of which we find acceptable. First, it presupposes "the need to agree upon a single rational and workable system for naming and classification of igneous rocks, which geoscientists throughout the world will use," and, second, the classification incorporates "igneous and igneous-looking rocks, in the sense of Anglo-Saxon authors, irrespective of the genesis of the rocks."[4] In the face of wide divergence of purpose and attitude among students of igneous rocks, unanimity and precision of terminology are impossible to achieve. (We do not even admit that they are desirable objectives.) To group with truly igneous rocks superficially similar products of metamorphism and metasomatism can only confuse an already difficult and complex natural situation by obscuring genetic implications otherwise obvious in common igneous assemblages.

To demonstrate the aims and qualities of the IUGS scheme, Figure 2-13 illustrates in greatly simplified form the basis of primary subdivision and nomenclature of plutonic igneous rocks, depicted on two triangular (three-phase) projections sharing a common boundary *AP*. Oversaturated rocks are encompassed by the alkali-feldspar–plagioclase–quartz triangle *APQ*; feldspathoidal rocks fall within the *APF* triangle, with feldspathoids represented at the *F* apex; the locus of all saturated rocks and of nonfeldspathoidal (olivine-bearing) undersaturated types is the common boundary *AP*.* Secondary partition of each of the named fields is accomplished by means of accessory three-phase projections. Clearly ambiguity has been eliminated, and limits of most familiar rock types (granite, granodiorite, syenite, gabbro, and others) have been precisely defined.

---

*A similar basic classification and nomenclature of volcanic rocks has also been proposed (*Geology,* vol. 7 (1979): p. 332).

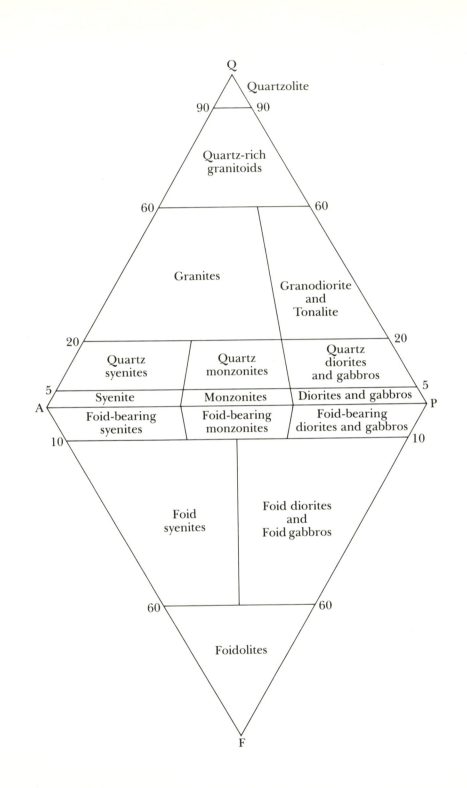

This classification and other pigeonhole classifications having precise nomenclature as the primary objective are largely artificial, although they may be well adapted for compiling and storing in orderly fashion the essential data of igneous mineral assemblages. The geometrical symmetry with which divisional boundaries are drawn and their rectilinear character are at variance with general asymmetry in patterns of natural petrogenetic processes. Furthermore, once the divisional boundaries have been drawn every "pigeonhole," regardless of its relative importance in nature, is assigned a rock name with the result that the classification fails to bring out, and may even conceal, the natural associations and relative abundances among rock types.

Bowen closed his classic exposition, *The Evolution of the Igneous Rocks* (1928), with an eloquent, two-page defense of flexibility in classification as exemplified in the Rosenbusch system. An apt analogy is drawn from the progress of ideas on classification in the organic kingdom. Early classifications based purely on plant and animal morphology, Bowen points out, were not abandoned when the fact and principles of organic evolution became established. Instead,

. . . for the most part establishment of evolution gave merely a fresh impetus and a new meaning to existing classification.

The situation will not be found very different in the case of rocks when a general understanding of the process of their development has been reached. The minerals of which rocks are composed are the expression of a response to the conditions of their genesis. A genetic classification cannot transcend a classification based on . . . modal mineral composition [p. 320].

The inference is that particular emphasis should be placed on those mineralogical characteristics of igneous rocks that clearly have genetic implications—even though the evolutionary processes and trends themselves may as yet be only imperfectly understood. As magmatic evolution becomes perceived with greater clarity, a well-designed mineralogical classification can be modified here and there in detail, but will never be superseded. Such a classification, endowed with the necessary potential

---

Figure 2-13. Basis of IUGS General Classification and Nomenclature of Plutonic Igneous Rocks. (Greatly Simplified from Figure 1, *Geotimes*, October 1973, p. 26.) Q = quartz; A = alkali-feldspar; P = plagioclase; F = feldspathoid ("foid").

for genetic interpretation, is that of Rosenbusch (as modified to suit present needs). It is characterized by looseness and flexibility in its quantitative aspects.

Bowen concluded his argument and his book with this statement:

> The quantitative element is not lacking, but is relatively unimportant. The looseness of the classification is not due to our lack of knowledge of rocks but to our very knowledge of them. . . . With increase in our knowledge of rocks as members of series . . . there will come a greater appreciation of the desirability of loose classification [p. 322].

Paradoxically, this statement is the exact antithesis of that which opens the preamble to the IUGS classification (p. 85). Choice between the two schemes just outlined is clear-cut; it is a matter to be decided individually according to personal attitude and purpose. In our view,

> . . . all classifications of igneous rocks are open to criticism, and none more so than those designed for the beginning student. Because serious difficulties arise in trying to be consistent in the use of a single criterion, it is best to make use of several criteria, emphasizing one or another as the purpose justifies. Let every student bear in mind what has been stressed before: that all rocks are transitional members of series and that clear-cut boundaries are not to be expected.*

The Nature of Lithologic Transitions

When we speak of a transitional or gradational relation between rock types or broader lithologic divisions—for example, between basalt and andesite, or between trachyte, quartz trachyte, and rhyolite—the connotation of the phrase varies according to the purpose of discussion and the viewpoint of the individual.

Use of continuously variable criteria in classification—for example, SiO_2 percentage or *An* content of plagioclase—automatically imposes one type of transitional relation between adjacent categories so defined. Thus, by the very criteria of definition the sequence basalt-andesite-latite-trachyte (as the terms are defined here and in the IUGS scheme)

---

*This admonition, taken directly from the first edition of this book (p. 36), is just as appropriate today as when written three decades ago by our late coauthor, Howel Williams. It applies to format of presentation of data as much as to design of classification of the rocks themselves.

represents a continuous compositional spectrum. Yet continuity so implied is purely artificial. The sequence, while convenient enough for purposes of data storage, museum arrangement, or teaching at the elementary stage, pertains to neither lithologic associations as seen in the field nor to evolutionary lineages in the parent magmas. Use of pigeon-hole diagrams such as Figure 2-13 accentuates the artificial element in the transitional relation. Any line, rectilinear or otherwise, traced on such a diagram defines a series of transitions at successively encountered interdivisional boundaries. Henceforth, any sequence of the above kind will be termed a *lithologic spectrum* without implications as to possible natural relationships.

However convenient such treatment may be for comprehensive presentation of petrographic data, it is essential to seek and recognize natural transitional relations of a deeper significance (as emphasized by Bowen in the passages just cited). Only thus does petrography come to life. Rocks of what will henceforth be termed a *natural transition series* not only show continuous mutual chemical and petrographic gradation as seen in the laboratory; but especially they have a tendency for repeated mutual association in both space and time. One such example is the series alkali olivine basalt, hawaiite, mugearite, trachyte, encountered in many basalt-dominated provinces. Another is the upward generalized sequence peridotite, norite, gabbro, ferrogabbro, displayed from bottom to top in many layered basic plutons. Field relations and radiometric dating show that each of the above series, in the order given, represents a chronological sequence of relatively short span. Natural transition series in some instances are of more limited extent. Thus either typical granites (by decrease in content of quartz) or typical syenites (by increase in quartz content) may grade into an intermediate type (quartz syenite in Figures 2-13 and 5-1), but complete gradation between granites and syenites in the field, if such do exist, must be rare.

ADOPTED PLAN OF TREATMENT

What follows (as summarized in Figure 2-14) is not of itself a system of rock classification. Rather it is a synopsis of the plan adopted for presentation of the data of igneous petrography. Derived largely from the time-honored, familiar classification of Rosenbusch, the terminology is consistent with widely accepted usage. Current conflicts and ambiguities are resolved and arbitrary limits defined as far as possible in a manner con-

	Acid	Intermediate	Basic	Ultrabasic
	Chapter 5 Oversaturated rocks; CI—0 to 40 Alkali feldspar ↔ Plagioclase An_{10-30}	**Chapter 4** Saturated rocks; CI—0 to 40 Alkali feldspar ↔ Plagioclase An_{20-50}	**Chapter 3** Saturated and undersaturated; CI usually > 40 Plagioclase An_{50-100}	**Chapter 8** Undersaturated; CI—90 to 100 Plagioclase 0–10%
Plutonic	Quartz > 20% Granite │ Granodiorite │ Tonalite alkali granite │ adamellite Quartz 5–20% Quartz syenite │ Quartz monzonite │ Quartz diorite	Syenite Monzonite Diorite	Gabbro Norite Troctolite Anorthosite Mg and CaMg pyroxenites Alkaline gabbro	Feldspathic peridotite
Volcanic	Rhyolite Dacite	Trachyte Latite Andesite	Tholeiitic basalts and diabases Alkali olivine basalts hawaiite mugearite	

	Chapter 7		Chapter 6		
	Chapter 7 Feldspathoidal rocks; CI—low to medium Alkali feldspar		**Chapter 6** Feldspathoidal rocks; CI—low to high Plagioclase	Feldspar lacking	
Plutonic	Feldspathoidal syenites	shonkinite	Feldspathoidal gabbros	Ijolite	
	nepheline syenite sodalite syenite		essexite theralite Analcime diabase	Alkaline pyroxenite	
Volcanic	Phonolite		Trachyandesite Trachybasalt	Tephrite → Basanites Leucitites →	
			Wyomingite	Nephelinite Limburgite	

Chapter 8
Miscellaneous ultrabasic rocks

Lamprophyres
  biotite and hornblende lamprophyres*
  camptonite
  monchiquite

Melilite-rich rocks
  melilitite
  alnöite
Carbonatite
Kimberlite

Volcanic or quasi-volcanic

Nonfeldspathic peridotite (plutonic)
Komatiite (volcanic)

*Feldspar-rich types are basic.

Figure 2-14. General Character and Organization of the Principal Igneous Rocks as Treated in This Book.

sistent with the IUGS scheme of classification. Where terminology differs significantly from that of IUGS appropriate cross-references or footnotes are inserted.

Rock types are defined and grouped primarily on the basis of petrographically determined criteria, with the main emphasis on mineralogy. Chemical parameters play no part in the rock definitions, although they play an underlying role in the choice of definitive mineralogical criteria. Principal chemical characteristics—notably weight percentage SiO_2 and values of Q, hy, ol, and ne in the norm—are added, in parentheses as it were, to facilitate broader application of descriptive data in other fields of geology. And largely for the same reason, notes on field occurrence and mutual association are also included.

The format of presentation is dominated by the concept of gradational lithologic spectra. It is influenced, to different degrees in different sectors, by factors relating to relative global abundance, bulk SiO_2 content, and varying degrees of silica-saturation (and implied SiO_2 activity) expressed in mineralogical terms. The descriptive material is presented in a manner deemed conducive to genetic interpretation in terms of concepts relating to "evolved" versus "primitive" magmas and the existence of magmatic lineages (cf. Figure 2-12). But such concepts are necessarily of somewhat speculative nature and they play no direct part in the overall format we have adopted.[5]

Basalts are by far the most extensive of exposed crustal igneous rocks. Second in global abundance are rocks of the granite-granodiorite spectrum. The two corresponding magmas represent opposite ends of a continuous, broad lithologic spectrum that accounts for well over 90% of exposed igneous rocks and extends from basalt and gabbro (basic) through saturated rocks of intermediate SiO_2 content to oversaturated acid rocks (Figure 2-12). This spectrum is given first priority in treatment and is covered in Chapters 3 to 5. Then follows in Chapters 6 and 7 a spectrum of undersaturated feldspathoidal rocks whose total compositional range extends from less than 40% to 55% SiO_2 (ultrabasic to intermediate). There remains an assortment of other rocks, mostly ultrabasic, insignificant in combined extent but mineralogically and genetically of great interest. These are treated in Chapter 8; although some of these groups show mutual genetic affinities they have little in common as to detailed petrochemical pattern. In order of treatment they include lamprophyres, melilitites, carbonatites, kimberlites, and peridotites (and their serpentinized and volcanic equivalents). Appropriately the final emphasis falls on rocks (kimberlites and peridotites) that yield valuable clues as to the petrography of materials composing the upper mantle. Since the *ultimate* source of magmas lies here, a note on upper mantle petrography has been appended at the end of Chapter 8. Volcanic rocks of fragmental (clastic) texture are covered in Chapter 9.

GENERAL REFERENCE

See the works listed at the end of Chapter 1, p. 35.

REFERENCE NOTES

[1]Data from H. R. Shaw, *Journal of Geophysical Research*, vol. 68 (1963): pp. 6337–6343.

[2]For full details, see J. P. Iddings, *Igneous Rocks*, vol. 1, New York: Wiley, 1909, or A. Johannsen, *A Descriptive Petrography of the Igneous Rock*, vol. 1. Chicago: University of Chicago Press, 1931. The latter also describes a somewhat comparable treatment of rock chemistry by P. Niggli, A. Osann, and others.

[3]S. J. Shand, *The Eruptive Rocks*, 2nd. ed. New York: Wiley, 1943.

[4]*Geotimes*, October 1973, pp. 26, 27.

[5]A. Joplin, *A Petrography of Australian Igneous Rocks*. Sydney: Angus and Robertson, 1964. Presents a beautifully illustrated and systematic survey of Australian igneous petrography with a primary emphasis on magma types (a generalized concept of lineage).

3

Nonfeldspathoidal Basic Rocks

GENERAL MINERALOGICAL CHARACTER

Largely liquid basic magmas as directly observed in volcanic flows erupted at the earth's surface range in silica saturation from strongly undersaturated to slightly oversaturated types. They crystallize as simple combinations of a few main minerals—plagioclase (An_{80}–An_{50}), pyroxenes (augites and magnesium varieties), magnesian olivines, iron-titanium oxides, and (only in highly undersaturated types) nepheline or some other feldspathoids. Basic igneous rocks are simple combinations of the above minerals, dark in hand specimen, with a color index typically above 40, and a silica content of 45–52%. They fall naturally into two divisions: one, by far the more extensive, is nonfeldspathoidal (commonly termed calcalkaline); the other carries significant quantities of modal feldspathoid (usually nepheline). The present chapter covers nonfeldspathoidal rocks, with which are included rocks of transitional character in which modal nepheline is not obvious though small amounts of *ne* appear in the norm. Basic feldspathoidal rocks will be treated separately in Chapter 6.

Volcanic representatives of the nonfeldspathoidal division are grouped together as *basalts*, their plutonic counterparts as *gabbros*. Apart from a variable content of phenocrysts most basalts are rather fine-grained rocks. But there are also rocks that, by reason of slow cooling in thick flows or, more commonly, in shallow dikes and sills, have developed distinctive medium-grained textures, typically ophitic or subophitic. These are *diabases*.°

VOLCANIC DIVISION: BASALTS AND DIABASES

Global Abundance

By far the most abundant of all volcanic rocks are basalts. Indeed, their total surface exposure, neglecting thin sedimentary covers, exceeds that of all other igneous rocks combined. Spreading from fissure systems situated along the mid-oceanic ridges, they have covered the entire ocean floor beneath its veneer of deep-sea sediments; where copiously discharged from continental fissure swarms, as in southern Brazil, the northwestern United States, and the Deccan of western India, they have flooded entire landscapes to build basaltic plateaus of enormous extent; they are by far the dominant components of major shield volcanoes—continental or oceanic—of the Hawaiian type; and along with andesites and rhyolites they are conspicuous among lavas of island arcs and mobile belts situated along plate junctions.

Diabases, spectacularly developed in massive sills and dikes invading upper levels of the oceanic crust, likewise predominate overwhelmingly among shallow intrusive rocks.

Chemical and Mineralogical Delineation of Basalt Types

Since R. A. Daly assigned to basaltic magmas a global primitive role in petrogenesis, attention has focused increasingly upon variety in chemical pattern among basaltic magma types as expressed in the basalts themselves, and as a potential parental influence on the evolution of chemically divergent magma series. Different major basalt types have thus come to be delineated in terms of chemical criteria, particularly as

---

°The term *diabase* as used in North America and Germany is synonymous with *dolerite* as used in England. When English petrographers employ the term diabase they refer to "altered dolerites," ophitic or subophitic rocks in which feldspars are albitized or otherwise conspicuously altered, and pyroxenes tend to be replaced by aggregates of fibrous amphibole and chlorite.

expressed in normative parameters. Corresponding rock names—tho-leiite, alkali olivine basalt, and so on—have been coined and have won general acceptance by petrologists and geochemists. So deeply have these terms now become ingrained in petrological nomenclature that system-atic petrographers are faced with a special problem: They must seek and define mineralogical criteria to fit a system of terminology already estab-lished on the basis of chemical pattern.

With this problem in mind, we present a slightly simplified version of a system for chemical classification of basalts developed from a synthesis of petrographic, chemical, and experimental data by H. S. Yoder and C. E. Tilley along lines consistent with contemporary thinking among petrologists.[1] Major basaltic rock types, in order of increasing undersa-turation in SiO_2, will be named as follows:

1. Tholeiites:
    a. slightly oversaturated; normative Q and abundant *hy*.
    b. saturated; abundant normative *hy*.
2. Olivine tholeiites: undersaturated; normative *ol* and *hy*.
3. Olivine basalts: undersaturated; normative *ol*; *hy* insignificant or com-pletely lacking.
4. Alkali olivine basalts: strongly undersaturated; normative *ol* and minor *ne*.
5. Basanites: strongly undersaturated; *ol* and significant *ne* in the norm.

In both groups of tholeiite, SiO_2 tends to be higher and $(K_2O + Na_2O)$ distinctly lower than in alkali olivine basalts and basanites. Some writers recognize a group of "high-alumina basalts," mostly tholeiitic and with high Al_2O_3 ($\sim$ 18–19%).

Our descriptive treatment of basalts (and, in Chapter 6, basanites) employs as far as possible the same terminology as that just presented. But the criteria adopted, as befits the scope of this book, are mineral-ogical. We must remember, however, that the mode is by no means a precise mirror of the norm; so that some anomalies, especially in the case of transitional types, must be expected.

Oversaturated tholeiites cannot be distinguished sharply from satu-rated tholeiites. Quartz, which is the critical index of oversaturation in the norm, can seldom be distinguished as such in the mode. Magnesian olivine, because of its stability in slightly oversaturated melts at near-surface pressures (see Figure 1-2), can appear as sparse phenocrysts even in slightly oversaturated tholeiites. Abundance of olivine is important:

Great abundance of olivine phenocrysts (in picrite basalts) strongly suggests accumulation by gravitational sinking; and presence of olivine in two generations—phenocrysts and groundmass—implies pronounced undersaturation of the magma in SiO_2. The nature and composition of the pyroxenes is particularly significant: Magnesian pyroxenes (hypersthene or pigeonite) reflect high values of normative *hy*; augites close to $Ca(Mg, Fe)Si_2O_6$ when unaccompanied by magnesian pyroxene indicate an absence of *hy* in the norm; titanium becomes concentrated in augites crystallizing from alkaline basic magmas, generally imparting to the pyroxene crystals a distinctive pale violet hue.* Finally, in the groundmass of many alkali olivine basalts, especially those transitional to basanites, plagioclase laths are rimmed with alkali feldspar, and small crystals of brown hornblende or wisps of biotite may be visible.

Using purely mineralogical data of this kind, we shall describe the main basaltic rock types under four headings, tholeiitic basalts, tholeiitic olivine basalts, alkali olivine basalts, and picrite basalts. This last group includes basalts with abnormally high phenocrystic olivine ("oceanite"), augite ("ankaramite"), or both. Olivine basalts with little or no normative *hy* as defined chemically by Yoder and Tilley cannot be distinguished petrographically from olivine tholeiites with abundant *hy* in the norm. Solely on petrographic criteria both would be identified as tholeiitic olivine basalts.

Textural Characteristics

Generally, basalts are fine grained and diabases are medium grained (Figure 3-1). Most are holocrystalline except for small amounts of interstitial glass in some basalts (imparting to the groundmass an intersertal texture as in Chapter 2, Figure 2-9B). Intergranular texture (Figure 2-9A) is common among basalts, ophitic and subophitic textures typical of

---

*Chemical analyses of clinopyroxene cited in this chapter are reduced (by neglecting other ions such as Al, Ti, and Fe^{3+}) to a simplified formula $(Ca_xMg_yFe_z^{2+})Si_2O_6$, where $(x+y+z) = 2$. In what are loosely called "normal," "calcic," or "diopsidic" augites, x approximates 1.

Distinction between pigeonites, subcalcic augites, and calcic augites is possible by normal optical means. The principal criterion is the value of 2V (positive), which can be estimated roughly by standard convergent-light technique on grains nearly normal to an optic axis, or more readily and accurately by means of a universal stage (augites, $2V \approx 50°–60°$; somewhat magnesian augites, $2V \approx 40°–45°$; subcalcic augites, $2V \approx 20°–40°$; pigeonites, $2V \approx 0°–20°$, in convergent light nearly uniaxial).

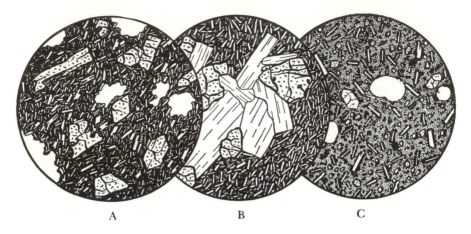

A   B   C

Figure 3-1. Basalts and Basaltic Andesite

A. Basaltic andesite, Parícutin, Mexico. Diam. 2.5 mm. Phenocrysts of olivine, some elongated parallel to the base, and microlites of labradorite in a vesicular matrix of black glass.

B. Glomeroporphyritic olivine-augite basalt, Copco Dam, northern California. Diam. 2.5 mm. A cluster of bytownite and olivine phenocrysts lies in a groundmass of labradorite laths, granular augite, and interstitial black glass.

C. Olivine-augite basalt, Craters of the Moon, Idaho. Diam. 2 mm. From the vesicular, glass-rich crust of a recent pahoehoe flow. Small crystals of olivine, augite, and labradorite, accompanied by abundant granular opaque iron oxides, in a base of clear, brown glass.

diabases (Figure 3-2). Glassy varieties of basalt, though encountered less frequently than in acid lavas, are nevertheless typical of local rapidly chilled variants: on borders of narrow shallow intrusions, in lava crusts and pyroclastic fragments, in pillow margins of lavas extruded beneath sea or lake waters, and in the peculiar environment of extrusion beneath terrestrial ice (as exemplified in Pleistocene flows of Iceland). Basaltic glasses, being highly unstable, not only tend to devitrify with the passage of time but also are particularly vulnerable to chemical alteration by percolating waters. Very commonly, therefore, they are suffused with dark pigment (cryptocrystalline) or have been partially replaced by a variety of brownish or greenish alteration products.

Porphyritic textures are widespread in basalts, much less so in diabases. Where olivine is present it is almost invariably phenocrystic; but phenocrysts of augite also are common, and many rocks contain microphenocrysts of plagioclase (Figures 3-1 and 3-4). The latter mineral,

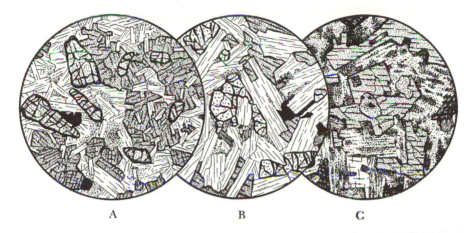

A   B   C

Figure 3-2. Diabases

A. Tholeiitic diabase, West Rock, New Haven, Connecticut. Diam. 2 mm. Colorless pigeonite, marginally altered to serpentine; fresh ophitic plates of pale-brown augite; laths of labradorite; granules of opaque minerals; and interstitial chloritic material. Not shown in this section, but found elsewhere in the sill from which this specimen came, are a little interstitial biotite and micropegmatite.

B. Alkali olivine diabase, Pigeon Point, Minnesota. Diam. 3 mm. Laths of calcic labradorite; olivine; ophitic, purplish augite; opaque minerals; reddish-brown biotite; and chlorite.

C. Tholeiitic diabase, Pwllheli, North Wales. Diam. 3 mm. A single plate of subcalcic augite (2V = 40°) ophitically encloses calcic plagioclase, which is almost entirely altered to calcite and prehnite and heavily stippled with granular leucoxene. The opaque grains close to the edge of the section are composed of exsolution intergrowths of ilmenite and magnetite; near the center are two round patches of talc and serpentine after olivine; near the lower edge is an area of calcite.

which is conspicuous as phenocrysts in many andesites, is seldom obvious in hand specimens of basalt. But there are exceptions: Oversaturated high-aluminum basalts in the island of Mull contain labradorite phenocrysts in such abundance that they have been designated by British petrologists as typifying a special porphyritic-central magma type of the Hebridean volcanic province. Finally, flotation of plagioclase crystals and sinking of augite and olivine may result in clustering of early-formed crystal aggregates that impart glomeroporphyritic textures to tops and bottoms of thick flows and sills (Figure 3-1B).

Petrographic Character of Tholeiitic Basalts and Diabases

Tholeiitic Basalts

As in all basalts, pyroxenes and plagioclase (labradorite) are the principal constituents of tholeiitic basalts. Magnesian olivine is almost always present, typically as phenocrysts making up less than 5% of the mode. These commonly show incipient serpentinization or deuteric marginal alteration to brown pleochroic iddingsite (Figures 2-7B, 3-4B). Colorless or pale greenish augite ($2V_\gamma = 45°–55°$) is abundant, especially as phenocrysts, in many tholeiites; but it is the presence of one or more magnesian pyroxenes—particularly in the groundmass—that is the hallmark of the tholeiitic basalt in the strictest sense of the term, reflecting in the mode the chemical pattern that is also expressed in high values of normative *hy*. In the groundmass of many tholeiites there is a residue of interstitial, pigmented, partially devitrified glass, imparting to the rock an intersertal texture (Figure 2-9B). Chemical analysis of this glass reveals a relatively high content of SiO_2, which explains why many tholeiitic basalts containing modal olivine nonetheless show minor normative Q ($< 4\%$). In some rocks one of the polymorphs of crystalline SiO_2 can indeed be identified in a fine-grained interstitial residue.

It is widely held that certain other chemical traits are characteristic of tholeiitic basalts in particular tectonic settings: low K, for instance, in basalts from the ocean floor; high Al in basalts of some island arcs and mobile belts. These peculiarities, however, have no unique diagnostic expression in tholeiitic mineralogy. Thus while many high-alumina tholeiites carry conspicuous phenocrystic plagioclase, others (such as those of Medicine Lake Highlands in the northwestern United States) are nonporphyritic. And some "high-alumina" basalts even conform in all respects to the alkali-olivine-basalt chemical pattern.

Tholeiitic basalts (in the strict sense of the term) are the principal lithologic components of the great active shield volcanoes Mauna Loa and Kilauea, and of some of the older eroded volcanoes, in the Hawaiian Islands.[2] The basalts of Mauna Loa and Kilauea carry phenocrysts of magnesian olivine ($2V_\gamma = 85°–90°$) which in the undersaturated types tend to be more numerous and larger than in quartz-normative rocks. Plagioclase ($An_{60}–An_{50}$) and augite usually are present as microphenocrysts. Groundmass minerals are subcalcic augite ($2V = 45°–20°$), plagioclase, iron-titanium oxides, and in many rocks a little dark-colored interstitial glass (intersertal texture). Textures of other Hawaiian tholeiites may be intergranular (subcalcic augite grains interspersed with plagio-clase laths), subophitic, or in coarser rocks truly ophitic. Exemplifying this last is a basalt from the island of Kauai with coarse subcalcic augite enveloping laths of labradorite and sparse granules of olivine (Figure 2-7B).

Much rarer among Hawaiian basalts are hypersthene tholeiites; they have been recorded from Mauna Loa and in the eroded basement (Koo-lau Series) of a Pliocene shield volcano on the island of Oahu.[3] All are quartz-normative rocks, though most show sparse phenocrysts of olivine in the mode. Hypersthene typically occurs as prismatic phenocrysts in some cases mantled with pigeonite. These, along with microphenocrysts of subcalcic augite and sparsely scattered olivines, are set in a groundmass of interspersed laths of andesine-labradorite and granular subcalcic augite ($2V = 15° – 40°$). One specimen (figured by Muir and Tilley, 1963, p. 115) shows phenocrysts of olivine in a groundmass of columnar hypersthene, pigeonite, and plagioclase.

In another midoceanic setting basaltic lavas extruded from fissures and, building central shield volcanoes, blanket the whole 50,000 km^2 extent of Iceland—the largest landmass located on the mid-Atlantic Ridge. They date from the mid-Tertiary onward. Over 95% of chemically analysed Icelandic basalts fall in the broad category of tholeiites. But only some of them display the diagnostic petrographic characteristics of tholeiites in the strict sense. Such are the basaltic lavas that make up more than half the partially dissected late Tertiary volcano Thingmuli (the remainder being derivative ferroandesites and rhyolites).[4] The tholeiites of Thingmuli are quartz-normative rocks in which scattered small phen-ocrysts of labradorite or bytownite, augite, and in some a little olivine are set in a groundmass of granular pyroxene, zoned plagioclase laths (labra-dorite-andesine), plentiful iron-titanium oxides, and a little interstitial heavily pigmented glass. Pyroxene granules of the groundmass are augite rimmed with pigeonite. Chemically the Thingmuli basalts and associated lavas are characterized by a high Fe/Mg ratio.

Strictly tholeiitic basalts carrying pigeonite or subcalcic augite are thought to be the most characteristic and widespread of continental bas-alts. They are principal constituents of most basaltic plateaus; for exam-ple, Yakima basalts (120,000 km^3) of the Miocene-Pliocene Columbia River province of the northwestern United States; Cretaceous-Eocene Deccan province of western India; Lesotho plateau remnant, Karroo province (Jurassic) of South Africa. Moreover, they are the typical basic members of calcalkaline lava series (basalt-andesite-rhyolite) of active volcanic island arcs and mobile belts such as those that margin the Pacific Ocean. The petrography of island-arc tholeiites will be illustrated by three specific instances.

Two chemical types of tholeiitic basalt have been distinguished by H. Kuno in the eastern marginal volcanic belt of the Japanese and adjoining Izu island arcs.[5] The first type is concentrated along the oceanic edge of the belt and extends south (into the Pacific) in the Izu Islands. Basalts here are slightly oversaturated and conform chemically and petrographically to the strictly tholeiitic pattern. They include both aphyric and porphyritic types, the latter with phenocrysts of calcic plagioclase (anorthite, bytownite), augite, orthopyroxene, and olivine. Groundmass minerals are labradorite, subcalcic augite, rarely olivine mantled with pigeonite, and minor quartz (or some other form of SiO_2). The second type, localized along the inland (western) side of the marginal belt, is "high-alumina" basalt with 17–18% Al_2O_3. Some of these rocks are oversaturated and scarcely differ petrographically from the tholeiitic type just described, except that their olivine phenocrysts contain inclusions of picotite. But those that are undersaturated tend to have more groundmass olivine, and this is never mantled with pigeonite. There is a complete chemical and petrographic gradation between "high-alumina" basalts with tholeiitic affinities and others that are closely similar to alkali olivine basalts that occur still further westward (on the continental side of the arc).

In island arcs of the southwest Pacific, extending 3000 km from eastern New Guinea to Tonga, andesites and rhyolites are much the most voluminous products of currently active volcanoes; but basalts are also present in most sectors. In chemistry these basalts vary considerably, but the olivine-tholeiitic chemical pattern predominates (Carmichael, Turner, and Verhoogen, 1974, pp. 533–551). At some volcanic centers, recently erupted basalts such as the 1967 flow of Lopevi in the New Hebrides arc show all the characteristics of tholeiitic basalt: phenocrysts of bytownite, augite, olivine,and minor hypersthene set in a groundmass of plagioclase, augite, iron-titanium oxides, and brown glass; significant Q and hy in the norm. Basalts of Talasea in New Britain are quartz-normative rocks bordering in silica content on basaltic andesites. They are porphyritic, with phenocrysts of zoned plagioclase (An_{90}–An_{60}), diopsidic augite, and magnesian olivine; the groundmass is composed of plagioclase laths (An_{70}–An_{60}), augite, and either pigeonite or hypersthene, iron-titanium oxides, and interstitial glass, with accessory apatite and in some instances cristobalite.

Tholeiitic Olivine Basalts

Between tholeiitic basalts, such as have just been described, and alkali olivine basalts with their equally distinctive mineralogy, lies a sort of petrographic no man's land occupied by a widespread type of undersaturated basalt whose mineralogical character—more particularly the nature of the pyroxene—gives no certain clue as to its tholeiitic chemical pattern: significant normative *hy*, usually accompanied by *ol* but more rarely by minor *Q*. In the chemical classification presented earlier (p. 96) these rocks are olivine tholeiites. To conform as closely as possible to this usage we group them here as tholeiitic olivine basalts. Two examples suffice to demonstrate their rather ambiguous mineralogical character.

Many ocean-ridge basalts, perhaps the most extensively developed and certainly the most imperfectly sampled of terrestrial volcanic rocks, belong to this general petrographic category. Most are olivine tholeiites in the chemical sense. Tending to be fine-grained and in many instances (e.g., on lava-pillow rims) partially glassy, their principal recorded mineral constituents are diopsidic augite, magnesian olivine (microphenocrysts and groundmass), and labradorite.

From extensive regional sampling it would seem that most Icelandic lavas are tholeiitic olivine basalts.[6] In one well-sampled area, the western Reykjanes Peninsula, basalts of the shield volcanoes have about equal amounts of normative *ol* and *hy* (each $\sim 10–12\%$). They are even-grained rocks with ophitic texture and few phenocrysts of olivine (Fo_{80}) with small brown picotite inclusions, and rare plagioclase (An_{87-78}). The groundmass minerals are labradorite, a normal augite, minor olivine, and interstitial iron-titanium oxides. Chemically these rocks are interesting in that they duplicate the very low K of "abyssal" oceanic tholeiites. Throughout the whole of this area the fissure basalts are of a distinctly different chemical and petrographic type. They are slightly oversaturated or undersaturated in SiO_2 (normative *ol* < 6%). They are spotted with microphenocrysts of plagioclase (An_{80-70}), augite $Ca_{0.8}(Mg, Fe)_{1.2}Si_2O_6$, and olivine ($Fo_{80-70}$), the two latter in smaller quantity. Clots of all these phases are common, imparting to rocks in which they are plentiful a glomeroporphyritic texture. The groundmass is a fine-grained mixture of labradorite-andesine, augite, iron-titanium oxides, glass, and rare granules of olivine. A third type of lava erupted in subordinate quantity from shield volcanoes is a picrite basalt.

Tholeiitic Picrite Basalts

It has been repeatedly demonstrated in thick diabase sills that early-formed, relatively large, dense crystals of magnesian olivine tend to sink in static bodies of basic magma and to accumulate in the lower levels—the classic example of gravitational differentiation. In many basaltic provinces, localized flows crowded with olivine are thought to have such an origin, representing drafts of magma drawn from the lower levels of reservoirs in which differentiation was in progress. Such rocks are called *picrite basalts* (Figure 3-4B); and to those that are associated with tholeiitic lavas the adjective "tholeiitic" is also applied. They are but minor components of tholeiitic volcanic complexes.

The picrite basalt of the 1840 flow from Kilauea (at Nanawale Bay)[7] is a porphyritic rock with abundant phenocrysts and clusters of olivine (Fo_{88}), scattered phenocrysts of augite ($2V = 56°–52°$), and micropheno-crysts of labradorite (An_{60}). These are set in a fine-grained groundmass of zoned labradorite ($An_{60–52}$), subcalcic augite ($2V = 46°–32°$), and iron-titanium oxides. In spite of the high content of modal and normative olivine (*ol* 28%), *hy* also is high (18%) in the norm, and the groundmass pyroxene is that of typical tholeiites.

Tholeiitic picrite basalts comprise about 1% of the total volume of lavas in the basaltic field of western Reykjanes Peninsula in Iceland (p. 103), where they were erupted from shield volcanoes whose main products are tholeiitic olivine basalts. The picritic rocks are strongly porphyritic. Phenocrysts of magnesian olivine (Fo_{90}), some as large as 1 cm in diameter, make up between 25% and 50% of the mode; scattered small phen-ocrysts of chromite are ubiquitous and plagioclase phenocrysts (An_{90}) commonly present. The groundmass is a fine-grained aggregate of labra-dorite laths, clinopyroxene, minor olivine, and interstitial iron-titanium oxides. Only the field association and the consistent presence of nor-mative *hy* (mostly 5–15%) distinguishes these Icelandic picrite basalts as tholeiitic.

Tholeiitic diabases are generally subaluminous rocks, saturated or slightly oversaturated with silica, but near the bottoms of thick sills they may grade into olivine-rich diabases (Figure 3-3B). The quickly chilled margins of intrusions usually display either a fine intergranular or an intersertal texture; toward the centers of intrusions the texture generally

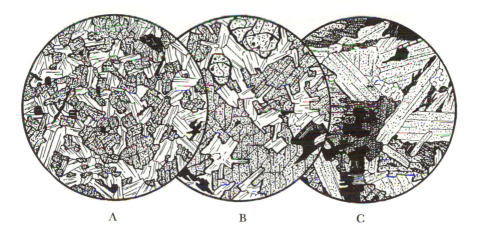

A   B   C

Figure 3-3. Differentiation in a Tholeiitic Diabase Sill, New Jersey

A. Specimen 3 m above the base. Diam. 3 mm. Composed of labradorite, clinopyroxenes, and a little hypersthene, ilmenite, and biotite.

B. Olivine-rich specimen, 15 m above the base. Diam. 3mm. Consists of olivine, ophitic pigeonite, labradorite laths, ilmenite, and, close together, accessory biotite and micropegmatite.

C. Specimen from upper part of sill. Diam. 3 mm. The chief constituents are pyroxene, altered labradorite, and iron-titanium oxides. Deuteric hornblende and biotite border the pyroxene and oxides; patches of interstitial micropegmatite near center and right edge of section; prism of apatite adjoins upper-right edge.

becomes coarser and more ophitic (Figure 3-2). There may be wide textural variations, however, even in single thin sections, and some diabases are notably porphyritic.

Because of the remarkable differentiation exhibited by most tholeiitic sills, the proportions and nature of the component minerals vary from layer to layer. An average sample is made up of plagioclase (40–55%), pyroxene (35–45%), iron-titanium oxides (8%), micropegmatite (5%), and olivine (3%), plus a little biotite, amphibole, and apatite.

Plagioclase generally varies in amount from a minimum of about 40% in chilled border facies and the lower parts of sills to a maximum of approximately 65% near the top. Most of it ranges in composition between An_{50} and An_{60}, though early-formed crystals may be as calcic as bytownite and the last to crystallize may be as sodic as oligoclase. Normal zoning is widespread, and generally the percentage of calcium in the feldspar diminishes upward from the olivine-rich layer, reaching a minimum in the late-crystallizing veins and streaks of granophyre. The last feldspar to form is usually orthoclase or anorthoclase, graphically intergrown with quartz in areas of micropegmatite. These intergrowths are present in small amount at almost all levels, and in the cores of thick sills they may constitute as much as a quarter of the bulk. They are a principal constituent of granophyric diabases.

Pyroxenes in tholeiitic diabases are generally of three kinds, augite, pigeonite, and orthopyroxene, and the later-formed crystals of each contain more iron than the earlier ones.Of the three, brown or purplish augite ($2V = 30°–60°$) is generally the most abundant and is present throughout the sills. Near sill margins it tends to be euhedral; elsewhere it is subhedral or subophitic. Most of the orthopyroxene forms microphenocrysts, and it is confined chiefly to chilled margins and the lower layers of sills. Near the bottoms it is generally bronzite; higher up it is hypersthene. It is absent from micropegmatite areas. Colorless pigeonite ($2V < 30°$) varies in habit. Early, magnesian crystals tend to be euhedral; later, more ferriferous ones tend to build ophitic plates. Most of the pigeonite crystallized after the hypersthene, and the two are seldom seen together in a single specimen. Pigeonite and augite are generally found side by side, for they crystallized mostly at about the same time, though ferroaugite continued to separate after formation of pigeonite ceased. Nowhere are the optical properties of pigeonite better displayed than in the diabase of West Rock, New Haven (Figure 3-2A).

Olivine is absent in most tholeiitic diabases. However a magnesian variety, Fo_{70-80}, makes up as much as 25% of the mode toward the bases of some thick sheets, as in the Triassic Palisades sill, 300 meters thick, of New Jersey. Varieties richer in iron may appear in much smaller quantities in central sectors of some sills (e.g., in the South African Karroo Province). Olivines close to fayalite in composition are minor but characteristic constituents of granophyric quartz diabases in the uppermost levels of some intrusions.

Biotite and brown or greenish-brown hornblende are late reaction products. They are almost ubiquitous, but usually in trifling amount save in areas of micropegmatite, where they may be accompanied by scraps of bluish-green alkali amphibole. Most of the biotite is molded on pyroxene or iron oxidies and is mantled, in turn, by hornblende.

The oxides are intergrowths of ilmenite and magnetite or hematite. They are rare in chilled border facies, become increasingly common in the more siliceous parts of sills, and are most abundant in patches of micropegmatite. They may form euhedral crystals or skeletal grains molded on or intergrown with pyroxenes, or occur as jagged rods and plates. Finally, most tholeiitic diabases carry slender needles of chlor-apatite, some of early but much of late formation; these latter are most abundant in areas of micropegmatite.

Alkali Olivine Basalts and Diabases

General Modal Character

The distinctive chemical pattern expressed in the norm by high *ol*, minor *ne* (< 5%), and corresponding absence of *Q* and *hy* is reflected in the mode principally by abundance of magnesian olivine, commonly in two generations, and presence of a single pyroxene, an augite close to diop-side-hedenbergite in its relative proportions of Ca, Mg, and Fe. Typically the augite is significantly titaniferous, characteristically but not invariably pale violet in thin section, and except in iron-rich varieties has 2V in the range 50°–60°. Nepheline normally cannot be identified in the mode of basalts, but in some diabases its chemical equivalent may appear in the form of minor interstitial analcime. Additional signs of alkalinity in the groundmass of some basalts may be a little alkali feldspar (rimming pla-gioclase laths) or sparse wisps of brown biotite. Occasionally alkaline bas-alts may carry prisms of brown amphibole (kaersutite) rimmed with titan-augite.

Distribution

Alkali olivine basalts have been recorded from almost all tectonic envi-ronments—even in close association with the much more typical tholeiitic basalts of continental plateaus and the deep ocean floor. But their pref-erential locations are in areas of oceanic or continental rifting, for exam-ple, in islands that dot the mid-oceanic ridges and in continental rift valleys such as the lower Rhine Graben and those of eastern Africa. They may be closely associated with tholeiitic basalts, as in the Hawaiian Islands and the Hebridean volcanic province situated along the rifted Atlantic edge of northern Scotland. In many provinces their associates may be basic and even ultrabasic lavas. While atypical of mobile fold belts, alkali olivine basalts may be erupted over wide continental provinces situated on the landward side of island arcs. A classic example is provided by Quaternary volcanism along the western margin of the main Japanese islands and still farther west in Korea and Manchuria.

Petrography of Basalts

Alkali olivine basalts are much the same the world over. We take as an example the Late Pleistocene and recent lavas of a small local volcanic field, the site of the city of Auckland, New Zealand.[8] Here, within 100 km of the much more extensive Taupo zone of contemporary andesite-rhyolite volcanism—where the Tonga-Kermadec island arc cuts New Zealand—local eruptions have occurred every few hundred years for 50,000 years or more. With one minor exception (a few blocks of basanite debris), the Auckland lavas are alkali olivine basalts and subordinate but widespread chemically related picrite basalts.

The Auckland basalts are medium-gray even-grained rocks with small phenocrysts of olivine (up to 3 mm long) making up about 20% of the mode. Augite and labradorite (An_{80}–An_{50}) may appear as microphenocrysts; but both minerals along with abundant iron-titanium oxides occur principally in an intergranular or intersertal groundmass. The pyroxene is everywhere the same: a pale yellowish or purplish titaniferous diopsidic augite with high 2V (commonly ~ 60°). Olivines are highly magnesian, with 2V close to 90°, and for the most part are unaltered. Some olivine phenocrysts show corroded outlines; others are euhedral; and in most sections a few are mantled with prismatic or granular augite (a feature generally considered highly atypical in alkali olivine basalts). Rarely the groundmass contains subhedral flakes of red-brown biotite.

The *picrite basalts* of the Auckland region conform to the same general mineralogical pattern, but are much richer in augite (about 40% of the mode) and poorer in plagioclase (< 25%) than the rocks designated basalt. One recorded specimen has relatively coarse biotite (0.5%). Augite-rich picritic basalts° are common in alkali-olivine-basalt provinces elsewhere. In the Auckland picritic basalts, augite is deeper in color than that of the other basalts and tends to be strongly zoned: Clusters and stellate aggregates are common. In picritic basalts with intersertal texture the interstices are usually filled with finely crystalline alkali feldspar rather than glass.

Alkali Olivine Diabases

In many provinces sills, dikes, and thick flows of diabase duplicate the alkali-olivine-basalt mineralogy. But the grain is coarser, and ophitic and

---

°Sometimes called *ankaramites* as distinct from *oceanites* (rich in olivine) of tholeiitic provinces.

poikilitic textures are typical. Augites, because of slow cooling, may enclose exsolved lamellae of a more magnesian pyroxene.

Rocks of some thick sills, whose chilled margins have the chemistry of alkali olivine basalt, contain minor interstitial analcime (which marks a transition toward feldspathoidal basic rocks treated in Chapter 6). An example from the Tertiary Hebridean province of western Scotland is a sill nearly 200 m thick exposed in the offshore Shiant Isles.[9] This shows marked vertical compositional variation, probably the result of differentiation in which the influence of gravity clearly played a main role. The modal composition in the main layer (total thickness > 100 m) is olivine 10%, augite 25%, labradorite 60%, with iron-titanium oxides and analcime (or other sodic zeolite) together accounting for the remaining 5%. The pyroxene, a calcic augite ($2V = 50°$–$55°$) approximates $(Ca_{0.85}Mg_{0.75}Fe_{0.4}) Si_2O_6$. Crystals of olivine show strong zoning, from core to rim $Fo_{80-70} \rightarrow Fo_{15-10}$. Plagioclase crystals likewise are normally zoned, $An_{80} \rightarrow An_{35}$. Textures are mainly ophitic. This diabase grades downward into olivine-enriched picritic diabases.° The mean modal composition of the picritic diabase over a thickness of 30 m is given as olivine (zoned $Fo_{80} \rightarrow Fo_{50}$) 30%, calcic augite ($2V = 50° - 53°$) 15–20%, zoned plagioclase ($An_{80} \rightarrow An_{50-30}$) 50%, plus opaque oxides and traces of zeolites and brown hornblende (barkevikite). The texture is ophitic near the top. The lowest differentiated layer—*picrite* of almost ultramafic composition—is less than 5 m thick. The mode is olivine (Fo_{85-80}) 60%, calcic augite ($2V = 48°$) 10%, bytownite (An_{80}) 25%, together with iron-titanium oxides (2%) and zeolites (3%). The texture is poikilitic, olivine grains being enclosed in coarse anhedral plagioclase and augite—much as in crystal cumulates, in the lower levels of major basic plutons (p. 119).

Hawaiites and Mugearites

In the late volcanic series of various Hawaiian shield volcanoes there is a natural gradational transition from alkali olivine basalts to associated basic lavas in which the plagioclase is more sodic than labradorite. These rocks, which habitually appear in other alkali-olivine-basalt provinces the world over, have always presented a problem in classification. In classifications in which plagioclase composition is the principal index they fall in the andesite class. In other classifications they are grouped as andesine or oligoclase basalts. Our preference is for this second alternative; and

---

°*Picrodolerite* (olivine > 30%) and *picrite* (olivine > 60%) in British terminology.

following a terminology widely used among those who work with such rocks we adopt the now well-established terms *hawaiite* and *mugearite* to designate, respectively, the andesine- and the oligoclase-bearing rock types. In support of this option (which involves rejection of plagioclase composition as the universally overriding classificatory index) two items of great genetic significance are noted. First, the natural field and chemical* gradational sequence basalt → hawaiite → mugearite invariably terminates at the intermediate SiO_2 level in trachyte, never in andesite. Second, another natural transition, tholeiitic basalt → basaltic andesite → andesite, is typical of the island-arc or mobile-belt tectonic environment, one to which the hawaiite → mugearite → trachyte series is alien.

The problem of the petrographer is to search for visible mineralogical and textural criteria by which to distinguish hawaiites and mugearites from andesites proper. Admittedly, petrographic distinctions in the absence of definitive plagioclase composition prove to be rather subtle. And they need to be backed by details of natural environment—without which, except possibly for museum purposes, petrography would in fact have little value.

Hawaiites from the summit crater of the great Haleakala volcano on the Hawaiian island of Maui are mostly fine-grained dark nonporphyritic or sparsely porphyritic rocks, slightly fissile in hand specimen as a result of fluxional alignment of groundmass feldspars. Where present, phenocrysts include labradorite, calcic and somewhat titaniferous augite ($2V_\gamma = 60°$), and olivine ($2V_\alpha = 80°-85°$). The groundmass is dominated by laths of andesine with interspersed granules of augite, olivine, and magnetite. Careful search reveals optically negative untwinned alkali feldspar (a calcic anorthoclase) rimming the twinned groundmass plagioclase and, along with minor glass, filling the interstices in some rocks. Most rocks are characterized chemically by low SiO_2 (45–50%) and normative *ol* (7–10%), *or* (8–10%), and *ne* (~5%). Normative plagioclase is about An_{40}.

Mugearites, first reported from the Hebridean volcanic province, are now known to be typical minor associates of alkali olivine basalts in many parts of the world–the Hawaiian volcanoes, the Azores, and the Miocene alkaline province of Dunedin, New Zealand, among them.[10] They are basic volcanic rocks ($SiO_2 = 48-51\%$) with about 10% each of *or* and *ol*,

---

*For example, in a simple plot of ($K_2O + Na_2O$) against SiO_2 (see Carmichael, Turner, and Verhoogen, 1974, p. 413).

and minor *ne* in the norm; normative plagioclase is close to An_{30}. Feldspar is the most abundant modal constituent (65–70%) and dominates the texture by subparallel alignment of its slender lathy crystals that comprise the greater part of the groundmass (Figure 3-4A). Microphenocrysts (when present) and the larger crystals of the groundmass consist of two optically distinguishable phases: cores of albite-twinned sodic andesine (An_{35-30}; $2V_\alpha = 80°–70°$) with clear, usually nontwinned rims of a Ca-bearing anorthoclase ($2V_\alpha = 55°–70°$). In some mugearites (e.g., those of New Zealand) small amounts of interstitial sodic sanidine ($2V_\alpha = 25°–40°$) have been identified in the groundmass. There are numerous micro-phenocrysts of ferriferous olivine (Fo_{60}—Fo_{35}; $2V_\alpha = 75° - 60°$) with an unusual crystal habit—elongated parallel to the [100] axis—and commonly showing marginal alteration to iddingsite. Pyroxene is relatively inconspicuous; it is a calcic variety ($2V_\gamma = 55°–60°$) approximately midway between diopside and hedenbergite. Magnetite is rather plentiful and in some rocks even appears as microphenocrysts. Apatite, a ubiquitous accessory, is conspicuous by reason of its smoky appearance (a feature also common in apatite of phonolites).

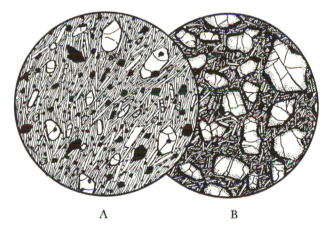

A   B

Figure 3-4. Basalts

A. Mugearite, Isle of Skye, Scotland. Diam. 3 mm. Essentially composed of oli-vine, oligoclase, and iron oxide, with accessory augite, apatite, and orthoclase. The smaller olivines are elongated along [100], the larger ones, terminated by domes, are elongated along [001].

B. Picrite basalt, Kauai, Hawaiian Islands. Diam. 3 mm. Abundant large grains of olivine, rimmed with iddingsite and magnetite, in an intergranular matrix of labradorite laths, subhedral augite, and magnetite.

Postmagmatic Elements in Petrography of Basaltic Rocks

In some basalts (and basaltic andesites) there are petrographic elements that clearly relate to postmagmatic processes in which circulating aqueous (and CO_2-rich) fluids played a significant role. There are two independent categories that merit present attention: development of *amygdules* in vesicular rocks, and partial overprinting of a new mineral assemblage upon the truly igneous mineralogy of an important group of basalts called *spilites*.

Amygdules

Amygdules are spheroidal or ellipsoidal bodies representing vesicles that have been partially or completely filled by minerals deposited from aqueous fluids either deuterically or, more commonly, much later during diagenesis or low-grade metamorphism. The mineral assemblages so formed tend to be simple, frequently monomineralic. Among those most widely encountered are the following: cryptocrystalline quartz (chalcedony) in some cases macroscopically color-banded (agate); clear isotropic opal (with very low refractive index); yellowish, greenish, or brown chloritic minerals (cryptocrystalline or in fibrous spheroidal aggregates); epidote, prehnite, or pumpellyite; zeolites of all kinds (mostly with very low refractive index), lining the larger cavities with euhedral crystals; carbonates, notably calcite, aragonite, and spherulitic siderite.

Spilites

The term *spilite* in today's usage covers a broad group of basic and semibasic volcanic rocks texturally resembling basalts or diabases, but with plagioclase of highly sodic composition (commonly pure albite), and an assemblage of associated secondary minerals familiar elsewhere as products of low-grade metamorphism. Their usual development is in submarine pillow lavas and dikes in the eugeosynclinal environment (see p. 521). Spilites in fact comprise the most voluminous of the three lithologic elements in *ophiolites**—composite rock units now generally considered to be segments of oceanic crust, flooring the geosyncline or transported

---

*The other two are serpentinite and deep-sea chert, underlying and overlying, respectively, the spilitic unit.

tectonically to present sites in mobile belts. Not all spilites are pillow lavas, and some indeed may not even be of submarine origin. Nor are all pillow basalts spilitic.*

Long thought to be strictly volcanic rocks crystallizing from a special highly sodic type of basic magma peculiar to the geosynclinal environment, spilites are generally interpreted today as normal basalts (mostly tholeiitic) whose present composition is the result of pervasive reaction with circulating waters at temperatures of perhaps 200 or 300 degrees. The transformation, it is believed, is effected mostly within piles of submarine lavas through the agency of heated sea water circulating in intracrustal geothermal systems such as are known to operate today in the vicinity of some oceanic ridges—for example, the Reykjanes Ridge off the southwestern coast of Iceland.

In details and variety of texture spilitic rocks resemble normal basalts and diabases. Chilled once-glassy dike selvages and pillow rims are commonplace; enclosed within these the diabases typically are ophitic, and spilitic basalts display the usual textures—microporphyritic, intersertal, and so on—familiar in normal unaltered basalts. Vesicles and amygdules usually are abundant (Figure 3-5B,C), the latter being filled with chalcedony, chlorite, calcite, opal, and in many cases a hydrated Ca-Al silicate—laumontite, pumpellyite, prehnite, or epidote. Original glass has been largely replaced by brownish isotropic products of combined hydration and oxidation (palagonite). In some rocks scattered radial growths of slender albite crystals (*variolitic* texture) suggest devitrification (Figure 3-5C). The overall microscopic appearance conveys an impression of more or less severe mineral degradation. Groundmass materials are largely replaced by chloritic minerals, pumpellyite (or other hydrous Ca-Al silicates), densely granular sphene ("leucoxene" replacing original iron-titanium oxides) and fibrous amphibole (after pyroxene). While olivine almost invariably has been converted to "serpentine," clear, coarse-grained augite commonly has survived the degration process.[†] Plagioclase phenocrysts and groundmass laths, while retaining their crystal form and patterns of internal twinning, have been replaced by highly

---

*Tertiary submarine lavas near Oamaru in southern New Zealand consist of perfectly shaped ellipsoidal pillows with glassy rims, the intervening spaces filled with bryozoan detritus; but both mineralogically and texturally these are normal basalts. Similar rocks have been recorded among products of recent eruption on the ocean floor.

†The striking combination of primary augite and ophitically enclosed clear albite was long cited in support of the concept of primary igneous origin of spilites—a view still not unanimously discarded.

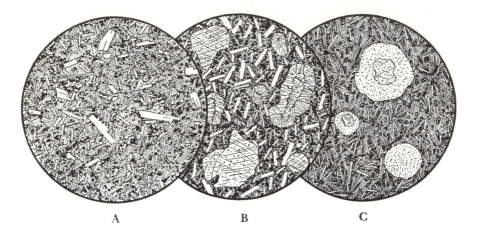

A   B   C

Figure 3-5. Spilitic Rocks

A. Spilitic diabase, Weilburg, Lahn, Germany. Diam. 2 mm. Cloudy laths of oligoclase in an intersertal matrix composed of chlorite, calcite, granular ilmenite, and leucoxene.

B. Amygdaloidal basalt, Coast Ranges, California. Diam. 2mm. Laths of cloudy oligoclase and a few of albite, with relic granules of augite, in a matrix of chlorite, calcite, ilmenite, and leucoxene. Amygdules filled by calcite and chlorite.

C. Variolitic basalt, Mount Tamalpais, California. Diam. 2 mm. Specimen from a pillow sill. Subradiating laths of albite and slender prisms of augite, in a groundmass of calcite, chlorite, and leucoxene. Amygdules of calcite and chlorite.

sodic plagioclase—commonly pure albite but in many rocks albite-oligoclase or even oligoclase in the range An_{10}–An_{20}. It is in the low-intermediate structural state, $2V_\alpha$ typically 90°–80°. The albite crystals, in some cases water-clear, more commonly are charged with dusty particles of chlorite or hydrous Ca-Al silicates. Some may enclose ragged relics of remnant labradorite. Clearly spilitization must be a complex process involving cation exchange between primary minerals, especially plagioclase and sea water, and also a general redistribution of elements within the rock itself. The most obvious resulting chemical effect shown by bulk analyses of spilites is a marked increase in Na_2O and H_2O. Rock permeability plays a major role in facilitating spilitization. Effects are especially obvious in originally vesicular, now amygduloidal flows.

In the coastal ranges west of Santiago, Chile, a thick sequence of tilted Cretaceous flows of porphyritic basaltic andesite in the Andean geosyncline shows a consistent pattern of selective spilitization related to permeability and ready access to water along flow tops.[11] Individual flows are 10–40 m thick. The lower part of each is nonvesicular and relatively fine-grained (except for feldspar phenocrysts) as a result of rapid cooling. In the middle and upper segments the rock becomes increasingly vesicular and correspondingly more permeable. Large phenocrysts of plagioclase are conspicuous, especially in rocks of the middle and upper levels. Near flow bottoms, feldspar is complexly twinned labradorite An_{58} to An_{63}, groundmass laths slightly less calcic. Higher in every flow, plagioclase has mostly been replaced by albite (An_{0-10}) or in some cases by sodic oligoclase; an imperfectly ordered structural state is reflected in anomalous values of optic axial angle ($2V = 90°-81°$). Crystals here retain their original euhedral form and twin pattern but are opaque white in hand specimen, cloudy beneath the microscope. Vesicles in the middle and upper levels of each flow are filled with secondary minerals— laumontite, prehnite, pumpellyite, or epidote accompanied in some cases by quartz, K-feldspar, a colorless mica, actinolite, chlorite, or calcite. Primary diopsidic augite ($2V_\gamma = 55°$) has been largely replaced, except in the lower segments of flows, by one or more of the above minerals, especially by fibrous actinolite. Rock analyses show marked upward increase in Na_2O within each flow.

Apart from relict augite, the minerals that typify spilites in general comprise an assemblage that is both metasomatic and metamorphic. To appreciate its implications as to pressure, temperature, and fluid composition we must compare it with comparable assemblages long familiar in rocks affected by low-grade metamorphism (see p. 520). But spilites, especially as components of ophiolites, have assumed special significance in modern concepts of volcanism in relation to tectonics. The student of purely igneous petrography therefore needs to be familiar with the mineralogical characteristics of spilites that reveal their peculiar status in the spectrum of basic volcanic rocks.

Cognate Xenoliths in Nonfeldspathoidal Basalts

Volcanic flows and associated tuffs commonly carry scattered rock fragments and crystals clearly of foreign origin. These are respectively termed *xenoliths* and *xenocrysts*. Some, having no obvious genetic relation to the enclosing rocks, are called *accidental xenoliths*. Many of these are of sedimentary or metamorphic origin, and they especially are likely to show striking effects of reaction with the magma in which they became immersed. Quartz xenocrysts rimmed with hypersthene, partially melted glass-streaked graywackes, and aggregates of aluminous minerals (spinel, cordierite, corundum, etc.) representing refractory residues from fractional melting of shale, are all common in basalts.

Students of igneous petrography are particularly interested in xenoliths of another kind—*cognate xenoliths*, so called since they appear to be related, either directly or indirectly, to the host magma. It is not surprising to find some degree of correlation between basalt lithology and the mineral composition of cognate xenoliths that they habitually carry. Take, for example, the basaltic lavas of the Hawaiian volcanoes. Extensive outpouring of tholeiitic lavas commonly has been followed, after a significant time interval, by eruption of alkali olivine basalts (and even ultrabasic feldspathoidal lavas) in the closing stages of the volcanic cycle. Each of the three lava types carries its distinctive set of cognate xenoliths.[12] And this lithologic correlation on the whole holds good from one volcano to another, regardless of relative age and location in the Hawaiian chain.

1. Xenoliths are much less numerous in tholeiitic basalts than in either of the other two types. Almost exclusively they are gabbros resembling those of the great continental layered intrusions. The typical mineral assemblage is olivine-augite (diopsidic)-orthopyroxene-plagioclase. Cumulate structures and small-scale layering are common. These xenoliths would seem to have come from the floors of differentiating reservoirs of basaltic magma deep beneath the sites of surface eruption.

2. In flows of alkali olivine basalt, hawaiite, and mugearite, cognate xenoliths are more frequently encountered and show more lithologic variety. Some are composed of gabbro, but many are of ultramafic composition and mineralogically simple, consisting essentially either of magnesian olivine alone (dunite) or of olivine plus augite (wehrlite). The historic alkali-olivine-basalt flow of 1801 from Hualalai on the island of Hawaii, encloses thousands of tons of angular xenoliths of gabbro and dunite presumably torn from an underlying layered intrusion.

The same pattern of correlation is repeated in other basalt-dominated provinces, both oceanic and continental. Thus in the Galapagos, tholeiitic pyroclastic accumulations on the two large western islands contain a varied spectrum of xenoliths—not only hornblende-gabbro cumulates but more evolved rock types such as ferrogabbro (with olivine Fo_{80}–Fo_{30}) and plagioclase-rich diorites; ultramafic xenoliths (olivine-pyroxene-spinel) are confined to alkali olivine basalts of the eastern islands. In the Holocene alkali-olivine-basalt province of Auckland City, New Zealand, cognate xenoliths are exclusively ultramafic: most are dunite (olivine Fo_{88} plus accessory chromite), but a few carry minor enstatite, En_{94}–En_{88}. Ultramafic xenoliths of this kind are generally thought to be refractory residues from melting of magma source rocks in the upper mantle.

PLUTONIC DIVISION: GABBROS, NORITES, TROCTOLITES, ANORTHOSITES, PYROXENITES

Field Occurrence

By far the most extensive basic plutons with the overall chemical composition of tholeiitic basalt are massive stratified complexes of all ages invading the continental crust. Classic examples are provided by immense Proterozoic sheetlike complexes of the South African Bushveld and the Stillwater province of Montana, and the smaller but intensively investigated Eocene Skaergaard intrusion of eastern Greenland. All such layered plutons appear to be of tholeiitic parentage. But internal differentiation involving crystal settling, sorting, and "sedimentation" in convecting magma bodies has led to large-scale heterogeneity, as evidenced by conspicuous compositional layering of the solidified mass, on every scale from millimeters to hundreds of meters. So although the parental composition in every case was basic (tholeiitic), the solidified end products cover a wide range of silica content: ultrabasic (peridotites, pyroxenites) at the base, basic (gabbros, norites, troctolites, anorthosites, pyroxenites) in the lower and middle levels, and even acid (ferrogranophyres) in the upper portions of the plutons. In this section we concern ourselves only with rocks of the basic class, with examples drawn mainly from layered complexes. And it will be noted that since these latter represent a sequence of crystalline fractions drawn from melts of continuously changing composition, none of the individual rocks has any precise chemical counterpart among the successive liquid fractions (magmas) from which they were generated. In fact, rocks of the gabbro family in general cannot safely be equated chemically either with basic magmas or with basalts (which more nearly reflect melt compositions).

Plutons composed mainly of gabbro or norite occur also in other geological situations. Some layered basic plutonic complexes appear to be components of the suboceanic crust. Again in compound batholiths dominated by acid plutonic rocks there may be component plutons of basic composition. Such is the San Marcos gabbro which makes up 14% of the total areal outcrop ($\sim$2000 km^2) of the granodioritic Southern California batholith.[13] There are also smaller plutons, such as the great gabbro ring dikes of the Hebridean province, that represent virtually undifferentiated basaltic magma.

Anorthosites—rocks with gabbroic affinities but composed largely or entirely of plagioclase—are prominent in most layered basic plutons. But their most spectacular development is elsewhere and takes the form of large independent plutons, most of mid-Proterozoic age and some of batholithic dimensions. They are exemplified by anorthositic massifs in the Adirondacks of New York State, throughout the Grenville province of eastern Canada, in southern Norway, and in Greenland.

Definition and General Petrographic Character

Typical members of the gabbro family have three characteristic features in common. With regard to SiO_2 content they are of basic composition; plagioclase, the most abundant single mineral, generally is more calcic than An_{50}; and the color index varies between 40 and 70 (whereas in most diorites it is between 10 and 40). Individual rocks, however, especially those on the borderline between gabbro and diorite, may depart in any of these respects from the normal pattern. Some rocks are classed as gabbro on account of high color index even though the plagioclase is less calcic than An_{50}. Then there are gabbros, as in the Duluth complex of Minnesota and on the Isle of Skye, with color indices below 40; these accordingly may be termed *leucogabbros*. At one extreme, rocks consisting largely of plagioclase (*anorthosites*), the color index indeed is usually less than 10. At the opposite extreme are *melagabbros* (and *melanorites*) transitional toward feldspathic peridotites. These have color indices exceeding 70 and are close to ultrabasic in composition. Here is another instance of natural field and chemical transition between two major classes of rocks: gabbros (or norites) and peridotites (or pyroxenites).

The principal mineral components of all members of the gabbro family are combinations of calcic plagioclase (An_{50}–An_{90}), diopsidic augite, hypersthene, and olivine. Hornblende, a mineral typical of diorites, nevertheless is plentiful in some gabbros. Biotite, alkali feldspar (orthoclase or microcline), or quartz may be present in some rocks, though always in minor quantity. An opaque mineral (ilmenite, magnetite, an iron sulfide, rarely chromite) and apatite are ubiquitous accessories.

Slow cooling in the plutonic environment is responsible for widespread exsolution phenomena in both varieties of pyroxene. In crystals of augite thin lamellae of magnesian pyroxene have exsolved parallel to {100} or in some cases {001}; the variety diallage has planar arrays of exsolved iron-oxide particles aligned parallel to {100}. Hypersthene crystals commonly enclose lamellae of exsolved diopside oriented parallel to {100}. In some rocks similar lamellae in hypersthene are inclined at about 74° to [001] in the zone of [010], showing that the orthopyroxene host has inverted from the high-temperature monoclinic polymorph pigeonite during the cooling process. Pigeonite itself is not found in basic plutonic rocks.

Rocks of the gabbro family are holocrystalline, varying from coarse- to medium-grained types (the latter sometimes called *microgabbros*). They display considerable textural variety. In undifferentiated gabbros and norites in which crystal accumulation has played but an insignificant role in the consolidation process the common textures are ophitic (Figure 3-6B) and anhedral granular (Figure 3-6A). In rocks of layered intrusions, especially at the lower levels, poikilitic textures are common:[14] small grains of magnesian olivine are enclosed in coarse, optically continuous crystals of plagioclase or of pyroxene (as many as 1000 olivine grains have been counted in a single crystal of labradorite or augite several centimeters in diameter).

Poikilitic and allied textures in gabbros of layered intrusions have been explained by Wager, Brown, and Wadsworth (1960) in terms of a two-stage model of crystallization yielding rocks that they term *cumulates*. In the first stage a loose aggregate of crystals of one or several mineral species forms by precipitation (a kind of "sedimentation") on the floor of the magma chamber; the interstices in this aggregate (the *cumulus* fraction) at this stage are occupied by a melt phase which may or may not be sealed off from effective diffusive exchange with the overlying magma. The second stage involves crystallization of the interstitial melt to yield the *intercumulus* fraction of the finally solidified rock. Olivine grains poikilitically enclosed in coarse plagioclase or pyroxene are envisaged as a single-phase cumulus fraction, the host crystals representing the intercumulus. In a similar fashion interstices in a preponderantly plagioclase cumulus become filled with skeletal crystals of olivine and augite that together constitute the intercumulus. Where the cumulus consists of several phases the individual grains may simply enlarge by accretion at the expense of the interstitial melt until the whole mass is crystalline, yielding an anhedral granular texture as the ultimate product. Or again if an entrapped liquid fails to maintain equilibrium with the cumulus crystals during the accretion process, the intercumulus fraction may maintain its identity through compositional zoning in the final product: labradorite rims surrounding cores of bytownite; more ferriferous rims to crystals of more magnesian olivine and hypersthene.

With falling temperature and decreasing pressure the typical basic plutonic mineral assemblage (labradorite-olivine-pyroxene) becomes unstable, particularly in the presence of an intergranular aqueous fluid. If cooling and unloading are sufficiently slow, adjoining grains of different phases may react under favorable conditions to produce kelyphitic rims (coronas) along the interfaces (Figures 2-8B,C; 3-6A). Around olivine grains the minerals constituting these concentric and usually fibrous coronas are pyroxene, spinel, garnet, actinolite, anthophyllite, and cummingtonite, all of which are common products of metamorphism. Reaction usually begins with formation of pyroxene or amphibole around the olivine; if a second shell is formed, it consists of amphibole and spinel; and if garnet is present, it is generally in the outermost parts of the rims, adjoining the plagioclase. Presence of an amphibole as one of the newly generated phases in such rims surrounding pyroxene grains testifies to participation of water in the reaction process. Another widespread effect of incipient hydration is partial replacement of olivine (or less commonly magnesian orthopyroxene) by a serpentine mineral (Figure 3-6C) clouded with specks of magnetite.*

Classificatory Terminology

According to long-established usage several different types of basic plutonic rocks are identified in terms of specific combinations of the four principal mineral constituents.

> *Gabbros*: plagioclase plus diopsidic augite; olivine or hornblende are prominent in correspondingly designated varieties.
> *Norites*: plagioclase plus hypersthene, typically with subordinate augite; olivine present in some rocks.

---

*The serpentinization process is considered in greater detail on pp. 247–248.

Troctolites: predominant plagioclase plus olivine.

Anorthosites: plagioclase the sole or greatly predominating mineral; minor hypersthene or augite are common.

Magnesian and calcmagnesian pyroxenites: magnesian orthopyroxene or diopsidic clinopyroxene the sole or principal constituents.

British petrologists have coined additional special terms to denote gabbros and norites with highly calcic plagioclase (bytownite or anorthite).[°]

Lithologic Variation in Layered Basic Intrusions

The nature of successive rock types, as seen in hand specimens and outcrops, and hence the pattern of lithologic variation in a vertical sense within a major layered basic pluton, depends on two simultaneously operating factors: the overall chemical trend in the fractionating magma, and the process of crystal settling and sorting as the solidified floor builds upward at the expense of the remaining liquid fraction. The first of these determines the compositions of mineral phases separating from the magma at any stage and follows a generally consistent pattern. The second, which varies according to local conditions of cooling and convection, is more capricious, and its detailed imprint on the lithologic succession has individual characteristics unique to each pluton.

The most striking overall upward chemical trends are progressive increases in FeO/MgO and in Na_2O/CaO. From the base to the top of each intrusion, over vertical distances commonly exceeding 5000 m, the mafic crystalline phases become increasingly ferriferous, the plagioclase increasingly sodic. As the activity values of key chemical components (SiO_2, FeO, etc.) in the fractionating magma change continuously with fractionation, the identity of the crystallizing solid phases likewise tends to follow a rather consistent pattern. Magnesian olivine begins to separate early and continues to do so, becoming increasingly ferriferous ($Fo_{80} \rightarrow Fo_{50}$) for many hundreds, even a few thousand meters, in the vertical section. Then olivine ceases to crystallize—only to reappear, however, far up in the section, where it is now a highly ferriferous type, $\sim Fo_{40}$, becoming nearly pure fayalite Fo_5 in the roof rocks. Orthopyroxene is typical

---

[°]*Allivalite*: bytownite or anorthite troctolite. *Eucrite*: noritic gabbro with bytownite (a term unfortunately borrowed in the first place from the long-established terminology of stony meteorites).

of the lower and middle segments of the section. It is soon joined or in some cases is preceded by diopsidic clinopyroxene; and through much of the section the two phases, both becoming increasingly ferriferous, crystallize together. Then orthopyroxene cuts out and clinopyroxene, now ferroaugite, is the sole pyroxene (associated with iron-rich olivine in the uppermost rocks). The mineralogical and lithologic imprint of the vertical chemical trend is too subtle to be recognizable in a hand specimen or an outcrop. It is sometimes described therefore as the pattern of *cryptic* layering within the pluton. Its nature is revealed only through thorough petrographic and chemical investigation of a suite of specimens covering a substantial vertical section (usually thousands of meters).

The processes of crystal sorting and settling, though perhaps still incompletely understood in some respects, determine the detailed pattern of macroscopically visible lithologic layering, that is, the succession of recognizable rock types displayed on a local map, in the individual outcrop, or even in a single large specimen. At any level in the pluton there may be conspicuous *rhythmic* layering of this kind. Individual minerals then become regularly segregated into alternating internally graded bands during the sorting process: pyroxene or iron oxides from olivine; plagioclase from mafic phases. Layers of almost pure plagioclase rock (anorthosite) 100 m or more in thickness have originated thus in some intrusions; and near the base of some plutons layers of almost pure chromite or of metallic sulfides have a similar origin.

Vertical zonation on the largest possible scale (1000 m or more per vertical unit)—the ultimate expression of the overall chemical trend—tends to conform in many plutons to some part or all of a general pattern. Successive zones numbered from the base upward may be designated broadly as follows:

5. Ferrodiorites and ferrogranophyres
4. Ferrogabbros
3. Normal gabbros or norites
2. Olivine gabbros
1. Pyroxenites and peridotites

Here we are concerned only with zones 1 (in part) to 4, noting that marked lithologic variation (rhythmic layering) within any zone follows a detailed pattern unique to every individual pluton.

Relevant experimentally investigated phase equilibria indicate several possible alternative sequences in the order of crystallization of the principal phases. Some of these, reflected in variations within the general upward lithologic sequences displayed by different layered basic plutons, are as follows (normative values referring to parent magmas):

Favored by high normative *hy/di* and *hy/ol*: olivine, bronzite (at expense of olivine), hypersthene + plagioclase.
Favored by high normative *an* and *di*: olivine, plagioclase + olivine, plagioclase + olivine + augite (± hypersthene*).
Favored by low SiO_2 and high normative *ol/an*: olivine, olivine + spinel, olivine + plagioclase, olivine + plagioclase + diopside, plagioclase + diopside + hypersthene* (with elimination of olivine).

Detailed Petrography

Gabbros and Norites

For the most part gabbros and norites are devoid of phenocrysts and except for ophitic and poikilitic rocks are medium- to coarse-grained rocks with subhedral or anhedral granular texture. Specimens from layered intrusions may show a pronounced tendency for primary segregation of alternating feldspathic and mafic bands as a result of crystal sorting during the process of accumulation on the floor of the magma chamber. A secondary foliation due to internal deformation after consolidation may be conspicuous in other gabbros. In these the feldspar may be granulated to an average diameter of less than 1 mm; and much of the pyroxene may be replaced by fibrous pale-green amphibole ("uralite").

In *normal gabbros*—so called simply because they are common members of the gabbro family—the principal constituents in subequal quantity are plagioclase and augite, either building a mosaic of equant, mutually interfering, anhedral grains (Figure 3-6A) or much less commonly ophitically related (Figure 3-6B). Plagioclase typically is labradorite or bytownite, with prominent albite and pericline twinning, nonzoned, or, particularly in cumulates, showing simple normal zoning. In many gabbros the feldspar crystals are water clear; in others they are heavily charged with minute inclusions of rutile, ilmenite, hematite, amphibole,

---

*Or, when $FeSiO_3$ exceeds ~35% (molar), pigeonite subsequently inverting to hypersthene on cooling.

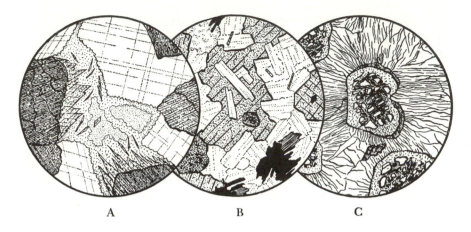

A   B   C

Figure 3-6. Gabbros and Troctolite

A. Gabbro, Volpersdorf, Saxony. Diam. 3 mm. Labradorite and diallage are the chief primary minerals; the latter shows kelyphitic fringes of tremolite. The remainder consists of serpentine and talc.

B. Gabbro, Glen More ring dike, Mull, Scotland. Diam. 3 mm. Chiefly composed of labradorite and augite ophitically intergrown. Accessory constituents include serpentinized olivine, needles of apatite, flakes of biotite bordering plates of ilmenite, and, in the upper-left portion, a micrographic patch of quartz and K-feldspar.

C. Troctolite, Volpersdorf, Saxony. Diam. 6 mm. Essentially an olivine-labradorite rock. The olivine is almost entirely converted to serpentine, and the surrounding feldspar is criss-crossed by expansion cracks. Accessory augite is partly embedded in the feldspar and also forms fringes around the olivine.

or pyroxene; in others again they are thoroughly altered, either deuterically or through incipient metamorphism, to cloudy masses of "saussurite" (white and opaque in hand specimen)—a finely divided aggregate of clinozoisite, zoisite, or epidote plus albite, quartz, and in some cases calcite, chlorite, or even a calcic garnet. Minor orthoclase or microcline may be present as narrow rims to plagioclase in some rocks, but more often as a constituent of interstitial patches of micrographically intergrown quartz and feldspar. The dominant mafic mineral is augite (or diallage), commonly with thin {100} lamellae of exsolved magnesian pyroxene. Pyroxene crystals may be rimmed with late-magmatic hornblende or in some cases are partially replaced by fibrous amphibole (uralite) of deuteric origin. Dark-brown biotite may appear in small quantities usually as reaction rims around grains of iron oxides. Other possible accessories include rutile, apatite, pyrope, zircon, and green spinel, brown picotite, or chromite.

Biotite is never a major component of gabbros. But some gabbros carry sufficient biotite (5–10%), along with abundant titaniferous or slightly sodic augite and in some cases a little K-feldspar, to point to close chemical affinities with alkali olivine basalts. Such are the gabbros of eroded older plugs in volcanic centers of the Permian alkaline province that occupies the Oslo graben of Norway. Typically these are rocks with color indices beteen 30 and 66 whose chief mineral components are pale violet or greenish augite and andesine-labradorite. Biotite is consistently present, small amounts of olivine almost always so; and in the lighter-colored rocks there is usually K-feldspar (5–15%).* Ubiquitous accessories are iron-titanium oxides and apatite. Chemically most of the Oslo gabbros are characterized by SiO_2 between 45% and 48%, with normative *ne* (2–3) and *ol* (5–10).

In *hornblende gabbros* the chief mafic mineral is brown or greenish-brown hornblende, and the plagioclase is generally more calcic than An_{50}. Some of these rocks are unusually mafic, having color indices of more than 60. Quartz, alkali feldspar, and biotite are more abundant than in normal gabbros, but olivine is much less abundant. Two examples must suffice. In the hornblende gabbro of the San Marcos complex of southern California, anorthite makes up two-thirds of the volume, hornblende a quarter, and the rest is composed of pyroxene, apatite, and iron oxides. In the Gilford gabbro of New Hampshire, labradorite makes up half the volume, hornblende a fifth, augite another fifth, and magnetite and other minor accessories form the remainder. In both of these examples, as in most hornblende gabbros, the hornblende occurs as large ophitic and poikilitic plates enclosing the other constituents.

In most *norites* the texture is anhedral granular; but in some rocks (e.g., Figure 3-7C) it is subhedral granular. Their typical feldspar is labradorite or bytownite, twinned as in normal gabbros; but in some specimens (usually cumulates) it is normally zoned from cores of anorthite to rims of andesine. Among the mafic constituents bronzite or hypersthene predominates. Hypersthenes show notably stronger pleochroism than orthopyroxenes of identical composition in volcanic

---

*Suggesting transition toward feldspathoidal gabbros called *essexites* (see p. 209). Indeed they were so named by the pioneer Norwegian petrologist W. C. Brøgger, and to this day are loosely referred to in Norwegian literature as "Oslo-essexites."

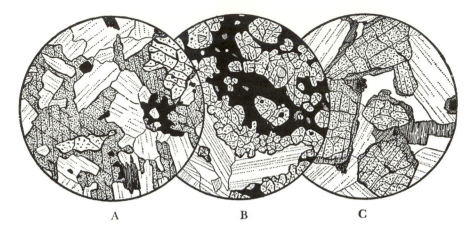

Figure 3-7. Norites and Ferrogabbro

A. Olivine norite, Aberdeen, Scotland. Diam. 3 mm. All the visible hypersthene is optically continuous; it encloses grains of olivine and is intergrown ophitically with calcic labradorite. Iron ore and biotite are accessory constituents.

B. Ferrogabbro, Iron Mine Hill, Rhode Island. Composed of labradorite, iron-rich olivine, and opaque oxides containing specks of green spinel. The opaque grains are exsolution intergrowths of magnetite and ilmenite.

C. Quartz norite, Sudbury, Ontario. Diam. 3 mm. Around the large hypersthene crystals are reaction rims of green hornblende and brown biotite. Biotite also envelops accessory iron oxides. The rest of the rock is composed of subhedral laths of labradorite and anhedral quartz. Elsewhere, but not shown here, bluish-green arfvedsonite forms fringes around some of the hornblende.

rocks—an effect that has been traced to crystallization at lower temperatures in the plutonic environment.* All orthopyroxenes of norites tend to show lamellar exsolution of clinopyroxene parallel to {100}; but it is only in iron-rich hypersthenes (with Fs>30) that exsolution lamellae also develop at about 74° to {100} in the zone of [010] indicating inversion from primary pigeonite. The inference is that more magnesian pyroxenes (bronzites) are stable in the orthorhombic form at temperatures extending up to the liquidus. In many norites a magnesian diopsidic

---

*Hypersthene of metamorphic rocks may be even more strongly colored. Pink crystals with inclined extinction—γ′ to a single set of cleavage or lamella traces—have sometimes been misidentified as clinohypersthene. A stereoplot of a section (SV_α = 75°) cut normal to the (110) cleavage and inclined at ~45° to (1$\bar{1}$0), which is thus invisible, shows γ′ inclined at 25° to the visible (110) trace. One cut at 60° to {100} lamellae—i.e., at 30° to α—extinguishes with γ′ at almost 10° to the lamella trace.

augite accompanies the orthopyroxene. The two appear to have crystallized side by side, thus demonstrating a broad immiscibility break in the pyroxene series at temperatures on the plutonic liquidus. Accessory minerals in norites are much the same as in gabbros. In addition, however, certain norites in the border zones of basic plutons invading pelitic metasediments (as near Huntly and elsewhere in Aberdeenshire, Scotland) contain significant amounts of aluminous minerals—cordierite, garnet, spinel, sillimanite—formed by reaction between the basic magma and the envelope. Indeed it has been shown experimentally that assimilative reaction of this kind tends to induce crystallization of hypersthene rather than augite from the contaminated magma. The result is a border zone of norite partially margining an essentially gabbro pluton.

Augite and hypersthene coexist so commonly in subequal amounts in silica-saturated basic plutonic rocks that many of these could equally well be called "noritic gabbros" or "gabbroic norites." Thus the average mineral composition in the lower gabbro zone of the Stillwater complex, Montana (zone 3 of the general sequence, p. 122), is plagioclase An_{79}, 60%; augite $(Ca_{0.82} Mg_{0.96} Fe_{0.22}) Si_2O_6$, 23%; hypersthene En_{77}, 17%.[15]

Hornblende norites are well represented in the small layered basic pluton of Bluff, southernmost New Zealand. A typical mode is labradorite, An_{65} (65); hypersthene, Fs_{50} (20); green hornblende (12) rimming pyroxene and in separate prisms; accessory oxides and apatite. Minor biotite and quartz, and in a few cases epidote and green spinel, may be present. The hornblende rims are probably deuteric, but prismatic hornblende may be a late-magmatic phase. In some other hornblende norites occurring in metamorphic complexes, such as the Cortland complex of New York, the amphibole is probably of metamorphic rather than deuteric origin.

As an example of quartz-bearing basic plutonic rocks, we select norites and gabbros of the Sudbury intrusion, Ontario (Figure 3-7C). In the lower levels of this pluton norite contains plagioclase with cores of An_{50} and thin, clear rims of calcic andesine; the mafics, in order of abundance, are ortho- and clinopyroxenes, amphibole, and biotite. Quartz occurs interstitially, replaces the margins of some of the plagioclase crystals in graphic intergrowths, and also forms larger plates enveloping the feldspars. Other more basic variants of the norite contain dusty-brown plagioclase (An_{55}) as the principal constituent. Next in order of abundance are clinopyroxenes and stumpy euhedral prisms of orthopyroxene, some of which contain lamellae of clinopyroxene. Reaction rims of greenish hornblende are common, and some of them are fringed by bluish-green, somewhat more sodic amphibole. Biotite occurs as a late reaction product, particularly around iron oxides. A little quartz is present interstitially. Higher in the intrusion the content of quartz tends to increase; there the rocks are gabbros rather than norites, for all the pyroxene is monoclinic. In some other quartz norites the quartz is intergrown with orthoclase in interstitial patches of micropegmatite.

All of the foregoing kinds of gabbro and norite have very coarse-grained pegmatitic variants, most of which carry abundant iron-titanium oxides, are rich in apatite, and have a slightly more sodic plagioclase.

In *olivine gabbros* and *norites* that dominate the lithology of the lower and middle segments of many plutons (zone 2, p. 122), olivine is a major though not the predominant mineral constituent. Toward the base of the pluton its composition is highly magnesian (Fo_{90} to Fo_{85}; $2V_\alpha > 90°-85°$); but the iron content increases upward in the section, so that at the top of the olivine-gabbro zone the composition may be Fo_{65} to Fo_{55} ($2V_\alpha = 75°-70°$). The corresponding upward range of plagioclase is from $\sim An_{85}$ to An_{60} or An_{50}. Orthopyroxene likewise varies from bronzite in the lower, to hypersthene (inverted pigeonite, $En_{<65}$) in the higher levels. Accompanying clinopyroxene (a diopsidic augite) similarly becomes increasingly richer in iron toward the top of the olivine-gabbro zone. A typical *olivine gabbro* from Pigeon Point, Minnesota, consists of labradorite laths (60%), augite (27%), olivine (10%), and apatite plus iron-titanium oxides (3%). The olivine is subhedral and poikilitically enclosed in coarser augite, which may itself be mantled with hornblende. The most abundant rocks in the lowest exposed levels of the Skaergaard intrusion (all within zone 2 of our general sequence) are olivine gabbros of simple and typical mineral composition. Labradorite (An_{55}) and less abundant olivine (Fo_{65-55}) (together constituting the cumulus fraction) are poikilitically enwrapped by augite ($Ca_{0.76}Mg_{0.82}Fe_{0.42}$)$Si_2O_6$ accompanied by a little opaque oxide and apatite.

The layered ultrabasic and basic complex on the isle of Rhum (Hebridean province) represents a basal remnant of a pluton now largely removed by erosion. It consists of a sequence of rhythmically banded layers, each 15 m to 150 m or so thick and each grading upward from peridotite through olivine gabbro to troctolite. A typical olivine gabbro from the middle of such a layer consists of highly calcic plagioclase An_{85}, 65%; diopsidic augite ($Ca_{0.88}Mg_{0.98}Fe_{0.14}$) Si_2O_6, 20%; olivine Fo_{85}, 15%; and accessory chrome spinel.

Olivine norites are less commonly found than olivine gabbros in the lower levels of basic plutons. One example from near Aberdeen, Scotland, is illustrated in Figure 3-7A. Coarse bronzite En_{80} encloses euhedral bytownite An_{75} and olivine Fo_{80}; there is also interstitial augite $(Ca_{0.84}Mg_{1.0}Fe_{0.16})Si_2O_6$ and accessory biotite and apatite. More widely distributed are noritic olivine gabbros in which hypersthene is subordinate to clinopyroxene. These are of two kinds. Just above the basal ultramafic zone of some intrusions (e.g., in the Cuillin remnant of Skye in the Hebridean province) are olivine gabbros and noritic gabbros ("eucrites") in which highly calcic plagioclase An_{80}–An_{70} is accompanied by magnesian olivine Fo_{80}–Fo_{70}, augite, and in some rocks bronzite or hypersthene. Much higher in the general zonal sequence (upper part of zone 2), just below the level at which olivine cuts out, are noritic olivine gabbros notably richer in iron than those lower down in the sequence. One such rock from the Skaergaard intrusion consists of labradorite An_{55}, augite $(Ca_{0.76}Mg_{0.82}Fe_{0.42})Si_2O_6$, inverted pigeonite $En_{\sim 65}$, and olivine Fo_{55}.

Olivine ferrogabbros are typical of the uppermost zones of the Skaergaard, Bushveld, and some other basic plutons. About 1000 m below the roof of the Skaergaard intrusion an average mode is calcic andesine An_{45}, 56%; augite $(Ca_{0.70}Mg_{0.74}Fe_{0.56})Si_2O_6$, 20%; olivine Fo_{39}, 16%; iron-titanium oxides 7%, with inverted pigeonite $En_{\sim 54}$, apatite, and quartz together totaling 1%. Above this lies a hortonolite* ferrogabbro with andesine An_{38} 45%; ferroaugite $(Ca_{0.8}Mg_{0.42}Fe_{0.78})Si_2O_6$, 28%; olivine (hortonolite) Fo_{20}, 17%; iron-titanium oxides, 6.5%; apatite, 2.5%; and a little interstitial micropegmatite. The uppermost rocks in the Skaergaard pluton are quartz-bearing fayalite ferrodiorites: andesine An_{30}, 24%; ferroaugite $(Ca_{0.86}Mg_0Fe_{1.14})Si_2O_6$, 37%; fayalite Fo_{0-4}, 21%; opaque oxides, 9%; quartz, 6%; apatite, 3%. The plagioclase is rimmed with perthite and the quartz for the most part is intergrown with augite or with alkali feldspar (as micropegmatite). These andesine-bearing rocks, because of their high color index and basic composition (SiO_2 ~45%), are classed as gabbros rather than as diorites. Note that in these rocks, which represent the last stages of fractional crystallization, the concentration (and activity) of P_2O_5 in the liquid residues had increased to a point where apatite was precipitated freely.

Troctolites

Troctolites are basic, in some cases almost ultrabasic, rocks composed almost entirely of calcic plagioclase and less abundant olivine. They are

---

Hortonolite refers to olivines ranging in composition from Fo_{50} to Fo_{10}.

found in the lowest levels of some layered basic plutons as in the isles of Skye (Cuillin complex) and Rhum in the Hebridean province of Scotland. They are plagioclase-olivine cumulates in some cases (as on Rhum) consisting of feldspathic segregation layers alternating rhythmically with bands of peridotite (dunite) in a basal zone of overall ultrabasic composition. Such rocks represent that crystallization trend in high-alumina olivine-tholeiitic magma where the first silicate phase to join olivine is calcic plagioclase.

The troctolites° of Rhum, as represented by the upper layer in each rhythmic unit, consist of plagioclase An_{85}, 84%; olivine Fo_{86}, 15–16%; and chrome spinel, 0.5%. The plagioclase is in subhedral laths tabular parallel to {010} with a tendency to subparallel alignment. Comparable rocks from the Cuillin complex of Skye are almost identical, but contain minor diopsidic augite as an additional cumulus phase. In other troctolites (Figure 3-6C) the texture is equigranular, crystals of both the main phases being anhedral and equant. The texture of troctolites at the base of the anorthosite zone (see below) of the Stillwater complex is different again. These rocks consist of bytownite An_{80}, 65%, and olivine Fo_{80}, 35%. The olivine occurs as isolated coarse grains 5 mm to 15 mm in diameter enclosed in an aggregate of subhedral plagioclase crystals of much smaller size. Some of these penetrate or are enclosed in the outer zones of the olivine crystals. It would seem that only the olivine cores constitute a true cumulus fraction.

Olivine of troctolite commonly shows effects of post-consolidational reactions. Slow cooling in an almost dry environment may permit development of kelyphitic reaction rims or coronas around the olivine consisting of orthopyroxene, spinel, and in some cases fibrous amphibole. Almost universal too is partial alteration of olivine grains to a serpentine mineral, greenish or brownish in thin section—a process involving introduction not only of water but also of SiO_2 (or alternatively loss of MgO by outward diffusion). As required by addition of water and SiO_2, the serpentinization process in troctolites is accompanied by significant expansion; for the serpentinous product not only replaces part of the olivine in each grain but also spreads outward filling expansion cracks in the enclosing plagioclase (Figure 3-6C). In hand specimens the darkening effect of serpentinization brings the olivine grains into sharp contrast

---

°Troctolites of this kind, with plagioclase in the bytownite range, are sometimes called *allivalites*.

with the enclosing pale-colored plagioclase, thereby producing a characteristic spotted appearance in hand specimen (as implied by the German name *Forellenstein*—literally "trout-stone").

Anorthosites

Anorthosites are plutonic rocks consisting largely or entirely of plagioclase of uniform composition somewhere in the range An_{35} to An_{80}. As regards silica content the bytownite and labradorite anorthosites fall in the basic, andesine anorthosites in the intermediate category. But since they are all closely associated with gabbroic and even with ultramafic rocks, even though often in only minor amount, and because no sharp line of distinction can be drawn between the more sodic and the more calcic rocks, all are conventionally treated as a single class.

Anorthosites occur as cumulate layers from a centimeter or so to a few hundred meters thick at different horizons in layered basic plutons. Their most spectacular occurrence, however, is as independent plutons of great size—1,000 km² to 20,000 km² in surface outcrop—in Precambrian (especially middle Proterozoic) terranes. In anorthositic massifs of this kind, accompanying anorthositic gabbros play only a minor role. Those of southern Norway, western Greenland, the Grenville province of Quebec and Labrador and its southern offshoot in the Adirondacks of New York State are classic examples. They are sometimes referred to as anorthosites of the Adirondack type. In layered basic plutons anorthosites occur at all levels as segregation bands in mafic rocks. They reach their most striking development in the anorthosite zone (2000 m thick) of the Stillwater complex of Montana. Here three separate anorthosite units each about 400 m thick are separated by norite bands of comparable thickness.

Anorthosites of layered basic plutons (Stillwater type) are mostly bytownite or labradorite rocks depending on the level at which they occur. Those of the Adirondack type, by contrast, typically are composed of andesine. But this generalization is not without exceptions. A typical anorthosite from the lower unit of the Stillwater anorthosite zone has an average modal composition plagioclase An_{79}, 90%, orthopyroxene (inverted pigeonite) En_{60} and diopsidic augite $(Ca_{0.8}Mg_{0.8}Fe_{0.4})\,Si_2O_6$, 5% each. Both pyroxenes occur in skeletal crystals ophitically related to neighboring grains of plagioclase. Anorthosites of the rhythmically banded "critical zone" low down in the Bushveld complex consist of greatly predominant bytownite (An_{76} to An_{78}) partially enwrapped by skeletal crystals of bronzite (En_{75} to En_{85}). Plagioclase of thin anorthosite bands in the lower olivine gabbros of the Skaergaard intrusion is labradorite An_{60}–An_{53}.

The great anorthositic massifs of the Adirondack type are thick lenses or sheets in which anorthosite proper with 5% or less mafic constituents (cf. Figure 3-8A) is concentrated in the cores. (Border-zone rocks are anorthositic gabbro with 10% to 30% pyroxenes.) Plagioclase typically is andesine An_{45} to An_{40}. The principal mafic minerals are a faintly brownish augite and hypersthene. A little hornblende or biotite may sometimes be seen as reaction rims surrounding pyroxene crystals. Common accessories are spinel, ilmenite, magnetite, rutile, and sulfides of iron and copper. Aluminous minerals, corundum (Figure 3-8B) or kyanite, are occasionally present—in some cases at least as products of superposed metamorphism. An average modal composition carefully computed by A. F. Buddington for Adirondack anorthosites with SiO_2 = 53–55% is calcic andesine An_{50}–An_{40}, 94%; augite, 2.5%; hypersthene, 0–2%; hornblende, 1%; magnetite-ilmenite, 0.5%; plus accessory apatite and rarely a little quartz.

Texturally anorthosites of the Adirondack type are generally coarse grained, anhedral-granular rocks with conspicuous evidence of strain

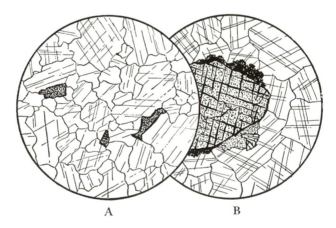

A   B

Figure 3-8. Anorthosite of the Adirondack Type

A. Anorthosite, Frontenac County, Quebec. Diam. 1 cm. An anhedral granular intergrowth of labradorite and accessory green hornblende.

B. Andesine anorthosite from same locality. Diam. 1 cm. Interlocking anhedra of calcic andesine; large crystal of corundum fringed with iron oxide, green spinel, talc, and clinozoisite.

incurred either during emplacement in a nearly solid state or after intrusion was complete. Large grains with internally curving twin lamellae and undulatory extinction show marginal "granulation" (or annealing recrystallization), and may even be reduced to ovoid remnants by marginal attrition. In some rocks there are intervening irregular veinlets of postdeformational minerals—quartz, microcline, sodic plagioclase, garnet, and others—all of metamorphic origin. Finally it should be noted that even in very coarse anorthosites of the Adirondack type the interstitial pyroxene typically is ophitically related to plagioclase. In one such case, the San Gabriel anorthosite of southern California, individual skeletal crystals of hypersthene can be traced (in fact mapped!) for distances as great as 50 cm in the outcrop.

Magnesian and Calcmagnesian Pyroxenites

Pyroxenites are coarse-grained ultramafic rocks composed solely or principally of pyroxene—a single species usually greatly predominant. Easily identified beneath the microscope, they can readily be subdivided into *orthopyroxenites* and *clinopyroxenites*—with of course the inevitable two-phase transitional types. Both these groups, however, cover wide ranges of SiO_2 content, orthopyroxenites ranging from basic to intermediate (mostly 47–55% SiO_2), while clinopyroxenites cover an ultrabasic-to-basic span (~42–50% SiO_2). This fact raises a problem regarding systematic treatment in terms of the general plan adopted in this book. The dilemma is compounded, but paradoxically in a way resolved, by a pronounced tendency for rocks with each pyroxene species to be genetically affiliated and in the field associated with one or more well-recognized groups of ultrabasic or basic plutonic rocks. This important characteristic is readily brought out if we treat pyroxenites in three separate chemical, rather than crystallographically defined, groups.

1. Magnesian: orthopyroxenite; basic-intermediate.
2. Calcmagnesian: diopsidic clinopyroxenites; basic.
3. Alkaline: clinopyroxenites, mostly peralkaline, in which the pyroxene species are a titaniferous augite, an aegirine-augite, or both; ultrabasic.

In this chapter we consider only groups 1 and 2. Alkaline pyroxenites will be discussed in Chapter 6 in the context of their habitual associates, ultrabasic feldspathoidal rocks.

By far the most extensive of all pyroxenites are magnesian *bronzite pyroxenites* occurring as massive cumulate layers in the lower levels of stratified basic plutons (p. 122). They are coarse-grained aggregates of anhedral or subhedral bronzite crystals, usually with internal {100} exsolution lamellae of diopside. Where grains of other minerals—olivine, chromite, magnetite, diopsidic augite, or calcic plagioclase—are present these typically are enclosed poikilitically in coarse bronzite which then plays the intercumulus role. Rarer minor and accessory components may be hornblende, iron sulfides or graphite. *Olivine-bronzite pyroxenites* occur as transition rocks in massive internally graded units in which pyroxenite passes downward through harzburgite to dunite (p. 245), as in the "Great Dike" of Zimbabwe. In the "Critical Zone" toward the base of the Bushveld complex *chromite-bronzite pyroxenites* occur in similar fashion as transition rocks between pyroxenite and bands of pure chromite. Further up in the same pluton there is comparable downward gradation from bronzite pyroxenite through *magnetite-bronzite pyroxenite* to pure magnetite rock. In most of these transitional types bronzite is the intercumulus phase and poikilitic textures are general. But there are also chromite pyroxenites in which some of the chromite is interstitial.

Rather more magnesian *enstatite pyroxenites* occur chiefly as pods and irregular dikelets within peridotite bodies of the "alpine type"—including ophiolitic peridotites (p. 242). The principal component is enstatite. Chromite or picotite are widespread accessories, and some rocks contain minor chrome-diopside (distinguished by its pale emerald-green pleochroism). Postmagmatic partial replacement of enstatite by a fibrous serpentine mineral ("bastite") is common. Secondary tremolite where present is a metamorphic product.

There is a mineralogical and in some instances natural gradation from orthopyroxenites through orthopyroxene-diopside rocks (Figure 8-5C) to calc-magnesian clinopyroxenites in which the sole pyroxene is diopside. The transitional two-phase rocks *(websterites)*—some of them very coarse-grained—are by no means uncommon in peridotite bodies of the "alpine type," as in the Dun Mountain ophiolite complex of southern New Zealand and again in North Carolina.

In calcmagnesian *diopside pyroxenites* closely associated with gabbros, the pyroxene is the lamellar variety diallage; and common accessories are hypersthene, plagioclase, titanomagnetite, and apatite. There are diopside pyroxenites, associated with peridotites in the Pyrenees, in which the pyroxene is chromiferous and accessory phases include garnet, spinel, and opaque oxides. In a completely different environment diopside pyroxenites with accessory hornblende and biotite are locally developed where bodies of more siliceous rocks invade limestone. An unusual rock of yet another kind is the "Pusey iron ore" of Glamorgan, Ontario, essentially an augite titanomagnetite rock in which the pyroxene grains are rimmed by reaction zones of brown hornblende that in turn are surrounded by the opaque oxide phase.

Finally diopsidic pyroxenites of varied character are principal components of remarkable concentrically zoned cylindrical ultramafic plutons in southeastern Alaska and in the Urals. The principal rock type in the Alaskan plutons is a *hornblende clinopyroxenite* consisting essentially of an aluminous diopside (in simplified form, $Ca\ Mg_{0.7}\ Fe_{0.3}\ Si_2\ O_6$), green hornblende, and accessory magnetite. These rocks may grade inward through successive zones of *magnetite-pyroxenite* and *olivine-pyroxenite* to a core of peridotite (the variety dunite). In the olivine pyroxenites ($SiO_2 = 49\%$) the pyroxene is aluminous diopside virtually free of iron, and the olivine is Fo_{75}. In other parts of the world, as in the Scottish Garabal Hill mafic complex, hornblende pyroxenites contain ortho- as well as clinopyroxene, and grade, probably by postmagmatic replacement of pyroxenes by amphibole, into pure hornblende rocks.

GENERAL REFERENCE

See the works listed at the end of Chapter 1, p. 35.

REFERENCE NOTES

[1] H. S. Yoder and C. E. Tilley, *Journal of Petrology*, vol. 3 (1962): pp. 342–352 (especially pp. 353–358). See also G. A. Macdonald and T. Katsura, *Journal of Petrology*, vol. 5 (1964): pp. 82–133.

[2] For example, see H. S. Yoder and C. E. Tilley, *Journal of Petrology*, vol. 3 (1962): pp. 357–360; I. D. Muir and C. E. Tilley, *American Journal of Science*, vol. 261 (1963): 111–128; G. A. Macdonald, *Geological Society of America Memoir*, vol. 116 (1968): pp. 477–522.

[3] I. D. Muir and C. E. Tilley (1963).

[4] I. S. E. Carmichael, *Journal of Petrology*, vol. 5 (1964): pp. 435–460; *American Mineralogist*, vol. 52 (1967): pp. 1815–1841.

[5] *Journal of Petrology,* vol. 1 (1960): pp. 121–145; *Bulletin Volcanologique,* vol. 29 (1966): pp. 195–222.

[6] For example, see E. E. Sigvaldson, *Journal of Petrology,* vol. 15 (1974): pp. 497–523; S. P. Jakobsson, J. Jonsson, and F. Shido, *Journal of Petrology,* vol. 19 (1978): pp. 669–705.

[7] For example, see I. D. Muir, C. E. Tilley, and J. H. Scoon, *American Journal of Science,* vol. 255 (1957): pp. 241–253.

[8] E. J. Searle, *New Zealand Journal of Geology and Geophysics,* vol. 4 (1961): pp. 165–204.

[9] F. Walker, *Quarterly Journal of the Geological Society of London,* vol. 86 (1930): pp. 355–398; R. J. Murray, *Geological Magazine,* vol. 91 (1954): pp. 17–31.

[10] For detailed chemistry and petrography, see I. D. Muir and C. E. Tilley, *Journal of Geology,* vol. 69 (1961): pp. 186–203.

[11] B. Levi, *Contributions to Mineralogy and Petrology,* vol. 24 (1969): pp. 30–49.

[12] R. W. White, *Contributions to Mineralogy and Petrology,* vol. 12 (1966): pp. 245–314; E. D. Jackson, *Proceedings of the 23rd International Geological Congress,* vol. 1 (1968): pp. 131–150.

[13] E. S. Larsen, *Geological Society of America Memoir,* vol. 29 (1948): pp. 42–43.

[14] For example, see L. R. Wager, G. M. Brown, and W. J. Wadsworth, *Journal of Petrology,* vol. 1 (1960): pp. 73–85.

[15] H. H. Hess, *Geological Society of America Memoir 80* (1960): p. 77.

4

Silica-Saturated Intermediate Rocks

Igneous rocks of intermediate silica content fall naturally into two general divisions: essentially saturated nonfeldspathoidal rocks and undersaturated feldspathoidal rocks. By the IUGS classificatory scheme, rocks with a feldspathoid content greater than 10% of the felsic components are grouped in the second category; rocks of the first category are largely devoid of feldspathoids but may contain minor amounts.

This chapter focuses on the rocks of the first category, the silica-saturated rocks. Volcanic and plutonic types are treated separately, and each is subdivided into three groups on the usual basis of relative proportions of alkali feldspar and plagioclase in the mode. The general scheme of nomenclature to be followed is illustrated below, without arbitrarily drawn numerical limits.

	Volcanic Types	*Plutonic Types*
Plagioclase the greatly predominant or only feldspar	Andesite	Diorite
	Latite (Trachyandesite)	Monzonite
Alkali feldspar the greatly predominant or only feldspar	Trachyte	Syenite

VOLCANIC TYPES: ANDESITES, LATITES, TRACHYTES

Andesites

Andesites are the most abundant volcanic rocks of island arcs and mobile continental margins, especially in belts overlying Benioff zones. They are regionally associated with tholeiitic basalts and rhyolites, or both. Around the Pacific Ocean today, for example, andesites predominate in composite volcanoes of the Andes, Central America, the northwestern United States, and the Aleutian, Japanese, Javanese, and southwestern Pacific island arcs.

Texturally, most andesites are porphyritic rocks with prominent phenocrysts of both plagioclase and mafic minerals. These phenocrysts are typically enclosed in a pilotaxitic groundmass (Chapter 2, Figure 2-10B) or, in some rocks, in a groundmass of small plagioclase laths set in glass (Figure 2-10C). Intergranular, intersertal, and ophitic textures that are so typical of basalts and diabases are exceptional among andesites save in olivine-bearing types transitional to basalts.

Plagioclase in most andesites has a mean composition around An_{40}, as compared with An_{55} in average tholeiitic basalts. But precise estimation of plagioclase composition, because of widely prevalent zoning in phenocrysts and discrepancy between respective compositions of groundmass plagioclase and phenocrysts, is not possible by microscopic technique. A more reliable index is the normative ratio $100an/(an + ab)$. Moreover, in many andesites, such as those of active volcanoes of the northwestern United States, careful microscopic measurements (especially with the universal stage) reveal the presence of more than one crop of phenocrysts each with its own pattern of internal zoning and twinning. Thus in a single thin section one may find separate phenocrysts of bytownite, labradorite, and oligoclase-andesine. Zonary structure, highly prevalent in phenocrysts, can be either normal or oscillatory; groundmass plagioclase is usually homogeneous and decidedly more sodic than most phenocrysts. Small crystals of K-feldspar only rarely can be detected microscopically in the groundmass, but analyses of cryptocrystalline or glassy groundmass material may show normative *or* values up to 10. Increase in normative *or* beyond 10 and appearance of discrete crystals of K-feldspar in the groundmass mark transitions toward latites, or in more siliceous rocks toward dacites.

Mafic minerals include various combinations of olivine, augite, hypersthene, hornblende, and biotite. Specific combinations tend to correlate generally with SiO_2 content as it increases along the andesite-dacite spectrum; and various types of andesite may be conveniently distinguished on this basis, with special emphasis on mafic phenocrysts. In all of these andesites the proportion of phenocrystic plagioclase is variable (see Figures 4-1 and 4-2). The petrographic character of some typical varieties of andesite, in the general order of increasing SiO_2 content, are reviewed in the following paragraphs.

Olivine Andesites

Olivine is a typical phenocryst mineral in the more basic andesites ($SiO_2 = 52–55\%$). Many such rocks, in which the principal microscopically recognizable minerals are olivine and labradorite, are transitional to tholeiitic basalts, from which they are distinguishable only by a higher content of SiO_2 ($>52\%$) and somewhat higher normative Q. These rocks are generally designated *basaltic andesites* (or, alternatively, *andesitic basalts*), for natural transition between the two classes is completely gradational and any boundary must be arbitrarily drawn.

Typical of the olivine-bearing basaltic andesites that are so prominent in the circum-Pacific belt is a lava (Chapter 3, Figure 3-1A) extruded from the central Mexican volcano Paricutin, which came into being in 1943. The modal composition of this rock is phenocrystic olivine (5), hypersthene (4), and augite (1); labradorite microlites (50); dark glass with refractive index approximately 1.54 (40). The SiO_2 content is approximately 55%. Other olivine andesites of Paricutin (Figure 4-1 A–D) carry olivine in much the same proportions, but many of them contain less glass, and microlites of plagioclase accompanied by a good deal of finely crystalline hypersthene are correspondingly more abundant.

Pyroxene Andesites

Andesites in which the sole or dominant mafic components are pyroxenes are especially common on composite volcanoes in orogenic belts. Most of them carry abundant phenocrysts of zoned plagioclase with cores as calcic as anorthite and rims mainly of oligoclase. Zoning, though generally of the normal or normal-oscillatory type, in some cases may be reverse. Plagioclase microlites generally are andesine or oligoclase. In some rocks a potassic feldspar forms thin shells around plagioclase phenocrysts or may be intergrown with quartz in the microcrystalline groundmass; as the content of K-feldspar increases beyond 10%, however, andesites grade into latites.

Most of the hypersthene in pyroxene andesites occurs as phenocrysts (Figure 4-1F–H), and although it never exhibits the fine lamellar inclusions of clinopyroxene prevalent among the hypersthenes of more slowly cooled intrusive rocks, its vertical faces are often enclosed by jackets of augite. The dominant clinopyroxene is diopsidic augite with optic angles of more than 40°. Pigeonite rarely occurs as phenocrysts, and then only in rapidly cooled andesites, but it is common as a groundmass constituent along with augite. Occasionally these pyroxenes, like the plagioclases, show reverse zoning. Rims of hypersthene and augite may be more magnesian than the cores, and rims of pigeonite may be more calcic than

---

Figure 4-1. Basalt, Olivine Andesites, and Andesites of the Parícutin Region, Mexico. (Reproduced from figs. 87 and 88 in H. Williams, "Volcanoes of the Parícutin region, Mexico," *U. S. Geological Survey Bulletin 965B*, 1950, pp. 165–279.)

The number at the lower right of each diagram is the analyzed weight percentage SiO_2 for the rock illustrated.

A. Olivine-bearing basaltic andesite. Olivine phenocrysts in a fluidal, intergranular groundmass of plagioclase, augite, and opaque oxides.

B. Olivine-augite basaltic andesite. Large phenocrysts of olivine and augite and smaller ones of labradorite in a glass-rich matrix stippled with microlites of plagioclase, augite, and opaque oxides.

C. Olivine augite basalt, or basaltic andesite. Phenocrysts of olivine, augite, and calcic labradorite in a glass-rich matrix charged with granular opaque oxides, augite, and microlitic plagioclase.

D. Olivine-augite tholeiitic basalt. Phenocrysts of olivine and augite; groundmass contains labradorite, augite, iron-titanium oxides, and patches of black interstitial glass. Intersertal texture.

E. Pyroxene-hornblende andesite. Oxyhornblende, hypersthene, and labradorite phenocrysts in a base carrying plagioclase, augite, and opaque oxides.

F. Hypersthene andesite. Phenocrysts of hypersthene and labradorite in a glass-rich base stippled with augite, opaque oxides, and laths of andesine.

G. Pyroxene andesite. Microphenocrysts of augite and hypersthene, in about equal amounts, and phenocrysts of labradorite in a glass-rich base carrying granular opaque oxides, augite, and microlitic plagioclase.

H. Hypersthene andesite. Phenocrysts of labradorite and hypersthene in a glass-rich matrix containing microlites of andesine and rare granules of augite and opaque oxides.

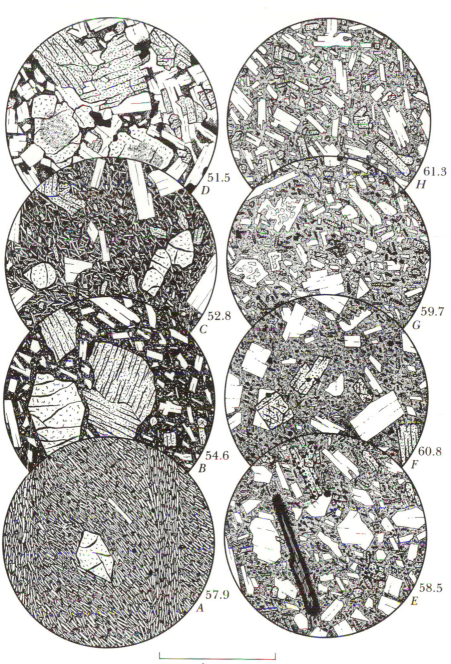

51.5
D

61.3
H

52.8
C

59.7
G

54.6
B

60.8
F

57.9
A

58.5
E

1 mm

the interior parts. Aegirine-augite and aegirine of deuteric origin are to be seen in a very few andesites, generally in association with sodic plagioclase, zeolites, or anorthoclase. The usual opaque oxide is magnetite; titanomagnetite is not uncommon, but ilmenite is rare. Hornblende, biotite, apatite, garnet, and cordierite are sporadic accessories, while tridymite and cristobalite, as linings and fillings of pores, are widespread but only in small amount. If glass is present, it is colorless or pale buff, with a refractive index close to 1.54.

The wide variety of olivine and pyroxene andesites that may be displayed within a small eruptive province (the Parícutin region) is partially illustrated in Figure 4-1.

Hornblende and Biotite Andesites

These generally form thick, short flows, steep-sided domical protrusions, or intrusive plugs and dikes. They are usually more siliceous and alkaline than pyroxene andesites, and hence grade into dacites and latites. Their hornblendes are seldom fresh and green except in quickly chilled, glass-rich types; otherwise they are brownish or reddish oxyhornblendes or are partly or completely replaced by granular mixtures of augite and magnetite (Figures 4-1E, 4-2C). Their biotites likewise are often reddened by oxidation and more or less replaced by granular opaque oxides. Fresh pyroxenes almost always accompany the hornblende and biotite, and occasionally a little olivine is to be seen. The plagioclase of these rocks tends to be somewhat more sodic than that of the olivine andesites and pyroxene andesites, and potassic feldspar tends to be more abundant.

Propylites

Andesites of all kinds are particularly susceptible to the kind of alteration known as propylitization. In most instances this alteration is to be ascribed to hydrous, deuteric fluids rich in carbon dioxide and sulfur, but in others it appears to result from introduction of similar fluids from extraneous sources into andesites that were already solid. Many propylitized rocks are found close to metallic ore bodies, but it is not certain in all cases that this relationship is genetic.

The characteristic features of propylites in early stages of alteration are as follows: a drab-green color in hand specimens; replacement of hornblende and biotite by chlorite, calcite, sphene, and iron oxides; and

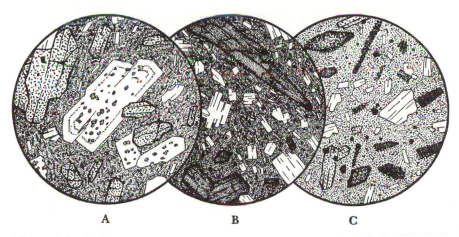

Figure 4-2. Andesites

A. Pyroxene andesite, Crater Lake, Oregon. Diam. 3 mm. Phenocrysts of zoned labradorite-andesine, with inclusions of glass and of hypersthene and augite, in a groundmass composed of oligoclase microlites, specks of opaque oxide and pyroxene, and interstitial cryptocrystalline material.

B. Hornblende andesite, Black Butte, Mount Shasta, California. Diam. 3 mm. Phenocrysts of oxyhornblende, pleochroic from gold to russet, fringed with granular magnetite; also phenocrysts of zoned labradorite. Pilotaxitic groundmass of microlitic andesine and interstitial cryptocrystalline material stippled with magnetite and fumarolic hematite.

C. Hornblende andesite, Stenzelberg, Siebengebirge, Germany. Diam. 3 mm. The hornblende phenocrysts are completely replaced by granular opaque oxides and augite. These, together with phenocrysts of diopsidic augite and calcic andesine, lie in a cryptocrystalline groundmass.

development of bastite serpentine, chlorite, and calcite from pyroxenes. Further alteration may result in albitization of the original plagioclase, sometimes with concomitant crystallization of epidote. Pyrite may or may not be present. A few propylites contain much quartz, with or without actinolite. Rather rarely, deuteric piedmontite may be found as veinlets and amygdule-fillings in propylites, the requisite manganese probably coming from the original mafic constituents.

Disequilibrium and Magmatic Mixing in Andesites

Over 40 years ago, E. S. Larsen and coauthors concluded from an exhaustive study of the Tertiary San Juan volcanic province, Colorado, that widespread disequilibrium in andesitic mineral assemblages of that region is evidence of preextrusive large-scale mixing of magmas of common parentage.[1] This now seems to be a common, even a characteristic, feature of andesites. Among the most widespread indications of departure from equilibrium within individual thin section are these: strong divergence in chemical composition between plagioclase phenocrysts, within the phenocrysts themselves (zonary structure), and between phenocrysts and groundmass plagioclase; presence of corroded phenocrysts of olivine, mantled in some instances with hypersthene; partial conversion of hornblende phenocrysts to dark aggregates of pyroxene and iron oxides; presence of three or more pyroxene phases in a single thin section (e.g., in an andesite from Hakone volcano, Japan); phenocrysts of hypersthene (mantled with augite), pigeonite, and augite in a groundmass containing both bronzite and augite; less commonly, presence of corroded xenocrysts of quartz, sanidine, or even nearly pure anorthite.

Latites

Greatly subordinate to andesites in global abundance, but widely distributed nevertheless in continental provinces, is a group of lavas of intermediate SiO_2 content, *hy*-normative, and carrying abundant andesine like the andesites but more potassic in that *or* is prominent in the norm and an alkali feldspar (unless occult in cryptocrystalline or glassy material) is visible in the mode. These rocks are generally termed *latite* in North American literature. They are roughly comparable with the less alkaline members of the plutonic monzonite family. (Indeed, in the IUGS scheme latites and monzonites are precisely equated in terms of essential minerals.) The term *trachyandesite* is partly synonymous with latite, but it carries a connotation of a greater degree of undersaturation in SiO_2, usually expressed by minor normative *ne* instead of *hy*. Trachyandesites, in contrast with typical latites, tend to be associated with lineages of alkali-olivine-basalt parentage; and they are not uncommon in oceanic islands as well as in continental provinces. Current terminology is confused not only by the partial overlap of latites and trachyandesites but also by differences in nomenclatural fashion (as between British and North American petrologists), and also by the transitional relation between trachyandesites, mugearites, and hawaiites. Here we first consider typical *hy*-normative latites such as are associated in the field with andesites of young orogenic belts—notably in the San Juan province of Colorado and in the Sierra Nevada of California.

The dominant plagioclase of latites is andesine or oligoclase, but cores of some phenocrysts may be labradorite or occasionally even bytownite. Shells of sanidine or anorthoclase commonly rim the plagioclase phenocrysts; but more generally alkali feldspar is confined to the groundmass or is an occult component of interstitial glass. In quartz-bearing latites transitional to dacites there may be considerable groundmass quartz. The usual pyroxene is pale-green diopsidic augite. Hypersthene is absent except in types transitional to andesite (as in the San Juan province). Aegirine-augite accompanies augite in some of the more alkaline types. Many of the more basic latites, like their andesitic counterparts, carry phenocrysts of olivine. But in fully saturated latites lacking olivine and with less abundant augite the phenocryst minerals include hornblende or biotite, or both. Minor accessories such as apatite, zircon, and iron-titanium oxides are less abundant than in most andesites. Finally we note that in a few latites, the appearance of small quantities of leucite marks a transition into highly undersaturated alkaline rocks, feldspathoidal trachyandesites (pp. 197; 201).

A type of latite with an unusual and striking texture, widely developed as lava flows and dikes in the Oslo province of southern Norway, has long been known as *rhomb porphyry*. Large phenocrysts of plagioclase (An_{20} to An_{40}) make up between 30 and 45% of the mode. In most specimens they average between 5 mm and 1 cm in diameter; and in some specimens they tend to show rhomboidal outlines due to a crystal habit dominated by {110} faces in the zone of [001]. These phenocrysts tend to be mantled with alkali feldspar, which also occurs internally as exsolved (antiperthitic) irregular inclusions and as the principal or sole feldspar of the groundmass. The color index is between 10 and 20. The usual mafic component is clinopyroxene (typically aegirine-augite) accompanied by minor chlorite, calcite, opaque oxides, and apatite. Subordinate alkaline amphibole, biotite, or ferriferous olivine may be present in similar rocks from other provinces. The rhomb porphyries of Norway are associated with chemically equivalent pyroxene monzonites and with quartz-bearing alkali syenites in a region of block faulting (p. 154).

Trachytes

Trachytes are volcanic rocks close to saturation in silica, consisting mainly of alkali feldspar. They are light in color, and the content of SiO_2 is mostly between 58 and 66%. Magmas of this composition are highly viscous, so that most trachytes occur in dikes, plugs, breccia pipes, and short, thick flows. There are, however, rare records (notably in the Pleistocene volcanic fields of Kenya) of outpourings of trachytic lava on a very large scale. Trachytes occur principally in close association with much more voluminous alkali olivine basalts and trachybasalts in oceanic islands and continental volcanic provinces associated with crustal rifting. Most petrologists consider the lithologic spectrum alkali olivine basalt (or trachybasalt) → hawaiite → mugearite → trachyte to represent a genetic lineage.

Nearly all trachytes are porphyritic rocks with large euhedral tabular crystals of feldspar set in a finely crystalline groundmass of slender feldspar microlites with subparallel fluxional orientation (trachytic texture, Figures 2-10A, 4-3B,C). The principal constituent of all trachytes, typically over 80% of the mode, is alkali feldspar, usually a sodic sanidine or

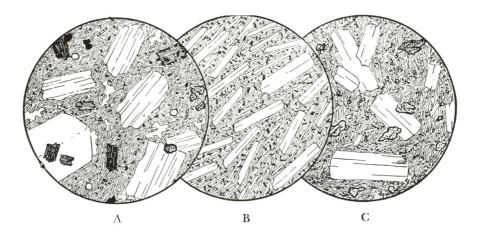

A   B   C

Figure 4-3. Trachytes and Keratophyre

A. Biotite trachyte, Drachenfels, Siebengebirge, Germany. Diam. 3mm. Phenocrysts of sanidine, andesine-oligoclase, and partly resorbed biotite, in a trachytic matrix of sanidine microlites with accessory iron oxide, apatite, sphene, and a little interstitial quartz.

B. Aergirine trachyte, Rosenau, Siebengebirge, Germany. Diam. 3 mm. Subparallel phenocrysts and microlites of sodic sanidine, divergent needles of aegirine, and a little interstitial sodalite.

C. Keratophyre, near Oroville, California. Diam. 3 mm. Phenocrysts and microlites of albite; anhedral granules of pale-brown hornblende and uralitized augite; abundant specks of epidote, ilmenite, and sphene.

anorthoclase.* Some trachytes also contain plagioclase—either oligoclase or andesine (Figure 4-3A). Rocks in which plagioclase of this composition becomes comparable to or exceeds alkali feldspar in the mode are transitional to mugearites. Mafic constituents, always greatly subordinate to feldspar, are combinations of fayalitic olivine, clinopyroxene, amphibole, and biotite. Sparse accessories are opaque oxides and smoky apatite.

Magnesian olivine, a mineral rather widespread in the more basic members of the latite family, is not found in trachytes except occasionally as xenocrysts; or rarely it may appear along with titanaugite in rocks (like certain flows in the east Otago province, New Zealand) of hybrid origin— possibly formed by mingling or reaction with partially crystalline basaltic magma.

Pyroxene Trachytes

In many trachytes the principal or only mafic constituent is diopsidic pyroxene or aegirine-augite, and the dominant sanidine is accompanied by a little sodic plagioclase (andesine or oligoclase) and quartz. In some varieties, such as those of Trans-Pecos, Texas, there are phenocrysts of anorthoclase, and in the groundmass aegirine-augite is accompanied by blue arfvedsonite. In others, such as those of Banks Peninsula, New Zealand, there are abundant phenocrysts of oligoclase, mantled with alkali feldspar; and the groundmass may contain arfvedsonite and a little tridymite. Others again (now known as *benmoreites*) have subequal alkali feldspar and oligoclase (or andesine) marking the transition to mugearites; these transitional rocks fall within the broad latite spectrum of the IUGS classification but are less potassic (lower normative *or/ab*) than typical latites described in this book (see p. 144).

Hornblende and Biotite Trachytes

Some trachytes are characterized by large phenocrysts of sanidine and fewer of oligoclase, hornblende, biotite, and diopside in a matrix composed chiefly of sanidine laths with a little interstitial quartz or tridymite or both (Figure 4-3A).

---

*The ratio Na/K in high-temperature alkali feldspars cannot be determined precisely by optical means. Where the characteristic patchy triclinic twinning can be detected in some sections it is safe to conclude that all the alkali feldspar is anorthoclase.

Peralkaline Trachytes

Peralkaline trachytes are characterized by dominance among the mafic minerals of aegirine, riebeckite, arfvedsonite, or cossyrite (Figure 4-3B). The preponderant or only feldspar is a sodic alkali feldspar—sanidine or anorthoclase. If aegirine-augite is present, it tends to form microphenocrysts; but other mafic minerals are restricted to the groundmass, where they occur as moss- and fernlike aggregates partly or entirely enveloping the feldspars in pseudopoikilitic fashion. These late-crystallizing minerals almost invariably show considerable oxidation to murky, brownish patches of hydrated oxide. In some peralkaline trachytes there may be a little fayalite, and many contain micropoikilitic patches of interstitial quartz.

Degrees of Silica Saturation in Trachytes

All trachytes, as implied by their mineralogical definition, are close to saturation in SiO_2. But some (quartz-bearing trachytes) are slighly oversaturated, some slightly undersaturated (feldspathoid-bearing trachytes). The intervening spectrum of exactly saturated trachytes (as seen from Figure 2-12) covers a wide range of silica activity and thus includes terminal products of many different magmatic lineages. In *quartz-bearing trachytes* quartz (less than 10% of the mode) rarely is phenocrystic; it typically occurs as small grains intergrown with alkali feldspar in the groundmass. In some rocks, such as those of dikes exposed in the eroded volcanoes of Banks Peninsula, interstitial tridymite, probably of early deuteric origin, takes the place of quartz. Plugs, domes, and short flows of quartz-bearing trachyte are familiar subordinate components of ocean-island volcanoes built largely of olivine tholeiites. One example is the mid-Atlantic island of Ascension, where a typical peralkaline trachyte ($SiO_2 = 66\%$; normative Q, 8) consists mainly of anorthoclase and minor aegirine. Associated lavas are trachyandesites (*hy* normative *hawaiites* and *mugearites*) and, at the siliceous end of the natural spectrum, peralkaline rhyolites (pantellerites) with $SiO_2 = 71\%$ and normative Q as high as 23. The same spectrum (olivine-tholeiite to pantellerite) is duplicated in the south Pacific on Easter Island. Elsewhere quartz trachytes of less alkaline character (metaluminous in Shand's terminology) have biotite—partially replaced by dusty opaque oxides—or hornblende as the mafic consituent.

Feldspathoid-bearing trachytes, at the opposite end of the silica-activity scale, have less than 10% modal feldspathoid and merge into phonolites, with which they tend to be associated in provinces where the principal magmas are alkali olivine basalts and trachybasalts (as on St. Helena and Tristan da Cunha in the south Atlantic). Typically, such rocks are peralkaline, *ne*-normative, and less siliceous than their quartz-bearing counterparts—a specimen from St. Helena has $SiO_2 = 60\%$, normative *ne* about 6. The feldspathoid may be nepheline, a sodalite mineral, or leucite—rarely as phenocrysts, usually in the groundmass; the sole feldspar is sodic sanidine or anorthoclase; and mafic minerals are sodic pyroxenes or amphiboles.

Keratophyres

Most keratophyres seem to bear much the same relation to andesites and dacites as spilites do to normal basalts, and, like spilites, most of them are interbedded with or intrude marine sedimentary rocks. Quartz-free types generally have phenocrysts of sodic plagioclase and hornblende or diopsidic augite set in a felted or trachytic groundmass made up mainly of albite or albite-oligoclase, chlorite, epidote, magnetite, sphene, and apatite (Figure 4-3C). Some varieties contain subordinate potassic feldspar, and others have sodic mafic constituents—riebeckite or aegirine.

Possibly some keratophyres (albite or oligoclase trachytes) are primary products of sodic magmas. But most appear to have formerly been more calcic rocks (andesites or dacites) that have become albitized by metasomatic processes. Presence of albite veinlets and albitic rims on relict, more calcic plagioclase and of much epidote and chlorite and fibrous amphibole (uralite) replacing original pyroxene all suggest activity of postmagmatic (actually metamorphic) processes.

PLUTONIC TYPES: DIORITES, MONZONITES, SYENITES

Diorites

Typical diorites are medium- or coarse-grained rocks in which andesine or calcic oligoclase is the predominant mineral and hornblende and biotite are the chief mafic constituents. Their textures are notably variable. Most diorites have granitic texture, but poikilitic and porphyritic textures are characteristic of many. Furthermore, there is often a segregation of light-colored and dark-colored minerals in irregular clots or lenticular layers, suggesting a hybrid origin for the rock as a whole.

Although diorites are essentially saturated rocks, many of them are somewhat oversaturated and contain small amounts of quartz. With increasing quartz content, diorites are transitional into *quartz diorites*; and finally, into *tonalites* (see Figures 4-4 and 5-1), which are here considered as among acid plutonic rocks to be discussed in the next chapter. More basic diorites, in contrast, usually carry pyroxenes, lack quartz, and cannot be sharply distinguished from gabbros.

Petrographic distinctions between the foregoing rock types thus are necessarily arbitrary. Gabbros are characterized by more calcic feldspar and are usually more mafic (color index >40) than diorites. Pyroxene is an essential mineral in most gabbros, while the commonest mafic constituent of diorites is hornblende. Quartz-bearing diorites are usually less mafic and richer in biotite than quartz-free diorites. In the field, diorites

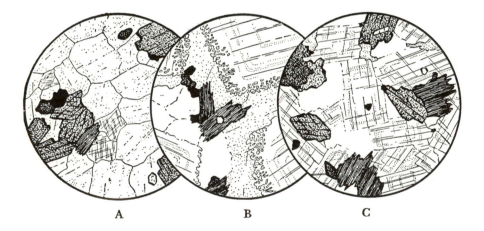

| A | B | C |

Figure 4-4. Diorite-Tonalite Spectrum

A. Hornblende diorite, near Stockholm, Sweden. Diam. 3 mm. Roughly equant subhedral crystals of andesine-oligoclase; a little microcline, hornblende, and biotite; accessory iron oxides, apatite, and sphene.

B. Felsic tonalite (trondhjemite), Castle Towers batholith, British Columbia. Diam. 2.5 mm. Main constituent is oligoclase showing oscillatory zoning and borders of myrmekite; next in abundance is quartz, then orthoclase. Accessory constituents are biotite, apatite, iron oxides, and sphene.

C. Tonalite, Adamello, Italy. Diam. 2.5 mm. Subhedral and euhedral zoned crystals of andesine-oligoclase, locally rimmed with orthoclase; anhedral patches of quartz; green hornblende and brown biotite; allanite partly fringed with epidote (lower right); accessory magnetite, apatite, and sphene.

and quartz diorites are generally found as member bodies of composite batholiths—both as small discrete plutons and as cupolas or marginal facies of larger more siliceous component plutons. Gabbros, especially hornblende gabbros, may occur in much the same fashion. But by far the greatest development of gabbros is in the great layered plutons of the Skaergaard and Bushveld type.* Thus although petrographically and chemically diorites and gabbros are mutually gradational, field transitions are the exception rather than the rule. The sequence gabbro-diorite-tonalite is essentially a lithologic spectrum that does not reflect, except in a crude manner, the genetic linkage that characterizes the magmatic sequence tholeiitic basalt → andesite → dacite. Finally, we note that some diorites have developed as subordinate marginal facies to granodioritic plutons and are probably of hybrid origin—the product of assimilation reaction between acid magma and adjoining or subjacent country rock.

The commonest kind of diorite consists essentially of oligoclase or andesine, hornblende with subordinate biotite, and traces of interstitial quartz and orthoclase. Zoning in plagioclase crystals is commoner and more variable than is the case in most gabbros, granodiorites, or granites. Normal zoning predominates, but both oscillatory and reverse zoning are common, especially in hornblende-rich diorites and in rocks of hybrid or contaminated origin. Cores of zoned crystals most commonly are labradorite. Deuteric alteration tends to accentuate zoning. Sodic rims may remain clear while the more calcic cores are changed to white or grey semiopaque microcrystalline aggregates of hydrous silicates ("saussurite" of early petrographers); or certain zones may become dusted with tiny flakes of colorless mica or clay minerals, while others remain clear. Plagioclase crystals of many diorites contain abundant inclusions of primary phases—mafic minerals, apatite, and opaque oxides.

Potassic feldspar, where present in diorites, is almost invariably orthoclase, and either is interstitial or mantles the plagioclase. Not uncommonly it is bordered by myrmekite, particularly where in contact with plagioclase. Microcline is scarce, and perthite, common in more alkaline plutonic rocks such as syenite and granite, is rarely present. Quartz is always interstitial and is often in micrographic intergrowth with orthoclase.

---

*Rocks in the uppermost 1000 m of the Skaergaard layered series, sometimes called *ferrodiorites* are here termed *ferrogabbros* for reasons given on p. 129.

Among the mafic constituents of diorites, green or, less commonly, brown hornblende predominates. In some rocks it forms stout prisms; in others it is acicular; and in still others it has a spongy or sievelike appearance owing to abundant inclusions of partially replaced clinopyroxene. Hornblende, in turn, may be surrounded by reaction rims of biotite. Typically, biotite is greatly subordinate to hornblende but becomes increasingly important in diorites with minor modal quartz. It is usually brown, rarely green, and characteristically shows some degree of deuteric alteration to chlorite and minor granular sphene.

Pyroxene when present in diorites is a colorless or pale diopsidic augite and is generally surrounded by reaction rims of hornblende. However, in the more basic rocks transitional to gabbro, augite may form discrete crystals, along with hypersthene; and one or the other may even exceed hornblende in amount. Only rarely, even in basic diorites, is there any olivine. Apatite, zircon, sphene, and iron-titanium oxides are all common accessories. Much more rarely minor amounts of garnet, cordierite, spinel, and andalusite or sillimanite (minerals typical of contact-metamorphosed shales) point to contamination of the diorite magma with sedimentary rocks of the envelope.

Diorite Porphyries

Porphyritic variants of the diorite family have precisely the same mineral content as that of diorites, from which they differ only in texture. They contain phenocrysts of zoned plagioclase, hornblende, and biotite (and occasionally quartz) in a dense anhedral-granular groundmass, almost devoid of mafic minerals and containing a little K-feldspar and quartz in addition to the dominant sodic plagioclase.

Mafic Diorites and Hornblendites

The color index of typical diorites is less than 40 (see Figure 4-4A). But darker rocks consisting of hornblende and subordinate andesine or oligoclase can be thought of, in spite of their low SiO_2 content (45%) as mafic extremes of the diorite family (meladiorites of IUGS). They occur as minor intrusions, are not widespread in their global occurrences, and typically grade into and are associated with dikes of virtually pure hornblende rock—*hornblendites*. Other associates are dikes of hornblende lamprophyre.

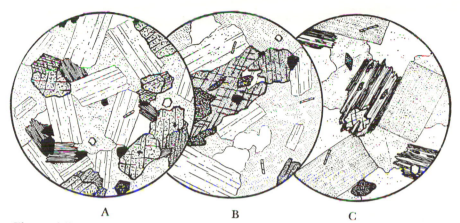

A   B   C

Figure 4-5. Monzonites and Plagioclase-Rich Granite (Adamellite)

A. Monzonite, Monzoni, Tyrol. diam. 2.5 mm. Euhedral laths of andesine; anhedral, turbid sodic orthoclase, and a little interstitial quartz. Diopsidic augite, partly bordered by green hornblende and brown biotite. Accessory minerals are opaque oxides, apatite, and sphene.

B. Quartz-bearing hornblende monzonite, Pine Nut Range, Nevada. Diam. 2.5 mm. Euhedral crystals of andesine, large anhedra of altered orthoclase, and smaller ones of quartz. Dark constituents are hornblende, sphene, and opaque oxides. Accessory needles of apatite.

C. Granite (adamellite), Shap Fell, Westmorland, England. Diam. 2.5 mm. Euhedral, altered crystals of oligoclase; anhedral quartz and slightly altered orthoclase. The flakes of biotite show alteration to chlorite with liberation of secondary sphene. Accessory constituents are primary sphene, apatite, fluorite (near center), and allanite (near bottom).

Monzonites

Monzonites occupy an intermediate position in the classificatory spectrum between diorites and syenites; hence, they are sometimes called syenodiorites. They are characterized by subequal amounts of K-feldspar and plagioclase, neither of which is less than one-third or exceeds two-thirds of total feldspar. Quartz is usually present but not in amounts exceeding 5% of the mode. Color indices generally lie between 15 and 30, but there are more mafic types (color index 50–60) of basic composition that may contain olivine and augite with or without biotite.[°] With increase in quartz, normal monzonites grade into a widely prevalent type of plagioclase-rich granite (Figure 4-5C). In fact monzonites (which are

---

°*Kentallenites.*

not common rocks) typically occur as border facies or satellitic offshoots of larger bodies of just such granites, or of diorites.

The type monzonite of the Tyrol is illustrated in Figure 4-5A. Its mode is as follows: euhedral or subhedral normally zoned plagioclase (An_{68-35}), 33; orthoclase, 32; augite plus a little hornblende, 24; biotite, 6; quartz, opaque oxides, apatite, zircon, and sphene, 5. The texture is often poikilitic, large anhedral tablets of orthoclase enveloping all other constituents except quartz.

In the more siliceous quartz-bearing monzonites, the dominant mafic constituents are biotite and hornblende; in the more basic varieties, augite takes first place among the dark minerals and may be accompanied by hypersthene and olivine.

Many monzonites contain patches of myrmekite and micropegmatite of deuteric origin, just as many diorites and granodiorites do, and these are molded upon, are inserted between, or exhibit fingerlike intergrowths with, the more abundant constituents. Also to be ascribed to deuteric action is the widespread alteration of augite to fibrous amphibole (uralite) seen in many monzonites.

The most extensively developed rocks of the Oslo province, Norway, are pyroxene monzonites of almost basic composition ($SiO_2 = 52-55\%$).° They cover an aggregate area of 1600 km²—one-third of the total mapped outcrop of shallow plutonic complexes which in the same region are associated with almost equally extensive effusive porphyritic latites (rhomb porphyries) of similar chemical composition, and with subordinate alkaline basalts. The mode of a typical Oslo augite monzonite is given by T. F. W. Barth as alkali feldspar (perthitic orthoclase), 48.5; cryptoperthitic oligoclase, An_{27}, 33; titanaugite, 9.5; biotite, 4; quartz, 1; iron-titanium oxides, 3; apatite, 1. In other specimens the clinopyroxene is diopsidic augite; and bronzite, olivine, barkevikite, and sphene are possible additional phases. The color index is about 12; but hand specimens have a characteristic blue-gray color, imparted by the schiller luster of the feldspars, that has made the rock celebrated as an ornamental stone. The grain size is medium to coarse, both feldspars tending to occur as subhedral to euhedral crystals several millimeters or even centimeters across, rather commonly showing the rhombic outline that characterizes the phenocrysts of some equivalent effusive latites.

Monzonite porphyries carry phenocrysts of plagioclase with andesine

---

°*Larvikites* of Norwegian literature.

cores and outer zones of oligoclase, which in turn may be enveloped in alkali feldspar. Orthoclase and perthite may also form phenocrysts; but large mafic crystals are rare. The groundmass consists of finely inter-grown sodic plagioclase and orthoclase, stippled with augite, hornblende, biotite, opaque oxides, apatite, and sphene. Texturally the Oslo rhomb porphyries, here termed latites because of their effusive character, could be called monzonite porphyries.

Syenites

The medium- and coarse-grained equivalents of trachytes are micro-syenites and syenites, in which alkali feldspars make up at least two-thirds of all the feldspar. Quartz or feldspathoid may be present as a minor constituent, but commonly both are absent. The color index seldom rises above 30. Two broad types of syenite are distinguished on the basis of the ratio $(K_2O + Na_2O)/CaO$.

Alkali Syenites

Some syenites in which alkali feldspars make up no less than 95% of the total feldspar occur as small plutons closely associated with trachytes and phonolites, as on the oceanic islands of Réunion east of Madagascar (Figure 4-6C) and Fuerteventura in the Canaries. More extensive alkali syenite plutons tend to be associated with others of peralkaline granite, as in the White Mountain magmas series of the New England Appala-chians and the Oslo province of Norway. In the latter region, quartz-bearing alkali syenites (nordmarkites)* outcrop over an aggregate mapped area of 1400 km² and are second in extent, among the shallow plutonic rocks of the region, only to the augite monzonites described on p. 154; associated bodies of peralkaline granite and others of biotite granite each cover 800 km.

Between 80% and 90% of a typical specimen of Oslo nordmarkite (Figure 4-6A) consists of micro- or cryptoperthite in which orthoclase or microcline plays host to albite or sodic oligoclase. Perthitic intergrowths of this kind are common in almost all syenites, and especially in alkali

---

*Some of these rocks contain enough quartz to qualify as quartz syenites (cf. Figures 2-13 and 5-1).

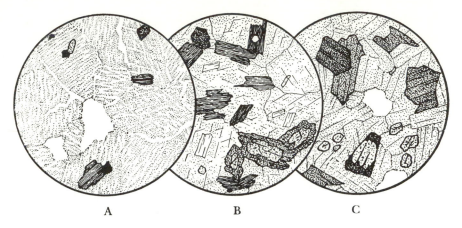

A   B   C

Figure 4-6. Syenites

A. Quartz-bearing syenite (nordmarkite), Oslo, Norway. Diam. 2.5 mm. Large
    crystals of microperthite, locally veined and fringed with albite; a little quartz
    and biotite; accessory opaque oxides, zircon, and sphene.

B. Syenite, Ymir, British Columbia. Diam. 3 mm. The main constituents are
    biotite, uralitized augite and altered orthoclase. Minor constituents are small
    euhedral andesines and apatite.

C. Alkali syenite, Cilaor, Reúnion Island. Diam. 2.5 mm. The feldspar is altered
    perthite; and there is a little interstitial quartz. The mafic minerals are aegi-
    rine-augite (palest), aegirine (darkest), and barkevikitic hornblende.

syenites. The two phases are often so minutely intergrown, however, that
the precise composition and relative amounts of the component feldspars
cannot be determined. The content of interstitial quartz usually ranges
from about 5% to 8%. In some varieties the chief mafic constituent is
iron-rich biotite; in others it is hornblende or hastingsite; in still others
one of the sodic varieties may preponderate, either aegirine-augite,
aegirine, arfvedsonite, or riebeckite. Sphene, apatite, zircon, and iron
oxides are ever-present though minor accessories.

Some alkali syenites are peraluminous, being composed almost
entirely of microperthite, biotite, and muscovite. A few, indeed, are so
rich in aluminum that they also carry abundant corundum and accessory
aluminous minerals such as spinel, pyrope, and chrysoberyl. Other alkali
syenites are peralkaline, in that all the mafic minerals accompanying the
microperthite are varieties rich in iron and sodium, like arfvedsonite and
aegirine. Such syenites commonly contain a little nepheline; these generally occur as border facies of nepheline syenite-plutons.

Peralkaline alkali syenites, like trachytes (see Figure 2-12), represent the full span of silica activity appropriate to saturated intermediate magmas. For in contrast with the quartz-bearing alkali syenites of the Oslo district, other syenites° composed essentially of alkali feldspar may carry minor nepheline or sodalite as well as titaniferous biotite, sodic pyroxenes, and a variety of such accessories as apatite, fluorite, zircon, and zirconium silicates.

Augite alkali-syenites are exemplified by rocks in the upper levels of the classic Shonkin Sag sheet of Montana, which solidified during differentiation of a thin shallow body of highly potassic undersaturated basic magma. The mode of typical specimens of syenite has been given by W. P. Nash and J. F. G. Wilkinson as: sanidine, 51; augite, 15.5; biotite, 11; accessory opaque oxides and apatite, 2.5; zeolites and carbonate, 20.[2] Rocks in the lower levels of the sheet, *shonkinites*, are much more basic and have high normative *ne* (15). Shonkinites are therefore treated in Chapter 7 as mafic variants of feldspathoidal syenite.

Alkali syenites and alkali syenite porphyries emplaced at shallow depths, as in the case of the Shonkin Sag sheet, may contain sanidine instead of orthoclase or microperthite. At Port Cygnet, Tasmania, there occurs an interesting sanidine-rich alkali-syenite porphyry in which large phenocrysts of sanidine (with inclusions of hauyne), numerous grains of melanite garnet, and phenocrysts of iron-rich hornblende, aegirine-augite, and aegirine lie in a trachytic groundmass of the same minerals. Melanite garnet may be abundant in alkali syenites, especially in mafic varieties transitional into shonkinites, but it is rare in alkali-lime syenites, where the characteristic garnet is andradite.

Alkali-Lime Syenites

Syenites of a less alkaline nature (alkali-lime syenites) are marked by a higher content of sodic plagioclase (5−30% of the modal feldspar). Usually this is oligoclase or andesine, but it may be as calcic as labradorite. In the coarser-grained types of alkali-lime syenite, the principal feldspar is micro- or cryptoperthite, the host mineral being orthoclase or, rarely, microcline. Some of these perthitic intergrowths are enveloped by thin shells of albite, and some enclose cores of oligoclase. In microsyenites and syenite porphyries, however, perthitic intergrowths are less common, most of the alkali feldspar being sodic orthoclase or sanidine.

°*Pulaskite.*

Quartz, if present, is found either as interstitial anhedral grains between the feldspars, in micrographic intergrowth with potassic feldspar, or in myrmekite. A well-known example of quartz-bearing syenite from Plauen, near Dresden, is illustrated in Figure 5-3A. This rock, like the type syenite from Assouan, Egypt, and many other syenites, is extremely variable in content of both quartz and plagioclase, so that it grades into granite on the one hand and into monzonite on the other.

The characteristic mafic minerals of alkali-lime syenites are green hornblende and brown biotite. These are almost always found together, the former generally predominating, but in quartz-bearing types biotite tends to preponderate. Diopsidic augite may be present as discrete crystals or as cores inside hornblende, and it tends to become more abundant as the rocks become more basic. Titaniferous augite, orthopyroxene, and olivine are scarce except in varieties transitional to monzonites and alkali gabbros. Minor accessories are generally less abundant and varied than in alkali syenites; among them are apatite, sphene, zircon, and opaque oxides, the last of which usually carry less titanium than those of alkali syenites.

Syenites of the alkali-lime type tend to occur as local variants of granitic plutons; but less commonly quartz syenites of this general group (10–15% modal quartz) form independent large plutons. A well-known example of the latter is the Diana complex in the Proterozoic massif of the Adirondacks, New York State, a plutonic body about 700 km² in outcrop area consisting essentially of quartz-bearing syenite.[3] The finer-grained rocks of the border zones are alkali-lime quartz syenites ($SiO_2 = 66\%$) with an average mode approximately microperthite, 65; oligoclase, 14; quartz, 13; augite, 5; hornblende, 1; accessory magnetite, apatite, and zircon, 2. Coarser variants within pluton are alkali syenites with either hornblende or augite as the main mafic constituent and a content of SiO_2 around 58–60%. Local variants have enough plagioclase to be classed as monzonites.

REFERENCE NOTES

[1] E. S. Larsen et al., *American Mineralogist*, vol. 23 (1938): pp. 237–257.

[2] W. P. Nash and J. F. G. Wilkinson, *Contributions to Mineralogy and Petrology*, vol. 25 (1970): pp. 241–269.

[3] A. F. Buddington, *Geological Society of America Memoir* 7 (1939): pp. 73–111.

5

Acid Rocks

GENERAL CHARACTER AND TERMINOLOGY

All acid rocks, defined as having a silica content higher than the arbitrarily drawn limit of 66%, are strongly oversaturated in SiO_2. In coarse- and medium-grained rocks this condition is expressed by high modal quartz—20% or more of the felsic fraction. For glassy and cryptocrystalline rocks we must fall back on the corresponding normative parameter $100Q/(Q + or + ab + an) > 20$.

As with intermediate rocks, compositional variation among acid rocks is expressed primarily in terms of the relative proportions of alkali feldspar, plagioclase, and quartz; with mafic constituents playing a subordinate but still significant role in framing more detailed rock terminology. The color index is generally low, ranging from almost 40 in some tonalites to less than 10 in typical granites (and rhyolites). A general scheme of classification is summarized below, without sharply drawn interdivisional boundaries, and illustrated in Figure 5-1.

Alkali Feldspar	Plutonic Types	Volcanic Types
<10% total feldspar	Tonalite	
10–35% total feldspar	Granodiorite	Dacite
>35% total feldspar	Granite	Rhyolite

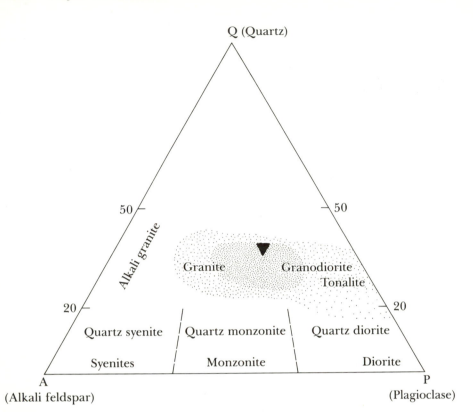

Figure 5-1. Essential Mineral Composition of Quartz-Bearing Plutonic Rocks
The most abundant rocks are indicated schematically by relative density of stippling. The solid triangle represents the approximate locus of minimum melting temperatures, at 1000 bars H_2O pressure, in the system albite-orthoclase-quartz.

Thus, in accordance with the IUGS classification (Chapter 2, Figure 2-13), *granite* is defined here to include not only rocks with greatly predominant alkali feldspar but also even more plentiful rocks with subequal plagioclase and alkali feldspar, which in other well-known systems of nomenclature have been termed "quartz monzonites" or *adamellites*.*

It must be genetically significant that in any chemical-mineralogical plot, such as Figure 5-1, the most abundant acid rocks cluster around the locus of minimal melting temperatures (varying with pressure and H_2O content) in the experimentally investigated system $KAlSi_3O_8$–$NaAlSi_3O_8$–SiO_2.

---

*Here, as in the IUGS scheme (cf. Figure 5-1), the term *quartz monzonite* is restricted to less siliceous rock with modal quartz between 5% and 20% of the felsic components.

PLUTONIC TYPES:
GRANITES, GRANODIORITES, TONALITES

Global Distribution

Acid plutonic rocks are virtually confined to the continental crust.* Here they comprise the great bulk of immense composite batholiths in orogenic belts; and in the ancient Archaean cratons they have contributed to a major degree to the primitive continental crust. Rocks of the tonalite-granodiorite-granite spectrum indeed rank collectively as the most voluminous and widespread of plutonic igneous rocks. Many Archaean granites and granodiorites have been deformed and internally reconstituted during later episodes of metamorphism. These are the *granitic gneisses* of general geological literature, and their igneous (plutonic) origin is beyond doubt.

Textures of Acid Plutonic Rocks

Most acid plutonic rocks are of medium to coarse grain, showing the typical *granitoid (subhedral granular)* texture illustrated in Figures 5-3 and 5-6. Both plagioclase and mafic minerals tend to form subhedral (less commonly euhedral) crystals. Anhedral quartz occupies intergranular spaces. Alkali feldspar when abundant also commonly forms subhedral crystals; but when in minor quantity it is usually found as anhedral grains associated with the interstitial quartz. Some granites and granodiorites are coarsely porphyritic, with large phenocrysts of orthoclase or microcline (both typically perthitic), in some cases as much as 5 cm to 10 cm in diameter (e.g., in the Cathedral Peak granodiorite of Yosemite, California). Locally these phenocrysts may be very abundant and most of them are euhedral, commonly with Carlsbad interpenetration twinning. They usually enclose numerous small crystals of other minerals, especially biotite flakes, concentrated in spaced arrays trending parallel to the phenocryst margins.† In a few porphyritic granites, the phenocrysts of

---

*A trivial exception is the occurrence of cognate xenoliths of roughly banded peralkaline granite in some ocean-island basaltic lavas (notably on Ascension Island) that presumably have been torn from upper levels of a subjacent minor body of differentiated tholeiitic magma. Another is the peralkaline granite of Rockall Island in the North Atlantic.
†These large K-feldspar crystals have been variously interpreted, both as true phenocrysts and as porphyroblasts formed by metasomatism after consolidation of the granitic rock. Mostly, in our opinion, they are true igneous phenocrysts.

K-feldspar, ovoid in shape, are encased in shells of oligoclase, thus displaying what has been termed *rapakivi texture*. In hand specimens of such rocks it is usually evident from the contrast in color between pink cores and white rims. In some instances this phenomenon is clearly a result of reaction between the partially crystalline magma and adjacent, more basic or pelitic rocks in the envelope. In the Dartmoor granite of Cornwall, for example, rapakivi texture is restricted to hybrid facies; in the near-by Bodmin Moor and St. Austell granites, also, it is best developed in contaminated rocks carrying andalusite and cordierite; and likewise in the granites of Jersey and Brittany it is commonest in basified portions.

A widespread minor textural feature among granites is *graphic* (or *micrographic*) intergrowth between a single K-feldspar crystal and enclosed small, often angular skeletal grains of quartz. Though appearing mutually isolated in a thin section, these patches of quartz are optically and presumably crystallographically continuous within each host crystal of feldspar. Graphic texture may be conspicuous in hand specimens of coarse-grained pegmatitic granites. On the smaller scale, micrographic texture is a microscopic characteristic of many granites, especially those of the type called *granophyre*, which is the typical final product of differentiation in the upper levels of tholeiitic diabase sills and of layered basic plutons. So the terms *micrographic* and *granophyric* have become synonymous in the textural sense. The texture is generally thought to represent rapid simultaneous growth of feldspar and quartz, each phase growing from a single nucleus of its own species. Closely similar, but almost certainly of postmagmatic and even metamorphic origin, is an intergrowth of finely vermicular, optically continuous quartz bodies in outer fringes of K-feldspar crystals where these impinge directly on grains of plagioclase. The intergrowth—known as *myrmekite*—encroaches into the plagioclase as lobate protuberances, clearly a result of localized metasomatism.

It is now generally recognized that in acid plutonic rocks, more especially in granites, grain boundaries among the felsic phases are not strictly the direct product of igneous crystallization. During slow postmagmatic cooling, when aqueous residual fluids still permeate the rock and internal exsolution and inversion processes are active in the crystals of alkali feldspar, primary grain boundaries tend to become modified. Myrmekitic lobes may encroach from K-feldspar into adjoining plagioclase at this stage. And around other K-feldspar crystals thin rims of albite may develop, probably as a product of exsolution from the potassic host crystal. In some rocks, fingerlike outgrowths of biotite may encroach upon alkali feldspar crystals; and less commonly localized graphic intergrowths of biotite and quartz may develop.

A special and rare local feature of some granites and granodiorites (and even of more basic plutonic rocks) is *orbicular texture*. The rock in question is crowded with large ovoid bodies—several centimeters or even a meter in diameter—made up of concentric shells alternately rich in the felsic and the mafic components of the rock. In some cases at least, such bodies seem to have formed by rhythmic selective crystallization of minerals around a nucleus of foreign rock.

Rocks in the cores of many large acidic plutons show little or no preferred orientation of the constituent minerals. But near the margins there is often a marked tendency for elongated minerals to be aligned with their major axes more or less parallel, so as to produce a distinct lineation; also, slight concentrations of light- and dark-colored minerals in alternating streaks may produce a crude planar banding, and mafic schlieren and basic inclusions often become abundant. Some of these inclusions are segregations of early-formed minerals; others are reconstituted xenoliths. They consist of the same minerals as the enclosing rock, but in different proportions; in particular, their content of mafic minerals, plagioclase, apatite, sphene, and magnetite tends to be higher (Figure 5-6C).

The marginal facies of some plutons have been sheared during emplacement, producing what has been called *protoclastic* texture in peripheral belts of gneissose granitic rock. In the earlier stages of deformation grains of quartz, and to a lesser degree those of feldspar, develop undulatory extinction—the optical expression of severe internal strain. Individual grains may begin to develop ovoid outlines, and in advanced stages of deformation, phenocrysts of K-feldspar become worn down to lenticular *augen* that swim in a matrix of finely recrystallized quartz-feldspar "debris." Much of this fine-grained material, once generally referred to as "granulated" or "comminuted" (as if the product of brittle crushing), is now thought to result from recrystallization (annealing) of strained, still-heated crystals to finer-grained aggregates of unstrained (and hence more stable) grains. The annealing process, by analogy with experimental annealing of metals (and now of some rocks), is driven by the strain energy thus released.

Finally, a word should be said as to the significance of porphyritic structure of some acid plutonic rocks with respect to environment of emplacement and so to rock terminology. Rocks of small, relatively shallow plutons are commonly porphyritic with finely granular or microgranular groundmass; but mineralogically they are identical with coarse-grained equivalent plutonic rocks. The phenocrysts may be numerous or sparse, and they may include quartz along with feldspars and mafic minerals, or even quartz alone. These rocks are designated by the name *porphyry*; typical examples are illustrated in Figure 5-2. Intrusive porphyries have affinities with both plutonic and volcanic rocks, and unless field relationships are well established, it can be a matter of choice whether a plutonic or volcanic rock name (e.g., granite porphyry or rhyolite porphyry) might be the more appropriate.

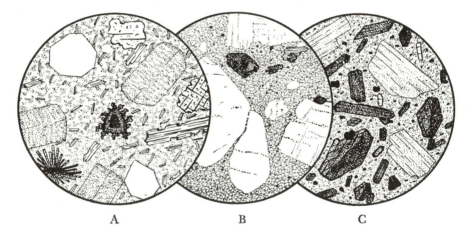

A   B   C

Figure 5-2. Porphyries

A. Pneumatolyzed granite porphyry, Cornwall, England. Diam. 5 mm. Euhedral phenocrysts of quartz and altered perthite in a microgranular groundmass of the same minerals accompanied by abundant muscovite, topaz (near top), fluorite (right edge), and two generations of tourmaline.

B. Granodiorite porphyry, Paiyenssu, northwestern Yunnan, China. Diam. 3 mm. Large crystals of quartz and calcic oligoclase, with smaller ones of hornblende and biotite, in a microgranular matrix of quartz and alkali feldspar with accessory sphene and epidote.

C. Hornblende diorite porphyry, Carrizo Mountain laccolith, northeastern Arizona. Diam. 3 mm. Phenocrysts of andesine, partly altered to calcite and clay minerals, and of green hornblende, some of which are twinned on the front pinacoid. The groundmass consists chiefly of microgranular feldspar with minor quartz and accessory grains of apatite and zircon. This rock might also be called *andesite porphyry*.

Granites: General Mineralogy

More than a third of the feldspar in granites is alkali feldspar, typically orthoclase or microcline. However, as in syenites, the initial high-temperature alkali feldspar usually contains appreciable sodium, and exsolution of a sodic phase during the slow cooling process that occurs in plutonic masses leads to the widespread formation of perthite. Perthitic intergrowths are often rendered conspicuous under the microscope in ordinary light by selective alteration of their potassic and sodic phases. For instance, either the sodic or the potassic phase may have been partially altered to powdery material—milky by reflected light, pale buff in transmitted light—while the other phase remains fresh or shows a different style of alteration. Plagioclase, except in peralkaline granites, is seldom antiperthitic. Discrete crystals of albite, usually euhedral, likewise appear only in highly alkaline granites—and in some of these (e.g., the albite granites of Oregon) the albite may be essentially a metamorphic product as in associated keratophyres and spilites. The plagioclase of most granites is oligoclase, in some instances enclosing cores of andesine.

Quartz makes up about 20–40% of most granites. Only in granite porphyries and in some granophyres does it form phenocrysts; normally it is anhedral, occupying irregular spaces between the other constituents, though even then one may sometimes detect a preferred orientation of the crystal axes. The quartz may contain gas or liquid inclusions; far more often it carries abundant dusty and acicular inclusions, as of rutile and apatite. Undulatory extinction resulting from strain is particularly prevalent in granites taken from the marginal parts of plutons.

In normal or calcalkaline granites the characteristic mafic mineral is dark-brown biotite, the iron content of which tends to increase with the silica content of the enclosing rock. Inclusions of apatite, zircon, allanite, monazite, and iron-titanium oxides are common, and around those that are radioactive, especially zircon, there may be pleochroic haloes. The typical alteration of biotite is to chlorite and granular sphene.

Muscovite and lithium micas are confined to peraluminous and at the same time highly siliceous varieties of granite—especially granite pegmatites and aplites and rocks affected by pneumatolysis (see p. 172). The molar ratio $Al_2O_3/(K_2O + Na_2O)$ is much higher for muscovite than for alkali feldspar—3 compared to 1; the ratio for lepidolite is somewhere in between. So these two micas are diagnostic indices of peraluminous rock composition, which is expressed by the appearance of corundum (C) in the norm.

In calcalkaline granites the usual amphibole is green hornblende in stout subhedral prisms. It generally increases in amount at the expense of biotite as the content of plagioclase increases. In alkali granites, in contrast, the typical amphiboles are hastingsite, riebeckite, and arfvedsonite, and these are usually anhedral.

Pyroxenes are rare in calcalkaline granites, and most of them are relict cores of diopsidic augite enclosed by reaction rims of hornblende. In alkali granites, late-crystallizing aegirine-augite and aegirine are common, as acicular crystals, as spongy aggregates partly enclosing quartz and feldspar, or filling angular interspaces. In some basic granophyres the pyroxene is hedenbergite, and in these, and also in acid granophyres and pegmatites, there may be a little iron-rich olivine.

Accessory minerals always present in granites include apatite, sphene, zircon, and magnetite. The first three are usually more abundant in alkali granites than in granites of adamellite type; and the first two are particularly abundant in granites contaminated by assimilation of basic rocks. Epidote, zoisite, clinozoisite, and allanite, though generally secondary, may be primary constituents. In the granites of Victoria, Australia, allanite is especially common where abundant xenoliths and basic clots suggest contamination.

A number of other minerals, much more commonly found in contact-metamorphosed shales than in acid plutonic rocks, nevertheless may appear as accessory constituents of certain granites. Typical of these are almandine garnet, cordierite, andalusite, and sillimanite. In some granitic plutons these minerals have been shown by careful field and petrographic observation to be xenocrysts set free by mechanical dismemberment of pelitic hornfels during the course of partial assimilative melting of adjacent metasediments. In others—for example, certain granites of Victoria, Australia—large euhedral or subhedral crystals of cordierite have crystallized directly from magma that became enriched in Al_2O_3 by assimilation of pelitic rocks; here cordierite crystallized relatively early and later became embayed by quartz and alkali feldspar. Spessartine-almandine is widespread as an accessory primary mineral—in small euhedral crystals—in certain granitic pegmatites and aplites. Almandine of the Dartmoor granites, Cornwall, may even form graphic intergrowths in phenocrysts of feldspar. Elsewhere almandine is clearly of xenocrystic origin. Andalusite or sillimanite occasionally appear as what seem to be primary accessories in certain peraluminous muscovite-bearing pegmatites and aplites. In other granites the same phases, accompanied by topaz or tourmaline, appear to be products of late pneumatolytic reactions within the already solid rock. In others again both andalusite and sillimanite are xenocrysts derived from incorporated fragments of hornfels. Green spinel, which occurs sporadically in some granites, likewise appears to be a by-product of assimilation of aluminous xenoliths.

Granites: Principal Types

Alkali Granites

All those granites in which alkali feldspar greatly preponderates over plagioclase are grouped as alkali granites. Chemically they range from peraluminous muscovite-bearing varieties to peralkaline granites in which all the mafic minerals are sodic.

The majority of alkali granites are micaceous. Strictly peraluminous varieties typically carry both muscovite and biotite; many also contain minor amounts of sillimanite, andalusite, cordierite, almandine, topaz, or tourmaline as well. Biotite granites (lacking muscovite and any of the other preceding minerals) are much more widespread and indeed are the commonest type of alkali granite. The Conway granite of New Hampshire (Figure 5-6A) contains about 70% microperthite and sodic orthoclase, in subhedral-to-anhedral oblong grains, and 20% quartz; other constituents are oligoclase (chiefly as cores inside the alkali feldspars), biotite, and the usual accessories. The Pikes Peak granite of Colorado is characterized by large phenocrysts of red microcline in a matrix composed principally of quartz, oligoclase, and abundant biotite, with accessory allanite, sphene, apatite, magnetite, and fluorite. In some alkali granites, biotite is accompanied by hornblende (Figure 5-3A) or augite; but most hornblende granites are more basic and belong to the adamellite group to be described shortly. The example illustrated in Figure 5-3A is transitional to quartz syenite; this, in fact, is a natural field transition.

In a very few granites both biotite and hornblende are lacking, the mafic constituents being ortho- and clinopyroxenes, occasionally accompanied by a little fayalitic olivine. Hypersthene granites with both orthoclase and andesine and carrying 20−30% quartz occur in the Blue Ridge batholith of Virginia. Hypersthene granites are acid members of the *charnockite* family—rocks typical of Archaean and earlier Proterozoic sectors of the major continental cratons, notably southern India and neighboring Sri Lanka, western Australia, southern Brazil (around Rio de

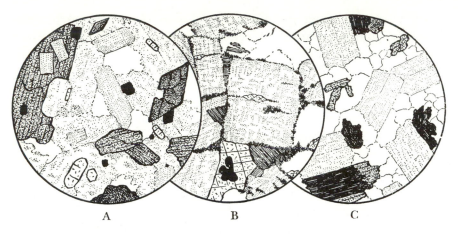

Figure 5-3. Granites

A. Hornblende "granite," Plauen, near Dresden, Saxony. Diam. 3 mm. Composed of green hornblende, orthoclase, oligoclase, and quartz, with accessory magnetite, apatite, sphene, and allanite. Note that some of the oligoclase is enclosed poikilitically by hornblende and orthoclase, and, left of center, there is a little myrmekite at the contact between two orthoclase crystals. With decreasing quartz, the rock grades into syenite.

B. Biotite granite, Rockport, Maine. Diam. 3 mm. Euhedral and subhedral crystals of microcline-perthite; strained anhedral crystals of quartz. Two generations of biotite; the earlier in large flakes; the later in radiating tufts occupying cracks and veins. The later biotite is darker and richer in iron and is associated with pneumatolytic fluorite.

C. Peralkaline riebeckite-aegirine granite, Quincy, Massachusetts. Diam. 3 mm. Euhedral and subhedral crystals of microperthite, and anhedral quartz; dark constituents are riebeckite, aegirine, and allanite.

Janeiro), and parts of west Africa. The mode of charnockite from Madras, India, is given as microcline, 48; quartz, 40; oligoclase, 6; hypersthene, 3; magnetite, 2; biotite, 1. Charnockites habitually occur in terranes of high-grade regional metamorphism dominated by rocks of the granulite facies (cf. p. 568 and Figure 20-8C). So arises this question: Could at least some charnockites themselves be metamorphic, completely dehydrated derivatives of acid volcanic rocks or of quartzo-feldspathic sediments? This question will be discussed more fully later (p. 184). In the meantime we note that most petrologists today agree that at least some charnockites are primary plutonic rocks.

Peralkaline granites are marked by amphiboles and pyroxenes rich in sodium and iron, and some also by much albite or antiperthite. The

Quincy granite of Massachusetts is an excellent example, though highly variable in mineral content (Figure 5-3C). In an average sample, a little more than half consists of sodipotassic feldspar, mainly perthite, and quartz makes up about a third. The mafic minerals are aegirine and riebeckite; common accessories include apatite, sphene, zircon, iron oxide, and astrophyllite. Cutting this granite is another peralkaline rock marked by microperthite, albite, and quartz, in which the principal mafic constituents are olive-green kataphorite and deep-brown biotite. Other peralkaline granites occur in the Kudaru Hills of Nigeria. Most of these are composed of sodic microperthite (58%), quartz (38%), and riebeckite (4%), though some carry a little iron-rich biotite and fayalite as well. In still other peralkaline granites, such as are found in the Oslo region of Norway, the characteristic dark minerals are arfvedsonite and aegirine.

Some of these peralkaline granites carry fine-grained inclusions much richer in mafic minerals than the host rocks. The coarse aegirine-riebeckite granite of the Atlantic island of Rockall, for instance, has abundant dark inclusions whose constituents, in order of abundance, are dark-green aegirine, quartz, albite, and microcline. Like many other sodic igneous rocks they are relatively rich in zirconium, chiefly in the mineral elpidite. Peralkaline granite porphyries are illustrated in Figure 5-4.

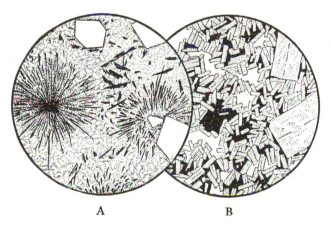

A   B

Figure 5-4. Peralkaline Granite Porphyry

A. Riebeckite granite porphyry, Lake Brunner, New Zealand. Diam. 3 mm. Phenocrysts of quartz and sodic orthoclase (latter not shown), in a graphic groundmass of the same two minerals accompanied by acicular riebeckite.

B. Riebeckite granite porphyry, Ailsa Craig, Scotland. Diam. 2 mm. Essentially composed of sodic orthoclase with interstitial riebeckite and quartz.

Adamellites

In adamellites plagioclase and alkali feldspar are present in subequal amount (each between 35 and 65% of the total feldspar). Thus, they have more alkali feldspar than do granodiorites into which they grade imperceptibly in nature— as in the great Sierra Nevada batholith of California. Adamellites have more quartz than monzonites; but natural gradation between the two is uncommon. Biotite and hornblende are the typical mafic components, for adamellites, like granodiorites and many monzonites, are typical metaluminous rocks.

The well-known Shap Granite of Westmorland, England, is a good example of adamellite (Figure 4-5C). Large pink phenocrysts of orthoclase make up about a third of the bulk; white oligoclase forms another third, and quartz about a quarter. The chief mafic mineral is biotite; then follows hornblende. Zircon, apatite, and iron oxides are the normal accessories. An interesting feature of this adamellite, especially in its early basic phases is the presence of oligoclase mantles around some of the orthoclase phenocrysts. The resultant rapakivi texture is attributed to basification of partly crystallized magma by reaction with andesitic wall rocks.

The Mesozoic Sierra Nevada batholith of California, over 500 km long by 50 km to 80 km wide, is a composite body made up of perhaps 200 separate plutons.[1] The great bulk of these consist of adamellite or of granodiorite; and between these types there is complete natural chemical and mineralogical gradation. Typical adamellites (e.g., the Half Dome and Cathedral Peak plutons) have a content of SiO_2 of about 72% to 74% and color indices around 10 or less. Quartz is uniformly abundant (normative Q ~30). Either alkali feldspar or plagioclase may predominate slightly over the other. The former, pink in hand specimen, is usually coarser in grain and in some rocks (e.g., the Cathedral Peak pluton) occurs in coarse, euhedral, internally zoned phenocrysts that poikilitically enclose small crystals of plagioclase and mafic constituents. Plagioclase, gray or white in hand specimen, is oligoclase typically in smaller, normally zoned, subhedral grains. Biotite is by far the most abundant mafic phase. Hornblende, where present, is much subordinate. Muscovite invariably is absent. Common accessories are sphene, iron-titanium oxides, apatite, zircon, and allanite.

Adamellites (and granodiorites, too), like alkali granites, include peraluminous varieties carrying minor amounts of andalusite (or sillimanite), cordierite, or almandine garnet in addition to more abundant biotite or hornblende, or both. One such rock is a cordierite-bearing porphyritic adamellite that makes up a significant portion of the composite Clouds Creek pluton in the Appalachian Piedmont of South Carolina.[2] An approximate average mode, neglecting deuteric alteration products, is quartz, 25; plagioclase (with oscillatory zoning An_{35} to An_{10}), 40; microperthite, 20; biotite, 10–12; cordierite, ~3; plus accessory ilmenite, sphene, and tourmaline. The cordierite takes the form of subhedral prisms up to 3 mm long, largely free of inclusions and showing sector twinning.

Granophyres

Granophyres are volumetrically insignificant, texturally and mineralogically distinctive members of the granite family whose mode of occurrence throws significant light on the process of differentiation. Most are sodic rather than potassic rocks with an unusually high FeO/(FeO + MgO) ratio. Quartz, one of the last minerals to crystallize, for the most part is micrographically intergrown with alkali feldspar (*granophyric texture*)—possibly reflecting cotectic or eutectic crystallization of the final residual melt. That the parent melt is indeed of such an origin is indicated too by the highly ferric nature of the mafic phases—hedenbergite, fayalite or, in a few peralkaline rocks, riebeckite.

Granophyres are commonly found as minor streaks and lenses in the upper levels of thick differentiated sills of tholeiitic diabase. The largest recorded bodies are *ferrogranophyres* developed near the roof regions in major layered basic plutons of the Skaergaard type (p. 117). In the Skaergaard pluton itself, ferrogranophyres consist mainly of sodic plagioclase that passes marginally into cloudy potassic feldspar micrographically intergrown with quartz. The mafic constituents are hedenbergite and fayalitic olivine (in some specimens with a little hornblende) and opaque iron oxides.*

Some granophyres are not strictly acid rocks: SiO_2 may be anywhere between 64% and 72%. There can be little doubt that where they are components of basic sills and layered plutons granophyres represent

---

*Some granitic plutons have local border variants with granophyric texture. These have commonly been called granophyres, although a preferable name for them is *micrographic granite*. Such rocks are chemically and genetically distinct from the granophyres described here.

residual melts developed in the final stages of fractional crystallization of parental tholeiitic magmas.

Deuterically Altered Granites

During an early stage of postmagmatic cooling many granites—still hot but now essentially crystalline—become altered by residual aqueous supercritical fluids that continue to rise through the intergranular pore spaces. As a result of reactions that thus ensue,* earlier crystalline constituents become partially replaced by minerals such as tourmaline, topaz, fluorite, muscovite, lithium mica, cassiterite, and wolframite. Some of the tourmaline found in granites as an accessory phase crystallized during the magmatic stage; but where it is plentiful most of it certainly developed later by attack of borofluoric fluids upon feldspars, a process that was often facilitated by prior shattering of the primary minerals. At the same time apatite, biotite, sphene, and ilmenite were partly or wholly destroyed, and their constituent elements contributed to the formation of more tourmaline and of anatase and brookite.

No tourmalinized granite is better known than that of St. Austell, Cornwall, illustrated in Figure 5-5A. Here one sees two generations of tourmaline, an older yellow variety of primary magmatic origin, and a younger greenish variety forming conical and spherulitic clusters of slender euhedral prisms. Many of these beautiful radiating aggregates ("tourmaline suns") grow from the edges of feldspars; others are embedded in quartz mosaics. And in some instances the diverging rays end abruptly along straight lines in the quartz, wholly unrelated to the present quartz boundaries. These "ghost boundaries" of the tourmaline rays mark the faces of vanished crystals of feldspar and quartz.

Development of tourmaline may be contemporaneous with that of topaz, cassiterite, and wolframite, or there may be an overlapping sequence, topaz forming first, then tourmaline, and the tin-tungsten minerals finally replacing both.

Other pneumatolyzed rocks that occur mainly near the margins of granite plutons are *greisens* (Figure 5-5B,C). They form bands and vein-like bodies, a few centimeters or meters in width, with indefinite margins

---

*Generally termed *pneumatolytic* since there can be no doubt that the active fluids are in the supercritical (gaseous) state.

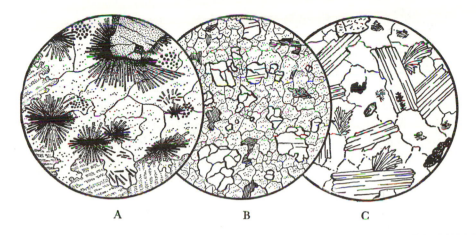

Figure 5-5. Pneumatolyzed Granites

A. Tourmalinized granite, Cornwall, England. Diam. 3 mm. Clusters of radiating blusih-green tourmaline needles, some of them bordering a corroded phenocryst of primary brown tourmaline. The remainder of the rock consists of microperthite and quartz, the latter invading the former. At the upper right are several tourmaline needles that terminate against a ghost boundary which marks the edge of a vanished quartz or feldspar crystal.

B. Greisen, Geyer, Erzgebirge, Germany. Diam. 5 mm. Composed of topaz, lithium mica, and dusty quartz.

C. Greisen, Grainsgill, Cumberland, England. Diam. 3 mm. Composed essentially of quartz and muscovite, with accessory rutile, apatite, and arsenopyrite. The large flakes of muscovite are relics from the original granite; the plumose muscovite is secondary after orthoclase; the minute, densely packed scales of muscovite are secondary after plagioclase. Other accessory minerals in this rock, not shown, are tourmaline and molybdenite.

that grade into unaltered granite. They were doubtless formed by flow of mineralizing fluids through fissures. Most of them are almost pure quartz-mica rocks, the feldspars of the original granites having been replaced completely by muscovite or lithium mica. Topaz is almost always present and occasionally may be the predominant constituent. Accessory minerals include tourmaline, fluorite, apatite, rutile, cassiterite, and wolframite.

Under yet other physical conditions pneumatolysis yields *china clay* by converting feldspar to kaolinite, muscovite, and quartz under the influence of fluids rich in H_2O, CO_2, and in some cases, F and B.

Granodiorites and Tonalites

Almost equal in abundance to granites are acid plutonic rocks in which alkali feldspar is greatly subordinate to plagioclase. These are grano-diorites and tonalites, mutually gradational rocks arbitrarily distin-guished by the content of alkali feldspar (<10% of total feldspar in ton-alites). With increase in proportion of alkali feldspar, granodiorite is transitional into granite of the adamellite type.

Granodiorites

Much of the huge Sierra Nevada composite batholith of California is granodiorite (Figure 5-6B). In fact, its prevalence there led W. Lindgren to introduce the name granodiorite for rocks intermediate between gran-ite and quartz diorite. According to him, the average mode of grano-diorite is as follows (weight percentages): sodic andesine, 40; quartz, 21;

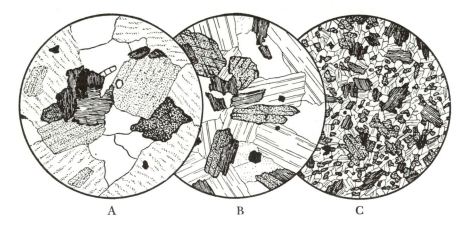

A   B   C

Figure 5-6. Granite and Granodiorites

A. Biotite granite, Conway, New Hampshire. Diam. 3 mm. The feldspars are microperthite and altered oligoclase; quartz is anhedral. Dark minerals are biotite, allanite, and a little magnetite. Two crystals of apatite near center.

B. Hornblende-biotite granodiorite, Yosemite, California. Diam. 3 mm. Approximately half the rock consists of normally zoned plagioclase (An_{50-20}), and a quarter of quartz. The remainder is composed of perthite, hornblende, and biotite, with accessory magnetite.

C. Basic inclusion in granodiorite from the same locality. Diam. 3 mm. Richer in hornblende, biotite, plagioclase, sphene, and apatite, but poorer in quartz and potassic feldspar than the enclosing rock.

orthoclase, 18; hornblende, 17; accessory biotite, iron-titanium oxides, apatite, and sphene, 4. In many granodiorites, however, biotite is about as abundant as hornblende, and in some it may be even more abundant. Augite may also be present in minor amount.

The plagioclase crystals in granodiorites are generally euhedral or subhedral except where orthoclase is sparse, and some of them are partly or wholly enclosed by orthoclase. Normal zoning is common, from cores approximating An_{50} to rims of about An_{25}. The orthoclase seldom forms phenocrysts and is perthitic more often than in tonalites, but less so than in granites and syenites, and it may be proxied by microcline.

In a segment of the Sierra Nevada batholith near Bishop, California, most of the component plutons are composed either of adamellite or of granodiorite;[3] and there is complete compositional gradation between the two arbitrarily distinguished rock types. Average modes of two extensive granodioritic plutons in this region are as follows:

Lamarck pluton: quartz, 24; alkali feldspar, 21.5; plagioclase, 44.5; biotite, 6.5; hornblende, 2.5; accessories, 1. This rock, it will be noted, is close to the granodiorite-adamellite boundary.

Round Valley Peak pluton: quartz, 26; alkali feldspar, 15; plagioclase, 46.5; biotite, 7; hornblende, 3.5; accessories, 2. The mode is well within the granodiorite range.

Both are acid rocks, according to the modal-quartz parameter (>20). But while the first rock has $SiO_2 = 67\%$ (normative Q, 23), the second has $SiO_2 = 63.5\%$ (normative Q, 16)—the silica percentage being well below the limit for rocks of the acid category as chemically defined—a pattern that is common enough in Sierran granodiorites. Adamellites from the same region by any criterion are acid rocks ($SiO_2 = 71–74\%$; Q, 25–30).

Tonalites

Tonalites resemble granodiorites except for significantly lower alkali feldspar and somewhat lower quartz in the mode. They tend to be more mafic than granodiorites (color index commonly 20 to 40); but there are felsic varieties with color indices as low as 5 or 10. The chief mafic minerals of tonalite are hornblende and biotite; augite, when present, is usually rimmed with one or the other of these minerals. The type rock, from Adamello in the Tyrol, is illustrated in Figure 5-7A; the approximate mode is quartz 15–20; alkali feldspar, 5; normally zoned plagio-

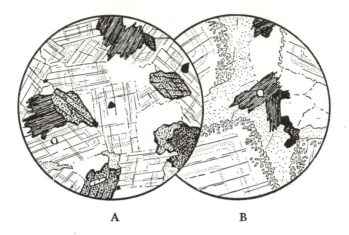

A   B

Figure 5-7. Tonalites

A. Tonalite, Adamello, Italy. Diam. 2.5 mm. Subhedral and euhedral zoned crystals of andesine-oligoclase, locally rimmed with orthoclase; anhedral patches of quartz; green hornblende and brown biotite; allanite partly fringed with epidote (lower right); accessory magnetite, apatite, and sphene.

B. Felsic tonalite (trondhjemite), Castle Towers batholith, British Columbia. Diam. 2.5 mm. Main constituent is oligoclase showing oscillatory zoning and borders of myrmekite; next in abundance is quartz, then orthoclase. Accessory constituents are biotite, apatite, iron oxide, and sphene.

clase (An_{63-41}), ~50–55; mafic minerals (biotite and hornblende), ~25; accessory apatite, sphene, and opaque oxides.

The most felsic variety of tonalite is *trondhjemite* (Figure 5-7B), a rock consisting principally of plagioclase (sodic andesine or oligoclase) and quartz, with K-feldspar and biotite as minor constituents. The color index is less than 10. An extensive pluton of this type is exposed as conspicuous white outcrops along the near-vertical glaciated walls of the northern arm of Lake Manapouri, southwestern New Zealand. The average mode is approximately quartz, 15–40; oligoclase, 55–80; microcline, <5; biotite, 3–5; with accessory epidote (with allanite cores), sphene, apatite, and rarely zircon. The microcline occurs mainly as grains of jagged outline between or enclosed within the larger subhedral crystals of plagioclase. In many rocks round grains of quartz are sparsely enclosed in the plagioclase crystals. The Manapouri trondhjemite probably represents a mobilized melt fraction separated from adjoining amphibolite migmatites.

Acid Aplites and Pegmatites

Many granodiorites and granites, together with their walls, are cut by light-colored rocks richer in quartz, alkali feldspar, muscovite, and pneumatolytic minerals than the associated plutons. These light-colored rocks develop from residual melts derived from the magmas that produced the plutons they transect, and hence are composed chiefly of the minerals that crystallized last in the interstices of the parent bodies. Some of these rocks, the aplites, are characterized by an even grain size that rarely exceeds 2 mm (Figure 2-6C). In hand specimens they appear "sugary." Contrasted with these are the pegmatites Figure 5-8), which vary much more in texture and coarseness. Some are as fine-grained as aplites; most are coarser, and a few carry crystals tens of centimeters or even meters in length.

Most aplite bodies are only a few centimeters up to a meter in width, but pegmatites range up to hundreds of meters across, and many are of

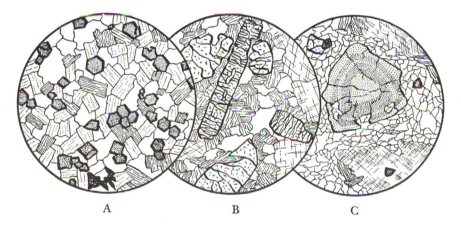

A   B   C

Figure 5-8. Granite Pegmatites

A. Garnetiferous fine-grained pegmatite, Pala, California. Diam. 2 mm. Composed of spessartine, lithium mica, albite, microcline, quartz, and a little deep-blue tourmaline.

B. Tourmaline pegmatite, Pala, California. Diam. 2 mm. Large crystals of colorless elbaite, scattered in a matrix of lithium mica, albite, and quartz.

C. Tourmalinized pegmatite, Tuolumne Canyon, Yosemite, California. Diam. 2 mm. Large crystal of zoned blue tourmaline; abundant granulated quartz and strained microcline; accessory muscovite and spessartine.

extremely irregular outline. Most pegmatites, moreover, contain symmetrical zones that differ in mineral composition, a feature seldom seen
in aplites; and they are generally much richer in rare minerals. Yet the
two rocks are found in intimate association, cutting each other or forming lenticular masses one within the other. Except for the greater abundance of pneumatolytic minerals in pegmatites, both rocks have similar
mineral composition but pegmatites vary more and tend to show more
extensive recrystallization and replacement of older by younger minerals. Some coarse pegmatities, indeed, appear to have been produced
by replacement of fine-grained aplites. Both rocks may occupy sharply
defined fractures, and both, but particularly pegmatites, may form
replacement bodies with indefinite margins. In both, the effects of strain
and crushing are usually exhibited by undulatory extinction of the
quartz, by curvature and faulting of the twin lamellae of the feldspars,
by crumpling of the micas, and by granulation or annealing of crystal
borders.

As we have noted, most aplites and pegmatites are composed principally of quartz, alkali feldspar (chiefly perthite, microcline-perthite,
microline, and sodic plagioclase), and muscovite. In aplites, the accessory
minerals include almandine and spessartine garnets, zircon, tourmaline,
topaz, lepidolite, spodumene, epidote, and allanite. Replacement of
potassic by sodic feldspars, though less widespread than in pegmatites,
is not uncommon. For pegmatites, the list of accessory minerals is very
much longer and some of them are very abundant (e.g., Figure 5-8).
Residual aqueous fluids in this case evidently become enriched, at various
stages, not only in boron, fluorine, and chlorine but also in elements such
as tin, tungsten, and beryllium otherwise present in granitic melts in only
trace concentrations. Some pegmatites therefore contain, in addition to
the common aplitic minerals just cited, such minerals as beryl, apatite,
eudialite, amblygonite, fluorite, wolframite, cassiterite, columbite, tan-
talite, and lithiophilite. Most of these are to be found in pegmatites transecting schists, gneisses, and quartzites, rather than in those cutting the
parent plutons.

In zoned granitic pegmatites the inward sequence of mineral zones
usually includes some combination of the following: (1) margins of plagioclase, quartz, and muscovite; (2) plagioclase, quartz; (3) quartz, perthite, plagioclase ± muscovite or biotite or both; (4) perthite, quartz; (5)
perthite, quartz, plagioclase amblygonite, spodumene; (6) plagioclase,
quartz, spodumene; (7) quartz, spodumene; (8) lepidolite, quartz; (9)
quartz, microcline; (10) microcline, plagioclase, lithium mica, quartz; (11)

quartz. In zones 1–3 the plagioclase generally ranges from andesine to median albite; in zones 5–10 it is more sodic albite. Crystallization of the successive zones proceeded inward from the walls, and potassic feldspars were generally replaced by sodic ones, with consequent formation of muscovite. Then garnets and tourmaline developed from the feldspars, and finally lithium minerals were produced. There are, however, many exceptions to this usual sequence; beryl, garnet, and tourmaline may crystallize early or by replacement of other minerals, or they may develop at various stages, and albite may crystallize before or after the lithium-bearing minerals. The borders of many pegmatites consist of graphic intergrowths of perthitic microcline and quartz, while the minerals of the interiors originated by their replacement.

The extraordinarily complex mineralogical and structural patterns displayed by zoned pegmatites reflect an unusually complex crystallization history that started with a silicate melt unusually rich in water and other fugitive components. If micas fail to crystallize in sufficiently large amount, concentration of water builds up to the saturation point and boiling ensues. Three sets of phases may now exist in mutual equilibrium—crystalline solids, water-saturated silicate melt, and a separate gaseous (supercritical) aqueous phase. Still later in the crystallization history, as temperatures drop below the critical level, the aqueous fluid itself will divide into a liquid and a gaseous phase. From now on solids, silicate melt, aqueous liquid, and the gas phase could coexist, each with its individual capacity to take up selectively the initially minor elements in the parent melt. This type of behavior has been experimentally investigated in the analogous but far simpler system $NaAlSi_3O_8$–H_2O.

Andalusite, sillimanite, kyanite, and corundum are not uncommon in pegmatites, where they are believed to have been formed in part by primary crystallization from highly aluminous fluids, and in part by reaction with aluminous wall rocks and xenoliths. Occasionally the mineral content of pegmatites varies according to the character of the wall rocks, but far more often there is no correlation.

Petrographic Interrelations Between Acid Plutonic Rocks and Contiguous External Rocks

Note on "Granitization"

The famous "granite controversy" of the 1940s and 1950s was essentially a prolonged debate on two diametrically opposed models of the genesis of large bodies of "granitic" (i.e., acid plutonic) rocks. Opposed to the orthodox view that such rocks are truly igneous—that is, of magmatic origin—was the alternative view eloquently argued by the most-informed student of "granites" in his time, H. H. Read, that many, perhaps most, acid plutons are products of metasomatism of preexisting rocks, the remnants of which now constitute the envelope of the pluton. The evidence on either side was the same set of facts: field relations at plutonic contacts and petrographic and chemical data obtained from rocks in the contact zone. We are here concerned with the petrographic data, in a way the most important type of evidence. But to avoid repeated qualification it will be presented from the "magmatist" viewpoint, which differs from that of the "granitizer" by invoking the presence in large volumes of a silicate-melt phase (magma) of acid composition. The magmatic model is strongly upheld by the present writers, and it has indeed now been accepted by the majority of petrologists familiar with plutonic and metamorphic phenomena.*

Contaminated Acid Plutonic Rocks

At well-exposed contacts between acid plutons and enveloping rocks, the plutonic rock very commonly is seen to have become strongly modified, presumably by chemical and mineralogical reaction between magma and wall rock. Among the countless examples of such reaction we select for illustration one from Lake Manapouri, New Zealand (Figure 5-9). Here the wall rocks are metagabbros that still retain a gabbroic texture and are composed chiefly of sodic andesine and green hornblende pseudomorphs after pyroxene. Quartz is a minor constituent, and biotite is present in even smaller amount. Apatite, iron-titanium oxides, epidote, finely divided white mica, and chlorite make up the remainder. Progressive alteration of this rock is depicted in Figure 5-9B,C. It involves change of andesine to oligoclase with resultant separation of epidote, replacement of hornblende by biotite, conversion of opaque oxides to sphene, formation of abundant apatite, introduction of quartz, and, in advanced stages, development of K-feldspar. These changes were brought about by introduction into the metagabbro of silica, alkalies, titanium, and phosphorus. The end product, represented by rocks within a few centimeters of the contact and by xenoliths within the granite, is a dark, fine-

---

*This broad consensus is, of course, no proof of the validity of our preferred model.

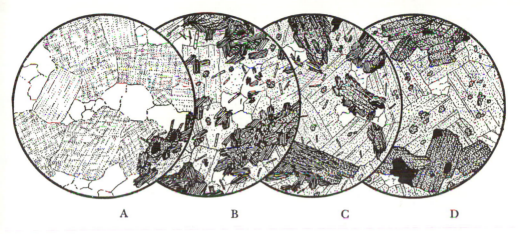

Figure 5-9. Granite-Gabbro Reaction Series, Lake Manapouri, New Zealand

A. Granite. diam. 3 mm. Composed mainly of microcline-perthite, quartz, albite, and biotite. The dark clot is a gabbro relic now composed of biotite, sphene-rimmed opaque oxide, and acicular apatite.

B. Transitional rock. Diam. 3 mm. The constituents, in order of abundance, are oligoclase, biotite, orthoclase, hornblende, quartz, sphene, apatite, epidote, and iron oxide. In this specimen most of the hornblende of the original gabbro has been replaced by biotite.

C. Transitional rock, nearer the gabbro contact. Diam. 3 mm. Chiefly andesine and hornblende, the latter in process of replacement by biotite. Iron oxide partly replaced by sphene, abundant apatite, and a little quartz and epidote.

D. Metagabbro. Diam. 3 mm. Least-altered material. Only difference from unaltered gabbro is the presence of a little introduced quartz. Bulk of rock consists of andesine and hornblende, with accessory epidote, sphene, white mica, chlorite, and opaque oxide.

grained rock like the parent metagabbro, but composed of minerals similar to those of the adjoining granite, though in very different proportions. The granite, also, has been affected by the contamination process. Its chief constituents are microcline-perthite, albite, and quartz of magmatic origin. But in the contaminated zone so much potassium has been withdrawn into the metagabbro to convert hornblende to biotite that the K-feldspar has become greatly subordinate to albite. Even at distances of several hundred meters from the contact, the dark constituents of the granite are mainly aggregates of biotite, granular sphene, and apatite, the texture of which indicates that they are microxenoliths derived from the metagabbro.

On the island of Alderney, in the English Channel, a hornblende-biotite granite contains xenoliths of diabase that show various stages of alteration. The least-modified inclusions are fine-grained and consist of zoned laths of plagioclase (An_{30-10}), hornblende, biotite, magnetite, and apatite. When these inclusions reacted with the granitic magma, more and more "phenocrysts," or porphyroblasts, of plagioclase (An_{10}), like those of the granite itself, were formed, and crystals of hornblende and biotite became more numerous; at the same time the groundmass became coarser, and patches of quartz and micropegmatite developed within it. At this stage the xenoliths had become clots of quartz diorite. By still further reaction the grain size of all the constituents increased until it was nearly the same as in the granite itself; then the inclusions scarcely differed from the enclosing rock except in containing somewhat more plagioclase, hornblende, and biotite.

Among the adjacent granites of Brittany there are fragments of older gabbro that also reveal all stages of replacement. The olivine, pyroxene, and calcic plagioclase of the gabbro were partly or wholly changed to later members of the reaction series then separating from the granitic magma—namely, hornblende, biotite, alkali feldspars, and quartz. When reaction was complete, all signs of the xenoliths had disappeared.

An excellent example of granitic rock contaminated by inclusion of sedimentary wall rock is the tonalite of Loch Awe, Scotland, described by S. R. Nockolds. The normal tonalite at this locality consists principally of oligoclase, microcline-perthite, quartz, biotite, hornblende, and diopsidic augite. It contains numerous xenoliths of argillaceous material, now composed chiefly of cordierite, spinel, corundum, and labradorite. The tonalite close to these is modified in several respects; its plagioclase is more calcic than usual; it has more biotite, but less microperthite and quartz; and it contains apatite and pyrite, but no hornblende, pyroxene, or sphene. Apparently the tonalite magma absorbed some Al_2O_3, MgO, and FeO from the xenoliths and, in exchange, enriched them somewhat in K_2O and SiO_2. The tonalite also contains inclusions of argillocalcareous material, now represented mainly by bits of diopside-hypersthene-plagioclase hornfels. Close to these, also, the tonalite is modified: its amphibole is colorless, and it contains very little augite or microperthite but more than the usual amount of magnetite, pyrite, and zircon. Besides, its plagioclase is more calcic. Greater chemical changes were involved in this reaction. Iron and magnesium were gained by the tonalite magma while some potassium and aluminum were lost. The tonalite also carries xenoliths of impure calcareous material, now represented by clots of diopside and plagioclase. Around these the tonalite consists essentially of diopside, plagioclase, hornblende, and quartz; here the magma gained some CaO and MgO and lost some Al_2O_3 and Na_2O.

In each of the foregoing reactions, entry of material from the magma into the xenoliths produced minerals identical with those in the surrounding tonalite. Clearly, the function of the reactions was not to change the chemical composition of the xenoliths to conform with that of the magma but to produce solid phases stable in contact with the magma at the time of incorporation, as predicted on the basis of experiment by N. L. Bowen.

At granite-limestone contacts, SiO_2 and Al_2O_3 are withdrawn from the magma to convert calcite and dolomite to silicates such as diopside and grossular. Here the contaminated granite may become dioritic; or in some cases a marginal variant rich in K-feldspar and containing sodic pyroxene develops.

Many of the dark basic clots seen in granites and granodiorites, such as those in the marginal zones of individual plutons in the Sierra Nevada batholith, are altered xenoliths composed of the same minerals as the granites themselves though in different proportions (Figure 5-6C). They tend to be richer in biotite and hornblende, and have more plagioclase and less K-feldspar, than the surrounding granites, and they usually contain more apatite, sphene, and magnetite. It should be stressed, however, that basic clots of similar composition may be not reconstituted xenoliths but autoliths formed by segregration of early-formed crystals from granitic magma. Inclusions characterized by a sugary or hornfelsic texture (see p. 477) may be considered without doubt to be xenoliths, generally of sedimentary material, and granites that contain cordierite, sillimanite, and andalusite can be produced by reaction with aluminous wall rocks.

Migmatites

Migmatites—the term signifies "mixed rocks"—are heterogeneous rocks developed on a large scale as migmatite complexes within regions of high-grade metamorphism (in the amphibolite or granulite facies, see p. 456), or alternatively, as broad migmatitic zones bordering major acid plutons. In hand specimens or within individual outcrops most migmatites consist of two lithologic components comprising mutually interspersed or interlayered streaks and bands of contrasted color. One of the lithologic components, light in color, is broadly "granitic" in mineralogical and chemical terms; its composition in different instances may be anywhere within the granite-granodiorite spectrum. The second and darker component is clearly metamorphic—commonly amphibolite or some similar rock.

The origin of migmatites, undoubtedly different in different individual cases, has been much debated; and of course this problem figured prominently in the granite controversy. The consensus today is that several processes may be involved, each dominant in specific cases.[4] One of these processes, of particular interest in the context of this chapter, is partial fusion (*anatexis*) of mixed sedimentary and volcanic rocks as the culmination of high-temperature regional metamorphism. Another is mechanical injection of intrusive magma along joints and fractures in the envelope of a pluton. In some instances selective metasomatism of rocks in process of metamorphism under the influence of pervasive aqueous fluids seems to be responsible for development of at least part of the light-colored lithologic component of the migmatite (this would be a true instance of granitization in the broadest sense of the term).

Whatever its origin, the lighter element in most migmatites conforms to the textural and compositional specifications of acid plutonic rock. It is here included in this category. The darker lithologic element will be considered as metamorphic—for in most cases it is identical with high-grade metamorphic rocks of nonmigmatitic terranes. But this brief statement must be concluded on a cautionary note of great interest. In some migmatites, such as those of the Rio de Janeiro district in Brazil and parts of southern Sri Lanka, the lighter streaks and bands are of charnockite (see p. 167); and these swim in a darker and predominant lithologic phase which is a quartzo-feldspathic hornblende-biotite gneiss. It has been suggested that such migmatites represent the product of partial fusion of high-grade, water-deficient metamorphic rocks;[5] and that the host gneiss is the product of partial fusion in the course of which all the water component (now held in mica and hornblende) was drained into the melt fraction, the *anhydrous residue* now being represented by charnockite. Such charnockites would not on any count be classed as igneous.

VOLCANIC TYPES: DACITES AND RYOLITES

There are two principal varieties of acid volcanic rocks, namely, *dacite* and *rhyolite*. Dacite has a bulk composition that corresponds generally to that of granodiorite or tonalite; the composition of rhyolite corresponds to that of granites. But while the coarser-grained plutonic rocks can be readily distinguished and identified by their visible mineral assemblages, dacites and rhyolites tend to be glassy or aphanitic and are often mutually indistinguishable by optical means. Reliable identification of them, therefore, commonly requires chemical analysis and normative parameters.

Like their plutonic counterparts, the acid volcanic rocks are almost exclusively found in mobile belts of island arcs and continental margins. But in oceanic islands, such as Ascension in the Atlantic and Easter in the Pacific, the end products of differentiated lineages of olivine-tholeiite parentage may be peralkaline sodic rhyolites.

Textures of Acid Volcanic Rocks

Dacites and rhyolites typically are either glassy or aphanitic; most are also porphyritic, although aphyric (nonporphyritic) varieties also are common. The highly siliceous magmas from which these rocks evolve are so extremely viscous that ionic diffusion and crystal growth are impeded and glassy textures are very common, much more so than in intermediate and basic volcanic rocks that evolve from less viscous magmas. Indeed, in many young dacites and rhyolites, the most distinctive and readily identified constituent is glass of low refractive index (Figure 2-11). Glass is metastable, however, and it seldom survives in ancient rocks. It may devitrify early as it cools through a moderately high temperature range, becoming a micro- or cryptocrystalline aggregate of feldspar and tridymite or cristobalite; or it may become finely crystalline later as a result of reheating or of reaction with hydrothermal or meteoric fluids. The alteration of glass by reaction with aqueous solutions is discussed in Chapter 9 (pp. 272–274).

Rapid undercooling (quenching) of dacitic and rhyolitic magmas produces rocks that consist wholly or almost wholly of glass. These are called *obsidian*—or *pumice* if they are highly vesiculated (Figure 5-10C). Because magma extruded onto the surface, or intruded at very shallow depth, at high temperature (usually on the order of 750°–950°C) can retain very little water in solution, obsidians contain less than 1% H_2O by weight. However, the glass usually becomes hydrated by later absorption of water, up to nearly 10%.[*] Hydrated glass is called *pitchstone* if it has dull, resi-

---

[*]That the glass is usually hydrated by meteoric water is shown by analysis of oxygen and hydrogen isotope ratios in the glass. See H. P. Taylor, *Contributions to Mineralogy and Petrology*, vol. 19 (1968): pp. 25–31.

nous luster in hand specimens; most of it is *perlite*, so called because it is characterized by perlitic structure consisting of numerous curving cracks roughly concentric around closely spaced centers (Figures 2-5B and 5-12A).

Flow structure is a conspicuous feature in many siliceous glasses. In hand specimen, thin planar or contorted bands of differing shades may be seen traversing the glass; on exposed clean surfaces, the banding may be accentuated by incipient weathering. In thin section, the fluidal structure is generally marked by curving trains of crystallites (Figures 5-10A and 5-12A) or by distinct boundary septa between adjacent flow bands.

Here a word of caution must be inserted concerning the chemical composition of glassy rocks. It is often assumed that glasses, because they form by rapid chilling, indicate better than other igneous rocks the composition of the original magmas, and in most instances the assumption appears to be valid. Some glasses, however, are modified in composition as they cool. In the selvages of a rhyolite dike in Iceland, for example, there are phenocrysts of anorthoclase that cannot have crystallized from a melt with the composition of the surrounding glass. Apparently the glass not only absorbed water but also was enriched in potassium at the expense of sodium.

Aphanitic (crypto- or microcrystalline) dacites and rhyolites consist largely of feldspar with quartz, tridymite, or cristobalite, and perhaps a little interstitial glass. The individual constituent grains are generally so small that they can seldom be identified optically, but the microscopic fabrics vary distinctly as a consequence of crystallization, or devitrification, under differing conditions of temperature and viscosity. Lavas undercooled substantially below the liquidus temperature for feldspar are essentially anhydrous hot glasses with exceedingly high viscosity. Devitrification of the hot glass typically results in slender fibers of feldspar intergrown with tridymite or cristobalite in feathery, finely fibrous aggregates. However, in lavas of comparable composition at temperatures near the liquidus for feldspar, viscosity is somewhat lower. Crystallization may then produce euhedral feldspar microlites enveloped in a residual glass matrix. (*hyalopilitic texture*, Figure 2-10C),* or it may produce microgranular aggregates of small, more or less equant, interlocking crystals of feldspar and quartz (*felsitic* or *microfelsitic texture*). The latter

---

*The ophitic, intergranular, intersertal, and pilotaxitic textures characteristic of basalts and basaltic andesites are exceptional in dacites and rhyolites.

is perhaps most characteristic of rocks in shallow intrusions, especially the groundmasses of acid intrusive porphyries (Figure 5-2).

Radially fibrous spherulites (p. 66 and Figure 5-12B) and spherulitic aggregates are the most easily recognized products of devitrification of hot glass, though where a thin section cuts across the spherulitic fibers near the periphery of a spherulite the texture may appear minutely granular. Separate spherulites may develop at widely spaced centers, or they may be concentrated in particular flow layers, while the surrounding rock remains glassy. Complete devitrification may produce a mass of coalescing spherulites or an aggregate of small, interlocking fibrous bundles or sheaves. Under crossed polarizers, the latter may at first glance appear microgranular; but when the stage is rotated, an extinction plume moves across each of the "grains" and indicates their fibrous structure. Devitrification fabrics often vary from one flow layer to the next, but the largest spherulitic structures typically cut across flow bands, a consequence of their late development after flow had effectively ceased. By contrast, feldspar microlites, being minute discrete crystals, may be oriented with their long dimensions approximately parallel to the fluidal structure, a consequence of their earlier formation at somewhat higher temperatures and before flow had ceased.

Dacites

Most dacites carry phenocrysts of plagioclase, quartz, or sanidine, and generally fewer of pyroxene, hornblende, or biotite. The groundmass is usually glassy, hyalopilitic, or felsitic. Most dacitic glass has a refractive index below 1.520.

Plagioclase phenocrysts may range in composition from labradorite or, rarely, even bytownite to oligoclase, but most of them consist of calcic andesine and sodic labradorite. The average composition, however, of all the plagioclase, both phenocrysts and in the groundmass, is at least as sodic as An_{50}. Normal, reverse, and oscillatory zoning are prevalent in the larger crystals, many of which may be spongy because charged with blebs of glass. Jackets of potassic feldspar are not uncommon.

Phenocrysts of quartz may be present as deeply embayed bipyramidal crystals; phenocrysts of sanidine are seldom as numerous as those of plagioclase and may be lacking. Mafic minerals are almost invariably phenocrystic. Diopsidic augite and hypersthene generally predominate, but lamellar intergrowths of these minerals, such as are common in slowly cooled, coarse-grained, intrusive rocks, are quite exceptional. In rocks transitional to rhyolites, hornblende and biotite may be the chief or only mafic constituents, but it is only in rapidly chilled, glass-rich dacites that these minerals remain unaltered; in aphanitic dacites they are partly or wholly replaced by granular iron oxides and pyroxene. Both biotite and hornblende are unstable under surface conditions at high temperatures. Thus, for example, the lava forming the plug-dome of Lassen Peak, California, is a hornblende-biotite dacite, whereas the lava of 1915, which was extruded into the summit crater from shallow depth, is a pyroxene dacite though identical in chemical composition. In a few dacites the phenocrysts of hornblende have cores of cummingtonite.

Olivine, usually fayalite, is only rarely a constituent of dacites, either associated with patches of tridymite or as reaction rims around hypersthene. It appears to be produced under the influence of hydrous residual fluids that are enriched in iron and silica. Where magnesian olivines are present in dacites, as they are near Clear Lake and Medicine Lake, California, they signify mingling of dacitic magma with more basic material. Minor accessory minerals, in addition to the usual apatite and iron oxides, may include garnet and cordierite.

Many dacites, particularly those found in steep-sided domes of Peléan type, carry abundant basic inclusions. These are fragments torn from the walls of the feeding conduits and from the roofs of the underlying reservoirs, and they are characterized by lamprophyric texture. Most of them consist of pyroxene prisms or slender crystals of oxyhornblende and laths of labradorite, arranged in a criss-cross fashion, with interstitial glass and much tridymite or cristobalite of pneumatolytic origin (Figure 5-10B).

Rhyolites

Rhyolites may be divided into potassic and sodic (typically peralkaline) types. In the former the chief mafic constituent is usually biotite, while in the latter it is sodium-rich amphibole or pyroxene or both, and the sodic rhyolites generally carry anorthoclase as one of their feldspars. In both types, corroded bipyramidal phenocrysts of quartz are seldom lacking and in microcrystalline rocks there is much granular quartz in the groundmass. Tridymite and cristobalite are also widespread, and in many specimens there is abundant opal or chalcedony or both, lining pores or filling amygdules (Figure 5-12B).

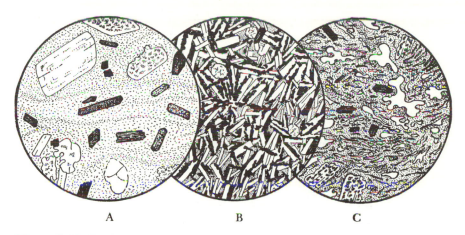

A   B   C

Figure 5-10. Dacites

A. Hyalodacite, near Lassen Peak, California. Diam. 3 mm. Phenocrysts of glass-charged, zoned andesine, quartz, green hornblende, biotite, and hypersthene, in a glassy groundmass stippled with crystallites.

B. Basic inclusion in dacite, Lassen Peak, California. Diam. 3 mm. Laths of labradorite and calcic andesine, and prisms of reddish-brown oxyhornblende largely replaced by magnetite and hematite. Interstitial colorless glass and cristobalite; some of the latter also occurs in spheroids.

C. Pumiceous dacite obsidian, Rock Mesa, near Three Sisters, Oregon Cascades. Diam. 2 mm. Microphenocrysts of hypersthene and corroded, glass-charged andesine, in a matrix of colorless vesicular glass.

In *potassic rhyolites*, sanidine is the principal feldspar. Microcline and perthite, which are abundant in granites, are not to be expected in rhyolites. Occasionally, as in rhyolites from Yellowstone and Skye, there may be phenocrysts of irregular or rectangular outline, in which K-feldspar and quartz are graphically intergrown. Plagioclase is seldom a constituent of the groundmass; it usually forms zoned phenocrysts of sodic andesine or oligoclase. Among the mafic minerals, dark-brown, iron-rich biotite is the commonest; next in abundance are green and brownish-green hornblendes. However, both of these mafics tend to be replaced by granular iron oxides and pyroxene. Phenocrysts of green diopsidic augite occur in some rhyolites; hypersthene is present only occasionally. In many rhyolites, lithophysae are lined with crystals of pneumatolytically deposited minerals—tridymite, cristobalite, hematite, fayalite, and less commonly, ferrosilite and Mn garnet. In some specimens topaz, fluorite, and muscovite are to be seen (Figure 5-11A), and tourmaline also is not uncommon in pneumatolyzed rocks. Cordierite has been noted in some

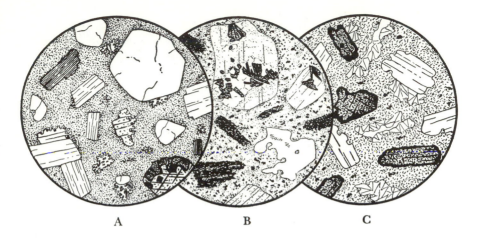

A   B   C

Figure 5-11. Rhyolite and Dacites

A. Rhyolite, Climax, Colorado. diam. 4 mm. Phenocrysts of quartz, orthoclase, oligoclase, and biotite, in a cryptocrystalline base stippled with minute flakes of white mica, larger, spongy granules of topaz, and (lower right) grains of fluorite and pink garnet.

B. Dacite, Sidewinder Mountain, near Barstow, California. Diam. 3 mm. Corroded phenocryst of quartz; other phenocrysts of andesine and of resorbed biotite and hornblende. Groundmass composed chiefly of quartz and K-feldspar (microfelsite). The feldspar is partly altered; piedmontite clusters occur inside the porphyritic andesine; and smaller specks are visible inside the hornblende and biotite crystals as well as in the felsitic groundmass.

C. Tridymite-rich hypersthene dacite, Crater Lake, Oregon. Diam. 3 mm. Phenocrysts of hypersthene rimmed with magnetite and hematite resulting from fumarolic oxidation; also phenocrysts of andesine. Cryptocrystalline groundmass stippled with hematite dust; irregular patches of tridymite with characteristic fan-shaped twins.

contaminated rhyolites; and graphite, apparently produced by reaction between iron oxides and carbon dioxide, is another sporadic constituent. Minor accessories common to almost all rhyolites are zircon, sphene, and iron oxides. The groundmass of rhyolites is generally glassy, microfelsitic, or spherulitic, and if crystalline consists of quartz or tridymite intergrown with K-feldspar. Rhyolitic glass usually has a refractive index below 1.50.

Sodic rhyolites are distinguished by the presence of phenocrysts of sodic sanidine, anorthoclase, or albite. As in potassic rhyolites, bipyramidal phenocrysts of quartz are usually present. The mafic constituents are chiefly sodic amphiboles and pyroxenes. Mafic phenocrysts, however, are exceptional and are generally composed of either diopsidic or aegirine

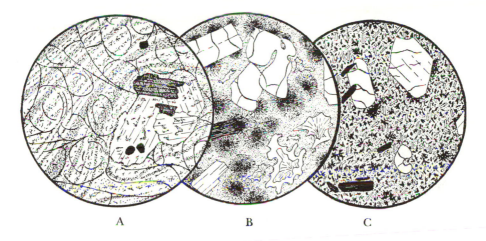

A   B   C

Figure 5-12. Rhyolites

A. Rhyolite pitchstone, near Shoshone, California. Diam. 2.5 mm. Phenocrysts of brownish-green hornblende and of andesine, in a base of banded glass showing perlitic cracks and abundant curved crystallites.

B. Spherulitic biotite rhyolite, Apati, Hungary. Diam. 3 mm. Phenocrysts of quartz, sanidine, andesine, and reddish-brown biotite in a devitrified spherulitic groundmass containing amygdules of opal and radiating chalcedony.

C. Sodic rhyolite (pantellerite), Santa Rosa, California. Diam. 2 mm. Phenocrysts of sodic sanidine or anorthoclase, corroded quartz, and deep-brown enigmatite. Groundmass of quartz and sanidine with needles and mosslike patches of arfvedsonite, subordinate needles of aegirine, and anhedral specks of enigmatite. In other specimens from this locality the rhyolite contains abundant opal and tridymite lining pores.

augite. The dark minerals are mostly restricted to the groundmass, where they occur as clusters of acicular and fernlike aegirine, and mosslike, patchy aggregates of riebeckite, arfvedsonite, cossyrite, and enigmatite, many of which are partly altered to brownish clots of iron oxide. A typical sodic rhyolite is shown in Figure 5-12C. This illustrates a widespread but never-abundant peralkaline variety (*pantellerite*) that represents the final product of fractional crystallization in some lineages of tholeiitic parentage.

REFERENCE NOTES

[1] P. C. Bateman and others, *U. S. Geological Survey Professional Paper*, no. 414-D, 1963.

[2]J. A. Speer, *Canadian Mineralogist*, vol. 19 (1981): pp. 35–46.

[3]P. C. Bateman, *Bulletin of the Geological Society of America*, vol. 72 (1961): pp. 1521–1538.

[4]See I. S. E. Carmichael, F. J. Turner, and J. Verhoogen, *Igneous Petrology*. New York: McGraw-Hill, 1974, pp. 572–575; K. R. Mehnert, *Migmatites and the Origin of Granite Rocks*. Amsterdam: Elsevier, 1968.

[5]See O. H. Leonardos and W. S. Fyfe, *Contributions to Mineralogy and Petrology*, vol. 46 (1974): pp. 201–214.

6

Feldspathoidal Mafic Rocks:
Basic and Ultrabasic

Our concern in this chapter is with mafic rocks of recognizably alkaline character reflected in the mode by the presence of a feldspathoid and significant alkali feldspar. The color index ranges from 40 upward and in a few extreme cases (certain alkaline pyroxenites) even exceeds the ultramafic limit (70). In sodic types in which the feldspathoid is nepheline or a sodalite mineral, normative *ne* is in excess of 5 (typically >10) and *or* is between 5 and 20. However, in the much more restricted potassic group in which leucite is the sole or principal feldspathoid, normative *ne* may be low while *lc* may reach values of 40 or 50. There is complete mineralogical and chemical gradation between mafic rocks with only minor modal nepheline (or leucite) and alkali olivine basalts (and equivalent gabbros) already described in Chapter 3. Volcanic and plutonic divisions as usual are distinguished mainly on the basis of texture. Differences in their respective environments commonly are trivial. Many of the classic descriptions of plutonic members indeed refer to shallow intrusions exposed in the lower levels of what are essentially volcanic complexes.

VOLCANIC DIVISION:
TRACHYBASALTS, BASANITES, TEPHRITES,
NEPHELINITES, LEUCITITES

General Character and Terminology

Most of the common volcanic representatives of the feldspathoidal mafic class are dark-colored rocks of basaltic aspect. The most abundant and widespread mafic constituent is diopsidic augite, commonly the pale-violet titaniferous variety. A magnesian olivine usually is also present except in one group (tephrites). Various combinations of calcic plagioclase, alkali feldspar, and feldspathoid provide the basis for generally accepted terminology presented here in synoptic form. In the following sequence, the rock types express a spectrum of decreasing SiO_2 and rising $(Na_2O + K_2O)$ that together reflect a falling gradient of SiO_2 activity in corresponding magmas.

> *Trachybasalts:* plagioclase abundant, alkali feldspar in significant quantity (typically 10–25% of total feldspar), feldspathoid absent or in small amount (<10%).
> *Basanites*: resembling trachybasalts but with conspicuous modal feldspathoid (>10%). *Tephrites* are similar to basanites but lack olivine.
> *Nephelinites and leucitites:* one or more feldspathoids abundant, to the point of virtual or complete exclusion of feldspar of any kind. The presence of perovskite rather than sphene in many rocks testifies to extremely low SiO_2 activity in the parent magmas.

In general mineralogical definitions of the feldspathoidal mafic class, and in major subdivisions set up within them, equal weight is generally attached to all feldspathoids. Thus nepheline- and leucite-bearing basic rocks with otherwise similar mineral composition are usually placed together in a single major subdivision as basanites, tephrites, and so on. Yet each feldspathoid has its own peculiar significance with respect to activities of such components as SiO_2 and H_2O in the magma from which it crystallizes. And such characteristics must sometimes be taken into account in framing major classificatory divisions in terms of SiO_2 content and mineralogy. In particular, leucite has a much higher content of SiO_2 (~ 54%) than nepheline (40–44%) and kalsilite (38.5%); and, moreover, in simple melts at low pressures leucite alone can coexist with melts *oversaturated* in SiO_2 (with high SiO_2 activity).* In consequence, whereas nephelinites and kalsilite-bearing rocks are all ultrabasic, most leucitites are basic rocks, some even bordering on intermediate; but some rare leucite-bearing types rich in olivine or melilite are truly ultrabasic (SiO_2 well below 40%).

The general terminology just presented, being consistent with broad current usage, and, in essence, with the IUGS scheme,† is adopted in our treatment. But in view of the markedly different respective roles of nepheline (or sodalite) and leucite, and to bring out natural mineralogical and field relationships, a nephelinic spectrum has been separated from a leucitic spectrum. To the latter, for convenience of treatment, have been appended a rare group of ultrapotassic rocks with leucite as the sole felsic ingredient. Accordingly the order of treatment adopted is: (1) trachybasalts; (2) nepheline basanites and tephrites; (3) nephelinites; (4) leucite basanites and tephrites; (5) leucitites; (6) ultrapotassic leucitic rocks (intermediate and ultramafic).

Mineralogical transitions between individual rock types juxtaposed in this list are so gradual that it is impossible to draw intervening boundaries except in the most arbitrary manner. Yet there is no field indication whatever that the complete list or even any section of it represents a magmatic lineage. Where several rock types are indeed associated within some volcanic province, as in various Hawaiian volcanoes and in sectors of the Birunga province of the eastern rift in Kenya and adjoining countries of east Africa, significant separation in space and in time suggests that each individual type, at least in most cases, represents an independently generated "primitive" magma.

Global Distribution

Across the globe volcanic rocks of the whole trachybasalt-basanite-nephelinite spectrum show much the same pattern of distribution as alkali olivine basalts. Both groups repeatedly appear among the products of the closing stages of the eruptive cycles of Hawaiian volcanoes. Then

---

*At room pressure sanidine—like forsterite (Chapter 1, Figure 1-2)—melts incongruently to yield crystalline leucite plus a melt with excess SiO_2.
†Except that we avoid rigid boundaries between rock types and contrived terms such as "foid" and "foidite."

there are other midoceanic islands in which the lavas are exclusively trachybasalts and basanites (e.g., Tahiti and Tristan da Cunha), or others again in which nephelinites predominate (e.g., Trinidad and Fernando de Noronha, both east of the northern Brazilian coast). Continental eruption of alkaline basic magmas shows some degree of correlation with more or less contemporaneous regional rift tectonics.* Classic examples are seen in the rift systems of eastern Africa and the Rhine Graben of Germany, in each of which particular alkaline rock types tend to dominate individual subprovinces. Elsewhere there is no obvious connection between effusion of alkaline basic magmas and regional rifting—as, for example, in the Miocene alkaline provinces of northern New South Wales and southeastern New Zealand.

Leucite-bearing basic volcanics are most uncommon among alkaline lavas of oceanic volcanoes, the notable exception being the leucite basanites of Tristan da Cunha. Extremely potassic lavas carrying abundant leucite (or even kalsilite) are exclusively continental. They are well known in the Roman province of Italy, the Leucite Hills region of Wyoming, and in subprovinces within more extensive regions where alkaline rocks are predominantly sodic: the Eifel region of the lower Rhine and the Birunga and Toro-Ankole fields of the east African rifts are examples.

Trachybasalts

Most trachybasalts are closely associated with trachytes and phonolites on the one hand, and with olivine basalts and hawaiites on the other. They grade into the former by increase of alkali feldspar at the expense of plagioclase, and into the latter by the reverse change. They are characterized by the presence of olivine and augite, by plagioclase more calcic than andesine, and by alkali feldspar (orthoclase, sanidine, or anorthoclase) amounting to more than 10%. The plagioclase, which may be as calcic as anorthite, generally forms phenocrysts; the alkali feldspars, on the contrary, rarely do so, but usually occur in the groundmass or form jackets around the plagioclase.

Characteristically the dominant mafic mineral is diopsidic augite—in many cases the purplish titaniferous variety. Phenocrysts of magnesian

---

*This pattern is by no means universal. For example, basalts and diabases invading Triassic sediments in a series of down-faulted basins along a 2000 km stretch of the Appalachian Piedmont province in the eastern United States are exclusively tholeiitic (cf. Figure 3-3A).

olivine are almost always present; and some rocks carry a little brown amphibole or red-brown biotite in the groundmass. Dark-brown, strongly pleochroic amphibole (variously identified as basaltic hornblende, barkevikite or kaersutite) occurs as scattered large crystals in some trachybasalts. In many cases these seem to be xenocrysts of cognate origin. Thus in trachybasalts and in related "trachyandesites" (cf. p. 198) of the east Otago district in southern New Zealand, the same coarse brown amphibole is seen in cognate gabbroid xenoliths and occurs also as isolated crystals. In either case it appears to have been in violent disequilibrium with the enclosing magma under near-surface conditions; for it shows all stages of replacement by opaque oxides, titaniferous augite and apatite (the last-named mineral suggesting perhaps a significant content of fluorine in the amphibole). Although trachybasalts are *ne*-normative, modal feldspathoid cannot be identified in many of these rocks—which are said to carry occult nepheline since *ne* is hidden as a minor component of glass, amphibole, or even augite. Leucite is a modal constituent of the potassic trachybasalts that build much of the south Atlantic island of Tristan da Cunha. Analcime, probably of deuteric origin, may replace part of the groundmass feldspar of some trachybasalts.

Basanites and Tephrites

Basanites and tephrites are plagioclase-bearing basic lavas in which feldspathoids make up more than 10% by volume. Tephrites either lack olivine or contain very little; basanites, in contrast, carry olivine in considerable amount, though rarely to the exclusion of pyroxene. Rocks of similar chemical composition, but containing sodic glass instead of feldspathoids (feldspathoid being occult), are called *basanitoids*.

Nepheline Basanites and Tephrites

In most nepheline basanites and tephrites laths of labradorite or bytownite make up between 20 and 40% by volume, and many of them have thin rims of more sodic plagioclase or of sodic sanidine. Nepheline, which is generally about half as abundant, rarely forms phenocrysts; it normally occurs in the groundmass as anhedral interstitial grains or as large poikilitic patches. Hauyne, nosean, and leucite may be found as small euhedral crystals, while analcime and natrolite may occur in irregular blotches or may fill amygdules. The typical pyroxene is purplish titanaugite, which forms euhedral phenocrysts, sometimes fringed with greenish aegirine-augite, and small anhedra in the matrix, where it is accompanied by sodic pyroxenes. Brown barkevikitic hornblende and bluish arfvedsonite are seldom more than minor accessories; but there are rare lavas in which amphibole is abundant enough to be counted as an essential constituent. Reddish-brown titaniferous biotite, also, is normally present only in minor amount, but occasionally it may form large and numerous poikilitic flakes. Olivine, the presence of which distinguishes basanites from tephrites, is almost invariably in phenocrysts; much of it is altered to a serpentine mineral. Finally, in common with most basic, sodium-rich rocks, these lavas contain considerable titanium, the characteristic oxide being titanomagnetite, usually accompanied by sphene. Apatite is an ever-present accessory.

Rocks that carry more alkali feldspar than most nepheline basanites and contain andesine rather than labradorite represent a mineralogical and chemical, though not a genetic, transition toward latites (trachyandesites). Like normal nepheline basanites, they are thoroughly undersaturated in SiO_2 with normative *ne* exceeding 10 and *ol* somewhat lower. The SiO_2 content is below 50 (whereas latites belong to the intermediate SiO_2 category). Such departures from typical basanites, slight as they are, are not trivial; they mark a more potassic character that will persist along a differentiated lineage leading ultimately to phonolites. Among rocks of this kind, representing an early stage in one of several alkaline lineages in the Miocene volcanic province of east Otago (Dunedin), New Zealand,[1] is a porphyritic type with a glassy groundmass (38% of the mode). Phenocrysts include olivine, $Fo_{82,4}$; diopsidic augite zoned outward from pale violet to pale green, 18; feldspar, 26; nepheline, 7; opaque oxides, 7; and occasional large crystals of brown kaersutite largely replaced by a spongy mixture of pyroxene, opaque oxides, feldspar, and nepheline. The feldspar includes both oligoclase-andesine (much of it distinctly potassic) and anorthoclase. Many of the olivine phenocrysts are rimmed with clinopyroxene and the sparse phenocrysts of nepheline are fringed with reaction rims of sodic plagioclase.

Analcime Basanites and Tephrites

In some sodic basanites, instead of nepheline, analcime, usually attended by other zeolites, occurs either as irregular patches in the groundmass or as ovoid clots, some of which may be surrounded by tangentially arranged prisms of pyroxene. *Analcime basanites*, in which the olivine is invariably serpentinized to some extent, are much more abundant than *analcime tephrites*. The typical pyroxene of both lavas is purplish titanaugite; colorless diopsidic augite is less common; orthopyroxenes are absent, as in all feldspathoidal rocks. Calcic plagioclase and iron-titanium oxides accompany the groundmass zeolites, and a little brown hornblende may be present.

Nephelinites

Nephelinites are mafic volcanic rocks, mostly of ultrabasic composition, in which nepheline is the sole or predominant felsic constituent. In some types there may be another feldspathoid as well as nepheline. But feldspars are totally lacking or present only in minor amount (<10% of the mode). Most nephelinites are porphyritic, with diopsidic or titaniferous augite and nepheline in euhedral phenocrysts (Figure 6-1A)—the pyroxene commonly bordered wtih aegirine-augite; but phenocrystic nosean,

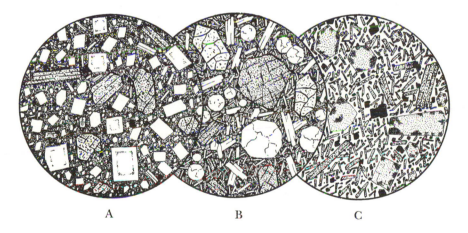

A   B   C

Figure 6-1. Feldspathoidal Lavas

A. Nephelinite, Mikeno, East Africa. Diam. 1 mm. Microphenocrysts of green augite and nepheline, in a matrix of dark-brown glass with granules of iron oxide, and slender microlites of sanidine.

B. Leucite basanite, Vesuvius, Italy. Diam. 3 mm. Phenocrysts of olivine, green diopsidic augite, and leucite, in an intergranular matrix of labradorite laths, iron oxide, and augite. Locally there are minute interstitial grains of sanidine.

C. Hauynophyre, Tahiti. Diam. 1 mm. Microphenocrysts of deep-sky-blue hauyne with webs of rutile; slender prisms of pale-green diopsidic augite and euhedral granules of iron oxide, in a matrix of pale glass.

hauyne, or sodalite also appears in some rocks. The groundmass typically is composed of the same minerals, accompanied by accessory iron-titanium oxides, apatite, perovskite (less commonly sphene), and needles of apatite. Small amounts of biotite or brown hornblende are occasionally to be seen. Deuteric zeolites are seldom absent. The *hauynophyre* illustrated in Figure 6-1C is a feldspar-free basic lava with hauyne instead of nepheline.

In the Hawaiian Islands, east Africa, western Japan (Nagahama), and near the Laacher See in the Eifel province, there are nephelinites of extremely basic composition (SiO_2 <40%) in which melilite is a prominent constituent. It takes the form of phenocrysts—tabular crystals showing the "peg structure" so characteristic of this mineral. Rocks of this kind mark a transition toward melilitites described in Chapter 8 (Figure 8-4B).

Finally, in provinces where associated rocks are strongly potassic we find nephelinites that carry leucite or even kalsilite as well as nepheline. In some of these the nepheline itself is high in the kalsilite "molecule" $KAlSiO_4$. All three feldspathoids may be mutually associated in the groundmass (e.g., in lavas of the east African Toro-Ankole field).

The xenoliths that so commonly are found in nephelinite lavas and breccias (and also in some basanites) conform to a lithologic pattern both striking and uniform. All are ultramafic. They include all possible combinations of magnesian olivine, diopsidic augite, enstatite, and chrome spinel—ranging from simple dunite (also found in tholeiitic basalts) to the full four-phase assemblage, lherzolite. All usually show textural evidence of strong internal deformation in the solid state. Almost certainly these have been derived directly from materials in the upper mantle. Other types of ultramafic xenolith peculiar to nephelinite host rocks are garnet pyroxenites (including true eclogite whose pyroxene is omphacite) and biotite-augite pyroxenites. Some nephelinites—and also some basanites—carry coarse xenocrysts of augite unusually rich in aluminum or of kaersutite. Individual crystals of the latter in nephelinite breccias may reach several centimeters in average diameter.

Limburgites

Limburgites are largely glassy, ultrabasic sodic rocks akin in composition to nephelinites. An analyzed specimen from the island of Principe in the

Gulf of Guinea, west Africa, has an SiO_2 content of 39.5% with normative values *an* 22, *ne* 14, *di* 25, *ol* 21. Limburgites form flows, dikes, sills, and plugs, usually associated with basic alkaline rocks, particularly analcime basanites and ultrabasic lamprophyres (monchiquites, p. 232). In hand specimens they are dark-colored rocks, usually devoid of feldspar or containing at most a small amount of sodic plagioclase or nepheline or both. Euhedral phenocrysts of clinopyroxene are always numerous, and generally they are zoned from cores of colorless diopsidic augite to lilac and purple rims of titanaugite. Phenocrysts of olivine are also abundant, but seldom more so than pyroxene, and in some specimens there are a few microphenocrysts of biotite and barkevikitic hornblende. All these constituents are embedded in a sodic matrix of brown glass, liberally stippled with granules of iron-titanium oxides, which are sometimes accompanied by tiny needles of clinopyroxene. Calcite and analcime are almost invariably present, either as irregular patches or filling amygdules. In some rocks there are veinlets and amygdular fillings of dark "chloritic" material.

Leucite Basanites and Tephrites

Examples of leucite basanites and tephrites as well as rocks transitional to trachybasalts and to leucitites are drawn mainly from areas of late Tertiary and Holocene volcanism along the middle Rhine (e.g., the Eifel and the Kaiserstuhl), the Roman region extending from Naples to north of Rome (Alban Hills), and the Bufumbira province situated on the western rift of equatorial east Africa. Rocks of this kind cover the whole range of SiO_2 content encompassed by the basic category.

Among the most basic types are *leucite basanites* of the Bufumbira area°—dark-colored rocks (color indices as high as 60) with SiO_2 about 46%. They consist of diopsidic augite, magnesian olivine, labradorite, leucite, and in some cases a little glass. A rather more siliceous, strongly porphyritic leucite basanite from Vesuvius is illustrated in Figure 6-1B. In most leucite basanites (and also in tephrites and related transitional rocks) crystals of leucite occur in two generations: large, feebly birefringent phenocrysts with characteristic obliquely intersecting sets of twin

---

°*Kivites* and *murambites* of east African literature.

lamellae, and small isotropic grains in the groundmass. Radial and concentrically disposed inclusions of pyroxene and opaque oxides are plentiful, particularly in the smaller grains. Leucite crystals may be completely replaced by analcime or by "pseudoleucite" (a fine-grained mixture of K-feldspar and nepheline or zeolites). Serpentinization of olivine is widespread. Pyroxene is usually greenish diopsidic augite, but in some cases is titaniferous; either may be rimmed with jackets of green aegirine-augite or aegirine. Plagioclase is usually labradorite, in phenocrysts of some rocks bytownite. Alkali feldspar rims plagioclase phenocrysts or appears as small microlites and anhedral grains, locally intergrown with nepheline in the groundmass. Some rocks carry minor biotite or hornblende. Prevalent accessories include other feldspathoids (nepheline, hauyne, or sodalite), melilite, sphene or perovskite, apatite, and opaque oxides. Basanites with leucite and nepheline both prominent in the mode are by no means unusual. Very basic rocks of this kind from the Korath volcanoes on the eastern rift in southern Ethiopia consist of phenocrystic olivine and titaniferous augite enclosed in a groundmass of plagioclase, leucite, nepheline, olivine, and phlogopite.

Leucite tephrites resemble basanites but lack significant olivine. Examples described by W. Wimmenauer from the Kaiserstuhl complex in the Rhine Graben, west of Freiburg, are dark-colored porphyritic rocks with SiO_2 around 45–46%. Phenocrysts (of titaniferous augite, 29; magnetite, 5; plagioclase, 3; and leucite, 5) are set in a groundmass of plagioclase (An_{60}–An_{55}), leucite, zeolites, and glass; some rocks have a few phenocrysts of olivine. Tephrites from the Korath volcanoes of Ethiopia have large phenocrysts of brown amphibole (kaersutite). Leucite tephrites of other regions may be lighter in color, more siliceous ($SiO_2 = 50$–51%), and carry notable alkali feldspar (phonolitic tephrites of IUGS). One such from the Roman region (Rocamonfina, northwest of Vesuvius)[°] has the following mode: leucite phenocrysts, 40; plagioclase An_{65}, 20; sanidine, 20; augite plus minor olivine, 15; accessories (among them nepheline), 5. A tephrite with a low feldspar content, and thus transitional toward leucitite, is currently being erupted from Vesuvius.[°°] It has this mode: leucite phenocrysts, 40; plagioclase An_{70}, 18; augite, 34; olivine, apatite, and opaque oxides, 6; nepheline or sodalite, 2.

[°] *Vicoite.*
[°°] *Vesuvite* (tephritic leucitite of IUGS).

Leucitites

Leucitites are basic volcanic rocks, strongly undersaturated in SiO_2, in which leucite is the sole or dominant felsic mineral, while feldspars either are absent or do not exceed 10% of the mode. High values of *lc* are characteristic of the norm. Color index and total SiO_2 content range widely from dark ultrabasic rocks to light-colored, leucite-rich types verging on intermediate.

Leucite typically occurs both as phenocrysts and in the groundmass; other feldspathoids—nepheline or a sodalite mineral—may appear in minor amounts. The usual mafic constituent is diopsidic or titaniferous augite, also in two generations. With it olivine appears only in the more basic leucitites such as those of the volcano Mikeno in the Bufumbira province of east Africa. Possible minor constituents are melilite, biotite, or melanite garnet. Apatite, sphene or perovskite, and opaque ores are common accessories. Melilite is sometimes abundant as large plates poikilitically enclosing leucite and augite.

The rock type *leucitite* as thus defined in general terms covers a wide lithologic spectrum whose total SiO_2 range is about 35–53%. This is an excellent example of a mineralogically gradational spectrum that is purely artificial, depending simply on the classic definition of leucitite. In individual provinces the compositional range is much narrower. Thus in leucitites of the Roman province SiO_2 ranges from 47% to 53%. In the Arizona lamprophyre province the span is 45–47%; in the Toro-Ankole and Bufumbira provinces of east Africa, leucitites are typically ultrabasic ($SiO_2 \sim 35$–40%).

Leucitites of the Roman province show considerable variety. They range from thoroughly basic ($SiO_2 \sim 47\%$) fine-grained lavas of the Alban Hills, consisting principally of leucite and augite in subequal quantities, to leucite-rich blocks in which leucite makes up as much as 90% of the total composition.° A mode for this extreme type ($SiO_2 \sim 53\%$) is given as leucite, 90; hauyne, 2; melilite, 2; augite, 4; melanite, biotite, magnetite, and apatite, 2.

Olivine leucitites (leucite basalts of some writers) are dark, very basic rocks ($SiO_2 = 45$–47%) in which leucite, augite, and olivine are all essential constituents. The mode of a dike rock from the Navajo lamprophyre

---

°Respectively *albanite* and *italite* of H. S. Washington.

province, on the Arizona-New Mexico boundary, is leucite (in granules up to 0.5 mm), 28; phenocrysts of diopsidic augite, 26, and of serpentinized olivine, 12; and a groundmass of microlitic augite, 16; phlogopite, 4; sanidine, 8; titanomagnetite, 4; and apatite, 2.

Wyomingites and Related Rocks

A rare and remarkable suite of ultrapotassic leucite-rich volcanic rocks, represented in the Leucite Hills, Wyoming, and the west Kimberley area of western Australia, departs so strikingly from leucitites and phonolites as to merit special treatment.[2] There is a good deal of mineralogical variation in the suite and a correspondingly complex and rather outlandish terminology has become accepted. As no adequate group name uniquely covers the suite as a whole, we refer to them as "wyomingites and related rocks."*

Despite their high content of leucite, wyomingites and related rocks—unlike leucitites—are essentially silica-saturated, neither normative (lc + ne) nor Q exceeding 5 in most rocks. This anomaly is due mainly to a combination of two factors: presence of excess SiO_2 in the leucite itself (which can hold as much as 30% $KAlSi_3O_8$ in solution) and significant normative Q (~15%) in an interstitial glassy residue. Moreover, wyomingites and allied rock types, again unlike leucitites, are strongly peralkaline (Al_2O_3-deficient) as indicated in the norm by significant values of ac (acmite) and ks (potassium silicate). Also characteristic are very high K_2O (usually 10–12%) and almost uniquely high concentrations of certain trace elements, notably P, Zr, and Ba. Finally the oxidation state of iron is exceptionally high, $Fe_2O_3/(FeO + Fe_2O_3) \approx 0.6-0.8$.

Most of these chemical traits can be inferred from microscopically recognizable mineralogical characteristics: abundance of mafic potassic minerals, notably phlogopite, the potassic amphibole magnophorite (strikingly pleochroic from lemon yellow to pink), and the widespread unique accessory phase wadeite, $Zr_2K_4Si_6O_{18}$ (colorless, highly birefringent, uniaxial, positive); complete absence of iron oxides, all ferric iron being accommodated in leucite and sanidine (in both cases substituting for Al) or in residual glass.

---

*By some writers they are called *leucite lamproites*. They can be fitted to *phonolitic leucitites* and *leucitites* of the IUGS scheme.

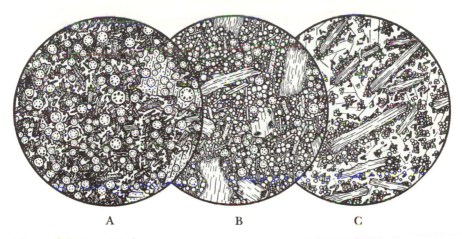

A	B	C

Figure 6-2. Leucitic Lavas

A. Phonolitic leucitite, Capo di Bove, Italy. Diam. 3 mm. Leucite crystals with concentric inclusions, greenish diopsidic augite, and flakes of biotite, in a matrix of anhedral poikilitic sanidines. Poikilitic tablets of melilite near right edge of section.

B. Wyomingite, Leucite Hills, Wyoming. Diam. 3 mm. Leucites with inclusions; slender prisms of pale-green diopside; large flakes of phlogopite with dark titaniferous borders and a few sagenite webs of rutile; interstitial colorless glass.

C. Orendite, Leucite Hills, Wyoming. Diam. 3 mm. Minute grains of leucite and diopside prisms and large flakes of phlogopite, in a groundmass consisting chiefly of divergent sanidine laths and much acicular apatite.

The principal rock types in the Leucite Hills are *wyomingites* and *orendites* (Figure 6-2B,C). Wyomingites are porphyritic rocks with abundant phenocrysts of phlogopite (one generation only) and microphenocrysts of diopsidic augite set in a groundmass of euhedral or subhedral leucite, magnophorite, diopside, and apatite. Magnophorite invariably is interstitial, apparently the last phase to crystallize. In some rocks there is a good deal of glass (Figure 6-2B). Orendites are more varied rocks with many of the same characteristics as wyomingites; but they carry abundant sanidine (Figure 6-2C) as well as leucite, in some rocks in the form of large phenocrysts riddled with small inclusions of diopside. Wadeite and priderite are widespread accessories in orendites—the former usually molded on apatite, priderite (approximately $KBaTi_8O_{16}$) in brown, highly refractive needles closely resembling rutile.

Both wyomingites and orendites border on intermediate silica content ($SiO_2 = 50-55\%$). Related rocks from western Australia illustrated in

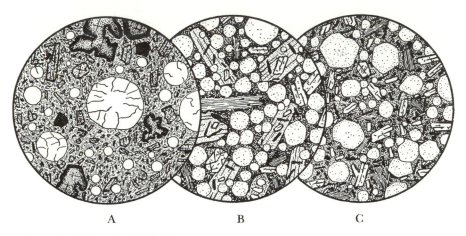

Figure 6-3. Leucite-Rich Lavas

A. Leucitophyre (leucite phonolite), Rieden, Eifel, Germany. diam. 9 mm. Phenocrysts of leucite, aegirine-augite, and embayed nosean, in a dense matrix of aegirine-augite, microlitic sanidine, a little nepheline, and melanite garnet.

B. "Wolgidite," West Kimberley, Australia. Diam. 2.5 mm. Composed of leucite, large prisms of magnophorite, smaller ones of diopside (some of which are included in the magnophorite), a little serpentinized olivine (lower edge of section), flakes of phlogopite, and small prisms of priderite, in a murky matrix of serpentine accompanied by a little zeolite.

C. "Cedricite," West Kimberley, Australia. Diam. 2.5 mm. The larger crystals are of leucite, diopside, and phlogopite; the matrix consists chiefly of priderite, perovskite, magnophorite, phlogopite, zeolites, and indeterminate materials. Not shown, but present in most cedricites, are serpentine pseudomorphs after olivine.

Figure 6-3B,C are truly basic (SiO_2 = 46–52%). In some of these latter magnophorite as well as phlogopite occurs in phenocrysts.

Analcime Diabases

Sills and other minor shallow intrusions composed essentially of analcime-bearing diabases have been described from a number of continental provinces where associated basic lavas have a recognizably alkaline character. Diabases equivalent to alkali olivine basalts for the most part carry only accessory amounts of analcime (p. 109). However, in the upper levels of differentiated sills of this same kind, and in other sills and minor intrusions comagmatic with trachybasalts and nepheline basanites, analcime becomes an essential (though still subordinate) constituent, and other mineralogical evidence of alkalinity also appears. Such rocks may truly be called *analcime diabases*. Of the various rock types that have been recognized—and separately named—on the basis of minor mineralogical and chemical variation, *teschenites* best typify the more conspicuously alkaline of analcime diabases. Further description refers mainly to rocks of this kind.° One example is illustrated in Figure 6-4B.

An average modal composition of teschenite may be given as purplish titaniferous augite, 45; olivine, 10–15; labradorite, 30; analcime, 5–15; minor alkali feldspar, barkevikite (or other alkaline amphibole), and green sodic pyroxene (rimming crystals of augite); accessory iron-tita-nium oxides and apatite. Plagioclase generally forms stout euhedral or subhedral laths, zoned outward from labradorite to andesine or oligo-clase, many of them with rims of K-feldspar, which may also appear in minor quantity as minute anhedra in the groundmass. In many teschen-ites the plagioclase is ophitically enclosed in augite. Deuteric minerals—zeolites, clay minerals, calcite—tend to be widespread. A minor constit-uent of some rocks is nepheline, more or less converted to analcime.

Most teschenitic sills show marked effects of internal differentiation. In the higher levels rocks of more felsic character with more abundant analcime and alkali feldspar may be locally streaked with veins and pods of analcime syenite (much more rarely mafic analcime-rich pegmatoids). Lower down, olivine and augite tend to increase markedly as olivine-rich teschenite merges downward into thoroughly ultrabasic but still recog-nizably alkaline *picrites* (p. 109). In these modal olivine is between 50 and 70, calcic plagioclase between 10 and 25, the remainder being titanaugite rimmed with aegirine-augite, and barkevikite (or arfvedsonite). One cumulate layer 10 m thick near the base of the Lugar sill of Ayrshire, Scotland,* even attains the composition of peridotite—dominant olivine together with titanaugite and brown barkevikite. Well-known examples of differentiated sills and minor intrusions of this nature are a suite of minor intrusions described by K. Yagi from the Morotu province of northern Japan, and the Mesozoic Prospect intrusion near Sydney, Aus-tralia.[3]

°*Crinanites* and *kylites* of British petrologic usage are teschenites respectively lower in both augite and analcime, and richer in olivine, than the average mode cited here.
*Made classic through the writing of G. W. Tyrrell, and site of the extreme nepheline-rich pegmatoid differentiate, *lugarite* (cf. F. J. Turner and J. Verhoogen, *Igneous and Metamorphic Petrology*. New York: McGraw-Hill, 1960, pp. 181–183.

Elsewhere in other varieties of analcime diabase, analcime is only a minor constituent, titanaugite less plentiful, and labradorite correspondingly more abundant than in typical teschenites. Others again contain sufficient alkali feldspar to be classed as *analcime trachydiabase*, a rock gradational in composition toward the latite group (cf. p. 144).

PLUTONIC DIVISION:
FELDSPATHOIDAL GABBROS AND IJOLITES

Under physical conditions prevailing at depths of greater than about 3 km, leucite has no stability field. So the feldspathoid, which is an essential constituent of all rocks in this general class, is almost invariably sodic, usually nepheline. In the more potassic rocks potassium is accommodated in alkali feldspar, or, especially in ultrabasic types, in biotite; some is taken up as the $KAlSiO_4$ component of nepheline. In potassic rocks with plutonic texture that crystallized at shallow depth, the feldspathoid may be leucite; but examples are rare.

There is a good deal of mineralogical variety within the division as a whole. Striking phase combinations have provoked unnecessary elaboration (and confusion too) in rock terminology—all the more unnecessary by reason of marked local variation displayed within relatively small intrusive bodies. Here we have simplified the situation by recognizing only two major groups, *feldspathoidal gabbros* and *ijolites* (the first with essential calcic plagioclase, the second completely lacking that mineral), and a minor group of related ultramafic rocks, *alkaline pyroxenites*.

Feldspathoidal Gabbros

In this section we consider rocks that resemble "normal gabbros" described in Chapter 3 in their basic composition, high color index ($_{>30}$), and presence of calcic plagioclase ($An_{>40}$), augite, and in most rocks olivine. They differ, however, in having a sodic feldspathoid as an essential constituent. The alkaline character thus imparted is petrographically accentuated by the nature of the pyroxene (titaniferous or sodic augite) and in many rocks by the presence of at least minor amounts of alkali feldspar, sodic amphiboles, or biotite. Accessory apatite and iron-titanium oxides are ubiquitous.

From the chemical standpoint feldspathoidal gabbros can be regarded as the plutonic counterpart of trachybasalts and basanites. By diminution of feldspathoid and of K-feldspar, they merge naturally into the more alkaline members of the gabbro family (such as the "Oslo-essexites" described earlier, p. 125). Here is one more illustration of the gradational nature of the alkali-feldspar–plagioclase join (*AP*, Figure 2-13) common to the *APF* and *APQ* triangles that figure so prominently in most classification diagrams.

The two most commonly encountered types of alkaline (feldspathoidal) gabbro are *essexite* and *theralite*.* Both carry essential nepheline. They are mutually gradational rocks that differ mainly with respect to the content of alkali feldspar, an essential constituent of essexites but minor or lacking in theralites. This difference is reflected in bulk content of SiO_2: theralites are the more basic of the two groups—many of them in fact being truly ultrabasic ($SiO_2 = 42–44\%$).

Essexites

The general composition of essexites is exemplified by the mode of the type rock from Salem Neck, Massachusetts: sodic orthoclase, locally perthitic, 20; plagioclase, 28; nepheline, 20; mafic minerals, 30; apatite, 2. The orthoclase, like the nepheline, is anhedral; in places it jackets crystals of plagioclase. The latter is normally subhedral or euhedral and varies in composition from about An_{60} to An_{35}. Among the mafic constituents of most specimens augite predominates. It is usually the purplish titaniferous variety, often as large phenocrysts, but may also be diopsidic. Aegirine-augite or aegirine is exceptional. Brown sodic or titaniferous hornblende is invariably present, and in some essexites it may be more abundant than pyroxene. Biotite is never more than an accessory mineral. Olivine, which may be altogether absent, is usually found in subordinate amount, but occasionally may be even more plentiful than pyroxene. The usual minor accessories are apatite, sphene, and titanomagnetite. Some of the interstitial nepheline in essexites may be replaced by analcime.

It should be noted that most of the essexites from Massachusetts are much richer in nepheline than those from other regions. In the specimen illustrated in Figure 6-4C nepheline forms only a small percentage of

---

*In the IUGS scheme *essexite* and *theralite* are permitted synonyms of *nepheline monzogabbro* and *nepheline gabbro*, respectively.

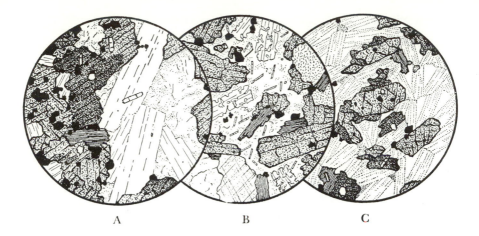

Figure 6-4. Theralite, Teschenite, and Essexite

A. Theralite, Duppau, Bohemia. Diam. 2 mm. Titanaugite is intergrown with dark-brown barkevikite and a little biotite; opaque oxides are titanomagnetite. Large laths of labradorite with thin shells of calcic andesine; interstitial, partly altered nepheline. Accessory prisms of apatite. Not shown, but seen elsewhere, are thin rims of sanidine around some of the plagioclase laths.

B. Teschenite, Inchcolm, Scotland. Diam. 2 mm. Large prisms of purplish titanaugite; flakes of deep-brown biotite; serpentinized olivine (right edge); and granular titanomagnetite. These dark constituents lie in a colorless matrix of corroded labradorite laths, deuteric calcite, and abundant analcime crowded with acicular apatite.

C. Essexite, Rongstock, Bohemia. Diam. 2 mm. The mafic minerals are titanaugite, barkevikite, and biotite. The chief felsic constituents are zoned plagioclase (An_{60-25}), wedge-shaped areas of orthoclase, and small interstitial patches of altered nepheline. The accessories are iron oxide, apatite, and sphene.

the volume, whereas orthoclase makes up 10% and plagioclase more than 40%. It is the significant content of plagioclase that distinguishes essexites from mafic nepheline syenites, also basic rocks, such as those illustrated in Figure 7-4B,C.

Finally there are very rare potassic essexites—such as one represented in ejected blocks on Monte Somma, Vesuvius°—that crystallized at sufficiently shallow depth for the pair leucite-sanidine to take the place of nepheline-orthoclase.

°*Sommaite.*

Theralites

Ideally theralites are nepheline gabbros that virtually lack alkali feldspar. But it is better perhaps to consider them as the end member of a natural transitional series that, with increase in alkali feldspar, passes gradually into essexites as we have just defined them. All theralites are mafic rocks of ultrabasic composition (SiO_2 typically 40–45%). Most contain olivine, a mineral sparsely present or lacking in most essexites.

A typical theralite described by G. W. Tyrrell, from a layer about 10 m thick in the upper section of the differentiated teschenitic Lugar sill, Scotland, has this mode: andesine, 20; nepheline, 15; titanaugite, 35; olivine, 15; biotite, 5; barkevikite, 5; plus minor analcime, iron-titanium oxides, and apatite. It passes downward via transitional rocks (through another 10 m) into barkevikite picrite.

In a similar shallow environment in the Miocene east Otago volcanic province of New Zealand, there is a thin differentiated alkaline sill whose lithologic components are mainly theralites. One representative specimen ($SiO_2 = 40.5\%$) with a color index 40–50, consists principally of zoned titaniferous augite (with thin, greenish sodic rims), andesine and nepheline, accompanied by significant amounts of distinctly ferriferous olivine $Fo_{65} \rightarrow Fo_{45}$ and anorthoclase, plus unusually plentiful accessory apatite. Crystals of andesine and alkali feldspar, both subhedral to euhedral, are poikilitically enclosed in coarse anhedral nepheline.

Ijolite Series

What may be called an *ijolite series* is a suite of plutonic rocks—mostly mafic but variable as to color index, all with $SiO_2 < 45\%$—that typically lack feldspar and whose principal constituents are a pyroxene and a feldspathoid (usually nepheline). Their total areal extent is extremely limited. They occur typically as bosses or annular intrusions that crystallized at shallow depth in provinces of alkaline volcanism. Their normal associates are carbonatites (cf. p. 240) and superjacent nephelinites—to which typical ijolites are chemically similar. There is a complex terminology based on color index.[°]

The typical *ijolite* (Figure 6-5A) is a holocrystalline subhedral-granular rock in which nepheline makes up about half the volume while aegirine-

---

[°]Arbitrary limits approved by IUGS are: *urtite*, 0–30; *ijolite*, 30–70; *melteigite*, >70.

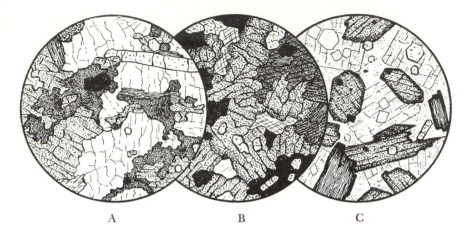

Figure 6-5. Mafic and Ultramafic Alkaline Rocks

A. Ijolite, Magnet Cove, Arkansas, Diam. 3 mm. Composed of nepheline, aegirine-augite, melanite garnet, and yellowish-green biotite, with accessory apatite, sphene, and pyrite.

B. Alkaline pyroxenite, Alnö Island, Sweden. Diam. 3 mm. Composed mainly of green aegirine-augite with purplish titaniferous borders, dark-brownish-green barkevikite, and titanomagnetite. Accessory minerals shown here are nepheline, apatite, and calcite; not shown, but common in alkali pyroxenites of Alnö, are melanite, pyrite, perovskite, and zeolites.

C. Mafic nepheline syenite, Crazy Mountains, Montana. Diam. 3 mm. Pale-green diopsidic augite with rims of dark-green aegirine; subhedral prisms of nepheline, partly altered to cancrinite and zeolites and including euhedral crystals of sodalite; interstitial barium orthoclase; zoned biotite; accessory apatite, sphene, and magnetite.

diopside or a member of the diopside-hedenbergite series makes up about a third. The long list of minerals normally found as accessories includes apatite, sphene, calcite, melanite, phlogopite, sodalite, perovskite, wollastonite, cancrinite, pectolite, and zeolites, any one of which may increase sufficiently in amount to become a major constituent. Much of the ijolite on Alnö Island, Sweden and at Iron Hill, Gunnison County, Colorado, is extremely rich in melanite, and some in Kenya carries abundant large crystals of wollastonite. To emphasize the variability of these rocks, it is enough to add that the ijolite series of Iron Hill ranges from almost pure diopside rocks to almost pure nepheline rocks.

Replacement features are widespread in all members of the series, but especially in the highly mafic ones, and usually the dark minerals formed at the expense of the light. Among the Kenya ijolites, for example, wollastonite has replaced nepheline, and then both it and pectolite have been partly replaced by pyroxene. In the melanite-rich ijolites of Iron Hill pyroxenes of the diopside-hedenbergite series formed first. These were then enclosed by large crystals of nepheline, and subsequently calcite, cancrinite, biotite, and melanite garnet were produced, the garnet replacing both the nepheline and the pyroxene. In other ijolites pyroxene has replaced wollastonite and pectolite.

At the felsic end of the series are rare rocks composed chiefly of feldspathoid. One such from the plutonic alkaline complex on Kola Peninsula, U.S.S.R., consists of nepheline (70−85%), aegirine, and accessory apatite and albite. Another type from Kenya has rather less nepheline (50−70%), abundant zeolites, aegirine-augite, and accessory calcic phases (calcite, melanite, pectolite, wollastonite). Another from eastern Ontario[°] consists principally of nepheline and a sodic amphibole, hastingsite; and another from the Kola Peninsula is a sodalite-aegirine rock with minor nepheline, cancrinite, natrolite, and eudialyte.

Finally, there are rare members of the ijolite group whose principal feldspathoid is pseudoleucite. In specimens from the Highwood Mountains, Montana,[°°] large pseudoleucites make up between half and two-thirds of the volume, aegirine-augite forms about a quarter, and the rest consists of iron oxides, olivine, apatite, and microlitic sanidine. The pseudoleucites are cloudy spheroids, consisting of granular or fibrous, radiating intergrowths of orthoclase or sanidine, nepheline, and zeolites. These minerals were partly evolved by reaction between original leucite and residual sodic fluids. Many of them are surrounded by prisms of pyroxene, which give the rocks an ocellar texture.

All kinds of extreme combinations of pyroxene, feldspathoids, iron-titanium oxides, perovskite, biotite, melilite, and calcite have been recorded from ijolitic complexes. Here we shall mention only the group of major variants that we can collectively term *alkaline pyroxenites*. These represent the ultramafic extreme of the ijolite series. Their chief component is titaniferous augite or aegirine-augite. In a well-known type (cf. Figure 6-5B) from Jacupiranga, southern Brazil,[°°] iron-titanium oxides come next in abundance to pyroxenes; biotite and melanite, normally

---

[°]*Monmouthite.*
[°°]*Fergusite.*
[°°]*Jacupirangite.*

minor constituents, locally rise to almost half the total volume; distributed sporadically in varying amounts are nepheline, apatite, perovskite, barkevikite, and olivine. Rocks of this kind are habitually associated, as in the Jacupiranga complex, with ijolites (into which they grade naturally) and carbonatites.

REFERENCE NOTES

[1]D. S. Coombs and J. F. G. Wilkinson, *Journal of Petrology*, vol. 10 (1969): pp. 440–501.

[2]See I. S. E. Carmichael, *Contributions to Mineralogy and Petrology*, vol. 15 (1967): pp. 24–66 (on the Leucite Hills); A. Wade and R. T. Prider, *Quarterly Journal of the Geological Society of London*, vol. 96 (1940): pp. 39–98 (on Kimberley).

[3]K. Yagi, *Bulletin of the Geological Society of America*, vol. 64 (1953): pp. 769–810 (on the Morotu province); H. G. Wilshire, *Journal of Petrology*, vol. 8 (1967): pp. 97–163 (on the intrusion near Sydney).

7

Feldspathoidal Felsic Rocks: Intermediate with Basic Variants

Most undersaturated rocks in the intermediate category have feldspathoids and alkali feldspars as principal constituents. Their SiO_2 content is usually between 52% and 60%. There are two main divisions: *phonolites* (volcanic) and *feldspathoidal syenites* (plutonic). Mineralogical-chemical transitions occur, with decreasing feldspathoid content and slight increase in SiO_2, into corresponding varieties of trachytes and syenites containing only minor feldspathoids. Some of the rocks treated in this chapter overlap into the basic category, by increase in the content of titanaugite, or amphibole, and incoming of olivine, or by virtue of an unusually high content of nepheline, analcime, or a sodalite mineral. Then again there are transitional basic rocks with significant plagioclase and olivine that form a natural link between phonolites and basanites, or between feldspathoidal syenites and alkali gabbros.

VOLCANIC DIVISION: PHONOLITE FAMILY

Distribution

Rocks of the phonolite family consistently occur in close association with more abundant basic or ultrabasic lavas such as basanites and nephelinites. So they appear on many oceanic islands—such as St. Helena, Tenerife, and Tahiti—and also in continental provinces of rift tectonics such as the German Rhine Graben and the east African rift region. Only in this continental setting do we find potassic varieties that carry leucite. Phonolites occur, too, in other continental provinces of alkaline volcanism where rifting is not conspicuous. Such are the east Otago field of southern New Zealand and various shallow alkaline ring complexes, composed principally of equivalent plutonic rocks and basic feldspathoidal associates, that dot large sectors of tectonically stable cratons. Very rarely, ♦ as in parts of Kenya, phonolitic lavas occur as extensive floods not immediately associated with basic lavas.

Sodic Phonolites

Most phonolites—including all those of oceanic islands—are sodic rocks with Na_2O/K_2O commonly as high as 2. Typically they are porphyritic with phenocrysts of sanidine or anorthoclase and sodic feldspathoids set in a microcrystalline trachytic groundmass dominated by alkali feldspar laths showing a marked tendency toward fluxional parallelism. Much more rarely the groundmass is largely or entirely glassy.* Many phonolites consist essentially of sodic sanidine or anorthoclase, nepheline and aegirine or aegirine-augite; in some a sodic amphibole may accompany or take the place of pyroxene. If nepheline is present in only small amounts (as in *trachytoid phonolites*), it is generally interstitial, anhedral, and difficult to detect optically. Where more abundant (Figure 7-1B,C) it tends to occur in euhedral crystals (recognizable by hexagonal and square outlines) both as phenocrysts and as small stout prisms scattered through the groundmass between swirling laths of feldspar. Deuteric alteration of nepheline to analcime or other zeolites or to powdery clay minerals is common and tends to conceal the true identity of the primary feldspathoid. Plagioclase is not a typical constituent of phonolites, and where present is usually a sodic variety, much subordinate in quantity to alkali feldspar.

---

*There are, however, exceptions. *Kenytes* from Mt. Kenya and from Antarctica) are black, glassy phonolites containing spectacular large phenocrysts of pink alkali feldspar—a potassic oligoclase optically resembling anorthoclase. The capacity of phonolitic magma to quench to glass during conditions of rapid cooling reflects the low entropy of crystallization, ΔS_m, of feldspathoidal melts (compare highly siliceous rhyolitic obsidians and ultrabasic alkaline limburgites).

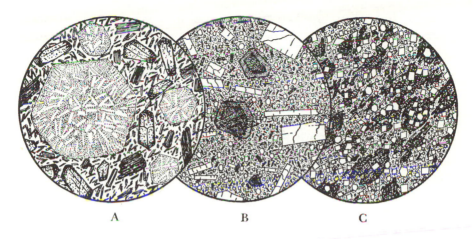

Figure 7-1. Phonolites

A. Mafic pseudoleucite phonolite, Bearpaw Mountains, Montana. Diam. 3 mm. Phenocrysts of pseudoleucite composed of sanidine, cloudy zeolites, and a little nepheline; also of biotite and diopsidic augite, the latter partly fringed with aegirine. Groundmass consists chiefly of aegirine needles, biotite, and anhedral sanidine.

B. Nosean phonolite, Wolf Rock, Cornwall, England. Diam. 2 mm. Phenocrysts of sanidine and zoned nosean, in a groundmass of euhedral nepheline, acicular aegirine, a few sanidine microlites, and a little interstitial turbid analcime.

C. Aegirine phonolite, Lead, South Dakota. Diam. 2 mm. Euhedral nephelines and poikilitic patches of aegirine, in a matrix composed mainly of sanidine microlites.

In a few sodic phonolites, nosean, hauyne, or sodalite accompanies or takes the place of nepheline (Figure 7-1B). Minerals of this group (isometric and thus completely isotropic) may be present as euhedral phenocrysts or as subhedral forms embayed by corrosion; they often exhibit strong color zoning (Figure 7-1B) in shades of gray, brown, or blue, or are rendered almost black by abundant minute inclusions arranged in dense grids. If they occur in the groundmass, they generally retain euhedral and subhedral shapes, but sodalite is commonly found in veins and poikilitic patches or in wedge-shaped spaces between the feldspar laths.

Analcime is seldom completely lacking in phonolites, and in some it is the principal constituent. In a variety found near Crowsnest Lake,

Alberta, analcime forms two-thirds of the bulk, occurring in reddish phenocrysts up to 2 cm across and as smaller, second-generation crystals in a fine matrix of nepheline, sanidine, aegirine-augite, and melanite garnet. In another Canadian phonolite, two-thirds of the volume is euhedral phenocrysts of analcime and anorthoclase in approximately equal amounts, accompanied by phenocrysts of augite and olivine in an interstitial matrix equivalent in composition to anorthoclase. And in the classic alkaline region of the Highwood Mountains of Montana, olivine-bearing analcime phonolites are among the most widely distributed lavas. They carry phenocrysts of olivine, of diopsidic augite fringed with green sodic pyroxene, and (a few) of biotite and barium sanidine, in a ground-mass of the same minerals plus abundant analcime.

The mafic minerals of sodic phonolites display unusual variety. In the more basic phonolites, a somewhat ferriferous olivine, diopside, and titanaugite are common. In most varieties, however, the principal pyrox-enes are aegirine-augite and aegirine; the former, which crystallizes first, forms stumpy phenocrysts or roughly equant granules; the latter either envelops the former or is grouped in ragged bundles of needles and mossy shreds that surround both the feldspar and the nepheline in pseudo-poikilitic fashion. The most characteristic amphiboles are the sodic varieties riebeckite, arfvedsonite, cossyrite, kataphorite, and, occa-sionally, barkevikite. Because all these crystallize at a late stage, they sel-dom form phenocrysts; they normally occur, like aegirine, in tufty inter-stitial clusters. And, like aegirine, they tend to be altered to brownish iron oxides. Biotite is not common in phonolites; it is never the only mafic constituent, but in rare instances it may be the predominant one.

Among the minor accessories, apatite (typically a "smoky" type crowded with minute dark specks) and zircon are scarce, but sphene is widespread, and the typical opaque mineral is titaniferous magnetite. Chemical analyses of phonolites commonly show high ZrO_2, much of which is accommodated in zirconium silicates such as eudialyte.

The term *tinguaite*, once used widely to cover "hypabyssal" dike rocks of phonolitic composition, still survives in modern literature. Many of these rocks (as exemplified in rare specimens from the east Otago prov-ince) are conspicuously porphyritic, with large phenocrysts of sanidine and feldspathoid (Figure 7-2) and in some rocks of sodic pyroxene as well. Other tinguaites of narrow dikes are essentially fine-grained rocks with few phenocrysts. In fact tinguaites differ in no respect (other than mode of occurrence) from sodic phonolites; the name has become redundant.

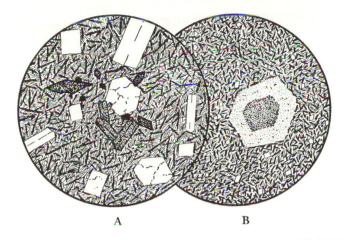

A   B

Figure 7-2. Tinguaites

A. Tinguaite, Serra de Tingua, Brazil. Diam. 2 mm. Phenocrysts of nepheline, sanidine, aegirine-augite, and sphene, in a groundmass composed chiefly of sanidine microlites and ragged needles of aegirine, with a little nepheline and analcime.

B. Sodalite tinguaite, Mühlörzen, Bohemia. Diam. 2 mm. Phenocrysts of zoned sodalite, in a dense matrix of sanidine, acicular aegirine, and euhedral nepheline.

Leucitophyres (Leucite Phonolites)

Leucitophyres are potassic rocks in which the principal or sole feldspathoid is leucite (or "pseudoleucite"). They are confined to and are typical of certain continental volcanic provinces where the dominant basic lavas are also potassic—leucite basanites and leucitites. No example is so familiar as the leucitophyres of the German Eifel province (see Figure 6-3A). The largest phenocrysts in these rocks are leucites showing sector twins; others consist of embayed dark-bordered nosean, zoned augite increasingly green and sodic toward the rims, and deep-brown melanite garnet. In some specimens, sphene also forms phenocrysts. The dense groundmass is composed mainly of sanidine microlites, leucite, aegirine-augite or aegirine, opaque oxides, apatite, and a little interstitial nepheline. Occasionally one may also see a small amount of hornblende and biotite.

Mafic leucite and pseudoleucite phonolites, such as that depicted in Figure 7-1A, are abundant in the Highwood and Bearpaw Mountains of Montana and in Murcia, Spain. They differ from orendite by having lower K_2O and darker color (CI about 50); they also have less sanidine and they contain olivine; and their typical sanidine is the barium-rich variety hyalophane. Their leucites are seldom fresh; generally they are completely or partly replaced by analcime or are altered to mixtures of sanidine, nepheline, and cloudy zeolites.

By increase in content of plagioclase, leucite phonolites grade into leucite tephrites and basanites; by disappearance of all feldspars they pass into leucitites.

PLUTONIC DIVISION: FELDSPATHOIDAL SYENITE FAMILY

Feldspathoidal Syenites

Feldspathoidal syenites, mostly of intermediate SiO_2 content, have alkali feldspar and a feldspathoid as principal constituents. They may form small plugs or ring dikes in eroded eruptive centers of essentially volcanic provinces dominated by basanite, nephelinite, and phonolitic lavas. But their best-known mode of occurrence is as principal components of sub-volcanic ring complexes scattered over stable continental cratons (notably in southern Africa and Brazil). Here their usual plutonic associates are rocks of the essexite, theralite, and ijolite families. Very rarely (as in the Ilimaussaq intrusion in Greenland and the great alkaline complex of the Kola Peninsula, U.S.S.R.) massive layers of nepheline-rich syenite with internal rhythmic layering are among the main components of large layered alkaline plutons.

Though among the least abundant of plutonic rocks, feldspathoidal syenites have always attracted special notice by reason of the great variety of their minerals, many of which are rare and hence of peculiar interest. Texturally, also, they are highly variable, and in addition, they exhibit a wide range in silica content, depending on whether feldspathoids or alkali feldspars predominate. It is not surprising, therefore, that feldspathoidal syenites have been given an excessive number of varietal names—here either omitted or relegated to footnotes.

Almost all feldspathoidal syenites are rich in sodium, for their commonest feldspathoids, in order of abundance, are nepheline, analcime, sodalite, and rarely nosean. Their sodic character is further reflected by predominance of sodic feldspars in many rocks—perthites, antiperthites, and albite—and the presence of such mafic minerals as aegirine-augite, aegirine, arfvedsonite, and barkevikite. Such rocks typically are peralkaline. In basic varieties titanaugite is common, and there may be olivine

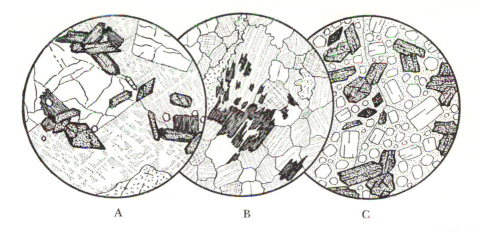

Figure 7-3. Feldspathoidal Syenites

A. Nepheline syenite (foyaite), Serra de Monchique, Portugal. Diam. 3.5 mm. Aegirine-augite with darker rims of aegirine; brown biotite; sphene, apatite, and opaque oxides. Large clear areas consist of nepheline; a little sodalite (bottom of section); the remainder is partially altered perthite.

B. Albite-rich nepheline syenite, Litchfield, Maine. Diam. 3 mm. Dark-green biotite; albite (with strip twinning); microcline (with grid twinning); orthoclase (in single twins); nepheline (clear); and cancrinite (stippled).

C. Nepheline syenite porphyry, Hot Springs, Arkansas. Diam. 3 mm. Euhedral and rounded crystals of nepheline, together with aegirine-augite and sphene, are enclosed poikilitically by crystals of sanidine up to 1 cm across.

as well; but as expressed in the saturation principle discussed on pp. 74–76, orthopyroxenes invariably are absent. Biotite, when present, is rich in both iron and titanium, and the usual opaque oxides also are titaniferous. In rarer peraluminous varieties, muscovite and corundum are fairly common. Apatite is almost always abundant, and zircon, sphene, and zirconium- and titanium-bearing silicates are far more plentiful than in most other plutonic rocks. Melanite garnet is also widespread, and many rocks carry deuteric cancrinite and calcite.

The most abundant feldspathoidal syenites have little or no plagioclase. Typical of these are nepheline syenites° with subequal amounts of alkali feldspar and nepheline (Figure 7-3A). In these rocks nepheline is usually euhedral or subhedral and is enclosed poikilitically by sodic orthoclase or microperthite. Many types contain much sodalite and

_____

°*Foyaites.*

analcime, and a few contain hauyne or nosean as well. Aegirine-augite is the principal mafic mineral in most specimens, but hornblende, biotite, or aegirine may predominate, and many varieties carry melanite garnet. Sphene, titaniferous iron oxides, and apatite are always present as accessories. Deuteric cancrinite and calcite, in part replacing nepheline, seem to be especially common in nepheline syenites close to contacts with carbonatite or limestone. A nepheline-syenite porphyry of similar chemical composition emplaced at shallow depth at Hot Springs, Arkansas, is illustrated in Figure 7-3C. In this the alkali feldspar is sanidine, occurring as large poikilitic plates enclosing corroded grains of nepheline and prisms of aegirine-augite.

In the Oslo province the principal feldspathoidal syenite (greatly subordinate in extent to both alkali syenites and peralkaline granites) is a basic type (SiO_2 <50%).° An average mode, compiled by W. C. Brøgger, is sodic orthoclase, 62; nepheline, 13; sodalite, 2; clinopyroxenes, 8; biotite, 9.5; opaque oxides, 3; apatite, 2. But the color index, it will be noted, is still low (<18).

A few feldspathoidal syenites rich in potassic feldspar have hastingsite as their chief mafic mineral. An example from Red Hill, New Hampshire, consists chiefly of microperthite, partly mantled with albite; approximately a third is composed of nepheline or sodalite or both, and the rest is made up of hastingsite, aegirine-augite, a little biotite, magnetite, pyrite, and pyrrhotite.

Other feldspathoidal syenites with K-feldspar are peralkaline, their chief mafic minerals being arfvedsonite, eckermannite, and aegirine. Remarkable rocks of this character are found in southern Sweden, some of which, especially those of coarse grain, are extraordinarily rich in apatite and sphene and in such zirconium-bearing minerals as eudialyte, catapleite, rosenbuschite, låvenite, and zircon. Indeed, in some specimens catapleite is a characterizing rather than an accessory mineral forming large phenocrysts as well as anhedral grains in the groundmass along with eudialyte. Rocks that carry many of these zirconium minerals are generally free from carbonates, and vice versa. Pectolite, which is secondary after eudialyte and aegirine in most feldspathoidal syenites, is here a primary constituent and is often abundant.

There are other nepheline syenites that locally display the same extraordinary concentration of zirconium minerals. Examples are laminated rocks from the outer rings of the Poços de Caldas ring complex

---

°*Lardalite* of Norwegian literature.

(dominantly normal nepheline syenite) of Brazil; in hand specimens, dark mafic laminae alternate with others that are pink by virtue of concentrated minerals of the eudialyte family. Similar rocks are conspicuous in the massive layered plutons of the Kola Peninsula, U.S.S.R., and Ilimaussaq in Greenland.

Among the otherwise rare minerals not uncommonly found in nepheline syenites (especially in pegmatitic varieties) is astrophyllite, a titanosilicate of K, Na, Mg, and Fe^{2+}. It occurs in tabular crystals with perfect {001} cleavage and striking pleochroism somewhat reminiscent of that in biotite: α deep orange-red, $\beta = \gamma$ pale yellow (parallel to the cleavage). Note, however, that the maximum absorption (parallel to α) is transverse to the cleavage.

Albite-rich nepheline syenites are well represented by rocks from Litchfield, Maine, illustrated in Figure 7-3B. About half of a typical specimen consists of albite, a quarter of orthoclase and microcline, and about 15% of nepheline. Green biotite is the only mafic mineral, and the usual accessories are cancrinite, sodalite, calcite, sphene, iron-titanium oxides, zircon, and apatite. Somewhat similar peraluminous syenites in the Blue Mountains of Ontario contain muscovite, corundum, hastingsite, and garnet. In other syenites, corundum may be present to the exclusion of nepheline, the accompanying minerals being microperthite and biotite. A few albite-rich nepheline syenites carry aegirine as well as or in place of biotite. In rocks transitional into nepheline-bearing monzonites, the plagioclase is oligoclase rather than albite and is accompanied by microperthite, the two together being far in excess of the feldspathoids.

Analcime syenites are especially common as late-formed streaks and veins within sills of analcime diabase and teschenite, though they also occur as discrete intrusive bodies. They are marked by interstitial analcime and by large tablets of anorthoclase. They also contain anorthoclase as irregular grains in the groundmass, along with laths of plagioclase that range in composition from labradorite to oligoclase. Slender prisms of grayish-violet titanaugite, fringed with green aigirine, predominate among the mafic constituents; less common are arfvedsonite and barkevikite. Ilmenite and apatite are always abundant. Some of the analcime appears to have developed during the magmatic stage while aegirine rims were forming round the titanaugites; some developed during the deuteric stage, along with calcite and zeolites, and partly replaced the plagioclase. In the Terlingua-Solitario region of Texas, where rocks of this kind are widespread, analcime is much more abundant in large intrusions emplaced at considerable depth than it is in smaller bodies intruded closer to the surface.

Pseudoleucite syenites are relatively rare rocks. In an example from Magnet Cove, Arkansas, some of the grains of pseudoleucite, composed of K-feldspar, nepheline, and zeolites, are several centimeters in diameter and make up as much as a third of the bulk. The groundmass in which they lie consists of abundant euhedral nepheline, melanite, diopside, aegirine, and a little alkali feldspar. A somewhat similar melanite-rich pseudoleucite syenite° comes from Loch Assynt, Scotland, but this is devoid of pyroxenes and much richer in orthoclase than the Magnet Cove rock, and carries a good deal of green biotite. Fragments of syenite that actually contain fresh leucite are found among the blocks ejected from Vesuvius. These have abundant sanidine and sodalite; the minor constituents are melanite, hornblende, augite, biotite, iron oxides, apatite, and sphene. The mineral assemblage is volcanic rather than plutonic. Two alternative modes of origin have been proposed for pseudoleucite. During crystallization of undersaturated potassic melts in the system $NaAlSiO_4–KAlSiO_4–SiO_2$ late fractions trend toward a ternary invariant point at which early-formed leucite reacts with the melt to yield nepheline plus anorthoclase (or sanidine). Alternatively, especially in plutonic rocks, pseudoleucites could form in the postmagmatic stage by reaction between leucite and aqueous sodic fluids.

Miscellaneous Basic Types

Certain rarer mafic feldspathoidal rocks that carry plentiful K-feldspar but lack plagioclase are best treated as variants of the feldspathoidal syenite family. The approximate mode of one such rock from Ontario°° (Figure 7-4A), is K-feldspar, 20; nepheline, 20; aegirine-augite, 50; apatite, melanite garnet, opaque oxides, and sphene, 10. The nepheline is mostly euhedral, and the K-feldspar tends to be coarse and poikilitically related to the other minerals.

Shonkinites (Figure 7-4B,C), as developed in the lower levels of the Shonkin Sag sheet, Montana (cf. p. 157), are more familiar. They are partially cumulate rocks, a typical mode being as follows: sanidine, 30; augite, 27; biotite, 18; olivine, 4; opaque minerals and apatite, 3; mixed zeolites and carbonates, 18.[1] Some petrologists consider shonkinites as mafic syenites; but because of their high normative *ne* (~15) and low

---

°*Borolanite.*
°°*Malignite.*

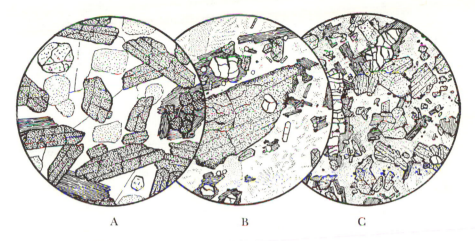

A   B   C

Figure 7-4. Malignite and Shonkinites

A. Mafic nepheline syenite (malignite), Poohbah Lake, Ontario. Diam. 3.5 mm. Deep-green aegirine-augite, green biotite, and melanite garnet are the dark constituents. Between them lie cloudy euhedral nephelines and grains of apatite embedded in poikilitic plates of orthoclase.

B. Shonkinite, Shonkin Sag, Montana. Diam. 5 mm. Approximately 8 m from the base of the sheet. Pale-green augite with rims of dark-green aegirine-augite; olivine, enclosed by both pyroxene and biotite; groundmass of sanidine, in part graphically intergrown with zeolitized nepheline; accessory apatite.

C. Shonkinite, Spanish Peaks, Colorado. Diam. 2 mm. The chief mafic minerals are diopsidic augite (partly fringed with aegirine), dark-brown biotite, and fresh olivine. The groundmass consists of large anhedral poikilitic plates of altered sanidine, together with a little analcime. Accessory opaque oxides and apatite.

content of SiO_2 (47–48%), they are here treated as mafic feldspathoidal syenites.

Finally, we note a suite of rocks from the Crazy Mountains, Montana, the average mode of which is augite, 38; barium orthoclase, 20; nepheline, 14; sodalite minerals, 10; biotite, 7; iron oxides, 6; olivine, 2; aegirine, 2; apatite, 1. Minor accessories include sphene, zeolites, and cancrinite. The pyroxene forms phenocrysts, in which cores of diopsidic augite are fringed with aegirine-augite and aegirine.* These rocks can

---

*Through misidentification of the feldspar as plagioclase, these rocks were originally designated as *theralite* (cf. p. 211).

be regarded as one of the mafic and ultrabasic extremes of the feld-spathoidal syenite family.

REFERENCE NOTE

[1]W. D. Nash and J. F. G. Wilkinson. *Contributions to Mineralogy and Petrology*, vol. 25 (1970): pp. 241–269.

8

Lamprophyres and Ultrabasic Rocks of Extreme Composition

In this chapter we discuss an arbitrarily grouped set of rocks having no unifying common trait except that none can be treated in the framework of Chapters 3 through 7. Five separate lithologic entities will be considered in the following order: (1) lamprophyres; (2) melilite-rich rocks; (3) carbonatites; (4) kimberlites; (5) peridotites and allied rocks. All but the last named are minor rocks of limited total volume and global extent.

There is no continuous lithologic spectrum. Nevertheless, certain of these rock types share some aspects of bulk chemistry, are rather commonly but by no means invariably associated in the field, and may have genetic affinities in common—for example, carbonatites with some melilite rocks and perhaps with kimberlites (but much more commonly with nephelinites and ijolites); ultrabasic lamprophyres with some melilite rocks (alnöites); kimberlites with mica peridotites. Notwithstanding such tenuous internal relationships, these five types have been brought together mainly as a matter of convenience to round off coverage of a set of phenomena—those of igneous petrography, which, as stated at the outset, are too complex for simple uniformly systematic presentation.

LAMPROPHYRES

Lamprophyres as a Lithologic Class

By long-established usage the term *lamprophyre* covers a diversified group of dark-colored basic and ultrabasic rocks habitually occurring in swarms of narrow dikes, less commonly as thin sills, flows, and plugs. All are relatively rich in alkalies and in (FeO + MgO). But their principal diagnostic character as a group is textural: lamprophyres are porphyritic,* with two generations of mafic minerals (combinations of biotite, amphibole, and augite), each generation in the form of euhedral crystals.° Feldspars where recognizable are confined to the groundmass. Olivine, however, occurs exclusively as phenocrysts and is seen mostly in ultrabasic varieties.

Lamprophyric magmas apparently are unusually rich in H_2O, CO_2, P_2O_5, and other fugitive components. This is reflected in the ubiquitous presence of hydrous primary phases (micas, amphiboles, analcime), abundant apatite, groundmass carbonates, and the presence of hydrous secondary minerals of the serpentine, chlorite, and zeolite families. Many lamprophyres are unusually rich, too, in certain minor and trace elements—Ti, Ba, and Sr being particularly noteworthy.

On the basis of field occurrence, texture, and gross mineralogical character lamprophyres are readily recognizable as such. But when chemistry and mineralogical detail as well as mutual association with other rock types are taken into account, it becomes obvious that the lamprophyre class is highly heterogeneous and that specific groups within it have totally different petrogenic significance. Not surprisingly, therefore, a profusion of rock names has developed, all well established in accepted specialized usage,† though some names have minimal value, and the diagnostic characteristics of others, though clearly intelligible in classificatory schemes, are petrographically unrecognizable in many rock specimens. This last statement applies particularly to identity of feldspar,

---

*The root *phyre*, as in *porphyry* or *leucitophyre*, designates a porphyritic character: *lamprophyre* = "glistening porphyry."
°*Panidiomorphic* texture of some writers.
†A clear discussion of lamprophyre nomenclature is provided by N. M. S. Rock, *Earth-Science Reviews*, vol. 13 (1977): pp. 123–169. In A. Strecheisen's summary of the IUGS scheme of classification of volcanic rocks (*Geology*, vol. 7 [1979]: pp. 331–335), approximately one-fifth of the discussion treats the nature and classification of this interesting, but insignificant, group of rocks.

which is highly susceptible to replacement by secondary minerals, especially carbonates.

We prefer to recognize two groups of lamprophyre, each characterized by abundance of one or more readily recognizable essential minerals (mostly mafic) and each covering a rather limited range of SiO_2 content. In IUGS terminology these are designated calcalkaline and alkaline respectively.

1. *Calcalkaline:* biotite and hornblende lamprophyres—approximately silica-saturated mildly potassic rocks (K > Na) ranging in SiO_2 content from about 50% to 54%.
2. *Alkaline:* ultrabasic lamprophyres ($SiO_2 = 38–44\%$), undersaturated in silica and distinctly sodic. The distinctive mineralogical feature is abundance of brown alkali amphibole. Principal rock types are *camptonites* and *monchiquites*.

Commonly placed with lamprophyres is a third group, *alnöites*, extremely basic in composition, with melilite an essential constituent. They show some affinities, and in some cases are associated in the field, with melilite-rich lavas, carbonatites, and even kimberlites. They are considered later in this chapter together with these rocks.

Calcalkaline (Biotite and Hornblende) Lamprophyres

Calcalkaline lamprophyres are most commonly associated with granitic and dioritic intrusive complexes. They are approximately saturated basic rocks made up essentially of biotite, green or brown hornblende, diopsidic augite and feldspar, with or without a little accessory quartz or altered olivine. Their color indices range from about 35 to 50. Either orthoclase or plagioclase may be the predominant felsic phase, although both are commonly altered beyond reliable identification; the predominant hydrous mafic mineral may be either hornblende or biotite.* Diopsidic augite is almost always present.

---

*Classic nomenclature, adopted too by the IUGS under the collective head "calcalkaline lamprophyres," may be summarized thus:

Biotite lamprophyres: *minette* (with K-feldspar); *kersantite* (with plagioclase). Hornblende lamprophyres: *vogesite* (with K-feldspar); *spessartite* (with plagioclase).

Most potassic are *biotite lamprophyres* in which the chief felsic mineral is K-feldspar. Typical specimens (cf. Figure 8-1A,C) are composed principally of biotite phenocrysts and subordinate euhedral and subhedral prisms of colorless or pale-green diopside, in a granular or poikilitic groundmass consisting mainly of K-feldspar and is generally rich in apatite, sphene, and iron-titanium oxides, usually with much calcite. The flakes of biotite are often characterized by "battlemented" ends (Figure 8-1C), and not infrequently they have dark-reddish-brown rims rich in iron and titanium enclosing pale-yellow cores rich in magnesium. Close to the edges of many flakes there may be sagenite webs of rutile; and secondary sphene is common in biotites that have been chloritized. Although diopside or diopsidic augite is the usual pyroxene, its place may be taken by pale-lilac titanaugite, and each of these may be partly or wholly replaced by mixtures of chlorite, calcite, and epidote. The chief

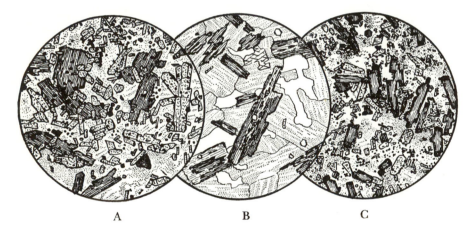

A   B   C

Figure 8-1. Biotite Lamprophyres

A. Biotite-augite lamprophyre Walsen dike, Spanish Peaks, Colorado. Diam. 1 mm. Diopsidic augite, reddish-brown biotite, and accessory iron oxides, apatite, and sphene, in a matrix of large anhedral plates of sanidine partly altered to zeolites.

B. Biotite lamprophyre (kersantite), Brest, Brittany. Diam. 1.5 mm. Flakes of chloritized biotite containing secondary sphene and ilmenite; altered laths of andesine; interstitial quartz and calcite; accessory prisms of apatite.

C. Biotite lamprophyre (minette), Trujillo Canyon, Spanish Peaks, Colorado. Diam. 2 mm. Biotite with "battlemented" ends, dark iron-rich rims, and pale magnesian cores; prisms of diopsidic augite; granular oxides and sphene; and needles of apatite. Groundmass composed of subhedral and anhedral plates of poikilitic sanidine, partly altered, with a little interstitial analcime.

felsic constituent is a potassic feldspar, sometimes rimmed with albite or oligoclase or accompanied by discrete crystals of those feldspars. A little interstitial quartz is present in most mica lamprophyres, and, in addition, many contain ovoid xenocrysts of quartz surrounded by reaction rims of slender diopside prisms. Olivine may be present, but is invariably altered to serpentine, chlorite, or carbonate. Some rocks of this class grade into olivine leucitites as biotite and K-feldspar give place to olivine and leucite, as happens for example in the Navajo region of Arizona and New Mexico.

Biotite lamprophyres in which plagioclase is the principal felsic constituent are illustrated by the rock in Figure 8-1B, which has the following mode: plagioclase (An_{30}), 53; chloritized biotite, 24; chloritized augite, 8; quartz, 9; calcite, 4; apatite and iron-titanium oxides, 2. In rocks of this general type, plagioclase typically occurs as subhedral or anhedral crystals in the groundmass. Usually it is andesine, though it may be oligoclase; almost invariably it is normally zoned and more or less altered to clay minerals and carbonates. Subordinate K-feldspar is commonly present filling interstices, and quartz may also occur interstitially, either as discrete patches or in micrographic intergrowth with K-feldspar. The characteristic mafic minerals are somewhat titaniferous biotite together with diopsidic augite. Subordinate hornblende may be present; altered olivine is rare.

In hornblende lamprophyres, the principal mafic mineral is green or brown hornblende, which typically occurs in euhedral prismatic crystals comprising a quarter to half of the rock. In some rocks it is in two generations, in others (Figure 8-2C) only one. Diopsidic augite is invariably present and may even be more abundant than hornblende. Biotite and altered olivine are common, though relatively minor components, along with the usual accessory minerals apatite, sphene, and iron-titanium oxides. Feldspar, either plagioclase or K-feldspar, or both, makes up half or more of most rocks. The plagioclase in hornblende lamprophyres is generally andesine but tends to be more calcic than that in biotite lamprophyres. Minor interstitial quartz is occasionally present. Hornblende-plagioclase lamprophyres (Figure 8-2C), apart from their "panidiomorphic" texture, could equally well be called microdiorites.

Alkaline Lamprophyres

Alkaline lamprophyres are at the same time both alkaline and ultrabasic, in that the norm shows significant *ne* (8–12) and the SiO_2 content is between 38% and 42%. Apart from higher H_2O and CO_2 these lamprophyres in fact are chemically similar to nepheline basanites and nephelinites. In the field they are commonly associated with these very rocks, or with related alkaline plutonic types such as feldspathoidal gabbros and nepheline syenites. Best known from continental occurrences, they also appear—as on Tahiti—in the ocean-island setting.

Mineralogically, alkaline lamprophyres are characterized by an abundance of titaniferous or sodic augite and of a brown alkaline amphibole (kaersutite or barkevikite); one or both, in accordance with the usual lamprophyre pattern, occur in two generations. Olivine and biotite are both possible additional phenocrystic minerals.

Classic terminology is both confused and needlessly elaborate; for it stresses minor mineralogical variations in a group of rocks that tend to be individually variable within a single limited geographical province or even in a single minor intrusion. Following classic terminology and recently sanctioned IUGS usage, we distinguish two general types of alkaline lamprophyre—*camptonites* and *monchiquites*. Even so, the distinction is by no means sharp and possibly trivial; for it may perhaps be determined largely, not by differences in fundamental chemistry, but by accidental factors such as different cooling rates.

Camptonites

Camptonites are alkaline lamprophyres with recognizable plagioclase in the felsic fraction of the groundmass. There is considerable variation among the rocks at the type locality, Campton Falls, New Hampshire. In some specimens (Figure 8-3C) barkevikitic hornblende occurs both as large, stout phenocrysts and as needles in the matrix, olivine and titanaugite being rare or absent. In other specimens the last two minerals and labradorite form the phenocrysts, while the groundmass is composed mainly of titanaugite, barkevikite, labradorite, and a little alkali feldspar plus the usual apatite and titaniferous iron oxides. Still other specimens contain scattered phenocrysts of biotite. In all varieties there are usually much deuteric calcite and zeolites, either filling amygdules or in irregular patches.

Monchiquites

By accepted definition, monchiquites are distinguished from camptonites by absence of feldspar, instead of which the felsic matrix of the groundmass is a colorless isotropic material—either sodic glass or anal-cime (itself recognizable in some cases as the product of devitrification). Typical specimens from the type locality in the Serra de Monchique, southern Portugal, have the following mode: phenocrysts of titanaugite, with fewer of barkevikite and biotite, 24; serpentinized olivine, 5; iron-titanium oxides, 4; the remainder (two-thirds of the rock) a glassy matrix (potentially a mixture of feldspar and nepheline) rich in microlites of pyroxene, barkevikite, opaque oxides, picotite, and apatite, and in deu-teric calcite, aragonite, and zeolites. Other varieties carry accessory mel-ilite, nepheline, hauyne, or leucite. Monchiquites from other localities are illustrated in Figures 8-2AB and 8-3B.

From these general descriptions it is clear that the only certain crite-rion that distinguishes camptonites from monchiquites lies in the nature

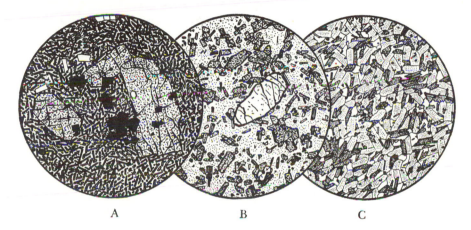

A   B   C

Figure 8-2. Lamprophyres

A. Monchiquite, near Wesseln, Bohemia. Diam. 2 mm. Phenocrysts of titanaugite and titanomagnetite, in a groundmass chiefly composed of barkevikite needles, specks of iron oxide, interstitial analcime, calcite, and a little glass. A few microphenocrysts of labradorite.

B. Monchiquite, Spanish Peaks, Colorado. Diam. 2 mm. Olivine, titanaugite, and dark-brown biotite, in a matrix composed mainly of colorless glass and anal-cime strewn with euhedral granules of magnetite and a few minute laths of sanidine.

C. Hornblende lamprophyre, Aschaffenburg, Bavaria. Diam. 2 mm. Euhedral and subhedral laths of zoned andesine-oligoclase; abundant prisms of green hornblende and a smaller amount of biotite. A little quartz and orthoclase occupy irregular interspaces, and needles of apatite are dispersed through-out.

of the felsic element in the groundmass: predominantly feldspars in the former, entirely isotropic (glass or analcime) in the latter. But what real significance can be attached to so finely drawn a distinction? This question will now be addressed with reference to a single well-documented instance.

Variation in a Suite of Alkaline Lamprophyres

Along a north-south belt 100 km in length, the Haast schists of southwestern New Zealand are cut sporadically by a swarm of alkaline (ultrabasic) dikes. The predominant rock type throughout is camptonite, but monchiquites also are well represented, particularly in the southern segment of the dike swarm, which has been described in detail by A. F. Cooper.[1] Locally, especially to the north, a few dikes of such diverse rocks as carbonatite, hornblende-mica peridotite, and tinguaite have also been recorded. But the whole suite of dikes is demonstrably cogeneric. And throughout the lamprophyre belt and far beyond, there is no trace of other igneous rocks of like age in an area of high topographic relief (> 2500 m).

The overall mineralogical pattern common to the two principal rock types is strikingly uniform. Strongly zoned phenocrysts of kaersutite and titanaugite and less abundant phenocrysts of olivine, titanomagnetite, rarely biotite or sphene, are enclosed in a groundmass in which a second generation of the same phases is set in a felsic base. In the southern segment of the lamprophyre belt, camptonites predominate. In these the felsic matrix consists mainly of plagioclase, An_{28} (now largely replaced by metamorphic albite and enclosed Ca–Al silicate grains), and minor quantities of alkali feldspar. But there may also be a small isotropic residue (glass or its devitrification product analcime), and interstitial patches of calcite are common. Monchiquites, in which the felsic fraction consists of pale-brown isotropic glass of phonolitic composition and is devoid of feldspar, also are well represented. They tend to occur as chilled marginal facies of dikes whose central lithologic component is holocrystalline camptonite. There is no perceptible chemical trait unique to either type. Both are ultrabasic ($SiO_2 \sim 40\%$), and both carry high normative *ne* (7–10) and *ab* (15–20); Na_2O in all cases greatly exceeds K_2O.

It seems that the classic distinction between camptonite and monchiquite is trivial. But it is part of generally accepted petrographic jargon and has received the sanction of IUGS nomenclature, and for these reasons is retained here.

Ocellar Texture

Characteristic of alkaline lamprophyres in many parts of the world is a striking textural feature that is carefully discussed by Cooper (1979) in his account of the New Zealand swarm. Scattered through the generally dark-colored rock (either camptonite or monchiquite) are light-colored spheroidal or ovoid bodies, *ocelli,* 1 or 2 mm in diameter. In the New Zealand rocks these tend to become concentrated toward the middle of individual dikes, where their aggregate may be as much as 15% of the total volume. Each ocellus, sharply bounded against the enclosing host, consists of the same minerals (olivine excepted) as the latter, and individually the crystals of each phase show identical zonal patterns in ocellus and host. But their respective proportions in the two are very different: In the ocellus, felsic phases greatly predominate, and slender prisms of kaersutite are much more plentiful than grains of augite and oxides. The ocellus boundary tends to be sharpened by internal tangential alignment of kaersutite prisms in the immediate vicinity. Enclosed in most ocelli is a core or eccentric globule of different composition, consisting in most cases of calcite and analcime. It must be emphasized that in monchiquites the felsic phase in the main body of the ocellus is glass, identical in composition with that in the matrix of the host.

A convincing case has been presented by Cooper (1979) for the origin of ocelli in both rock types by a two-stage process. When crystallization is well advanced the residual liquid boils to yield two contrasted immiscible fluid fractions: a silicate-melt phase saturated in H_2O and CO_2, and a gaseous phase consisting of H_2O, CO_2, and dissolved solids. Vesiculation results. With further cooling and condensation of the gas phase (to yield the calcite-analcime core or globule filling) the residual silicate-melt phase in the host rock is drawn in to fill the remaining space in the vesicle. It is the melt phase thus fractionated that ultimately solidifies to form the main body of the resulting ocellus.*

MELILITIC ROCKS

Melilite is by no means a common igneous mineral; but we have already noted its occurrence as an accessory in some monchiquites and in greater

---

*This model, consistent as it is with an impressive body of petrographic and chemical evidence, seems much more convincing than alternatives that invoke unmixing into two *melt* fractions, as favored by some petrologists.

abundance in certain nephelinites and olivine leucitites. Associated with and grading naturally into rocks of all three groups are rocks devoid of feldspathoid (and of feldspars) in which melilite assumes essential status (> 10% of the mode). There are three categories of these melilitic rocks: alnöites, melilitites, and melilite pyroxenites.

Alnöites

Alnöites, sometimes grouped with the ultrabasic lamprophyres, are among the most basic of all igneous rocks; some contain less than 25% SiO_2. They occur as dikes in intimate association with both volcanic and plutonic alkaline rocks of continental igneous provinces. Some are feeders to flows of olivine melilitite and of limburgite. But mostly they are associated with, and in some way apparently related to, nepheline syenites and especially ijolites.

The essential mineral assemblage comprises phenocrysts of biotite, with or without olivine, in a matrix of melilite, carbonates, and commonly perovskite. A typical variety at the type locality, Alnö Island, Sweden (Figure 8-3A), has a groundmass of ferruginous biotite, melilite, and carbonates with accessory perovskite, chromite, garnet, and pyrrhotite. This encloses phenocrysts of dark biotite (with reddish titaniferous rims) and smaller phenocrysts of olivine, augite, iron-titanium oxides, and apatite. Some rocks also carry a little barkevikite. Other alnöites, notably those of Isle Cadieux, Quebec, are remarkable for the presence of the calcic olivine monticellite (distinguished from common magnesian olivine by lower birefringence, $(\gamma - \alpha) < 0.02$, and $2V_\alpha = 72°–80°$). Here, associated minerals are melilite, biotite, marialite, perovskite, and titanomagnetite. Monticellite may be partially replaced by biotite, the extreme end product being seen in streaks of melilite-biotite rock as thick as 50 cm. More than 50 years ago N. L. Bowen proposed for the Isle Cadieux rocks a genetic model involving prolonged reaction between forsteritic olivine and fluids rich in H_2O, CO_2, and appropriate cations.

Only rarely alnöites carry minor amounts of interstitial feldspathoid—nepheline or hauyne.

By increase in forsterite and diminution of augite, some alnöites grade into kimberlites (essentially biotite-olivine rocks). Others, by increase in the content of carbonates, pass naturally into carbonatites that still carry olivine, melilite, melanite, apatite, and pyrite as accessories; and in the field, alnöites may be associated with either or both of these rock types.

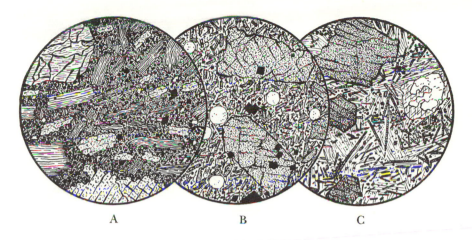

A   B   C

Figure 8-3. Alnöite and Alkaline Lamprophyres

A. Alnöite, Alnö Island, Sweden. Diam. 2 mm. Phenocrysts of phlogopite, melilite, diopsidic augite (bottom of section), and serpentinized olivine (upper left), in a dense groundmass of chromite, perovskite, sphene, and a little calcite. Not shown, but common in alnöites, are microphenocrysts of apatite.

B. Leucite monchiquite, Bohemia. Diam. 2 mm. Phenocrysts of purple titanaugite; microphenocrysts of titanomagnetite and leucite. Groundmass composed of slender prisms of titanaugite, ragged flakes of biotite, thin plates of ilmenite, and interstitial analcime.

C. Camptonite, Campton Falls, New Hampshire. Diam. 2 mm. Phenocrysts and small needles of dark-brown barkevikite, and euhedral granules of iron oxides in a matrix of subradiating slender laths of andesine, rich in acicular apatite. Amygdule filled with calcite and analcime. In other specimens from this locality there are phenocrysts of titanaugite and serpentinized olivine.

Melilitites

Volcanic rocks with abundant melilite and olivine but totally lacking either feldspathoids or feldspars have been termed *melilitites*. Very rare and extremely basic (SiO_2 = 30–35%), they are limited to a few volcanic fields—the best known in eastern and southern Africa—where the dominant rocks are nephelinites or olivine leucitites. Some of these latter contain melilite, at least in minor quantity, so that melilitites can be regarded as end members of spectra whose most abundant representatives are ultrabasic (and ultramafic) feldspathoidal rocks.

An example from Germany is shown in Figure 8-4C. Half of it consists of olivine and about a third of melilite. Accessory minerals include nepheline (typically absent from melilitites), apatite, perovskite, calcite,

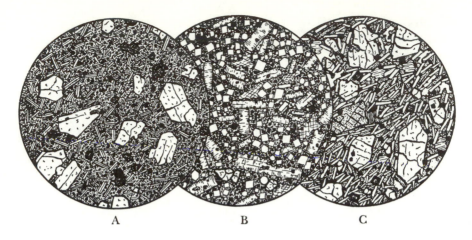

Figure 8-4. Melilitite-Rich Rocks

A. Olivine melilitite, Spiegel River, Cape Colony, South Africa. Diam. 2 mm. Phenocrysts of olivine and microphenocrysts of magnetite and perovskite, in a groundmass composed of melilite laths (some of which have axial rods and "peg structure"), iron-titanium oxide, perovskite, pyroxene, and a little interstitial glass.

B. Melilite Nephelinite, Bushwaga, Bufumbira volcanic field, East Africa. Diam. 2 mm. Phenocrysts of nepheline and melilite, in a groundmass of the same accompanied by iron-titanium oxide, olivine, leucite, apatite, serpentine, and calcite.

C. Olivine melilitite, Hochbohl, Württemberg, Schwäbische Alb, Germany. Diam. 2 mm. Phenocrysts of olivine and diopsidic augite, in a groundmass of melilite laths, intergranular opaque oxide, and augite. Sporadic flakes of phlogopite and octahedra of perovskite; a little analcimized nepheline occurs interstitially.

and iron-titanium oxides. Another from South Africa is shown in Figure 8-4A. In this one olivine and augite each make up about a quarter of the mode; melilite, in tablets showing "peg structure," and opaque oxides each form about 15%; and the remainder consists chiefly of perovskite, biotite, and apatite. Most such rocks are sodic rather than potassic. In the Toro-Ankole field of Uganda, however, there are unusually potassic melilitites consisting largely of olivine (~ 40%), melilite, and potassic glass.° Some contain accessory kalsilite or leucite. Their bulk chemistry (except for somewhat higher CaO and lower MgO) is not far from that of kal-

---

°*Katungite* of A. Holmes.

silite-olivine leucitites from the same region. Both types are extremely basic ($SiO_2 = 36\%$ in melilitite, 39% in leucitite).

CARBONATITES

Carbonatites are igneous rocks composed principally of carbonates—most commonly by far, calcite, dolomite, or ankerite. In texture and general appearance they superficially resemble marbles—for which indeed they were long mistaken. But as revealed in numerous mining operations and by careful mapping, their mode of intrusive emplacement is beyond doubt. Most commonly they occur as central plugs and irregular dikes in eroded volcanic centers, their habitual associates being basic and ultrabasic alkaline rocks, particularly ijolites or alkaline pyroxenites. Much more rarely, but still in the same general petrogenic environment, sodium carbonatites build extrusive flows with Pahoehoe surface structure (e.g., in the Toro-Ankole field of east Africa).* Carbonatite ash is now known to be rather widespread in parts of east Africa.

The principal mineral component of most carbonatites is calcite, dolomite,° or ankerite; siderite carbonatites are rare (Iron Hill, Colorado). Silicates, for the most part subordinate or even accessory constituents, include biotite, olivine (in some cases monticellite), sodic pyroxene, alkali feldspars, melanite, and others. Opaque phases are iron-titanium oxides and a variety of sulfides.

As a group, carbonatites are exceptionally rich in certain normally minor elements—Ba, Sr, Nb, rare-earth elements, P, F, S, and others. This trait is expressed in a wide variety of additional minerals such as fluorapatite, rare-earth phosphates (e.g., monazite and bastnaesite), barite, strontianite, fluorite, perovskite, and pyrochlore. Local concentration of apatite, barite, pyrochlore, or rare-earth minerals, either by magmatic processes or by superficial weathering, may yield economically workable ores of P, Ba, Nb, or rare-earth metals.† Carbonatites are holocrystalline, usually medium- to coarse-grained rocks. Crystals of calcite invariably

---

*An eruption of a unique sodium-carbonatite magma from the Tanzanian volcano Oldoinyo Lengai was witnessed by J. B. Dawson in 1960.
°Respectively, in *sovites* and *rauhaugites (beforsites)*.
†The *average* modal composition of carbonatite in the Mountain Pass deposit, California, has been estimated as carbonates (calcite > dolomite), 60; barite, 20; rare-earth minerals, 10; remainder, 10. That of Jacupiranga, southern Brazil, is calcite, 80–86; apatite, 15–16; remainder, 1–2.

are anhedral, those of dolomite mostly so. As in marbles, calcite commonly shows prominent lamellar twinning on $\{01\bar{1}2\}$, while grains of dolomite are but sparsely twinned on $\{02\bar{2}1\}$. Only silicates and some accessories (notably pyrochlore) tend to euhedralism.[2]

Field observations in many parts of the world, backed by experimental data, suggest that carbonatite and ijolitic magmas, while separately emplaced, may be dual products of unmixing from a common parent magma under plutonic conditions. Both carbonatite and ijolite-pyroxenite intrusions are commonly partially surrounded by broad aureoles of alkali metasomatism. Granitic gneisses of the envelope thereby become converted to *fenites**—rocks largely composed of alkali feldspar, blue sodic amphiboles, aegirine, and in some instances, nepheline. The end products may be virtually indistinguishable from peralkaline syenites.

KIMBERLITES

Kimberlites are ultramafic rocks, at the same time magnesian and potassic, that are properly considered by some writers as mica peridotites. Here they are treated separately because of unique characteristics relating to source, genesis, and mode of emplacement. Like their frequent associates carbonatites, kimberlites are confined to stable continental cratons notably in eastern and southern Africa, the Siberian platform, Brazil, the Grenville province of Quebec, and an inlier in Arkansas. Most kimberlites are prominent constituents of breccias that fill diatremes (exemplified by the "diamond pipes" of Kimberley, South Africa) whose other lithologic components are fragments of eclogite (itself occasionally diamond-bearing), garnet peridotite, and high-grade schists. With these are associated single crystals of minerals that singly and collectively indicate origin at very great depth—aluminous and chromiferous diopside, ilmenite, magnesian garnet, and diamond itself. The last-named mineral can have reached near-surface levels only by explosively rapid transport from depths greater than 100 km—well within the mantle.

In the type region in South Africa, the filling of "diamond pipes" is deeply weathered to oxidized "yellow ground" passing downward into "blue ground." Still deeper down the unweathered kimberlite proves to be a holocrystalline rock consisting of magnesian olivine, phlogopite,

---

*Named for the Fen area of Norway where H. von Eckermann demonstrated, once and for all, the intrusive nature of carbonatite cone-sheets and dikes.

chromian diopside (with characteristic green pleochroism), enstatite, pink magnesian garnet, and ilmenite—all set in a matrix of serpentinized olivine and calcite. The matrix, in turn, encloses scattered small grains of ilmenite, phlogopite, and perovskite. Kimberlites from Lesotho (southern Africa) have a lamprophyric aspect due to presence of phlogopite in two generations. Another variety, from Bachelor Lake, Quebec, consists mainly of olivine (largely serpentinized), phlogopite, and calcite. In the eastern American province, kimberlite dikes are constantly associated in the field with alnöites. Some of these kimberlites indeed have alnöitic affinities in that they contain plentiful melilite and perovskite.

PERIDOTITES AND ALLIED ROCKS

By far the most widespread members of the ultrabasic class are highly magnesian rocks consisting predominantly of olivine or its hydration products, serpentine minerals. These are *peridotites* and *serpentinites*—the latter, strictly speaking—metamorphic rocks. By virtue of texture and mode of emplacement peridotites conventionally are classed as plutonic. Much rarer, and indeed until recently not recognized as such, are volcanic equivalents, *komatiites*, found chiefly in older Precambrian "greenstone belts" of continental cratons.

Field Occurrence of Peridotites and Serpentinites

Large bodies of peridotite are found typically in one or the other of two independent tectonic-geographical situations: (1) as massive basal cumulate layers grading into pyroxenites or into troctolites in stratified continental basic plutons, as exemplified by the Bushveld (South Africa) and Stillwater (Montana) complexes (cf. p. 117); (2) as pods and elongate sheets (commonly largely converted to serpentinite) in mobile geosynclinal belts typified by the Alpine chains of Europe, the Urals, the Appalachians, and the young circum-Pacific fold belts of the western United States, Japan, New Caledonia, New Zealand, and intervening tectonic arcs (these have long been known as ultramafic bodies of the "Alpine type"). Probably more extensive still, but largely hidden by a veneer of basalts and sediments, are serpentinites of a third category—suboceanic crust underlain at no great depth by peridotites of the upper mantle. Fragments of serpentinite are common in deep-sea dredge hauls; and at least in one locality—St. Paul's Rocks, just south of the equator on the mid-Atlantic ridge—mantle peridotites appear to have been pushed up through a thin crust to above sea level. Peridotites of the Alpine type in all stages of serpentinization are now believed to represent fragments of suboceanic crust and mantle rock that have been tectonically emplaced in the solid state at their present sites in mobile belts. In some situations, as in the Italian Alps and Apennines, they form the basal layers of compound rock units, *ophiolites,* whose lithologic components in upward sequence are peridotite (or serpentinite), spilitic basalts and diabases, and deep-sea radiolarian cherts. Some students of tectonics tend to see *all* Alpine-type ultramafics as fragments of ophiolites.

Much less extensive than peridotites and serpentinites of the kinds just mentioned are xenolithic fragments in alkaline basic and ultrabasic lavas and breccias, thin cumulate layers in diabase sills, and core components of rare, zoned, cylindrical bodies recorded from southern Alaska and the Urals.

Deformational Textures in Peridotites

Peridotites of the Alpine type, as might be expected from their mode of tectonic emplacement as solid bodies, very commonly show textural evidence of internal deformation. Olivine, as has been demonstrated experimentally, can yield rather readily to internal stress by plastic strain. Consequently, even where overall strain has been rather trivial, individual crystals of olivine show some degree of undulatory extinction; and thin "deformation lamellae" parallel to {100}, differing in extinction by two or three degrees from the host, are quite common. More intense strain is manifested by progressively more conspicuous foliation, streaked textures and reduction in grain size of olivine (so-called granulation) by recrystallization at relatively high strain rates. The ultimate result can be mylonitic texture (Figure 8-6B; cf. also p. 503).

Enstatite, a widespread constituent of peridotites, yields to stress in a different manner. Individual coarse crystals develop sharply bounded kinks at high angles to a crystallographic glide plane {100}—the direction of internal slip being [001]. In a mylonitized peridotite from St. Paul's Rocks, mid-Atlantic, kinking in remnant rounded crystals of enstatite is spectacular; and the total picture of internal deformation within crystals rotated within the swirling matrix of fine-grained olivine is most convincing. Kinking of enstatite during rapid strain of pyroxenite at high confining pressure and a few hundred degrees Celsius in laboratory

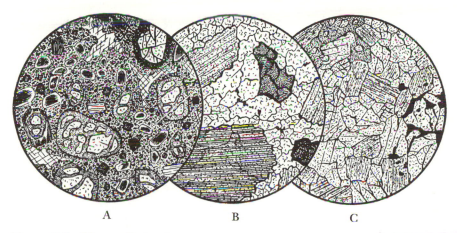

Figure 8-5. Ultramafic Rocks

A. Melilitite, Elliott County, Kentucky. Diam. 3 mm. Partly serpentinized phen-ocrysts of olivine, flakes of pale-brown phlogopite, plates of melilite with clear rims that polarize in ultra-blue, granules of perovskite and chromite, and, near top of section, a grain of pyrope garnet with a reaction rim. The dense matrix consists of iron oxide, perovskite, antigorite, and calcite, some of which is coarse grained and fills irregular pores.

B. Lherzolite, Haute Garronne, France. Diam. 3 mm. Diallage (at bottom), bron-zite, and granular olivine, with accessory green spinel (upper right) and picotite (lower right).

C. Pyroxenite, Hope, British Columbia. Diam. 3 mm. Approximately equal amounts of orthopyroxene and diopsidic augite. Some of the former contains lamellar inclusions of clinopyroxene. A little poikilitic hornblende (near top of section) and pyrrhotite.

experiments results in inversion (mainly within the kinks) to the poly-morph clinoenstatite (readily identified in sections normal to the kink axes [010]—α of the indicatrix—by inclined extinction with respect to {100} exsolution lamellae, and by higher birefringence than that of the enstatite host). Only rarely—notably in norites from central Australia and in highly strained peridotites from the Alps—has this phenomenon been recorded in naturally deformed rocks.

Peridotites: Terminology and Petrographic Characteristics

In current terminology several types of peridotite are recognized on the basis of additional essential mineral components other than magnesian olivine. All are holocrystalline, mostly medium- to coarse-grained rocks, with anhedral granular texture. While olivine of some rocks is completely unaltered, it usually shows some sign of deuteric or metamorphic hydration to a serpentine mineral (the fibrous variety commonly identified by petrographers as "chrysotile," the obviously flaky serpentine as "antigorite").* Pyroxenes—especially enstatite and diopsidic augite (often chromiferous)—are next in abundance to olivine in most peridotites. But calcic plagioclase ranks after olivine in many cumulate peridotites; and hornblende, biotite, and garnet-bearing peridotites as well as rocks locally rich in chromite or in magnetite are all well known. The petrographic character of individual types will now be reviewed.

Dunites

Dunites are almost pure olivine rocks. They take their name from Dun Mountain (named for the dun-brown color of the almost bare outcrop) located in the northern segment (100 km long) of a Permian ophiolite belt in southern New Zealand. The modal composition of dunite from this type locality is olivine, 97.5; chromite and picotite, 2.5; traces of chromian diopside and enstatite. Emplaced as a solid rock at temperatures too high to permit serpentinization, it consists of masses of strained, closely knit olivine anhedra. Strain has culminated locally in development of mylonitic bands in which finely recrystallized olivine and local streaks of chromite swirl around residual ovoid eyes of coarser material, each jacketed with a finely recrystallized ("granulated") rim (Figure 8-6B). Dunites from other regions may carry appreciable enstatite or diopside, or both, marking transitions to varieties of pyroxene peridotite. Dunites are much more widespread in ultramafic bodies of the Alpine type than in cumulate peridotite layers of basic intrusions.

While olivine of nearly all dunites is highly magnesian ($Fo_{>88}$), ferriferous *hortonolite dunites* with olivine $Fo_{<50}$ crowded with dendritic magnetite inclusions are known as cross-cutting pipelike bodies in the Bushveld complex. *Chromite dunites* with a high modal content of chromite (Figure 8-6C) are locally developed both in peridotites of the Alpine type (as at Dun Mountain) and in those of layered intrusions. Such rocks grade into *olivine chromitites*.

---

*Since much serpentine in peridotites displays neither habit clearly and since the true identity of a serpentine polymorph is readily revealed by x-ray powder pattern, it is no longer advisable to apply such names on the basis of microscopic observation alone.

Pyroxene Peridotites

Three mutually gradational types of pyroxene peridotites have been distinguished according to the nature of the pyroxene component. *Wehrlite,* not a common rock, consists essentially of olivine and diopsidic augite (diallage) in a ratio that approximates 3:1. Enstatite, picotite, chromite, and hornblende are the most common accessories. Wehrlites are well exemplified in the middle zones of cylindrical dunite-wehrlite-pyroxenite bodies in the Union Bay province of Alaska. *Harzburgite,* the most widespread type of peridotite in both major types of environment, consists almost wholly of magnesian olivine and orthopyroxene (enstatite or bronzite; both with thin exsolution lamellae of clinopyroxene parallel to {100}, i.e., normal to the β indicatrix axis. Common accessories are chromite, iron-titanium oxides, diopsidic augite, and, in cumulate rocks of layered intrusions, a little calcic plagioclase. Cumulate harzburgites commonly display a poikilitic relation between coarse skeletal bronzite and the dominant but smaller grains of olivine. Peridotites with subequal amounts of orthopyroxene and diopsidic augite are called *lherzolites* (Figure 8-5B). Some carry accessory magnesian garnet. All three types of

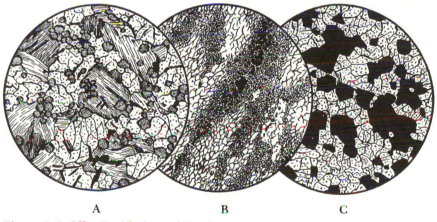

| A | B | C |

Figure 8-6. Mica Peridotite and Dunites

A. Mica peridotite forming lenses in norite, Kaltes Tal, Harzburg, Germany. Diam. 3 mm. Olivine, green spinel, reddish-brown titaniferous biotite, and plates of ilmenite.

B. Dunite mylonite, Dun Mountain, New Zealand. Diam. 3 mm. Augen of less-crushed olivine separated by winding bands of intensely granulated olivine. Trains of picotite and chromite granules.

C. Massive chromite-rich dunite, Dun Mountain, New Zealand. Diam. 3 mm. Virtually all fresh olivine.

pyroxene peridotite, like dunites, may carry secondary minerals of deuteric or metamorphic origin—serpentine, talc, fibrous amphibole, or carbonates.

Hornblende Peridotites

Hornblende peridotites are much less common than pyroxene-bearing types. They are characterized by large poikilitic crystals of hornblende (Figure 2-7A) usually of a patchy color ranging from pale green to pale brown, that enclose ortho- or clinopyroxenes or both, and rounded granules of serpentinized olivine. Pale phlogopite, occasionally with bright-green chromiferous rims, is a common accessory constituent; other minor minerals are magnetite, pyrrhotite, green spinel, apatite, and calcic plagioclase. In the hornblende peridotite layer of the well-known teschenite sill of Lugar, Scotland, the poikilitic amphibole is a reddish-brown barkevikite, enclosing olivine and titanaugite. Much of the amphibole in rocks such as these probably was formed by deuteric alteration of pyroxene and perhaps of olivine. In an interesting peridotite from St. Paul's Rocks, mid-Atlantic, the principal minerals are olivine, pargasite, enstatite, and spinel, with unusually varied accessories including chlorapatite, scapolite, allanite, biotite, and zircon.

Garnet Peridotites

A rare but genetically interesting kind of peridotite has olivine, pyroxene and garnet as principal constituents. One such is a body that outcrops some 250 km north of Bergen in southwestern Norway. The mode recorded by D. A. Carswell is olivine, 76; pyrope-rich garnet, 12; enstatite, 7; chromian diopside, 4. The pair olivine-pyrope is stable only at depths below about 60 km; at lower pressures the stable equivalent is enstatite-spinel (much more commonly found in peridotites):

$$Mg_2SiO_4 + Mg_3Al_2Si_3O_{12} \rightleftarrows 4MgSiO_3 + MgAl_2O_4$$
$$\text{olivine} \qquad \text{pyrope} \qquad \text{enstatite} \qquad \text{spinel}$$

The rate of inversion at temperatures of 500°C or so has been shown to be relatively fast. So the survival of garnet peridotites at the surface implies rapid upward tectonic transport from within the upper mantle.

Mica Peridotites

Mica peridotites, except for kimberlites, are rare varieties found only locally as segregations in biotite-olivine norites (Figure 8-6A). A rock from the Highwood Mountains, Montana, that has been identified as mica-peridotite contains magnesian olivine, monticellite, and poikilitic biotite.

Plagioclase Peridotites

In the lower levels of layered basic intrusions peridotites in rhythmically layered sequences or at the top of the ultramafic zone itself may grade into rocks of the olivine-gabbro series. Feldspathic peridotites consisting of olivine (with or without pyroxene) and subordinate calcic plagioclase, $An_{\sim 85}$, are typical of the transition zones. There is considerable textural variety due to poikilitic envelopment of cumulus olivine (with or without subordinate chrome spinel) by intercumulus plagioclase (with or without pyroxene). One type of many in the ultramafic zone of the Stillwater complex consists of equant grains of olivine ($\sim 90\%$) with interstitial skeletal crystals of bytownite. Another, from the Cuillin intrusion on Skye in the Hebrides of Scotland, has cumulus olivine and spinel poikilitically enclosed in subordinate augite and bytownite. A special type[°] from another Hebridean island, Rhum, consists of subparallel large olivine crystals with markedly tabular habit parallel to {010} oriented by upward growth normal to the plane of layering. The interstices are filled with intercumulus plagioclase and scattered crystals of chrome spinel.

A special type of feldspathic peridotite, *picrite*, occurs in cumulate layers near the base of some alkaline diabase sills, such as the Lugar sill, Scotland. These are hornblende-augite peridotites like that just described, but with calcic plagioclase or even analcime as additional minerals. Such rocks, with decrease in olivine, may grade into alkaline pyroxenites.

Serpentinites

Serpentinites are rocks composed largely or entirely of serpentine minerals. Field occurrence, the presence of relict partially serpentinized

---

[°]*Harrisite.*

grains of olivine and residual chromite, textures, and field transitions to peridotite show beyond doubt that most serpentinite bodies were formed by hydrothermal metasomatism of peridotites of the Alpine or ophiolitic type. In many cases the internally brecciated and slickensided condition of serpentinite lenses and sheets in mobile belts points, at least in some cases, to serpentinization prior to final emplacement at their present sites.

Experiment shows that at pressures corresponding to depths of a few kilometers, magnesian olivine in continuous contact with a flux of pure H_2O is unstable at temperatures below about 400°C. The stable product of hydration in a system closed except to aqueous fluid is serpentine (antigorite) and brucite. The corresponding transformation can be written:

$$2\ Mg_2SiO_4 + 3H_2O \rightarrow Mg_3Si_2O_5(OH)_4 + Mg(OH)_2.$$
$$\text{forsterite} \qquad\qquad \text{antigorite} \qquad \text{brucite}$$

Simple calculations show that such a reaction implies a great influx of water (16%, by weight, of the hydrated rock) and overall expansion of the rock by at least 50%. Expansion could, of course, be reduced to zero if the system were open to ions other than water and substantial amounts of MgO were removed. There is no way to ascertain the exact degree of volume change involved in a rock that has been deformed to the extent common in serpentinite bodies. And other reactions, involving introduction of SiO_2 and removal of MgO, can equally well be written; for brucite—if indeed it be present—has not been identified in many serpentinites. Then enstatite, when present, certainly plays a part in the serpentinization process. In the presence of water, the pair olivine-enstatite becomes unstable and could be replaced by antigorite and talc at temperatures as high as about 500°C. Thus it should be clear that serpentinites are metamorphic rocks. Yet—as with spilites and basalts—serpentinites are so closely associated with peridotites (especially those of the Alpine type) that from the petrographic standpoint they are most conveniently considered in the present context. New mineralogical patterns superposed during later metamorphic events will be considered in Part Three.

Microscopically, serpentinites are seen to consist of any combination of three morphologic varieties of serpentine, $Mg_3Si_2O_5(OH)_4$. A fibrous form (often identified as "chrysotile") tends to develop as an interlacing network of microveinlets, enclosing cores of a more weakly birefringent cryptocrystalline serpentine, within which relict remnants of olivine may still survive. The product has a distinctive overall *mesh texture*. It is accentuated by precipitation of iron, present in the parent olivine, as dusty, opaque oxide outlining the grain boundaries of the initial peridotite fabric (Figure 8-7A). Crystals of enstatite become pseudomorphed by fibrous serpentine ("bastite") whose fibers are aligned parallel to [001] of the parent pyroxene. Some or all of both olivine and enstatite—particularly in sheared serpentinite—may become replaced by a flaky (micaceous) variety of serpentine, usually identified, in most cases correctly so, as antigorite (Figure 8-7B)—the polymorph that is truly stable over the temperature range of serpentinization and of low-grade metamorphism. Primary diopsidic pyroxene also may become replaced in some peridotites by finely reticulated serpentine flakes. More commonly it is converted to a semiopaque mixture of indeterminate phases, white in reflected light. Chromite remains unaltered; or it may become reconstituted into secondary streaks and veinlets. Brucite, where present, easily

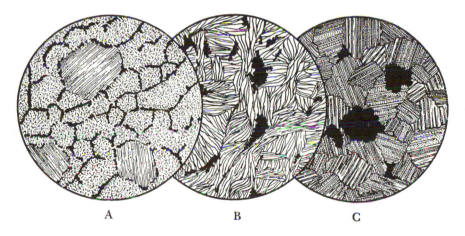

A   B   C

Figure 8-7. Serpentinites

A. Bronzite serpentinite, Kuhschnapper, Saxony. Diam. 2.5 mm. Bastite pseudomorphs after bronzite, in a matrix of antigorite showing mesh structure derived from olivine. Granules of secondary iron oxides outline the boundaries of the original olivine crystals. The parent rock was peridotite.

B. Flaky antigorite serpentinite, Sequoia Point, Oakland, California. Accessory pyrite, Diam. 3 mm.

C. Bastite serpentine derived from pyroxenite, near Forest Hill, Sierra Nevada, California. Diam. 3 mm. Chiefly antigorite, partially replaced by carbonates. The accompanying opaque mineral is magnetite.

remains undetected. But in a section of known thickness it can be recognized as dispersed flakelets of distinctly higher birefringence than that of the predominating serpentine minerals. Other secondary phases that may appear in minor amounts in some serpentinites are talc, actinolite, and a carbonate (dolomite or magnesite). Rocks composed largely of these minerals are metamorphic and as such will be considered later (pp. 497, 538).

While most serpentinites are hydrothermally altered peridotites, some are derivatives of magnesian pyroxenites. These typically consist mainly of bastite serpentine (Figure 8-7C).

Komatiites

The term *komatiite* was proposed in 1969 by M. J. and R. P. Viljoen for certain volcanic rocks of peridotitic or closely related composition in the Archaean of South Africa. These are rocks of great petrologic-tectonic interest; and they are now known from "greenstone belts" of Archaean age in other parts of southern Africa, in Western Australia, in India, and in Ontario.[3] The name komatiite, now in use only a dozen years, has already been redefined and broadened more than once. Here we consider only those rocks that are truly ultramafic. Their related, more siliceous, less magnesian, plagioclase-bearing associates can be considered as tholeiitic basalts with special textural characteristics.

Komatiites are volcanic rocks in the strictest sense of the term. Their occurrence as lava flows, some certainly extruded on the sea floor, and as thin sills has been documented in great detail. Some even show pillow structure. Devitrified glass is a characteristic constituent; and their most distinctive textural feature—*spinifex texture*—is attributed, by analogy with slags, to rapid quenching from unusually high temperatures (of the order of 1300°C or more).

Spinifex texture denotes an array of crisscrossing sheaves of closely spaced subparallel blades or plates (in many cases skeletal) of magnesian olivine. It tends to be localized toward the tops of flows with the mean direction of olivine elongation transverse to the trend of the flow surfaces. In this respect it shows some analogy with harrisitic texture of some Hebridean feldspathic peridotites (p. 247). Individual olivine crystals of the spinifex aggregates are commonly 1 cm or so long, but may be longer. They lie in a matrix of devitrified glass and acicular clinopyroxene, itself in some cases an obvious devitrification product.

There are two categories of ultramafic komatiites. In *peridotitic* flows or sills, whose overall composition is equivalent to that of peridotites, two textural variants are usually present: olivine cumulates toward the base, spinifex-textured rocks further up in the individual flow. The cumulates consist of magnesian olivine (60–80%) in closely packed equant grains, with minor chrome spinel, all set in a matrix of finely crystalline pyroxene (subcalcic augite) and devitrified glass. In the spinifex-textured variant, sheaves of skeletal platy crystals of magnesian olivine (35–60%) and scattered grains of chrome spinel lie in a groundmass of the same nature as that of the cumulates. In both types plagioclase is absent, SiO_2 is between 40% and 43%, and MgO usually 25–30%.

In *pyroxenitic komatiites* olivine is less abundant (typically 10–35%) and subcalcic augite correspondingly more plentiful. Again there are two variants: cumulates of close-packed olivine and augite granules, and rocks with spinifex texture. Typical analyses show SiO_2 about 48%, MgO about 14%. Again plagioclase is absent. But with incoming of plagioclase these rocks grade into a type of tholeiitic basalt ($SiO_2 = 50$–51%) that Arndt, Naldrett, and Pyke (1977) call "basaltic komatiite."

The great petrologic significance of komatiites is threefold: (1) They demonstrate beyond doubt that under some circumstances peridotitic liquids may rise high into the crust and be extruded as surface flows.* (2) They suggest that in early Precambrian times, and probably only then, the thermal regime of the upper-mantle-crust system permitted generation and eruption (at very high temperature) of peridotitic magmas, and that these played an important role in building the ancient crust. (3) Their relations to associated basaltic komatiites demonstrate that fractional crystallization of peridotitic (komatiite) magma at low pressures can generate magmas of tholeiitic composition—the complementary product being olivine-rich cumulates.

SUPPLEMENTARY NOTES ON THE PETROGRAPHY OF UPPER-MANTLE ROCKS

Nature of Available Data

Geological experience of prevalent lithologic heterogeneity on all scales within the accessible crust suggests that the upper mantle likewise may

---

*Thus refuting the persuasively argued thesis of N. L. Bowen (long convincing to many of us) that no magmas more magnesian than those of glassy or aphanitic basalt (normative *ol* not more than about 20) have ever been erupted as surface flows or injected as crustal intrusions.

well be heterogeneous, both laterally and vertically. Certain rock types, notably kimberlites, eclogites, and peridotites, have been noted in passing as rocks that are specially significant with respect to the composition of the upper mantle. With them are habitually associated individual xeno-crystic phases, which may have similar significance. But even if such assumptions are accepted and become substantiated, it must be admitted that direct natural sampling of the mantle must still have been sporadic and highly imperfect. Most rocks and many minerals that are stable at high pressures and temperatures prevailing in the upper mantle must become unstable, especially in the presence of water or liquid magma, as they ascend, by whatever vehicle of transport, through regimes of falling temperature and pressure en route to the surface. Survival of mantle samples, intact in mineralogical and textural detail, must surely be a matter of chance subject to the constraints of reaction kinetics in relation to velocity of transport. And in the course of ascent many samples, already notably unstable, must pass through a regime of temperatures still high enough greatly to facilitate reaction according to general prin-ciples of kinetics discussed in Chapter 2 (p. 46). Moreover, the chance for hydration and carbonation during and after ascent to shallower crus-tal levels must indeed be great—witness the altered states of many bodies composed of rocks of the types just mentioned. It is highly probable, therefore, that the upper mantle contains lithologic components that seldom or never survive ascent to the surface environment.

In spite of these qualifications, actual specimens of a few rock types— none of them voluminous in the crust on a global scale—provide the most tangible existing evidence of mantle composition. A highly signif-icant part of this evidence, independent of any geophysical, geochemical, isotopic, or cosmically based models of terrestrial composition, concerns the petrography of what are believed to be mantle-derived rock samples. The notes that follow are confined to microscopic evidence of this kind, considered within the context of field occurrence.

Mineral Indices of High Pressure

First, we must note thermodynamic and experimental evidence that has established a few critical minerals and phase combinations as reliable indices of crystallization at pressures greater than 10 kb.

First and foremost is diamond, a mineral whose natural primary crus-tal source is in kimberlitic breccias. Within the terrestrial environment diamond is the stable polymorph of carbon (with respect to graphite) only at pressures and temperatures prevailing at depths greater than about 150 km.[4] Minute inclusions of coesite (the high-pressure polymorph of SiO_2) recorded rarely in natural diamond indicate even higher pressures. Though rare in nature, diamond is the most valuable index of crystallization deep within the upper mantle.

The typical basalt-gabbro phase assemblage plagioclase-pyroxene becomes unstable at depths greater than about 25 km. Its place is then taken—in an anhydrous environment—by the eclogite pair omphacite-garnet, which in the absence of water melts only above about 1100°C. Stable assemblages in peridotites at depths greater than about 20 km consist of four phases: olivine + diopside + enstatite + an aluminous phase (spinel down to 30 km or 40 km, garnet from about 40 km downward). These are the familiar phase assemblages of lherzolites—in which the spinel, it will be remembered, is highly chromiferous (chromite or picotite).

The chemistry of both enstatite and diopsidic pyroxene also varies significantly with pressure. Enstatites of spinel lherzolites at first become notably richer in Al_2O_3 with increasing pressure up to about 10 kb or 15 kb; but the aluminum content then drops sharply at still higher pressures as associated spinels give way to garnet. Highly aluminous ("Tschermakitic") augites also are typical of the high-pressure environment.

Mantle Samples from Explosively Emplaced Breccias

Fragments of rock and individual xenocrysts believed to be of mantle origin are abundantly represented in volcanic breccias of two general kinds. These are kimberlitic pipes and tuffs, and diatremes whose volcanic component is nephelinite.

Kimberlite Paragenesis

Kimberlite pipes are restricted to specific, though extensive, sectors of ancient continental cratons. The information that they yield applies strictly therefore to some segments of the mantle immediately underlying continental crust. It is doubtful whether kimberlite itself, even rare fragments that actually contain diamond, represents mantle rock. Though it has certainly come from very great depths where diamond was stable, there remains the strong possibility that diamond may have been transported as xenocrysts caught in ascending kimberlite magma. Certainly there are rare allied diamond-bearing rocks from the Ivory Coast of west Africa that must finally have crystallized as dikes emplaced close to the surface; for these rocks contain plentiful leucite.

Kimberlitic breccias nevertheless habitually contain abundant mantle relics of several kinds—samples taken apparently over a wide range of depth. Chief among these are eclogite, spinel lherzolite, and garnet lherzolite—orthopyroxene greatly exceeding clinopyroxene in the two latter types. Rare specimens of eclogite (especially those from Siberian kimberlites) and of garnet lherzolite (from Lesotho in Africa, and Wyoming in North America) carry small euhedral crystals of diamond, apparently an integral primary phase. Spinel of kimberlitic lherzolites is a dark-colored picotite carrying notable Cr_2O_3 substituting for Al_2O_3.

Lherzolitic diopsides also commonly are chromiferous and are easily recognized as such by their characteristic emerald-green to colorless pleochroism. Single-crystal phases typical of kimberlite breccias include diamond and a suite of megacrysts (>1 cm in diameter) that include diopside, enstatite, garnet, ilmenite, and chromiferous spinel. Mineral inclusions in diamonds from many provinces, as recorded by H. O. A. Meyer and F. R. Boyd, conform to a general pattern: in order of abundance, olivine, highly chromiferous pyrope, chromite, enstatite, diopside; coesite, pyrrhotite, and pentlandite have also been recorded.

Nephelinite Breccias and Tuffs

The most carefully documented occurrences of rock fragments transported from the suboceanic mantle are scattered over a province 20 km in diameter northeast of Honolulu on the Hawaiian island Oahu.[5] These fragments are abundantly distributed in flows and explosively erupted breccias of late Pleistocene to Holocene age around nearly 40 centers of eruption. The volcanic rocks in nearly every instance are nephelinites or melilite nephelinites. At most centers the principal xenolithic types are dunite and spinel lherzolite, either of which may greatly predominate at specific centers. Most specimens of these rocks show strong textural evidence of internal deformation (cataclastic or even mylonitic texture, cf. p. 505) that was imprinted at the mantle source long prior to eruption. Salt Lake Crater, a few kilometers east of Honolulu, is one of several centers where the xenolithic paragenesis includes garnet pyroxenites and garnet lherzolite as well as various types of peridotite (mostly spinel lherzolite). The mode of one garnet pyroxenite recorded by Jackson and White is diopside 16; enstatite, 65; pyrope-almandine, 16; phlogopite, 0.5; remainder (oxides and sulfides), 2.5. The garnet pyroxenites of the Honolulu province must not be confused with eclogite, for the diopsidic pyroxene contains less than 2% Na_2O and only about 3% Al_2O_3—a composition far removed from that of omphacite. The mode of a garnet lherzolite from the same locality is olivine, 29; diopside, 12; enstatite, 33; pyrope-almandine, 22; oxide, 4.

To illustrate a xenolithic paragenesis of a nephelinite diatreme in a continental setting we have chosen a locality close to the village of Kakanui, southeastern New Zealand. Here a submarine breccia pipe of mid-Tertiary age is beautifully exposed by marine erosion. The matrix is nephelinitic glass, much of it fragmented. It is crowded with large xenoliths chiefly of peridotite and "eclogite" (but also including accidental inclusions of schistose rocks torn from the crustal basement). Peridotites are olivine-rich types with minor enstatite, chrome diopside, and picotite. The "eclogites"[6] more properly should be termed garnet pyroxenites. They consist of aluminous augite (with less than 2% Na_2O), pyrope-rich garnet, and kaersutitic amphibole. An average mode is garnet, 25; augite, 68; hornblende, 6; ilmenite, 1. A specimen unusually rich in kaersutite contains 62% of that mineral, 18% garnet, 18% augite, and 2% ilmenite.

A special feature of the Kakanui breccia is the presence of numerous large, polished, and rounded xenocrysts, often 10 cm in diameter, of pyrope, aluminous augite, and kaersutite.

Mantle Models Based on Breccia Petrography

Various tentative models for specific segments of upper mantle beneath major breccia provinces have been proposed from time to time. These, of course, are open to future modification or even ultimate rejection. But they illustrate the manner in which petrographic data must be incorporated into more sophisticated models that draw on all available data of many kinds—geochemical, geophysical, and mineralogical. Two petrographic models may be mentioned in outline.

For the kimberlite province of southern Wyoming and Colorado, M. E. McCullum and D. H. Eggler have constructed a model of layering in the following downward sequence.[7]

1. Depth 0 to 50 km: continental crust.
2. Depth 50 km to 180 km: lherzolite with profusely scattered pockets

of eclogite. Except for a thin upper layer of spinel lherzolite (and a patch or two of pyroxenite) the main mass of lherzolite is garnetiferous and very rarely carries diamond in the lowermost level.

The downward sequence of layering proposed by Jackson and Wright for the mantle beneath the Pleistocene-Holocene volcanoes of the mid-oceanic Honolulu province is this:

1. Depth 0 to 20 km: massive accumulations of older tholeiitic lava (Koolau shield, about 2 million years old) overlying thin oceanic crust of basalt and serpentinite.
2. Depth ~20 km to ~60 km: dunite and olivine-rich peridotite (= high-velocity seismic layer, P velocity ~8 km sec^{-1}).
3. Depth 60 to 90 km: garnet lherzolite (= low-velocity seismic layer, P velocity $<$ 8 km sec^{-1}).
4. Depth 90 km downward: garnet pyroxenite and garnet lherzolite (minimum-velocity seismic layer).

Xenolith Paragenesis of Basaltic Lavas

The nature of ultramafic xenoliths generally found in basaltic lavas of all types—tholeiitic to basanitic—has already been described (p. 115). Some of these have probably come from disrupted underlying layered basic intrusions. But many—especially the dunites and spinel lherzolites so abundantly represented in alkali olivine basalts and basanites—can, like their counterparts in nephelinite flows and breccias, be assigned with confidence to sources in the mantle. Again it is emphasized that textural evidence of extreme deformation is widespread in dunite and lherzolite nodules of this suite.

Significance of Tectonically Transported Blocks

Peridotites of the Alpine Type

In current fashion of thinking, large tectonically emplaced masses of peridotite in mobile belts are regarded as blocks of upper-mantle rock, or (if completely serpentinized, as in many ophiolites) slabs of suboceanic crust. There is much evidence to support this attractive hypothesis, but some of it is mineralogical and textural. To that extent, light thrown by such bodies upon mantle petrography is partly reflected from the initial premises.

Alpine peridotites in the unaltered condition are broadly comparable to the total suite of olivine-rich xenoliths in alkaline mafic lavas and tuffs. The lithologic range is from dunite through harzburgite and lherzolite to magnesian and calcmagnesian pyroxenite. Garnet lherzolites are rare;* but where they do occur they may be interleaved with garnet pyroxenites or eclogites. A few Alpine peridotites carry biotite, and others a pargasitic hornblende. Unlike peridotites of layered basic intrusions, those of the Alpine type typically lack cumulate textures and very commonly show textural evidence of intense penetrative deformation— even mylonitization (cf. p. 242).

Of particular interest with respect to our present topic are peridotites that make up the bulk of St. Paul's Rocks, a group of islets situated on the mid-Atlantic Ridge just south of the equator. The two principal assemblages here are olivine-enstatite-diopside-spinel (the familiar spinel lherzolite) and olivine-pargasite-enstatite-spinel. Both contain a highly varied suite of accessory phases—chlorapatite, biotite, scapolite, allanite, and zircon. If our general hypothesis is accepted (and in this particular instance it seems inescapable) St. Paul's Rocks furnish a unique large sample of suboceanic upper-mantle rock.

Eclogites

Some eclogitic blocks—commonly associated with blueschists in young mobile belts—are products of metamorphism of crustal basalts that were carried down into a deep-seated unusual environment of very high pressure and low temperature. These will be considered in detail in Chapter 20. But there are other blocks of eclogite, in terranes of high-grade metamorphism, that are generally considered to have been transported in their present state from the mantle. Besides the critical pair pyrope-rich garnet plus omphacite, many of these rocks contain kyanite or amphibole. Their bulk composition is that of tholeiitic basalt. Distinctive chemical differences between pyroxene-garnet pairs in these eclogites separate them from low-temperature–metamorphic eclogites of blueschist belts. In this respect eclogite blocks thought to be of mantle origin have some affinities with those of kimberlitic breccias.

---

*This is hardly surprising, since it has been shown experimentally that above 850°C, garnet lherzolite inverts rapidly to its spinel-bearing equivalent.

General Petrographically-Based Conclusions

On purely petrographic grounds with due regard to field associations we can now advance these tentative conclusions:

1. The upper mantle down to a depth of at least 200 km is both laterally and vertically heterogeneous on all ordinary geological scales—from that of a hand specimen to that of a quadrangle map.
2. Among its principal lithologic components are eclogites and garnet pyroxenites, and a wide range of magnesian ultramafic rocks—peridotites of all types, with dunite and lherzolites (both spinel- and garnet-bearing types) predominating. Peridotites containing significant biotite or amphibole (pargasite or kaersutite) or both conceivably could be locally abundant in parts of the upper mantle. Eclogites could well be much more widely represented in the mantle than is suggested by their relatively low abundance among accessible samples; for the total melting span of eclogites is very small, and they could easily become completely eliminated by melting during ascent from the depths.
3. Lherzolites, garnet pyroxenites, and eclogites are all potential sources of basaltic magma—by partial or even (in the case of eclogites) complete melting. Dunites, harzburgites, and perhaps some diopside-poor lherzolites, in contrast, must be virtually sterile in this respect. They are probably refractory residues left from partial melting of other mantle components.
4. It is possible that mechanical anisotropy may exist within layers of mantle peridotite even on a regional level. For in such rocks large-scale deformation accompanied by strong preferred orientation of olivine crystals is almost ubiquitous. It is conceivable—to students of fabric of deformed rocks (*Gefügekunde*) perhaps even likely—that a resultant seismic anisotropy, conforming to the symmetry of convective motion in the mantle, might persist on a very large scale—that of a convection cell.

REFERENCE NOTES

[1] A. F. Cooper, *Journal of Petrology*, vol. 20 (1979): pp. 139–163.
[2] For abundantly illustrated textural and mineralogical detail, the reader is

referred to E. W. Heinrich, *The Geology of Carbonatites*. Chicago: Rand McNally, 1966, pp. 156–219.

[3]For example, see D. R. Pyke, A. J. Naldrett, and A. R. Eckstrand, *Bulletin of the Geological Society of America*, vol. 84 (1973): pp. 955–978; N. T. Arndt, A. J. Naldrett, and D. R. Pyke, *Journal of Petrology*, vol. 18 (1977): pp. 319–369.

[4]See I. S. E. Carmichael, F. J. Turner, and J. Verhoogen, *Igneous Petrology*. New York: McGraw-Hill, 1974, pp. 102–105.

[5]For full petrographic and chemical detail, see R. W. White, *Contributions to Mineralogy and Petrology*, vol. 12 (1966): pp. 245–314; D. E. Jackson and T. L. Wright, *Journal of Petrology*, vol. 11 (1970): pp. 405–430.

[6]B. Mason, *Contributions to Mineralogy and Petrology*, vol. 19 (1968): pp. 316–327.

[7]M. E. McCullum and D. H. Eggler, *Science*, vol. 42 (1976): pp. 253–256.

9

Pyroclastic Rocks

PYROCLASTIC MATERIAL: GENERAL NATURE AND TERMINOLOGY

Pyroclastic rocks are volcanic rocks that have clastic texture; in other words, they are fragmental deposits formed as a direct result of volcanic action. The vast majority of them are products of explosive volcanic eruptions. But fragmentation of the material may also be caused by continuing growth of partially solidified volcanic domes, and it generally occurs where lavas are quenched by water.*

Pyroclastic material is classified primarily according to particle size. Pebble-size fragments between 2 mm and 64 mm in diameter are called *lapilli*. Smaller particles are called *ash*; larger ones are called *bombs* if during their formation they were partly or wholly molten, *blocks* if they were angular chunks of solid rock. Volcanic rocks consisting of ash and lapilli are called *tuff* if they consist largely of ash and *lapilli tuff* if lapilli predominate. Rocks composed chiefly of bombs are *agglomerate*, and those consisting chiefly of blocks are *volcanic breccia*. Those in which blocks are mixed with abundant ash are called *tuff-breccia*.

---

*The term *pyroclastic* is used here in the broad sense recommended by the IUGS Subcommission on the Systematics of Igneous Rocks (*Geology*, vol. 9 (1981): pp. 41–43; *Neues Jahrbuch fur Mineralogie* 1981, pp. 190–196). Many geologists, however, prefer to restrict the term to the material and deposits produced by explosive eruptions, in particular to fall, flow, and surge deposits.

Ash and lapilli deposited in and adjacent to low-lying areas where sedimentary detritus is accumulating are likely to become intimately mixed with mud, sand, and gravel. The resulting deposits are *tuffaceous sediments*, or *tuffites*. Such mixtures of pyroclastic and epiclastic material are to be distinguished, however, from wholly epiclastic deposits of volcanic debris derived by weathering and erosion of older volcanic formations. Discussion of these rocks is reserved for Chapters 10 and 13.

Ashes and tuffs may be classified further by their content of glass, crystal (mineral), and rock particles (Figure 9-1A, B). Those made up largely of particles of glass (glass shards) are termed *vitric ash* and *vitric tuff* and are said to have *vitroclastic texture*. The term *crystal ash*, or *crystal tuff*, signifies a deposit in which crystals are the predominant constituent, and the term *lithic ash*, or *lithic tuff*, refers to one in which rock fragments predominate. In this distinction between vitric, crystal, and lithic material: *lithic* refers to polycrystalline rock particles, mostly aphanitic; *crystal*

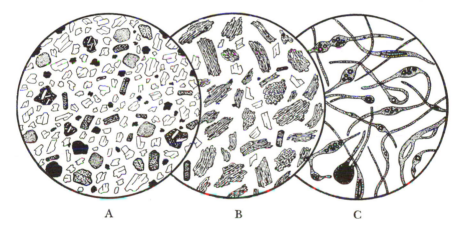

A   B   C

Figure 9-1. Volcanic Ashes

A. Andesitic crystal ash erupted from the volcano Santa María, Guatemala, in 1902. Diam. 2 mm. Broken crystals of plagioclase, dark-green hornblende, paler-green pyroxenes, rounded biotite flakes, magnetite, and a few lithic chips of andesite.

B. Dacitic vitric ash showing pumiceous texture. Diam. 2 mm. Product of the culminating explosions of Mount Mazama, which led to the formation of Crater Lake, Oregon. Shredded and cellular bits of pumiceous glass accompanied by fewer broken chips of plagioclase and small prisms of hypersthene.

C. Basaltic ash (Pele's Hair), Kilauea, Hawaii. Diam. 2 mm. Threads of brown basaltic glass containing bubbles of gas. Material discharged by lava fountains in the form of spray.

denotes single crystals and fragments of crystals; and *vitric* denotes particles that are essentially noncrystalline and glassy throughout. Lithic particles occur in all sizes from ash to the largest blocks, whereas single crystals and crystal fragments seldom are larger than coarse ash or small lapilli; vitric particles may be any size, but most of them are small.

Tephra is a general term that refers collectively to any or all pyroclastic material, regardless of size, that is ejected into the air during volcanic eruptions. Material produced by disruption of new magma is called *juvenile* or *essential* tephra. Fragments of older volcanic rock broken from the conduit and walls of the eruptive edifice are designated as *accessory*, and any derived from the subvolcanic basement are called *accidental*.

Most juvenile tephra is produced when gas-rich magmas vesiculate and become fragmented by the bursting of expanding gas bubbles. Extreme vesiculation of the very viscous silica-rich magmas produces the glassy foam called *pumice* and leads readily to violent disruption and to large volumes of vitric ash and pumiceous lapilli and bombs. Pumice is generally so porous that lapilli and bombs composed of it often will float on water and may drift for long periods before becoming sufficiently water-saturated to sink. Typically the vesicles are spheroidal or ellipsoidal, though they may be drawn out into slender tubes producing a silky, fibrous texture (Figure 9-3A). Comminution results in vitric ash composed of bits of the glassy septa between the pumiceous vesicles, chiefly curved glassy splinters and pointed chips with concave borders (Figure 9-2A).

Scoria is the general name applied to dark, highly-vesicular rock of basaltic composition; it is usually at least partially glassy but may appear aphanitic. Most basaltic lapilli and bombs are composed of scoria and are referred to as *cinders*. Extreme vesiculation of fluid basaltic magma may produce foamlike glass in which the vesicle walls are paper-thin, often threadlike septa of brown or black glass. This is called *thread-lace scoria*, or *reticulite pumice*, and is found commonly among the products of the spectacular lava fountains that play during the opening phases of Hawaiian eruptions; it also forms as crusts on fluid, gas-rich pahoehoe lava flows. If the lava is particularly fluid, as it tends to be in lava lakes of Kilauean type, bursting bubbles may hurl out the liquid as a fine spray that cools quickly to form small glassy pellets (*Pele's Tears*) or delicate glassy threads (*Pele's Hair*) like those illustrated in Figure 9-1C. These and other pyroclastic particles whose external forms are determined by the surface tension of the ejected lava, rather than by fracture surfaces, are called *achneliths*.[1]

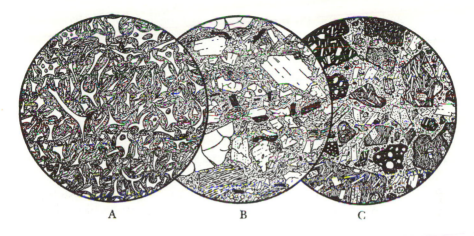

Figure 9-2. Tuffs

A. Rhyolitic vitric tuff, Shasta Valley, California. Diam. 2 mm. Shows typical vitroclastic texture. Arcuate shards of glass lie in a matrix of almost impalpable glass dust.

B. Rhyolitic crystal tuff, Etsch valley, Italy. Diam. 2 mm. Broken crystals of quartz and sodic plagioclase, together with small flakes of biotite, in a matrix of glass dust and pumice fragments.

C. Andesitic lithic tuff, near Managua, Nicaragua. Diam. 2 mm. Fragments of various kinds of andesite predominate; between these lies a matrix made up of plagioclase and pyroxene crystals and pale-brown glass dust.

Particles of scoria and pumice may be porphyritic (Figure 9-4C), the phenocrysts representing early-formed crystals that were suspended in the magma prior to eruption. Similar crystals usually occur along with vitric particles as constituents of finely comminuted ash; as might be expected, many of these have been broken and some have glass adhering to their margins. Also, some juvenile basaltic ejecta are aphanitic and may be classed as lithic fragments.

Accessory and accidental ejecta typically are either lithic or crystal fragments. Accessory ones are invariably angular chips blasted from the flow or dike rocks that constitute early-formed portions of the eruptive center; however, accidental fragments derived from the prevolcanic basement may represent any type of rock—igneous, sedimentary, or metamorphic. Both accessory and accidental ejecta sometimes exhibit the effects of contact metamorphism. Among the products of the 1924 steam-blast eruption of Kilauea, for instance, are fragments of plagioclase-hypersthene hornfels produced by reheating of basalt that formed the conduit walls; and on the flanks of Monte Somma, Vesuvius, are blocks of metamorphosed limestones and dolomites that were presumably torn from the roof of the magma chamber. Occasionally lithic fragments have been partially fused before ejection. For example, many lapilli and blocks of plutonic rock blown from the Parícutin volcano in Mexico contain vesicular glass formed by partial melting of feldspars.

Pyroclastic material may be deposited in a number of different ways, but two general types of deposits are most widespread: fall deposits and flow deposits.* *Fall*, or *airfall*, *deposits* consist of tephra that have been blown into the atmosphere and deposited by gravitational settling of individual particles. Deposits produced in this way usually show the effects of gravity sorting both in the size and in the composition of component particles, and they tend to be well stratified. The strata generally blanket an entire terrane covering hills and valleys alike, although they may be quickly removed from steep slopes by slumping and erosion.

In contrast, *flow deposits* typically are unsorted and unstratified. They are emplaced by relatively dense yet mobile mixtures of solid fragments and fluids that flow along the ground under the influence of gravity. Flows consisting largely of blocks, lapilli, and ash mixed with water are volcanic mudflows, or *lahars;* they may be either hot or cool. Other block and ash flows are avalanches that result from fracturing and collapse of newly formed domes or spines, or they may be produced by lateral blasts from the sides of developing domes. Flows of this kind may be very hot, as were those produced during the eruptions of Mount Pelée in 1902 and 1929–1932. Other pyroclastic flows are exceedingly hot masses of juvenile pumice lapilli and ash dispersed and fluidized by escaping magmatic gases. Flows of this type develop from rapid vesiculation of fresh magma and are hotter and more mobile than other types; the deposits they produce are called *ignimbrite*. They may be initiated when a relatively dense mixture of comminuted magma and gas in the lower part of an eruption column above a vent collapses altogether and flows rapidly downslope along the ground, impelled by gravity and by momentum from the initial collapse.[2] Small flows of this kind are generally directed by existing volcanic slopes and their deposits are more or less confined to valley areas. Very large flows, some of which probably are erupted from fissures, may surmount moderate relief and bury irregular terranes

---

*Surge deposits and various types of volcanic breccia are distinguished and identified primarily by their large-scale features and by their distribution and relationships as seen in the field; they are not considered separately in this book.

under extensive subhorizontal sheets of ignimbrite. Ignimbrites produced by some very large eruptions in the past have aggregate volumes of tens or hundreds of cubic kilometers. Most of these large effusions apparently occurred along ring fracture zones above relatively large, high-level bodies of silicic magma, and generally they were followed by collapse of the roof of the magma chamber and formation of a caldera.

VOLCANIC BRECCIA AND TUFF-BRECCIA

Coarse-grained pyroclastic rocks are invariably lithic; in other words, the blocks and lapilli that typically predominate are fragments of volcanic rock. Often it is newly solidified lava that has been fragmented to form the blocks, but accessory fragments of older lavas are also common. In most volcanic breccias and tuff-breccias, the rock fragments consist largely of a single general rock type, and the deposits may be designated according to the predominant lithic component (e.g., dacite breccia, andesite tuff-breccia, etc.). Although hand specimens and thin sections are necessary to identify individual components in breccias and tuff-breccias, the texture and bulk composition of such coarse-grained deposits is rarely well represented in a single sample and never in a single thin section. For this reason, breccias and tuff-breccias must be studied and described in the field prior to selection of small samples for laboratory analysis. Furthermore, the origin of a volcanic breccia or tuff-breccia is usually inferred from its field distribution and relationships; most of them have been produced by volcanic mudflows or by avalanches of blocks and ash.

AIR-FALL DEPOSITS

Ash blown high above eruptive vents may be dispersed afar by the wind, and it is cold before it falls back to earth. Air-fall deposits, particularly those derived from highly silicic magmas, invariably show effects of eolian differentiation. The largest and densest particles are not carried as high or as far as smaller and lighter particles, and they tend to accumulate closer to the parent volcano. The coarsest tephra falls relatively near the eruptive vent, whereas fine ash may be carried far away from it. But although tephra generally becomes finer away from the vent, at moderate range, lapilli composed of low-density pumice or scoria are often intermixed with much finer ash composed of nonvesicular glass and crystals. Other things being equal, crystal and lithic fragments tend to fall more quickly and closer to the source than vitric fragments, which are usually less dense. Thus in the deposit of a single ash fall, the denser crystal and lithic tephra tend to be relatively concentrated in the lower portion of the layer and the vitric tephra in the upper portion. Corresponding lateral transitions may also be observed as a layer of ash or tuff is followed away from the parent volcano. A single sample of air-fall tuff, therefore, may not exactly reflect the composition of the parent magma.

Successive air-fall strata also reflect variations in successive eruptive phases. At many volcanoes, accessory lithic fragments are more abundant during initial eruptions than during later ones, and the composition of juvenile ejecta may differ as successive eruptions tap different portions of the magma reservoir. Also, corresponding to changes in the force of eruption and in wind conditions, the different strata in any one locality may differ markedly in grain size.

The most widespread air-fall tuffs have highly silicic compositions and consist largely of vitric ash and pumice lapilli. Unaltered acid glass tends to be clear and colorless and has a refractive index that is usually substantially less than 1.51. A few crystals of quartz, sanidine, sodic plagioclase, or biotite are commonly scattered among the glass shards, and they may occur also as phenocrysts in pumice lapilli. In some tuffs, crystals are relatively abundant, either because of an abundance of early-formed crystals in the erupting magma or because of sorting of the erupted particles (Figure 9-2B).

Air-fall deposits of basic composition are numerous but generally less widespread than more silicic ones. In the coarse-grained deposits, lapilli composed of scoria are usually the principal constituents, as for example on typical basaltic cinder cones. Finer-grained basaltic tuffs consist largely of fragments of clear brown glass (*sideromelane*) or of black glassy particles that are opaque except along thin edges (*tachylite*); most of the latter, although they may appear glassy, are cryptocrystalline as revealed by X rays. Basaltic glass has a refractive index distinctly higher than 1.54, and the glass particles in basaltic ash tend to be more granular (less splinterlike) than typical shards of acid glass. Among the crystals occurring in basaltic tuffs, calcic plagioclase usually predominates, but olivine, pyroxene, and hornblende are characteristic and may be abundant.

When ash falls through moisture-laden eruptive clouds, some fragments may gather successive layers of very fine wet ash forming spheroidal ashy "mudballs" that are called *accretionary lapilli*. These form most commonly as a result of phreatomagmatic eruptions, many of which involve eruption through ponds or water-saturated ground. Most accretionary lapilli range from 2 mm to 10 mm in diameter, although larger ones are not uncommon, and they tend to accumulate relatively near the eruptive center. Consolidated deposits of accretionary lapilli are referred to as *pisolitic tuff*.

IGNIMBRITE

The term *ignimbrite* refers in general to the deposits laid down by very hot pyroclastic flows made up of comminuted viscous magma and gases.* Flows of this type move very rapidly, and the deposits are emplaced quickly at temperatures in the general range of 600° to 900°C. Some ignimbrites are the product of a single flow. The largest ones, however, consist of multiple flow units and are emplaced by several flows in such rapid succession that the deposit does not cool appreciably between them and the entire accumulation constitutes a single cooling unit. The rate at which an ignimbrite sheet cools is, of course, largely a function of its initial temperature and its extent—especially its thickness (cf. p. 43). The dimensions of the sheet depend on the volume of magma erupted and on the underlying topography. A few ignimbrites are no more than 1 m thick, but the largest ones—typically of acid composition—are very extensive and may reach thicknesses of several hundred meters, particularly where they fill valleys. Indeed, rhyolitic and dacitic ignimbrites of mobile belts are probably the most extensive surface manifestation of acid magma.[3]

Ignimbrites are unsorted mixtures of vitric ash and pumice lapilli. Most have rhyolitic or dacitic composition; some are andesitic. The lapilli may be small and sparsely scattered through a matrix of ash, but usually they are abundant and some reach the size of bombs. Crystals and crystal fragments are sparse in some ignimbrites and abundant in others, and most of them represent minerals that crystallized early and were suspended in the magma at the time of eruption (Figure 9-3B,C). A few crystal and lithic fragments of accessory and accidental origin are also to be expected in most ignimbrites.

The general composition of an ignimbrite is reflected both by the refractive index of the glassy fragments and by the assemblage of crystals

---

*The deposits produced by such flows are commonly referred to as *ash-flow tuffs*.

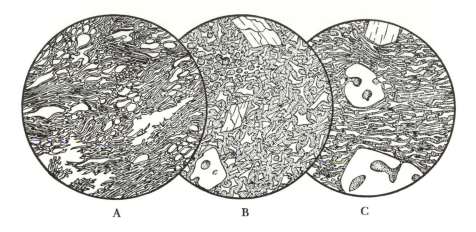

Figure 9-3. Rhyolitic Pumice and Ignimbrite

A. Rhyolitic pumice, Lipari Island, Italy. Diam. 3 mm. Entirely composed of extremely vesicular glass.

B. Incipiently welded ignimbrite, near Bishop, California. Diam. 3 mm. Specimen from the unwelded top of an ignimbrite. Crystals of quartz and sanidine, in a matrix of undeformed glass shards and dust, with well-preserved vitroclastic texture.

C. Welded tuff, from same locality. Diam. 3 mm. Specimen from the welded interior portion of the same ignimbrite. Constituents as in B, but here the glass shards are deformed and flattened.

included in the deposit. In highly silicic ignimbrites, the glass typically has a refractive index less than 1.50, and associated feldspars are usually sanidine and oligoclase; in the more alkaline types, the feldspar may be anorthoclase. In andesitic ignimbrites, the glass has a refractive index approximating 1.54, and andesine or labradorite is the feldspar to be expected. Quartz and biotite are commonly present in rhyolitic and dacitic ignimbrites. Hornblende, augite, and hypersthene may be present but are common only in the less silicic ones. Minor accessory minerals, such as magnetite, ilmenite, fayalite, zircon, apatite, and perhaps a few others, are commonly present but are invariably sparse.

Phenocrysts in pumice lapilli are indicative of the minerals that crystallized in the magma before eruption. Typically these phenocrysts are euhedral or subhedral, but similar crystals scattered through the ash matrix are likely to be broken. Crystals of feldspar and quartz are commonly rounded and embayed (Figure 9-3B,C). Biotite may be reddened by oxidation, and if temperatures were very high, hornblende crystals may have been altered to oxyhornblende. Accessory and accidental crystal and lithic inclusions alien to the magma that was erupted usually can be recognized by forms and compositions not consistent with those of the magmatic mineral assemblage present as phenocrysts in the lapilli.

Where large volumes of magma have been erupted by a rapid succession of flows, the later flow deposits are commonly less siliceous than the early ones. The final flows from Mount Mazama (Crater Lake, Oregon), for example, were andesitic and in marked contrast to earlier dacitic flows (Figure 9-4C). Presumably such changes reflect compositional differences in different portions of an erupting magma body, more basic portions perhaps being deeper and tapped by later eruptions. In any case, distinctly bounded compositional zones must exist in some magmas, for the changes in sequential ignimbrite compositions may be abrupt, as was the case at Mount Mazama. Also, a few ignimbrites contain pumice lapilli of two distinct compositions, as shown both by differences in the refractive indices of the glasses and in the phenocrysts, and some single lapilli may be hydrid and show streaks of contrasting composition.

As they cool, a few ignimbrites may remain masses of incoherent fragments, but usually the hot glassy particles are bonded at their points of contact and the deposit as a whole becomes coherent.* Ignimbrites bonded in this fashion without compaction or distortion of the glass particles are said to be incipiently welded. They have typical vitroclastic texture (Figure 9-3B) and, although coherent, they are soft porous rocks that can be sawed easily into blocks. Often, however, the vitric fragments have been deformed and flattened by compaction and then they become more firmly welded together, forming hard rocks of relatively low porosity that are called *welded tuff*. While still hot and under appreciable overburden, the glass shards are squeezed and flattened and may be bent between and around more rigid crystal fragments (Figure 9-3C). At the same time pumice lapilli and bombs are flattened into discoid lenses; when viewed edgewise, these commonly appear as thin lenticular streaks of dense glass, called *fiamme*.† As seen in thin section, most fiamme are traversed lengthwise by dark streaks or septa, which are traces of original pumice vesicles that have collapsed and coalesced during compaction and

---

*The bonding of hot glass particles is the result of a process called *sintering* in glass technology; it is dependent on temperature, composition, and viscosity of the particles.
†These glassy lenticular streaks are called *fiamme* [fee-ah'-me], Italian for "flames," because many of them are shaped somewhat like tongues of flame. The term is generally extended, however, to include all flattened lapilli in welded tuffs.

welding. The degree of welding in some ignimbrites has been so extreme as to reduce an original pumiceous lapilli tuff to nonporous, dense glass resembling obsidian, in which the flattened vitric fragments produce a streaky lamination deceptively like the fluidal banding seen in many lava flows. Additionally, as they cool, they may develop columnar jointing and crystalline spherulitic structures, so that their resemblance to lavas is increased. Little wonder that welded tuffs have often been wrongly identified as lava flows; the fact is that extensive flows of silicic composition usually are firmly welded ignimbrites. Features that aid in their recognition are the presence of relict vitroclastic textures and transitions of firmly welded tuff into less-compacted tuff of obvious pyroclastic origin. Such transitions occur near the tops, bottoms, and sides of most ignimbrites.

The welding of glassy fragments in ignimbrites is a function principally of temperature, composition, thickness, and rate of cooling. In many thick ignimbrites, firmly welded tuff occurs only in the lower-central portions of the deposit. However, if the vitric fragments have relatively low viscosity, perhaps because of unusually high temperature or concentration of H_2O and other fugitive components, even thin ignimbrites may become firmly welded throughout. In some of these rocks, small spherical vesicles occasionally appear in the densely welded fiamme, indicating continued emission of gas and the formation of bubbles even after compaction and welding, and structures produced by postwelding mass flowage have also been reported.[4]

As they cool, ignimbrites commonly lose their initial glassy texture and become wholly or partially aphanitic. The devitrification involves crystallization of the hot glass to a microcrystalline, generally fibrous, aggregate having the same bulk composition as the original glass, and also vapor-phase crystallization induced by hot gases escaping through the deposit. The effects of vapor-phase crystallization are most evident in the less-compacted, porous portions of ignimbrites.

Vitric fragments usually devitrify individually, so that original vitroclastic texture, although somewhat hazy, tends to be preserved in devitrified rocks. Typically in firmly welded ignimbrites, the glass is altered to fine-grained crystal aggregates that have distinctly fibrous texture. In thin section, the fibrous texture is clearly apparent and usually shows positive elongation under crossed polarizers, but the individual crystals are so slender that they usually cannot be discriminated by optical means. X-ray diffraction analyses generally identify the constituents as cristobalite and feldspar. Beginning at the fragment margins with fibrous crystals oriented at high angle to the margin, the crystallization progresses inward until the glass is completely devitrified. In the larger flattened shards and in fiamme, devitrification commonly produces *axiolitic texture* in which the crystal fibers are oriented normal to opposing walls and meet to form a septum along the central axis of the fragment. In some cases, radial fibrous plumes and spherulitic growths develop from separate centers along fiamme margins, and occasionally these may extend across the fragment margins. In a few welded ignimbrites, large spherulites have been formed; some of those, called *lithophysae*, contain central cavities that may be encrusted with crystals formed from a vapor phase or filled with chalcedony. Commonly, where devitrification has begun, it has progressed essentially to completion, but devitrified and unaltered glass may occur side by side within single fiamme, or the smaller fragments may be devitrified completely and the larger fiamme remain glassy.

Where crystallization occurs in porous, uncompacted portions of an ignimbrite, the pumice lapilli lose their original glassy vesicular texture and are converted to porous aggregates made up of minute crystal rosettes, and cavity walls are encrusted with small crystals precipitated from a vapor phase. Here tridymite and feldspar, usually with cristobalite, are the most abundant minerals, although the individual crystals are generally too small for visual identification. Other minerals of similar origin commonly include aegirine, hornblende, and scapolite. As a consequence of vapor-phase crystallization, incipiently welded ignimbrites may become somewhat more coherent, but they remain generally porous and soft.

AQUAGENE TUFF (HYALOCLASTITE)

Volcaniclastic deposits resulting from the fragmentation of magmas when quenched by water are termed *aquagene tuff* or *hyaloclastite*.[5] Generally they are related to submarine eruptions, but they also may form where lavas flow into the sea or into fresh water and where extrusion of magma occurs beneath glacial ice (common in Holocene volcanic deposits of Iceland). Most of them have basaltic composition and consist largely of sharply angular fragments of glass bounded by smooth fracture surfaces. Typical fragments are clear brown basaltic glass (sideromelane) in which vesicles either are lacking or are relatively sparse, in contrast to the highly vesicular glassy and cryptocrystalline fragments (tachylite) that characterize basaltic ash and lapilli produced in subaerial eruptions. But because basaltic glass reacts readily with water, the glass in most aquagene tuffs has been at least partially altered to a yellow or brown product called *palagonite* (p. 273).

Many aquagene tuffs are distinguished by their large-scale characteristics and associations in the field as well as by the physical character of the glassy particles. Perhaps the most common association is with pillow lavas. Indeed, the term hyaloclastite was coined originally for deposits produced by crumbling of the glassy crusts on pillow lavas, though it has since acquired a broader meaning. Particularly noteworthy are pillow breccias consisting of separate individual pillows and pillow fragments of various sizes scattered through a matrix of hyaloclastite.

DIAGENETIC ALTERATION OF VITRIC ASH

Volcanic glass is inherently unstable and it reacts readily and comparatively quickly with water in the near-surface environment. Calcic plagioclase, olivine, and pyroxene—the silicates that commonly occur in basic volcanic glasses—are also notably reactive. Vitric ash reacts with water to form a variety of minerals, the products depending primarily on the initial composition and texture of the ash, the physicochemistry of the environment, and the age of the deposit. The most common minerals produced are smectites* and zeolites. Many altered tuffs are so fine-grained that only general mineral identification is possible in thin sections; usually the constituents must be identified by x-ray diffraction.

Alteration products of silicic glasses are determined largely by aqueous chemistry and temperature. Near the surface, in either fresh or marine waters, silica-rich glass usually alters to smectite, sometimes with opal and zeolites. The term *bentonite* refers to rocks composed of relatively pure smectite derived by *in situ* alteration of vitric tuff. When bentonite absorbs water, the bulk volume increases, and it beomes a sticky, plastic mass of wet clay. Some bentonites are characterized by scattered euhedra of quartz, alkali feldspar, or biotite, and, in thin sections, vague relict outlines of vitroclastic texture may be apparent.

A vertical zonation of alteration products is exhibited by silicic tuffs that are interbedded in thick sedimentary sequences. In the upper, near-surface zone, phyllosilicates and opal are the principal products devel-

---

*The crystal structures and compositions of smectites and other phyllosilicates are outlined on pp. 317–319.

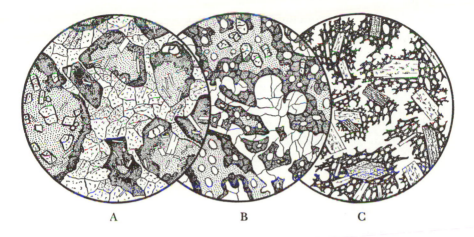

A   B   C

Figure 9-4. Basaltic Tuffs

A. Palagonite tuff, Oamaru, New Zealand. Diam. 4 mm. Fragments of palagonite, pale buff within and deep gold at the margins, including crystals of olivine and labradorite. Between these fragments is a matrix of calcite.

B. Palagonite tuff, Oahu, Hawaiian Islands. Diam. 4 mm. The cores of the vesicular fragments consist of fresh pale-buff palagonite including crystals of olivine; the rims of the fragments are fibrous and birefringent and largely composed of smectite. Between the fragments is a matrix of zeolites.

C. Hornblende andesite scoria, product of the last ash flows from Mount Mazama (Crater lake), Oregon. Diam. 4 mm. Phenocrysts of hornblende and labradorite, embedded in extremely vesicular, brown-to-black andesitic glass.

oped from fresh glass. Below 1 km to 2 km, glass alters to alkali-rich zeolites, often with opal and a phyllosilicate. Analcime, heulandite, quartz, and K-feldspar are characteristic of a succeeding lower zone, and below that, laumontite, albite, and prehnite may be expected. This vertical sequence of mineralogic changes represents reactions that occur in response to changes in water chemistry and to increasing temperature with depth.

Where pH and salinity are high, as in saline lakes and in the playa deposits of arid regions, alkali-rich zeolites are the typical near-surface alteration products of vitric ash. With aging and burial of the deposits, these become replaced by analcime and by quartz and alkali feldspar.

Glass of basaltic composition alters by reaction with water at low temperature, in both subaerial and subaqueous environments, to a resinous-appearing substance of brown or golden color that is called *palagonite*.[6] During the alteration process, K_2O, Na_2O, CaO, SiO_2, and not uncommonly Al_2O_3, are lost from the glass and much water is added. The chemical composition of the palagonite produced is variable, but invariably contains much ferric iron. Commonly the dissolved ions are redeposited as calcite and zeolites, cementing the rock and forming palagonite tuff (Figure 9-4A,B). With increasing age and depth of burial, palagonite is readily transformed to smectite, and this in turn may alter to chlorite at greater depth. Many palagonite tuffs also contain relict crystals of plagioclase, olivine, or pyroxene.

GENERAL REFERENCE

For general treatment of pyroclastic materials and deposits, see H. Williams and A. R. McBirney, *Volcanology*. San Francisco: Freeman, Cooper, 1979, ch. 6 and 7.

REFERENCE NOTES

[1] G. P. L. Walker and R. Croasdale, *Bulletin Volcanologique*, vol. 35 (1971): p. 308.

[2] R. J. S. Sparks and L. Wilson, *Journal of the Geological Society of London*, vol. 132 (1976): pp. 441–451.

[3] On the origin, character, and relationships of ash-flow tuffs and for an extensive review of the literature, see C. S. Ross and R. L. Smith, *U. S. Geological Survey Professional Paper*, no. 366 (1961). For a more recent symposium, see C. E. Chapin and W. E. Elston, *Geological Society of America Special Papers*, no. 180 (1979).

[4] H. U. Schminke and D. A. Swanson, *Journal of Geology*, vol. 75 (1967): pp. 641–664.

[5] For reference, see D. Carlisle, *Journal of Geology*, vol. 71 (1963): pp. 48–71; J. Honorez and P. Kirst, *Bulletin Volcanologique*, vol. 39 (1975): pp. 1–25.

[6] An excellent description of palagonite tuffs and discussion of the conditions under which palagonite originates are given by R. L. Hay and A. Iijima, *Geological Society of America Memoir*, vol. 116 (1968): pp. 331–376.

PART TWO
SEDIMENTARY ROCKS

10

The Origin of Sedimentary Rocks

INTRODUCTION

The distinction between igneous, sedimentary, and metamorphic rocks is basically genetic rather than descriptive. Sedimentary rocks are deposited in stratified fashion, layer upon layer, at the surface of the lithosphere, at relatively low temperatures and pressures. Most igneous and metamorphic rocks, in contrast, originated beneath the earth's surface, and were formed at much higher temperatures and pressures.

Each layer of sediment is buried ever more deeply as succeeding layers are deposited on top of it, and wherever deposition has continued more or less uninterruptedly for long periods of time, sediments may be buried very deeply indeed. Stratigraphic studies prove that sedimentary rocks may accumulate to thicknesses of thousands of meters. So long as deposition continues, each layer is subjected to continuously increasing pressure and temperature, and we can be sure that all deeply buried sediments have endured physicochemical conditions quite different from those existing at the surface where they were laid down. Postdepositional reactions can be expected to cause changes in the texture and composition of such sediments and tend generally to convert the initial loose sedimentary deposit into firmly lithified rock. Reactions that occur at relatively low temperature and pressure are considered to be part of the sedimentary process and are termed *diagenetic*; those occurring at higher temperature and pressure are designated *metamorphic*. Thus as temperature and pressure increase with increasing depth of burial, sedimentary rocks grade into metamorphic rocks. Some sedimentary deposits have been subjected not only to high temperatures and pressures but also to orogenic stress and have been so extensively recrystallized as to become typical high-grade metamorphic rocks.

The distinction between sedimentary and metamorphic rocks is arbitrary, however, and is not easily defined except where non-hydrostatic stress as well as heat has been a major factor producing the change. The classification of certain common rocks is therefore a matter of preference and may seem inconsistent. *Slate*, for example, is usually classed as metamorphic and *shale* as sedimentary, although there may be little visible difference in composition, whereas *dolomite* is generally classed as a sedimentary rock even when it has resulted from complete replacement of original calcite. In the field, however, dolomite is typically associated with other sedimentary rocks and is often undeformed, while slate occurs in highly deformed terranes where its tectonically induced cleavage, as distinct from sedimentary stratification, conforms to the regional structure and is easily recognized.

Rocks that are the direct products of volcanic processes are logically to be distinguished from sedimentary rocks, but in many places the two are interstratified. No confusion arises here except when a distinction must be made between *pyroclastic rocks*, produced by volcanic explosions, and *epiclastic rocks*, produced by subaerial weathering, erosion, and deposition of rock debris. Beds of ash and cinders (tuffs) that accumulate anywhere as a direct result of volcanic eruptions may logically be grouped with the igneous rocks; yet by normal erosion of older tuffs and flows similar products may find their way into widespread sedimentary basins, where they are deposited as more or less pure *volcanic sands*, *silts*, and *gravels*. These are epiclastic sediments in the strictest sense, but unless their exact mode of origin is known they are easily confused with pyroclastic rocks. Furthermore, mixtures of pyroclastic and epiclastic material are very common; these are called *tuffaceous sediments*. To assign a rock to one of these classes rather than another is often difficult and may be impossible where only a few specimens are available for examination. Collectively all of them may be referred to as *volcaniclastic rocks*.

To call a rock sedimentary, then, one must know or infer something about its origin. How is such knowledge to be gained from study of a specimen? What characteristics are peculiar to rocks that originate at low temperatures and pressures near the earth's surface? What visible or measurable properties of sedimentary rocks have genetic implications, and how are they to be interpreted? These questions outline a primary

problem of petrography, but they cannot be answered simply. Even stratification, which in one form or another is visible in nearly all sedimentary
deposits, is not a sufficient proof of sedimentary origin, for stratification
is simulated or inherited in many igneous and metamorphic rocks. And
organic remains, although abundant in many sedimentary rocks, are
entirely lacking in others and may occur in some pyroclastic and metamorphic rocks. However, the combined features of composition and texture when considered together do reflect more or less clearly the origin
and history of the rocks in which they occur. To interpret them requires
understanding of the processes whereby rocks originate. The brief outline of sedimentary petrogenesis in this chapter is intended to provide
a general background for the descriptive details that follow; for more
extensive coverage of sedimentary processes, several general references
are suggested at the end of the chapter.

A note of warning should be interjected here. No amount of laboratory study of hand specimens or of thin sections will supply a full interpretation of a sedimentary rock—or of any other. Significant as the
details of rock composition and texture are in themselves, their variations
and distribution within the sedimentary deposit as a whole are of still
greater significance, and these larger features must be examined in the
field.

FORMATION OF SEDIMENTARY ROCKS

In general, sedimentary rocks are formed in two different ways. Some
are *detrital*; that is, they are mechanical accumulations of mineral and
rock fragments (detritus) produced by disintegration of older rocks. The
vast majority of these are *terrigenous*, so called because they are derived
from the land by subaerial weathering and erosion. Typically the detritus
is transported and deposited far from its place of origin by water or wind
currents. Terrigenous deposits, including common sandstones and mudstones, are composed largely of quartz, feldspar, mica, and clay minerals.

A second type of sedimentary rock consists not of older rock debris
transported from afar but of minerals that originate by precipitation
within the depositional area. These minerals are principally calcium carbonates and opaline SiO_2 precipitated from surface waters, but many
others may form in this way; for example, in certain marine environments, calcium sulfates and phosphates, manganese oxide, iron silicates
and oxides, pyrite, and a few other metallic sulfides. Precipitation may

be a purely inorganic chemical process, as for example in saline brines that are concentrated by evaporation, or it may be caused by organisms. Some organisms, including bacteria, may cause precipitation indirectly by altering the chemical environment, whereas others, like coral, molluscs and foraminifers, secrete visible skeletons or shells that accumulate to form *biogenic (organic)* sedimentary deposits. Fossil coral and algal reefs, strata largely composed of fossil shells, and diatomite are representative examples of biogenic sedimentary rocks.

Sedimentary rocks of any kind—detrital, chemical, or biogenic—are usually modified after their initial deposition. Commonly they contain mineral components that have crystallized within the deposit itself, either precipitated in open pores in the deposit or replacing original constituents. Such minerals are termed *authigenic*. Quartz, calcite, and dolomite are the commonest of these, but the list of possible authigenic minerals is long. Postdepositional changes in rock texture also are to be expected. All the processes of postdepositional alteration at low temperature are referred to collectively as *diagenesis*.

The particular characteristics of a sedimentary deposit are determined by the entire setting within which the sediment has originated and accumulated and has later been lithified. The effects of *environment of deposition* are to be found in every sedimentary deposit regardless of age or kind. Evident also in most deposits, especially the older ones, are the effects of *diagenesis*. The influence of *source* and the effects of *transportation* are particularly noteworthy in detrital deposits.

Source Area

Parent rock controls to a large degree the composition of detrital sediment derived from it, a fact that is clearly evident in many sedimentary rocks. A *volcanic sandstone*, for example, consists of detritus necessarily derived from a volcanic terrane. Similarly, a sandstone composed essentially of feldspar and quartz is derived from coarse feldspathic rocks such as those composing large parts of the crystalline continental basement, and it is not formed unless such rocks are exposed in the source area from which detritus is derived. Among the grains of any sandstone, as Mackie pointed out long ago, varieties of quartz distinguished by the nature of their inclusions may indicate the type of parent rock from which the sand was derived. Or, occasionally, a particularly distinctive mineral in a sediment may lead one to the ultimate source from which it came; examples are the purple zircon (hyacinth) in the Paleozoic sand-stones of southern Scotland, which was derived from the Lewisian gneisses of the northern Highlands, and the glaucophane in Californian sediments, which comes ultimately from Franciscan blueschists. Few detrital sediments contain components that point so clearly to a particular source, but the composition of any detrital sediment reflects the character of its source in at least a general way.

Between parent rock and final sedimentary rock many minerals may be altered or completely destroyed, so that sediments never have exactly the same mineral composition as their parents. Some primary minerals may be destroyed or changed by weathering even before they are removed from the source area by erosion. If the environment is such as to induce intense decomposition and leaching, a selective destruction of minerals may take place that makes the weathered product differ greatly from its parent; it contains a concentration of relatively stable primary minerals such as quartz and muscovite among the sand and silt grains, and of clay minerals among the smallest particles. If, however, weathering involves chiefly physical disintegration, the weathered product is mineralogically like the parent rock; it contains both relatively stable and unstable primary minerals of various grain sizes, principally quartz, muscovite, feldspars, ferromagnesian silicates, and undisintegrated rock particles.

In general, chemical decay of minerals and rocks is fostered by warm, moist climates, particularly in areas that are well drained and heavily covered with vegetation, and is less active in cooler, drier climates. The climatic factor, however, is commonly overshadowed by relatively rapid erosion. All weathering processes are slow, but removal of the weathered material by erosion goes on at widely varying rates. Where erosion is very slow, a residual blanket of thoroughly weathered material—a *mature soil*—accumulates over the parent rock, and there the climate and biologic conditions of the area may have more influence than even the composition of the parent rock on the character of the weathered product. Where erosion is rapid, as in regions of high relief, steep slopes, and periodic heavy rainfall, rock waste may be removed almost as soon as it is loosened from the bedrock, and there is little decomposition before the material is carried away. Under such conditions, regardless of climate, the mineral composition of detrital sediment approximates that of the parent rock; even in the tropics, sediments that, like arkose, contain abundant unstable minerals are to be expected.

Transportation and Deposition

During actual transportation, chemical changes in detrital sediment are negligible, but two very important physical effects are produced. On the one hand, the individual particles are generally altered in *size, shape* and *roundness* by the abrasion and fracturing that result from rubbing and repeated impact of the particles on each other and on bedrock. On the other hand, *selective transportation* or *sorting* affects the total aggregate of grains, so that particles tend to be segregated according to *size, shape*, and *density*. Wherever a part of the detrital sediment load is deposited along the route of transportation, the larger and heavier particles tend to be deposited first. If the sediment is derived from a single source, the proportion of smaller particles in the sediment load progressively increases in the direction of travel as the larger and heavier grains are left behind, and the average grain size of deposits diminishes progressively in the direction of transport.

The combined effect of these processes on the sediment finally laid down is most clearly visible in the textural features, but it is evident also in the mineral composition. When, in a sediment consisting of particles of diverse size, the sand is segregated from the clay, as commonly happens in natural sorting, the resulting deposits are mineralogically as well as texturally distinct. The clay material contains a relative concentration of clay minerals (chiefly kaolinite, smectite, illite and chlorite), whereas the sand is composed largely of quartz, with or without feldspars, ferromagnesian silicates, and fine-grained rock fragments.

Very fine sand, silt, and clay material, together with larger micaceous grains, are apt to be carried in suspension in a transporting current and therefore suffer minimum physical attack by abrasion and fracturing. The coarser fragments, however, tend to be fractured and abraded by rolling, bumping, and sliding along at the bottom of the current. Cleavable grains, like feldspars, and also those weakened by incipient decay may be repeatedly chipped and broken, and thus tend to be gradually eliminated from the coarser sizes and added to the finer material in the suspended load.

Abrasion of the larger, heavier, and softer grains that are carried for long distances in the bottom load wears away sharp corners and surface projections, so that the grains tend to acquire a smoothly rounded surface, and at the same time to be somewhat reduced in size. The effectiveness of abrasion in rounding grains is a function of mode and distance of transportation and also of size and kind of grains. It is greatest on the larger and softer grains and is increased with distance of transportation. It is apparently greatest where wind or waves are the transporting agents. Well-rounded sand grains, however, should not always be attributed to a single cycle of sedimentation; many detrital grains have been involved in several cycles, being passed on from one sediment to another, and often the best-rounded sand grains are those inherited from older sedimentary rocks. Nor has every grain in a detrital sediment undergone the same history, for many deposits are of mixed derivation.

Some detrital deposits are very well sorted and contain particles of nearly uniform size, whereas others are unsorted or poorly sorted and contain particles that differ widely in size. Where sand, silt, and clay are all deposited together, sorting is poor and the deposits accumulate relatively rapidly. The best-sorted deposits, in contrast, result from selective transportation during which only a portion of the sediment load, with a restricted range of grain sizes, is deposited in any one place; some of these deposits are repeatedly winnowed by currents, as happens, for example, in shallow parts of the sea and on adjoining beaches and coastal dunes where sand is more or less continuously moved back and forth and winnowed by wind, wave, and tidal currents before being permanently buried.

But local variations in the competence of depositing currents are to be expected in most depositional environments, and they may produce, side by side, contemporaneous deposits of quite different texture and composition. Well-sorted sands are not the same grain size on all parts of a beach, and they are very different from the mud deposited in an adjacent lagoon. Also, the well-sorted sands of a river channel are but a small part of the total fluvial deposit, being associated with other deposits laid down on the channel banks and flood plain that are finer grained and less sorted. The variations themselves are significant in such formations, for they signify an environment that is not uniform from place to place, or from time to time. They also demonstrate clearly that no single sample can adequately represent a sedimentary formation.

A sure indication that conditions of deposition may vary with time is the fact that sedimentary strata in general are overlain and underlain by other strata of somewhat different texture and composition. The differences may appear to be random, but many sequences of strata become progressively coarser or finer grained or contain more or less rhythmic alternations of different sediment types. A noteworthy example of the latter is to be seen in the thick marine sequences known as *flysch*, in which beds of sandstone and mudstone or shale alternate repeatedly and with approximate regularity. The sandstone beds usually are thin but laterally extensive. Typically, each one is *graded*; that is, the average grain size diminishes upward within each sand bed, and each one rests with sharp contact on the mudstone below but grades without distinct boundary into the mudstone above. Such graded beds in many flysch sequences have been interpreted as deposits formed rapidly in deep water by periodic turbidity currents.

What has been said in general about the effects of transportation on detrital sediments applies to mineral and rock particles carried by wind, streams, and currents in oceans and lakes where the proportion of sediment to fluid is so low that each particle behaves as an individual. Under these conditions, selective effects on individual grains are to be expected. A large proportion of all detrital sediments, however, are transported more or less en masse by landslides, mudflows, and turbidity currents of high density. To these may be added transportation by glacial ice, though this is of more local importance. By such modes of transportation, masses of detrital sediment are moved quickly and with little or no selection, and many widespread thick deposits have been produced. Such accumulations are but poorly sorted, or totally unsorted, and the effects of abrasion on individual particles are at a minimum, except in some glacial deposits.

Environment of Deposition

Of all sedimentary environments the most extensive and the most enduring is the marine basin. In it most ancient sediments have been deposited and the great bulk of modern sediments is now accumulating. It is, moreover, the final destination of all sediment. Some very widespread and thick deposits, however, have been laid down above sea level on the continents, and much sediment is temporarily deposited in the continental areas en route to the sea. The major sedimentary environments to be distinguished are the oceans, the continents, and the marginal area between them, but each in turn includes many local areas having distinctive and varied characteristics that largely determine the particular kind of sediment that is deposited in each of them. Many environmental factors are important in determining the structure, texture, or composition of the sediments deposited. On dry land the important factors are climate, vegetation, and topography. In aqueous environments they include bottom topography, depth of water, temperature of water, salinity, acidity (pH), oxidation-reduction potential (Eh), freedom of circulation, turbulence, clearness or muddiness of the water, and organic population.

On the continents, some deposits accumulate on dry land and contain little besides the fragmented debris of older rocks. Some of these are *residual*; they have not been transported and lie as thin blankets of regolith over the parent rocks from which they were derived by weathering. Others have been moved and deposited by wind as sand dunes and loess and by gravity as talus and slide debris. The continental deposits that have been laid down in water also are largely detrital, but they often include some organic and chemically precipitated minerals as well. Some are *lacustrine* sediments, laid down in lakes; others are *alluvial* sediments (alluvium) deposited by streams in their channels and on their flood plains and fans. Glacial deposits include till and moraines deposited directly by ice; associated alluvial deposits, called outwash, are formed by melt-water streams.

Deposits in continental (nonmarine) environments, particularly those of alluvial and glacial origin, often lie at the surface exposed to subaerial weathering for long periods before being buried by additional sediment. During the period of exposure, soil-forming processes tend to produce changes in the original deposits, and if the time of exposure is long, distinct soil horizons may develop. In arid regions, vadose water tends to be drawn to the surface, and salts dissolved in it may be precipitated on and near the surface as the water evaporates, producing calcareous crusts called *caliche* or *calcrete*; these are essentially thin impure limestones. Elsewhere the surface horizon tends to be leached of some constituents that are carried downward in the vadose water. The leached material may be redeposited in a lower level of the soil, or it may be carried farther and added to the groundwater in the phreatic zone below the water table. Such processes may be interrupted when additional sediment is deposited, and any changes that have been produced in the original deposit may then become buried, preserved, and incorporated as primary characteristics in the total sedimentary formation.

The marine realm receives the bulk of terrigenous sediment, and in it the vast majority of chemical and biogenic deposits originate. It includes a shore zone consisting of subtidal, intertidal, and supratidal areas, and various zones of deeper water farther offshore.* In general,

---

*The shore zone is generally referred to as the *littoral zone: neritic, bathyal,* and *abyssal zones* correspond approximately to the continental shelf, the continental slope to a depth of about 2000 m, and the deeper sea, respectively.

coarser sediment and heavier shelled organisms tend to accumulate in the higher-energy environments that exist near shore and on shallow shelves and banks wherever the energy of large waves and strong currents is expended against the bottom and shore. The turbulence and the strong currents in these places keep the finer sediment in suspension and carry it away to lower-energy areas where it is deposited, both in sheltered bays and lagoons near shore and in deeper water farther seaward. Recent drilling, however, has encountered numerous layers of graded sand interbedded with mud on the deep-sea floor, indicating the periodic occurrence of currents competent to carry sand even to these depths. Most of these probably were turbidity currents that were generated on the outer continental shelf and upper slope and flowed down the slope and redeposited their load of sediment on the deep-sea floor. In this way, flysch-type deposits have accumulated at the base of the continental slope and vast submarine fans or aprons have been formed along most continental margins.

Chemical and biogenic deposits reflect the depositional environment most directly, for they consist of sediment that not only accumulated but also originated in the basin of deposition. In many parts of the open ocean where contamination by terrigenous and volcanic debris is minimal, the skeletal remains of free-floating organisms, collectively called *plankton*, have settled to the bottom from the surface waters where they live and have carpeted the sea floor with varying thicknesses of biogenic ooze. Chalk, diatomite, and radiolarian chert, for example, have originated in this way. Of the principal planktonic organisms, diatoms and radiolarians secrete tests of opaline silica, whereas foraminifers and coccoliths secrete calcium carbonate. The calcareous remains slowly dissolve as they settle through the deeper water, which is undersaturated with calcium carbonate, and at the carbonate *compensation depth* (a variable depth on the order of 4000–5000 m), carbonate sediment largely disappears from the sea-floor deposits. At greater depths, the biogenic oozes are siliceous; at shallower depths most of them are calcareous.

Vast deposits of calcium carbonate also are forming today, as they have in the past, in warm, shallow shelf seas and on shallow oceanic banks wherever there is a flourishing bottom-dwelling (benthonic) organic population supported by circulation of clear nutrient-laden water.* The sed-

---

*Such areas today include, for example, portions of the Persian Gulf, the shelf seas off southern Florida, Yucatan, and parts of northern Australia, the Bahama Banks, and numerous Pacific atolls.

iment deposited in these areas consists of the calcium carbonate secreted by organisms to form their shells and skeletons and additional fine-grained carbonate that is precipitated inorganically. In response to slow general subsidence, upward growth of the benthonic population and accumulation of biogenic sediment may continue for long periods, and where this has happened, the carbonate deposits are very thick. The character of ancient limestones indicates that most of them originated in similar environments and were largely biogenic.

In general, carbonate sediment tends to accumulate where it is pro-duced, and in this fundamental respect the formation of most carbonate deposits differs from that of terrigenous deposits. However, a portion of marine carbonate sediment is redistributed, abraded, and sorted by waves and currents just as terrigenous detritus is distributed after it enters the depositional basin. Well-sorted carbonate sands, like their ter-rigenous counterparts, are deposited, for example, on beaches and bars and in tidal channels and coastal dunes. Also, some of the carbonate sediment is transported by turbidity currents downslope to the deep-sea floor, where it is redeposited in graded layers of carbonate sand largely composed of skeletal debris of shallow-water origin. In some places, swift downslope transport has carried this carbonate sediment below the gen-eral carbonate compensation depth, and the carbonate sand layers are interbedded with deep-sea muds of noncarbonate composition.

In nonmarine environments, chemical and biogenic deposits tend to be thinner and more contaminated by terrigenous debris than their marine counterparts. Probably the most obvious examples of relatively uncontaminated, nonmarine, biogenic deposits are layers of high-grade peat and coal. Some of them extend over large areas, but all of them are comparatively thin. Undoubtedly they formed in swamps and marshes so situated that very little detrital debris reached the accumulating deposit of plant material. Such areas were perhaps comparable to the modern Dismal Swamp of the southeastern United States or to the swamps of northeastern Sumatra and the backwaters of the Ganges delta.

When we attempt to unravel the history of an ancient sedimentary formation by interpreting such features as composition and texture, our conclusions are likely to be uncertain or so general as to be vague. For this reason fossils in sediments are uniquely helpful: They are commonly a direct and positive indication of depositional environment. But fossils can be the basis of an accurate interpretation only when the organisms were buried without being moved from the area in which they lived and died. Some organic materials are detrital, and these cannot be directly interpreted; once deposited and buried, they are later eroded and washed into a younger sedimentary deposit in a different environment. Or the organic remains in shallow parts of lakes and seas may be carried by bottom currents to greater depths beyond their normal habitat. Some organisms that in life are free-swimming or free-floating—*pelagic* in contrast to *benthonic*, or bottom-dwelling, forms—may sink to the bottom and be included in various sediments in many different aqueous environments. Thus the presence of foraminifers in a sediment indicates a marine environment, but only certain benthonic genera and species suggest, to a practiced eye, particular depth zones and bottom conditions. Many other organisms are diagnostic, but many are not. Some molluscs, for example, are fresh-water forms and some are marine; others, including certain oysters, flourish in tidal and periodically brackish waters. To use the specific organic content of sediments as an indication of depositional environment requires some knowledge of paleontology, particularly of that branch of the science called paleoecology. Few petrographers are capable of interpreting this evidence for themselves, but they should, whenever possible, make full use of its interpretation by paleontologists, for organic remains are an inherent and important part of any sedimentary rocks in which they occur.

Specific inorganic criteria whereby the environment of deposition can be judged directly are relatively rare or poorly understood. One of the most useful is the mineral glauconite, which is widespread in sediments of all ages and apparently has formed only in marine sedimentary environments. It forms by diagenesis in muds on the sea floor, usually where the water is less than several hundred meters deep and under reducing conditions. Pyrite commonly forms in the same places, and the two minerals are often associated in sedimentary rocks.

Diagenesis

Both texture and composition are affected by changes that take place in sediments after deposition. Wherever these changes occur at relatively low temperatures, they are termed *diagenetic*, rather than metamorphic. Primary deposition and diagenesis are scarcely distinguishable when the post-depositional alterations occur at the site of original deposition and almost contemporaneously with it. Glauconite, for example, is usually classed as a diagenetic mineral, but it forms in sediments as they lie unburied on the sea floor (see pp. 409–411). Early diagenetic reactions in marine deposits are somewhat analogous to postdepositional changes in terrestrial deposits exposed to subaerial weathering before being buried; indeed, early marine diagenesis has often been referred to as submarine weathering or *halmyrolysis*. Just as organisms and organic products foster many reactions during subaerial weathering, so many of the reactions that take place during early marine diagenesis are induced by organisms, particulary by bacteria and the various scavengers which abound on the sea floor and inhabit the sediments that lie exposed there. An organic population is normally restricted to the uppermost part of a sedimentary deposit, and the longer sediments lie unburied and exposed to continued organic activity the more noteworthy will be the changes produced. The time available for these early reactions, therefore, is dependent on the rate of deposition, so that early diagenetic changes are at a maximum where deposition is very slow.

Other diagenetic reactions take place long after deposition and constitute what may be called *late diagenesis*. Some of these reactions occur after the sediment has been buried for some time, or even while it is being deformed. If they occur at great depth below the surface, the pressures and the temperatures are higher than those existing at the surface, and diagenesis then grades into metamorphism. The time available for such reactions may, of course, be very great, and the changes produced generally tend to produce compact, lithified rocks.

Processes that commonly occur during diagenesis include compaction, solution, authigenesis, and replacement. As sediments are buried, *compaction*** takes place; the solid particles of a sediment are pressed closer together by the weight of overlying material, and bulk volume is reduced. As a result, the fluid filling pores in the sediment is squeezed out and must generally migrate slowly upward through the deposit. The process of compaction is most noticeable in fine aqueous muds, either detrital or chemical, but it occurs to some degree in all sediments. The original, or *connate*, water with which marine and lacustrine deposits are saturated cannot be completely displaced by this process alone, however, and some of it may remain an integral part of deeply buried sediments for ages. Where it remains, it is the medium in which the later chemical changes occur, and its character exerts a considerable effect on many diagenetic

---

*The term *compaction*, as used by engineers who deal with soil mechanics, refers to reduction in volume produced by squeezing air, not water, out of a soil; *consolidation*, as they use the term, means reduction in porosity in general and is synonymous with the term *compaction* as generally used by geologists.

reactions. Elsewhere the connate pore water is displaced by slow general circulation of subsurface fluids through the rocks, and it is then the displacing fluid in which the later diagenetic changes occur.

The effects of solution during diagenesis are to be seen in many sedimentary rocks. Solution cavities may be formed or, if pressure closes potential openings as solution occurs within the rock mass, stylolites may develop. *Stylolites* are highly irregular surfaces within a rock and are typically marked by thin seams of insoluble residue; they commonly transect individual grains or fossils in a way that indicates removal of some material after deposition. Adjacent grains may interpenetrate along *microstylolitic* boundaries (cf. Figure 14-6B). Replacements and casts of original fossils are other indications that solution of original material has occurred. Also, various unstable minerals, particularly high-temperature ferromagnesian silicates such as olivine and pyroxene, tend to be eliminated from older terrigenous deposits by a process of *intrastratal solution;* these minerals are commonly seen only in relatively young rocks.

Authigenic minerals that are stable in the diagenetic environment commonly crystallize in sedimentary deposits. They may occur as authigenic cement precipitated in open pores between detrital grains or as replacements of original components of the deposit. Cements composed of SiO_2 or calcite are widespread, but many other cement minerals occur; dolomite and chert replacing calcite in limestones are widespread examples of authigenic replacement.

The effects of diagenesis are more noticeable and more extensive in some rocks, and in some environments, than in others. In general, however, the most visible effect is *lithification.* Some deposits are cemented by authigenic minerals; others have been firmly consolidated by pressure. Tight compaction of clay material, for example, produces a firmly bonded aggregate that cannot be easily disintegrated. Some coarser detrital sediments have been converted, by a process combining grain deformation and partial solution, into aggregates of interlocking grains bound together as in a mosaic (Figure 13-15B).

TECTONIC ENVIRONMENT

Various factors that play a part in the genesis of sedimentary rocks have thus far been considered separately, but these factors are not unrelated. Sedimentary rocks are products of composite general environments that exist during considerable periods of time throughout large areas and that vary within these limits from time to time and from place to place. All the genetic factors already mentioned are involved, but they are largely governed by the varying tectonic activity of the earth's crust.

The sedimentary deposits formed on the more stable parts of the crust are different from those formed in very unstable areas, and the differences are to be seen in the associations, the composition, and the texture and structure of the resultant sedimentary formations. It would be incorrect to assume, however, that a particular rock type is always deposited under one specific set of tectonic conditions.

Tectonic environments can be broadly classified in terms of several contrasting crustal settings. At one extreme are the cratons, those broad regions in which the continental platform is stable and has been relatively inactive for a long period. There the relief is generally low and the climate tends to be uniform. Decomposition of rocks by weathering tends to be relatively complete, because removal of weathered material is slow. When the detritus is removed, it may be carried for long distances, perhaps being deposited and reworked many times before it finally comes to rest. Sedimentary deposits are spread thinly over extensive areas and may lie exposed for long periods. Thus the regional configuration approaches a surface of equilibrium across which sediment is moved relatively slowly toward the continental margin. When slight epeirogenic subsidence occurs, broad areas of the craton may be flooded by very shallow epicontinental seas in which thin marine deposits are laid down. Only if gradual subsidence continues will a substantial thickness of strata accumulate, but such stratified sequences usually contain numerous discontinuities (disconformities) and do not attain the thickness of those laid down in unstable areas.

The terrigenous sediments deposited in cratonic settings tend to be concentrates of the relatively indestructible minerals. Pebbles and sand grains commonly become well rounded, and the deposits tend to be well sorted, sand and clay being deposited separately. In epicontinental seas, calcareous organisms may flourish and produce widespread sheets of limestone. The assemblage of deposits formed typically includes clean quartz-rich sandstones, clay shales, and where favorable marine conditions prevail, numerous limestones and dolomites. Such an assemblage is exemplified by the relatively thin and little-deformed sequence of Paleozoic strata in the midcontinental United States, which accumulated in epicontinental seas and on bordering lowlands entirely within the stable continental platform.

A less stable environment has existed since early Cretaceous time along the eastern margin of the North American continent. There downwarping has been essentially continuous, and rivers have supplied large volumes of terrigenous sediment derived from elevated parts of the continental interior. These two factors have combined to produce along the continental margin a sedimentary terrace 5 km or more thick and more than 100 km wide. Along the northern Atlantic coast, the strata are principally mudstones and quartzose and feldspathic sandstones. Southward, the terrigenous strata give way to limestones deposited in the warm, shallow seas of the south Florida–Bahama area, which has remained largely beyond the reach of terrigenous silicate detritus. Although both carbonate and terrigenous facies are very thick, the character of the strata indicates that they were laid down in the relatively shallow waters of the continental shelf, or on the adjacent coastal plain. Some detritus carried seaward across the continental terrace has been transported down the continental slope and has accumulated on the deep-ocean floor at the foot of the slope, forming the continental rise. The mineral composition of sandy layers in these deep-sea deposits corresponds in general to that of sediments on the adjacent shelf, but, as shown by deep-sea drilling operations, the texture and structure of the deposits are different and have the general character of flysch. Major sedimentary accumulations such as occur along the eastern margin of North America appear to be characteristic of rifted (trailing) continental margins elsewhere. The thick shallow-water sedimentary complexes that underlie broad continental shelves along rifted continental margins have been labeled *miogeoclines*.[1]

The most unstable environments are the active orogenic belts where continental and oceanic plates converge. Here, as exemplified in the circum-Pacific belt today, volcanic activity and mountainous relief are characteristic and are maintained by the continuing relative movements of the crustal plates. Large volumes of terrigenous sediment are deposited in geologically short periods and commonly in relatively restricted basins where the deposits pile up quickly, sometimes to extreme thicknesses. Along orogenic continental margins, most of the sediment carried to the sea is deposited in tectonically active, deep-water troughs and basins on the submerged margin of the continent and on the oceanic crust beyond it. These deposits usually are flysch sequences consisting of interbedded sandstones and mudstones; many of them also include interstratified submarine lava flows and volcaniclastic layers. On the elevated parts of the continental margin, thick alluvial fan-delta complexes, called *molasse*, are laid down in low-lying piedmont areas adjacent to recently uplifted mountain ranges.

Sandstones deposited along the orogenic (convergent) margin of a continental plate tend to be relatively rich in unstable detrital components, particularly plagioclase feldspars and fragments of aphanitic rock, and may include abundant volcanic detritus. Where quartzo-feldspathic plutons become exposed in highly elevated continental blocks, arkosic sandstones made up largely of feldspars and quartz may be formed. Furthermore, because tectonic activity and sedimentation continue concurrently for long periods, some early-formed deposits may be deformed and uplifted, and then they become, in turn, second-cycle sources contributing sediment to still younger deposits.

The terms *flysch* and *molasse* are used today in general reference to two contrasting sedimentary facies, both of which form in orogenic regions; they were coined originally for formations that occur in the Alps. There flysch refers to the deep-marine deposits that were interrupted and deformed by one of the orogenic pulses that produced the Alpine range; molasse refers to the piedmont deposits laid down along the northern base of the range during and after its elevation. Alpine flysch is typically thin bedded and sparsely fossiliferous, and its total thickness is great. It consists of alternating layers of argillaceous or calcareous graded sandstone and dark silty shale or marl. The molasse is also thick and sparsely fossiliferous, but it is younger and includes some debris eroded from the deformed and uplifted flysch exposed in the range. Also, its general aspect contrasts with that of the flysch, for it consists of conglomerates, sandstones, and mudstones of both fluvial and shallow marine types.

Recent intensive investigation of the ocean floor and the rapid development of plate tectonic concepts have focused attention on the great variety of tectonic settings in which sedimentary deposits may accumulate.[2] In contrast, earlier concepts concerning the relation between tectonics and sedimentation were developed essentially from studies of ancient sedimentary formations now uplifted and exposed to view on the continents. In the deformed belts of older mountain ranges, two general assemblages of sedimentary strata were recognized and distinguished. Exceptionally thick sequences characterized principally by limestones, quartz-rich sandstones, and shales, largely of shallow-water origin and abundantly fossiliferous, were referred to as *miogeosynclines*. Very thick flysch-type sequences of sandstone (commonly graywacke) and mudstone or shale with interstratified volcanic members (typically pillow lavas) and often with bedded radiolarian cherts were referred to as *eugeosynclines*. These two terms denote in a general way two contrasting types of thick stratified rock assemblages; unlike the newer term *miogeocline*, they refer to the rock assemblage without specifying a particular tectonic setting.

The tectonic settings and environments of sedimentation that can be directly investigated today are all geologically young, and therefore one cannot necessarily extrapolate directly from them to earlier, and especially to Precambrian, settings and environments. Consider, for example, the immensely thick marine-basin deposits of Proterozoic age, with their unique banded iron-formations (cf. pp. 408; 416) and economically valuable sulfide sediments. These are widespread on several continental cratons but their unique character, especially that of the metallic deposits, points to an early Proterozoic environment that has no younger counterpart.

GENERAL REFERENCE

For a discussion of the origin of sedimentary rocks, see H. Blatt, G. Middleton, and R. Murray, *Origin of Sedimentary Rocks,* 2nd ed. Englewood Cliffs, N.J.: Prentice-Hall, 1980; G. M. Friedman and J. E. Sanders, *Principles of Sedimentology.* New York: Wiley, 1978.

REFERENCE NOTES

[1] R. S. Dietz and J. C. Holden, *Journal of Geology,* vol. 74 (1966): pp. 566–583.

[2] For an illustration of the variety of tectonic settings in which detrital sediments are deposited and a summary of data relating sand composition to settings in the plate tectonic model, see W. R. Dickinson and C. A. Suczek, *American Association of Petroleum Geologists Bulletin,* vol. 63 (1979): pp. 2164–2182; W. R. Dickinson and R. Valloni, *Geology,* vol. 8 (1980): pp. 82–86.

11

Composition and Texture
of Sedimentary Rocks

SEDIMENTARY COMPONENTS
AS THE BASIS OF CLASSIFICATION

The physical components (individual particles) that make up a sedimentary rock are of two general kinds. The first includes all detrital (largely terrigenous) particles transported by physical means from extraneous sources to the site of deposition. The second, or nondetrital, category includes all materials formed by chemical or biochemical precipitation at the site of deposition, as well as chemical precipitates of somewhat later origin—authigenic cements and other products of diagenesis.

Detrital components, being solid particles that are transported by purely physical means, are classified primarily according to particle size as pebbles, sand, silt, and clay. Any mineral present in the source terrane that survives the processes of weathering and transportation may occur as a detrital constituent in a sedimentary deposit. Nondetrital components, being chemically or biochemically deposited, are restricted to minerals that may crystallize, or recrystallize, in aqueous solutions at surface and near-surface temperatures and pressures. Whether a constituent is detrital or nondetrital generally must be inferred from its mode of occurrence and its textural characteristics and relationships in a sedimentary rock. Many mineral species may have either origin. Indeed, the most abundant and widespread minerals in sedimentary rocks—clay minerals, quartz, alkali feldspars, and calcium-bearing carbonates—may be either detrital or authigenic, or both.

Most sedimentary rocks contain several different components, including both detrital and nondetrital types. For purposes of description and classification, only a few of the many potential constituents are quantitatively important, in particular the following kinds.

1. Detrital particles
   a. *Pebbles, sand,* and *coarse silt*: Relatively large grains that are carried largely as a bed load in transporting currents. They consist chiefly of quartz (including chert), other common rock-forming minerals (mostly silicates), and fragments of polycrystalline rock.
   b. *Mud* (clay and fine silt): Minute particles that are carried largely in suspension in transporting currents. They consist chiefly of clay minerals but usually include tiny fragments of quartz and mica.
2. Nondetrital minerals
   a. *Calcium carbonates*, chiefly calcite and aragonite (largely of biochemical origin) and dolomite.
   b. SiO_2 in the form of opal (largely of biochemical origin), chalcedony, or microgranular quartz.

The composition of the common sedimentary rocks can be described in terms of these four elements, which may occur either singly or intermixed in any proportions. Rocks of mixed composition are classified and named according to their principal constituents.

Because the variations between rock types are gradational, all classifications of sedimentary rocks are necessarily somewhat arbitrary. Thus, in Figure 11-1 rocks are classified broadly as mixtures of three components. In Figure 11-1 the corners of one triangle represent calcium carbonate, mud, and nondetrital SiO_2, and the corners of the other calcium carbonate, mud, and sand; points on either triangle represent all possible mixtures of the three components represented by its corners. Other mixtures involving sand with nondetrital SiO_2, with silt in place of sand or with a fourth component added, could be classified in similar fashion.

Complete classification of sedimentary rocks involves consideration of other chemically deposited mineral components in addition to those included in Figure 11-1, but the principles of classification are the same. *Phosphorites*, for example, are sediments in which a principal constituent is the cryptocrystalline variety of apatite, called *collophane*. This material may occur practically pure, but is generally mixed with sand, clay, or carbonates to form phosphatic sandstones, mudstones, and limestones. *Evaporites* are sediments containing the more soluble minerals, such as gypsum, anhydrite, and halite, precipitation of which is caused by evaporation. These, too, may occur nearly pure but are very commonly mixed

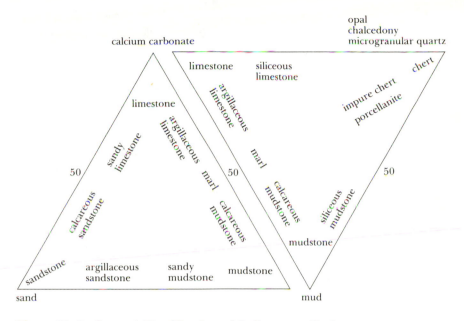

Figure 11–1. General Classification of Sedimentary Rocks

The transitional nature of the various rock types is indicated by omitting fixed boundaries from the diagram.

with sand, clay, or calcium carbonates. Other noteworthy components include iron-rich minerals such as hematite, pyrite, siderite, glauconite, chamosite, and greenalite, which may be concentrated to form iron-rich deposits.

Chapters 12 through 14 are devoted to the description of the most common sedimentary rocks—mudstones and shales, sandstones, and limestones and dolomites. Chapter 15 is allotted to a brief description of miscellaneous sedimentary types that are locally abundant. Before we begin detailed descriptions of these rocks, however, certain general features of rock fabric and mineralogy remain to be discussed.

OCCURRENCE AND STABILITY
OF MINERALS IN SEDIMENTS

Many common rock-forming minerals are *unstable* in the surface environment; they tend to be destroyed or altered during weathering, transportation, or diagenesis, and are therefore less common in detrital sedimentary rocks than in the source rocks. Minerals that remain essentially unchanged by sedimentary processes are considered *stable*. But stability in this sense is relative, and on this basis minerals do not fall readily into two groups. They resist destruction in varying degrees: at one extreme are such minerals as calcite and olivine, which dissolve or decompose very readily under some surface and near-surface conditions; at the other extreme are such minerals as quartz, which effectively resist change in all sedimentary environments. Between these extremes are many minerals of intermediate stability.

Consider, for example, the common ferromagnesian silicates of igneous rocks. Subjected to chemical weathering or to intrastratal solutions in buried sediments, olivine decomposes most readily, followed by pyroxenes, hornblende, and biotite, the same order in which they normally crystallize from cooling magmas. Among feldspars the calcium-rich varieties decompose more readily than those richer in alkalies; the most stable are albite, orthoclase, and microcline. Quartz and muscovite stubbornly resist decomposition. The stability series among these minerals, then, is similar to the magmatic reaction series (p. 19), the later-forming, lower-temperature igneous silicates being the most stable in sedimentary environments.

Rock-forming minerals may be arranged in a general stability series that expresses their relative resistance to destruction by normal processes of sedimentation; but the series does not remain the same under all conditions. Calcite, for example, is commonly dissolved in vadose water during weathering, especially in warm humid regions where vegetation is abundant; but in the depositional environment, and during later diagenesis, it may be stable and commonly crystallizes as an authigenic mineral. Alkali feldspars also, judged by their frequent formation as authigenic crystals, are stable during diagenesis of marine sediments; yet they are destructible and comparatively unstable under some conditions of weathering and transportation. Among the clay minerals formed by weathering, kaolinite is generally more stable under acid conditions and smectite under alkaline conditions, but both may change slowly to illite or chlorite during marine diagenesis.

The relative stability of minerals in various diagenetic environments, for example, in marine as compared with fresh or very saline waters, is uncertain, but the stability series is presumably different in each. Some common minerals, however, are apparently never authigenic and probably always tend to be eliminated by intrastratal solution during diagenesis. Such, for example, are the high-temperature silicates of igneous rocks (olivine, pyroxene, hornblende, and calcic plagioclase), which are not known to crystallize at the low temperatures of the sedimentary environment. Since these minerals are common in igneous and metamorphic rocks and scarce in the older sedimentary rocks, they must be largely eliminated during sediment history.

The purposes of this discussion can best be served by classifying the more common minerals into the following broad groups.

1. *Unstable minerals*
   a. Minerals that tend to be altered or destroyed during any or all stages of the sedimentary cycle—by weathering, transportation, or diagenesis. In approximate order of increasing stability the commonest are olivine, pyroxene, calcic plagioclase (An_{50-100}), hornblende, biotite, and oligoclase-andesine.
   b. Minerals that are generally stable during diagenesis, particularly marine diagenesis. These minerals tend to resist alteration once they are deposited and are commonly authigenic, but they tend to be destroyed by weathering and abrasion and are therefore considered unstable when they are detrital. In approximate order of increasing stability, the commonest are the various carbonate minerals, glauconite, pyrite, apatite, zeolites, and alkali feldspars.
2. *Stable minerals*
   a. Minerals that effectively resist alteration and destruction during all stages of the sedimentary cycle; these minerals occur both as detrital and authigenic constituents in sedimentary rocks. The commonest are clay minerals,* quartz (including chert), muscovite, tourmaline, zircon, rutile, brookite, and anatase.

All the minerals included in the foregoing list are commonly found in sedimentary rocks, but only a few are likely to be abundant. The most abundant are quartz, which is the chief component of sandstone, and the clay minerals, which are the chief constituents of mudstones and shales. Also abundant are the feldspars orthoclase, microcline, albite, and oligoclase-andesine, the carbonates calcite and dolomite, and silica in the form of opal, chalcedony, and microgranular quartz. Widespread but considerably less abundant are iron oxides. All other minerals together constitute probably less than 3% of the average sedimentary

---

*To call the clay minerals stable is to classify them as a group and is correct only in a very general sense. Different species of clay minerals are strictly stable, during either weathering or diagenesis, only under particular conditions (p. 420).

rock. Locally, however, some of the less common constituents may be so concentrated as to become principal components in particular deposits, and of these the most common are goethite (including limonite), hematite, siderite, glauconite, chamosite, gypsum, anhydrite, halite, and collophane.

It is noteworthy that the most widespread sedimentary minerals, and particularly the ones that may be abundant in sedimentary rocks, are among those that are stable during diagenesis. That is, they are minerals that may crystallize at low temperatures. The most common of all, quartz and clay minerals, are stable in all sedimentary environments.

TEXTURES OF SEDIMENTARY ROCKS

Texture refers to the physical make-up of a rock as distinct from its mineral or chemical composition, in particular to its crystallinity and grain size and to the mutual relationships of the individual components. Similar textures are to be found among rocks of very different mineral composition, and rocks of diverse texture may be similar mineralogically. Limestone, for example, has many textural varieties. All these consist essentially of calcite, but some are coarsely crystalline and others are so fine-grained that no individual crystals are easily visible. Some are composed of grains all of a single size, whereas others are mottled by large crystals scattered through a matrix of smaller ones. In some limestones crystals of calcite interlock tightly, producing a nonporous structure like a mosaic, but other limestones are porous. The calcite crystals may be arranged in a lamellar or fibrous fashion having a recognizably organic structure; they may have a crude columnar structure such as develops by crystal growth normal to the walls of an open cavity; more commonly they are equant grains and the rock has a sugary appearance. Some limestones are *oolitic*, consisting of innumerable small spheroids each usually having concentric and often radial structure. In still others the calcite has the character of sand grains, which may be either uncemented or very firmly cemented by calcite precipitated in the pores between the grains. All the features referred to in this example are *textures*, and they constitute a significant and useful basis of rock classification.

To decide how a sedimentary rock was deposited, one must study both the mineral composition and the textural relationships between individual components of the rock. The distinction between mechanically deposited grains and those that have been chemically precipitated or recrystallized is based primarily on textural features. It is therefore appropriate to classify textures as either *clastic*, consisting of mechanically deposited fragments, or nonclastic.

Nonclastic Textures

A typical nonclastic texture consists of interlocking crystals molded to each other as in a mosaic. Such a texture is generally referred to simply as crystalline. The individual crystals may have a variety of shapes and sizes and their mutual boundaries may be either smooth and regular or jagged and irregular. If the crystals are anhedral or subhedral grains of roughly equidimensional shape and approximately uniform size, the texture is said to be *crystalline granular*. Other crystalline textures may show a wide range of grain sizes and may include euhedral crystals, such as those of dolomite replacing calcite in limestones (Figure 14-10A). If the component crystals have fibrous, prismatic, or platy habits, they may be arranged in different ways to produce a variety of fabrics, for example, Figures 15-10 and 15-11.

Visibly crystalline textures in which the individual crystals cannot be clearly discriminated are called *microcrystalline*, a term usually applied to aggregates in which individual grains are less than 30 μm, the thickness of a standard thin section. If crystallinity is not evident except at exceedingly high magnification, or by x-ray diffraction, the texture is called *cryptocrystalline*. Strictly noncrystalline texture, such as that of biogenic opal (cf. p. 399), is to be expected only in relatively young sedimentary rocks and does not characterize large rock masses. The general term *aphanitic* covers both microcrystalline and cryptocrystalline textures.

The essential features of nonclastic crystalline textures arise by the growth or enlargement of crystals within an aggregate, and they can best be described by considering the stages of their development. The crystals first precipitated from a solution are minute, and are enveloped in a slightly oversaturated solution which nourishes them and in which they may grow freely. These initial crystals may be attached to any solid surface with which the solution is in contact, or they may form in suspension. However, the aqueous solutions in seas and lakes that produce most chemical sediments are so fluid and have such low density that any suspended crystals quickly settle to the bottom. The accumulated precipitate is a porous aggregate of crystals that do not interlock. Lime muds, for example, are of this sort, but they do not long remain in this condition.

The crystals are pressed closer together by the weight of additional sediment, and they may continue to grow by precipitation from the slightly oversaturated solution that remains in the pores of the deposit. In most chemical sediments, at least partial recrystallization of the original precipitate is induced by increased pressure and temperature resulting from burial.

Without any recrystallization whatsoever, precipitation from dilute aqueous solutions would be expected to result in rather porous fabrics. More compact deposits may result from crystallization of concentrated brine, but even the evaporites in desiccated lake basins have a large intergranular porosity so long as they remain unburied and unrecrystallized. Compact, nonporous textures do result, however, where crystal aggregates grow by slow accretion outward from a solid surface into a slightly oversaturated solution. To nourish these growths, there must be a continuing supply of ions in solution, and these must be brought to the site of crystallization by diffusion or by movement of the solution itself. If these conditions are met, a compact aggregate results, as for example, in many cavity and fissure fillings and in the cement that fills the pores of many rocks.

Since most sediments of chemical and biogenic origin consist of minerals that are at least appreciably soluble in water, they are particularly susceptible to recrystallization during diagenesis, especially after temperature and pressure are increased by burial. During simple recrystallization of original sediment, without addition of any new minerals, the grain size of an aggregate tends to be enlarged, so that individual crystals become tightly interlocked. Boundaries between the crystals may be relatively smooth, as in a mosaic, or they may be jagged and interpenetrating. In either case the fabric is without visible spaces between grains. Where *replacement* of the original substance by new minerals takes place, the result may be different. Reduction rather than increase of grain size may occur, as in dolomitization or silicification of coarse limestones, and porosity may be produced if the new materials occupy less volume than the original.

Clastic Textures

Rocks formed by accumulation of mineral and rock fragments have clastic texture. The particles may have any size, form, or composition, and they may be packed together in any fashion, loosely or very tightly. Most clastic aggregates, like sand and gravel, have some intergranular porosity, but they may become nonporous and impermeable by crystallization of authigenic cement in the pores. Or, by very tight packing and perhaps partial solution of grains under pressure of deep burial, a clastic rock may be converted into a firm aggregate that simulates crystalline granular texture. The term clastic, however, is as appropriate for such consolidated rocks as it is for incoherent porous sand; indeed, it may be applied to the texture of any aggregate whose original fragmental character is either clearly visible or can be inferred with assurance.

In poorly sorted, fragmental aggregates having a wide range of particle sizes, the material is conveniently subdivided, for purposes of description, into *grains* and *matrix*, which are roughly analogous to the phenocrysts and groundmass of porphyritic igneous rocks. No particular grain size is implied in the distinction; the terms have reference to the *relative* size of particles and to their disposition in the aggregate. Very small particles are normally packed between the larger ones, and in the aggregate they have the appearance of a matrix surrounding and containing the larger grains. The distinction between matrix and grains is based simply on this visual contrast, which can be seen only in the aggregate and results from the different form or darker color of the finer material. If grains compose more than about two-thirds of a rock, they will be in contact, and the texture is said to be *grain-supported;* the matrix simply fills in the potential pores between the grains. Where grains are much fewer, they "float" in the matrix; in other words, they are loosely packed and scattered through the matrix, in which they appear to be suspended. Such textures are *matrix-supported*, or *mud-supported.*

Except for those most recently deposited, few clastic sedimentary rocks are altogether lacking in authigenic *cement*. Minerals commonly crystallize in the voids between fragments, precipitating first on the grain surfaces and gradually filling the pores by continued growth. Cement is very common in fairly well-sorted silts, sands, and gravels, but it is much less common in argillaceous (muddy) sediments, perhaps because the relatively impermeable character of clay tends to prevent migration of solutions and the precipitation of large amounts of cement. Clay material alone is a firm bonding agent when compacted, and it commonly serves to consolidate sedimentary rocks that do not contain precipitated cements.

All detrital rocks have clastic texture; so also do many sediments composed of nondetrital components, such as organic skeletal structures (shells, etc.), which develop on the sea floor but may be broken and moved about and thus take on the character of detrital grains. As they move these grains may be sorted and current-bedded, or they may accumulate as ill-sorted aggregates. They may become mixed with terrigenous silicates, but often they are not. Diagenetic recrystallization and replacement, to which such rocks are particularly susceptible, may later destroy all trace of original clastic texture, but often the outlines of original clastic grains are discernible within the recrystallized aggregates.

SIZE AND SORTING OF CLASTIC GRAINS

Size, being one of the most readily visible properties of a particle, is the basis for classification of unconsolidated clastic sediments as gravel, sand, silt, and clay. Each of these terms has been given quantitative definition as follows (based on the mesh size of standard sets of sieves):

Names of Particles   *Diameters of Particles**

	boulders .... more than 256 mm
Gravel	cobbles .... from 256 mm .... to 64 mm
	pebbles .... from 64 mm .... to 4 mm
	granules .... from 4 mm .... to 2 mm
	very coarse .... from 2 mm .... to 1 mm
	coarse .... from 1 mm .... to 0.5 mm
Sand	medium .... from 0.5 mm .... to 0.25 mm
	fine .... from 0.25 mm .... to 0.125 mm
	very fine .... from 0.125 mm .... to 0.062 (or .05) mm
Silt .... from   0.062 (or .05) mm ..... to 0.005 mm
Clay .... less than 0.005mm

From this classification arises the familiar nomenclature of detrital deposits, *conglomerate, sandstone, siltstone*, and *claystone* or *mudstone*.[†] An aggregate composed mainly of sand-sized grains is called *sand*, or *sand-*

---

*The size of a particle can be defined in various ways—by volume, by diameter, by sieve mesh through which particles will pass, by weight, or by settling velocity in a fluid. The diameter of irregular, nonspheroidal grains is not easy to define; in this table, "diameter" refers to the intermediate diameter of a particle.

†Alternative names derived from Greek and Latin roots are psephite or rudite for gravelly aggregates, psammite or *arenite* for sandy rocks, and *pelite* or *lutite* for clay or mud rocks. *Argillite* is also a mud rock, but the term customarily refers to one that is very firmly indurated, usually somewhat recrystallized. Adjectives indicating clay material are *argillaceous, pelitic*, and lutaceous; those indicating sandy material are *arenaceous* and psammitic; those indicating gravel or conglomerate are psephitic and rudaceous. The terms in italics are those most often used.

stone if consolidated; one in which most of the particles are smaller than sand but coarser than clay is *silt* or *siltstone*; and aggregates of the finest particles are *clay* or *claystone* and *mud* or *mudstone*. If more than one-half of the grains are coarser than sand, the loose aggregate is *gravel* or *rubble*, and the consolidated rock is *conglomerate* or *breccia* (p. 306). This classification of detrital aggregates is simple and widely used. It also has genetic significance because grain size in a clastic aggregate is determined by the hydrodynamics of erosion and deposition.

No clastic sediment is perfectly sorted; many contain a wide range of particle sizes and are very poorly sorted indeed. Poorly sorted rocks are generally designated according to the predominant grain size present. For example, a rock that contains 55% sand, 15% silt, and 30% clay may be called a sandstone. The name "sandstone" alone, however, does not distinguish this rock from one composed entirely of well-washed sand grains. It is best, therefore, to call such a heterogeneous rock by a term that contains a qualifying adjective, or by some specific term. It may be called *argillaceous sandstone* or *wacke*.

Differences in sorting among clastic sediments have great genetic significance, and there is need here to roughly define the terms used in characterizing these differences. To say that the particles of unsorted and poorly sorted deposits range greatly in size means that the larger particles are *many times* larger than the smaller ones. Sand grains, for example, are larger than the finest clay particles by a thousand times or more, and a rock in which the two, together with all intermediate sizes, are intimately mixed is poorly sorted. The range in particle size would be far greater in an argillaceous conglomerate, but it would be less great in mixtures of sand and gravel without clay, which would therefore be somewhat better sorted. In the best-sorted clastic sediments, resulting from highly selective deposition, the largest particles have no more than five or ten times the diameter of the smallest. A rock that is well sorted may be fine or coarse, provided it consists of particles that are fairly uniform in size.

Shape and Roundness of Clastic Grains

The *shape*, or *sphericity*, of a grain refers to the relative dimensions of its three mutually perpendicular diameters, whereas its *roundness* refers to the angularity of its edges and corners. According to their shape, grains may be grouped qualitatively as *spheroidal* or equidimensional, *disk-shaped* or platy, *rod-shaped* or prismatic, and *blade-shaped*; according to their degree of roundness they are *angular, subangular, subrounded,* and *rounded.* These two properties, although frequently confused, are geometrically distinct and not fundamentally related. Particles of the same shape may have varying degrees of roundness, and those of similar roundness may have various shapes. Dodecahedral garnet crystals, for example, are spheroidal whether their interfacial edges are sharply angular or have been rounded by abrasion, and a prism of hornblende that was originally euhedral and sharply angular may become well rounded without losing its general prismatic shape (Figure 11-2).

Precise determinations of shape and roundness, involving, as they do, all three dimensions of a particle, are tedious and are not normally made in routine sedimentary analysis. Generally the shape and roundness of grains are estimated by visual comparision with a set of standards, from the outlines of grains as seen in two dimensions. Thin sections lend themselves to this sort of qualitative evaluation, but in them roundness is more reliably represented than shape. Being sliced along a random plane, each grain in a thin section reveals two random diameters that commonly do not adequately represent its shape but do give a good idea of its degree of roundness.

Shape and roundness differ in significance. Shape is largely inherited; it depends on the shapes of minerals in the parent rocks or on their fracture habits, and is relatively little altered during transportation. It does exercise a major influence, however, on the hydrodynamical behavior of a particle, affecting its settling velocity and mode of transport in a fluid current. Thus it affects selective transportation, for grains are commonly sorted by shape as well as by size and density. Roundness, in contrast, has little effect on the behavior of a particle in a fluid; it is, rather, a measure of the amount of abrasion during transportation and of the susceptibility of particles to abrasion. However, where an element of grain rolling enters transport (as in longshore drift in shallow water and in wind transport of sand) well-rounded grains tend to become selectively concentrated in the down-current direction.

Subdivision of clastic deposits according to the roundness of their particles cannot be applied to very fine-grained deposits because small particles are not abraded and are invariably angular; but it is readily applied to the coarser deposits. Thus *breccia* is a coarse-grained clastic rock composed of angular grains; a similar rock containing grains that have been abraded and hence are somewhat rounded is *conglomerate.*

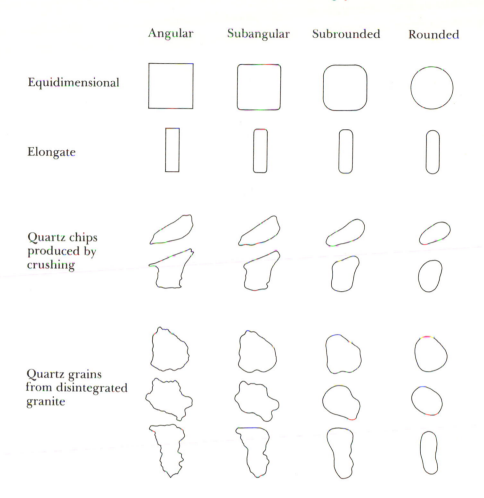

Figure 11–2. Two-Dimensional Shape and Roundness of Grains

As seen in two dimensions, grains are either *equidimensional* or *elongate*. Cross-sections of a sphere, the small section of a prism (rod), and the broad section of a disk are all equidimensional. The long sections through rod-, disk-, and blade-shaped grains are elongate. Typical quartz grains, shown in the lower and middle parts of the diagram, tend to be equidimensional rather than markedly elongate.

The four vertical columns represent four classes of roundness, which can be described qualitatively as follows. *Angular*: all corners sharp, having radius of curvature equal to zero; surface not abraded. *Subangular*: corners not sharp but have very small radius of curvature; most of surface not abraded. *Subrounded*: corners very noticeably rounded but surface not completely abraded. *Rounded*: entire surface abraded; radius of curvature of sharpest edges is about equal to radius of maximum inscribed circle.

PACKING OF GRAINS

The particles in a clastic sediment may be packed together so closely as to occupy the smallest possible total volume with little pore space between the particles. One might expect to find close packing of this kind in deposits that have been deeply buried and subjected to high pressure; indeed, in some such deposits, the grains have been deformed, or differentially dissolved at points of contact, and are locked together in an aggregate that has exceedingly low porosity. In contrast, the particles in newly formed, unburied clastic deposits tend not to be closely packed, and the aggregates are generally very porous.

The particles in a clastic aggregate can be arranged and packed together in many different ways. The possible packing schemes can be systematically analyzed where all particles are identical and no deformation of individual particles occurs. Such an analysis has been made by Graton and Fraser[1] for aggregates of uniform spheres, and their findings, although not directly applicable to natural aggregates, are nevertheless most instructive. If the spheres are systematically arranged in a cubic fashion and packed so as to touch each other, the aggregate texture is open and has a porosity of 47.5%. If the spheres are packed in a rhombohedral fashion* the aggregate has the smallest bulk volume that can be produced without distortion of the grains, and its porosity is only 26%. Between these extremes are systematic packing schemes that yield intermediate values of bulk volume and porosity. But the packing arrangement may also be haphazard, or nonsystematic, producing porosities as high as 45% or 50%. Chance packing, the arrangement normally to be expected, is a combination of systematically packed domains surrounded by, or alternating with, haphazardly packed domains. The average porosity of chance-packed aggregates of uniform spheres is slightly less than 40%.

Variations in size, shape, roundness, and sorting of the grains in a clastic aggregate complicate the possible packing schemes enormously. Irregular, nonspherical grains might be arranged so that the irregularities of one fit those of another in roughly complementary fashion, like the stones in a well-built stone wall. Such an arrangement reduces bulk

---

*In cubic packing the unit cell contains eight spheres each centered at one of the eight corners of a cube whose edge is the diameter of the spheres. Rhombohedral packing has a unit cell of eight spheres each at the corner of a regular rhombohedron whose edges equal the diameter of the spheres.

volume and porosity, whereas a loose, haphazard packing of the same irregular particles produces especially porous texture. Where an aggregate is poorly sorted, containing a wide range of particle sizes, the smaller are packed between the larger, and low porosity and bulk volume per unit of solid volume are to be expected. This is the well-known relationship between sorting and porosity—namely, that poor sorting produces lower porosity.

Thin sections, being random planes cut through a chance-packed aggregate of natural grains, do not represent precisely the actual packing and porosity of the aggregate. Petrographers examining clastic aggregates in thin section must therefore interpret their observations in the light of a general understanding of the relation between the thin section and the three-dimensional aggregate from which it is sliced. We are indebted to Graton and Fraser for an ideally simplified illustration of the problem. Figure 11-3 shows the distribution of voids and grains seen on a random plane cut through a chance-packed aggregate of uniform spheres, and it clearly reveals some of the shortcomings of thin sections when carelessly used to represent texture. The grains are not correctly represented as to size and sorting, and many of them do not appear to be in contact. Unless the section cuts exactly through the point of contact between two grains, they appear to be separated, and where a grain is sliced near its margin it appears smaller than it actually is. Thin sections, of course, are not infinitesimally thin; a small third dimension is visible

Figure 11–3. Distribution of Voids and Spheres on a Random Plane Section Through an Aggregate of Uniform Spheres Chance-Packed in a Container

Note that the correct size of spheres is rarely apparent and that many are apparently not in contact. (Figure after L. C. Graton and H. J. Fraser, *Journal of Geology*, vol. 43 (1935): p. 848, fig. 14.)

under a microscope, and where grain boundaries are not normal to the plane of the section, two grains may overlap and appear to be in contact when they are actually separated by a thin void. Such contacts are hazy rather than sharp and clear.

These shortcomings do not invalidate textural observations made in thin sections provided due allowance is made for them. Compare Figure 11-4A,B, which shows the textures of two natural unconsolidated sands as they appear in random thin sections, with Figure 11-3, representing a section through a chance-packed aggregate of uniform spheres. The aspect of the natural sands is modified by variations in particle form and size, but they as well as the aggregate of spheres have open textures in which most grains do not appear to be in contact.

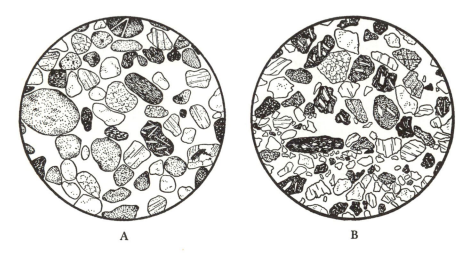

A   B

Figure 11–4. Recent Sands as Seen in Thin Section

A. Firm beach sand, Point Reyes, California. Impregnated with plastic before collection in order to preserve texture. Diam. 3 mm. Uncompacted subrounded grains very well sorted; porosity very high—about 30%. This is a lithic sand with high feldspar content; it contains abundant chert grains (heavily stippled), quartz (lightly stippled), feldspar (shown with cleavage lines), and various rock fragments.

B. Sand from channel of Jacalitos Creek, Coalinga, California. Impregnated with plastic before collection in order to preserve texture. Diam. 3 mm. Uncompacted subangular grains fairly well sorted; porosity very high; finer-grained layer at bottom. This is a lithic sand derived from a mixed sedimentary terrane including volcanic sandstones; it contains about 40% chips of andesite, argillite, shale, chert, and serpentine, 35% quartz, and 25% feldspar.

Most buried sands have, as might be expected, a more compact fabric than those just recently deposited. In thin sections this is recognized by closer packing of the grains (Figure 11-5), a feature indicating some postdepositional change in the packing arrangement (compaction), however it may have been accomplished. Many ancient and deeply buried sandstones are very tightly packed indeed, some so tightly that their grains interlock. Where little or no cement is visible in these rocks, one must assume that the detrital grains themselves have been forced into intimate contact. A somewhat similar appearance may also be produced by a form of cementation, common in quartzose sandstones, in which the detrital grains of quartz are enlarged by authigenic growth until they interfere and the pores are completely closed; such a process does not require tight packing of the grains.

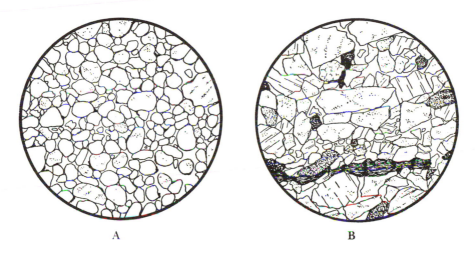

A   B

Figure 11–5. Uncemented Sandstones as Seen in Thin Section

A. St. Peter Sandstone (Ordovician), Beloit, Wisconsin. Diam. 2.5 mm. Very well-sorted sandstone consisting of subrounded quartz grains, a *quartz arenite*. The texture is very porous, but grains have been compacted until they are in close contact. Compare texture in Figure 11–4A.

B. Temblor arkosic sandstone (Miocene), 2500 m below surface, Kettleman Hills, California. Diam. 2.5 mm. Moderately sorted sandstone consisting of abundant subangular grains of quartz and feldspar (with cleavage), together with fewer biotite flakes (lined) and rock particles (heavily stippled). Texture very porous, but deep burial has caused rearrangement and compaction of grains. Compare the texture in Figure 11–4B. Note deformed biotite pinched between compacted grains.

"Packing" refers specifically to the arrangement of clastic grains, entirely apart from any authigenic cement that may have crystallized between them. Thus, in estimating the packing of cemented aggregates, the cement should be regarded as equivalent to the voids in an unconsolidated rock. Some completely cemented sandstones consist of very loosely packed detrital grains, which, except for the cement, have textures like those of unconsolidated sand. Cementation in these probably occurred soon after deposition, at the latest before deep burial of the detritus could produce tight packing. Other deposits were more tightly packed before cementation, and the proportion of cement is correspondingly less.

What has been said about packing applies specifically to the grain-to-grain relationships in relatively clean sands and sandstones. Where sand grains and mud are deposited together in poorly sorted deposits, the admixture of mud restricts close packing of the grains. Contacts between sand grains are few if, at the time of deposition, mud comprises 35% to 40% of the deposit; if mud predominates, the sand grains are surrounded by and suspended in it (mud-supported texture). Compaction of the interstitial mud does, of course, bring the sand grains closer together, but the sand grains themselves cannot be as tightly packed as is possible in clean sandstones.

STRATIFICATION AND GRAIN ORIENTATION

The conditions of deposition are rarely the same for successive strata in a sedimentary sequence, and as a consequence the strata vary in texture or mineral composition, or both. Because the strata are juxtaposed, even slight differences between them are usually evident, so that visible stratification typifies sedimentary formations. The mechanics of deposition vary, however, and so also do the types or modes of stratification. Much stratification is seen on a large scale and must be observed in the field, but the thinnest strata, called *laminae*, are less than a few millimeters thick and are significant features in many hand specimens and thin sections. Characteristic of numerous sandstones, for example, are plane parallel laminae that are distinguished by slight differences in grain size and mineral composition, which are produced by deposition from a traction load moving across a smooth bottom. Regular parallel laminae are also common in deposits of silt and clay, as in the seasonal layers in glacial varves. If the grain size varies progressively from top to bottom of individual laminae, the laminae are said to be *graded;* typically, the coarser sizes are at the bottom and the grading provides a useful criterion for distinguishing top and bottom in a sequence. *Cross-lamination* is produced by deposition from small ripples of fine sand or silt as they are moved by weak currents across the bottom. Because it represents deposition from the traction load transported by a current, however weak, cross-lamination is not to be expected in the finest muds and oozes that have settled to the bottom from suspension in quiet water; it is, however, common in fine sandstones and siltstones. *Convolute laminae* are typical of the fine sand and silt portions of graded layers deposited by turbidity currents.

Most sedimentary deposits of sand size or finer tend to be laminated when they are deposited, but after deposition the laminae are often disrupted and may be completely obliterated by active benthonic organisms. In marine environments especially, burrowing and scavenging organisms are usually abundant on and in the finer bottom sediments, and they effectively disturb and mix many of these deposits before they become buried beyond the limits of organic activity. This process, termed *bioturbation*, is widespread in the marine environment and is particularly effective wherever conditions favor an abundant active benthonic population; it also embraces the effects of nonmarine organisms, including the effects of roots, worms, and burrowing animals and insects in subaerial deposits. If a deposit has not been completely churned by organisms, their activity is usually indicated by casts of irregular burrows and trails filled with sediment of slightly different texture and usually of higher organic content. Movements caused by slumping and by the burden of overlying deposits also disrupt and deform original laminae and are common in unstable, water-saturated deposits of fine sediment.

In clastic sediments, mica, platy shell fragments, and other nonspheroidal particles generally lie with their longer diameters approximately in the bedding plane. With micas this tendency is so pronounced that the orientation of the flakes can be used to indicate the stratification planes when other evidence is lacking (Figure 13-7A,B). Preferred orientation of grains may go undetected in some sedimentary rocks, but few of them are without it.

Grains in sedimentary rocks may be oriented either by the process of deposition or by later deformation. Mica flakes are oriented as they are deposited. In many stream gravels disk-shaped pebbles have been deposited roughly parallel, with their broad surfaces dipping upstream, in what is known as *imbricate structure*, and in glacial till the long axes of pebbles and cobbles tend to lie parallel to the direction of ice motion.

The platy structure of some shales may develop during compaction or deformation by the parallel orientation of platy clay minerals, chiefly illite. An anisotropic fabric in quartz sandstone, which was probably produced by compaction coupled with differential solution of quartz grains, is illustrated in Figure 13-15B.

In a rock whose grains have preferred orientation, many properties as viewed in thin section vary with the plane of the section. When disk-shaped particles, for example, lie parallel to the plane of stratification, thin sections cut across and parallel to the bedding will reveal different shapes and sizes of grains, different packing arrangements and porosities, and perhaps even different mineral frequencies. Other mass properties of the rocks, such as permeability and conductivity, also vary with the direction in which they are measured. Because of this fact, thin sections used in quantitative petrographic studies of sedimentary rocks should usually be cut both normal to and parallel to the stratification. In very thin-bedded rocks, sections normal to the bedding cut across several laminae rather than a single one, and thus give a more representative picture of the rock.

GENERAL REFERENCE

A comprehensive description of sedimentary particles, textures, and rocks is offered by F. J. Pettijohn, *Sedimentary Rocks,* 3rd ed. New York: Harper & Row, 1975. An up-to-date, concise account of sedimentary rocks is given by M. Tucker, *Sedimentary Petrology: An Introduction.* London: Blackwell, 1981.

REFERENCE NOTE

[1]L. C. Graton and H. J. Fraser, *Journal of Geology*, vol. 43 (1935): pp. 785–910.

12

Mudstones and Shales

GENERAL COMPOSITION AND TERMINOLOGY

Mudstones and shales, the *argillaceous* or *pelitic* rocks,* are the most abundant and widespread of all sedimentary rocks. They consist essentially of silicate detritus having the grain size of clay, usually including some fine silt and abundant particles less than 1 or 2 μm in diameter.

The mineral constituents in mudstones and shales are more varied than the fine-grained texture of the rocks at first suggests. The principal detrital constituents are clay minerals (kaolinite, smectite, illite, and chlorite; see p. 317), and small particles of quartz, feldspar, and mica. Nondetrital components, including organic remains and a variety of authigenic minerals, are commonly present. The most widespread authigenic minerals are illite, chlorite, calcite, dolomite, opal, microcrystalline quartz, pyrite, glauconite, and zeolites. The clay minerals illite and chlorite may be detrital but they are commonly formed by diagenetic alteration of other detrital clay minerals, especially by alteration of smectite in deeply buried marine deposits. Illite, in particular, may thus become the

---

Argillaceous and *pelitic* are adjectives widely used in general reference to detrital materials consisting largely of clay. The term pelitic, however, is also used in reference to highly aluminous metamorphic mineral assemblages derived by recrystallization of clay-rich sedimentary rocks (cf., for example, pp. 453; 526); thus, pelitic specifies sedimentary rocks in which aluminum-rich clay minerals are abundant and which therefore have bulk chemical compositions showing high proportions of aluminum relative to other ions.

315

principal mineral in a deposit, as it is in a great many old marine shales and mudstones.

Some detrital clay minerals are inherited directly from older argillaceous rocks that are exposed to erosion; others are produced by decomposition of feldspathic parent rocks. Although any of the clay minerals may be derived from older mudstones and shales, this is probably the origin of most detrital illite and chlorite. Kaolinite and smectite are the minerals typically formed by decomposition of feldspars and ferromagnesian silicates during surface weathering.

Organic remains may be abundant in some argillaceous rocks but are lacking in others. The chief kinds of organic material in mudstones and shales are dark bituminous or carbonaceous matter, calcite in the tests of foraminifers and in larger shells and shell fragments, and opal in the tests of diatoms and radiolarians and in sponge spicules.

The term *mud* applies to unconsolidated argillaceous material in general, whereas *clay* refers specifically to material that is uniformly very fine-grained (clay size) and essentially free of silt. Also, most clay is distinctly plastic when wet and is composed largely of clay minerals. *Mudstone* and *claystone* are the consolidated equivalents of mud and clay, but these terms usually also imply rocks with a blocky habit of fracture. Shale, in contrast, is rock that splits parallel to bedding into thin plates or flakes. *Argillites* are very firmly indurated argillaceous rocks, and their formation generally involves some recrystallization of the original material; in short, they appear to be incipiently metamorphosed mudstones devoid of slaty cleavage.

To the foregoing names it is desirable to prefix appropriate adjectives indicating color or accessory components. *Black pyritic shale* is a good example. Not only is the name descriptive, but black color and abundant pyrite have genetic significance. Shales of this kind tend to form in water where poor circulation has resulted in depletion of oxygen and accumulation of organic matter, which darkens the rocks. *Red shale* and *red mudstones* contain ferric oxide, which implies oxidizing conditions during deposition and early diagenesis or, alternatively, rapid deposition of red muds derived from a source terrane undergoing oxidation. *Glauconitic shales* and *mudstones* contain the authigenic mineral glauconite, which generally indicates deposition in relatively shallow marine environments. Other shales and mudstones may be characterized, for example, as calcareous, siliceous, foraminiferal, or diatomaceous. So, by using appropriate adjectives, a variety of argillaceous rocks can be simply classified.

As the proportion of silt and sand intermixed with argillaceous material increases, shales and mudstones grade through *silty* or *sandy shales* and *mudstones* into *siltstones* and *sandstones*. Or by increase in the content of nondetrital constituents, they may grade into various predominantly chemical or biogenic deposits. Thus *marl* and argillaceous limestones result from an increase in the proportion of calcium carbonate, and as the content of authigenic silica increases, *siliceous shales* and *mudstones* may pass into *porcellanites* or impure *cherts* (Figure 11-1).

CLAY MINERALS

Minute crystals of a number of different hydrous aluminum silicates are the principal constituents of clays and are primarily responsible for the distinctive properties of clay and of most argillaceous rocks. Collectively these minerals are referred to as *clay minerals*. Most of them have layer lattices and, like micas, their crystals have a flaky or platy habit and a single perfect cleavage. A number of distinct patterns characterize the ionic structures of these minerals and define several different groups of clay minerals. *Illite* and *smectite* are composed of ionic sheets similar in general to those in micas. A different type of ionic sheet characterizes *kaolinite*, as well as other minerals of the kaolinite group (dickite, nacrite, halloysite). *Chlorite** is another common clay mineral and has yet another structural arrangment and composition; so also does *gibbsite*. The compositions and structures of these minerals and of other common phyllosilicates are depicted in Figure 12-1.

Clay minerals, especially smectites, have a pronounced capacity and tendency to adsorb various cations and anions from surrounding aqueous solutions. Loosely held on the outside of the structural units composing the clay minerals, the adsorbed ions are exchangeable and tend to be replaced by other ions if the surrounding chemical environment changes. The exchangeable-ion composition of an argillaceous rock

---

*Chlorite is not considered a clay mineral when it occurs well crystallized in veins, in metamorphic rocks, or as deuteric and hydrothermal alterations in igneous rocks. But chlorite of exceedingly fine grain size, often poorly crystallized, can be identified by x-ray diffraction in many recent marine muds and in most older shales and mudstones where it is typically associated with illite and smectite. Its widespread occurrence in such deposits and its close association with other clay minerals lead to inclusion of chlorite among the clay minerals.

Figure 12–1. Structure and Composition of Common Phyllosilicates

Schematic representation of the ionic structures in phyllosilicates, including the principal clay minerals. The structures consist of two types of ionic layers, namely, *octahedral* layers in which cations are in octahedral (sixfold) coordination with O^{2-} or OH^- and *tetrahedral* layers in which cations are in tetrahedral (fourfold) coordination with O^{2-}. Octahedral layers may occur alone or linked to adjacent tetrahedral layers. The various ions are shown in proportions corresponding to ideal chemical formulas for each mineral, using symbols for the various ions as indicated on the diagram.

---

is, therefore, a general reflection of the chemical environment of deposition, or diagenesis, and although it is not evident petrographically, it is a significant factor in the geochemistry of argillaceous rocks.

Clay aggregates commonly contain more than one clay mineral but are so fine-grained that individual crystals can rarely be discriminated with a petrographic microscope. Consequently, the different clay minerals usually must be identified by x-ray diffraction, although some general distinctions can be made in thin sections. Like mica, clay minerals are optically negative and have the high refractive index (the slow vibration direction) in the plane of the sheet structure and cleavage. Differing refractive indices and birefringence in the various clay minerals are the basis for most optical identifications.

Illite and smectite have high birefringence and even very small crystals can be expected to show second-order interference colors in thin section. But in both minerals, large variations in refractive indices are caused by different ionic substitutions in their lattices and, in smectite, by changes in the amount of interlayer water. The mean refractive index of hydrated montmorillonite is slightly below 1.54, the usual refractive index of the mounting medium,* but it is slightly higher for the dehydrated mineral. In nontronite, a smectite rich in ferric iron, refractive indices are much higher, in the range of 1.60. Most illites have a maximum refractive index between 1.57 and 1.58, slightly lower than that in muscovite.

In contrast, kaolinite has low birefringence and shows only first-order gray color in thin section. Its refractive indices vary slightly around a mean value of 1.56 and are a little higher than those of the mounting

---

*Various compounds may be used to mount thin sections. In this book, references to the refractive index of the mounting medium assume a value of approximately 1.540, similar to that of the classic mounting medium Canada balsam.

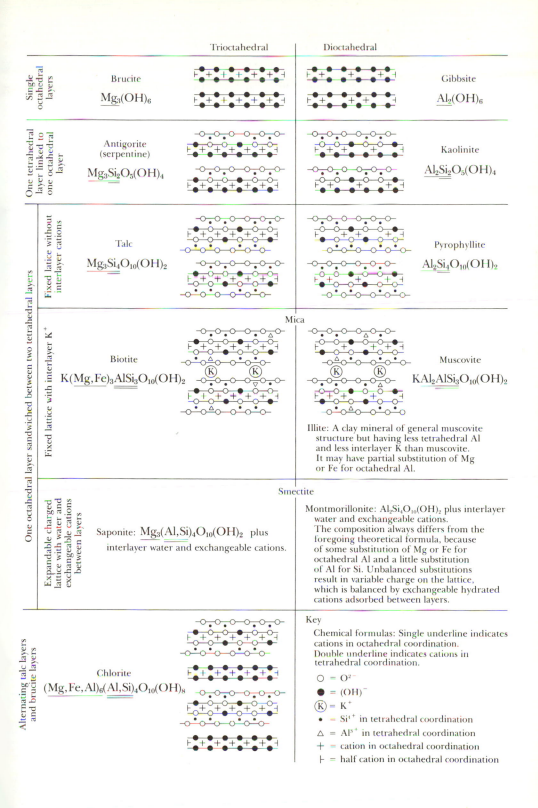

		Trioctahedral	Dioctahedral	
Single octahedral layers	Brucite $Mg_3(OH)_6$			Gibbsite $Al_2(OH)_6$
One tetrahedral layer linked to one octahedral layer	Antigorite (serpentine) $Mg_3Si_2O_5(OH)_4$			Kaolinite $Al_2Si_2O_5(OH)_4$
Fixed lattice without interlayer cations	Talc $Mg_3Si_4O_{10}(OH)_2$			Pyrophyllite $Al_2Si_4O_{10}(OH)_2$

One octahedral layer sandwiched between two tetrahedral layers

Mica

| Fixed lattice with interlayer K^+ | Biotite $K(Mg,Fe)_3AlSi_3O_{10}(OH)_2$ | | | Muscovite $KAl_2AlSi_3O_{10}(OH)_2$ |

Illite: A clay mineral of general muscovite structure but having less tetrahedral Al and less interlayer K than muscovite. It may have partial substitution of Mg or Fe for octahedral Al.

Smectite

Expandable charged lattice with water and exchangeable cations between layers

Saponite: $Mg_3(Al,Si)_4O_{10}(OH)_2$ plus interlayer water and exchangeable cations.

Montmorillonite: $Al_2Si_4O_{10}(OH)_2$ plus interlayer water and exchangeable cations. The composition always differs from the foregoing theoretical formula, because of some substitution of Mg or Fe for octahedral Al and a little substitution of Al for Si. Unbalanced substitutions result in variable charge on the lattice, which is balanced by exchangeable hydrated cations adsorbed between layers.

Alternating talc layers and brucite layers

Chlorite $(Mg,Fe,Al)_6(Al,Si)_4O_{10}(OH)_8$

Key

Chemical formulas: Single underline indicates cations in octahedral coordination.
Double underline indicates cations in tetrahedral coordination.

$\bigcirc = O^{2-}$
$\bullet = (OH)^-$
$\circledR = K^+$
$\cdot = Si^{4+}$ in tetrahedral coordination
$\triangle = Al^{3+}$ in tetrahedral coordination
$+ =$ cation in octahedral coordination
$\vdash =$ half cation in octahedral coordination

medium and of quartz. Halloysite, which is typically more hydrated than kaolinite, has a somewhat lower refractive index. Substitutions do not occur in the kaolinite lattice, and the kaolinite minerals show only minor variations in optical properties.

Chlorite-type clay minerals generally have low birefringence. Their refractive indices are above those of the mounting medium and of quartz but somewhat lower than those in the larger chlorite crystals seen in igneous and metamorphic rocks. Usually they show a faint greenish color in plain light.

Identification of specific clay minerals by optical methods is usually questionable. The optical properties seen in clay aggregates are often those of compound particles rather than of individual mineral crystals. Montmorillonite, for example, often occurs interlayered with illite or chlorite, and mixed-layer particles composed of illite and chlorite are also common. Furthermore, minute mineral flakes piled atop one another tend to produce aggregate optical effects that differ from those of individual crystals. Probably kaolinite is the clay mineral most often identified with confidence, for it tends to occur in larger crystals than the others and has more uniform optical properties. Smectite and illite, being optically rather similar, are difficult to distinguish from each other, but their high birefringence is usually apparent and serves to distinguish them from kaolinite and chlorite.

The clay minerals present in argillaceous deposits at the time of deposition depend on the character of the source rocks and also on the environment of the source area. They may be altered, however, by diagenesis in the depositional environment and, especially, after deep burial. During chemical weathering, formation of smectite is favored by volcanic parent rocks and by the relatively alkaline conditions to be expected in areas where rainfall is moderate to low and the soils are not thoroughly leached. Kaolinite is more likely to develop in somewhat acidic soils and where rainfall is heavier and leaching is more thorough. Gibbsite is formed only by intensive chemical weathering in tropical and subtropical climates where even silica tends to be leached from the soil. Kaolinite seems to be more stable in fresh-water than in marine environments, for it tends to occur in greatest abundance in nonmarine rocks. Illite and chlorite commonly form at the expense of smectite and kaolinite in marine deposits, especially after they have been buried to substantial depth. For this reason, illite and chlorite generally predominate in ancient marine shales and mudstones; in recent deposits, they are common detrital clay minerals derived from older shales and mudstones.

PETROGRAPHY

Microscopic techniques that are successfully applied to coarser rocks have been less fruitful when applied to shales and mudstones. Indeed, most of what we now know about clay is based upon chemical, thermal, x-ray, and electron-microscope studies. Ordinary microscopic examination of clay material, unless linked to these other types of analysis, can provide only general conclusions. In this book, therefore, which is devoted to the microscopic study of rocks and omits other means of identification, less space is allotted to rocks composed of clay minerals than to others. This brevity of treatment, however, is not intended to minimize either the importance of argillaceous rocks or the significance of whatever microscopic studies can be made of them. Microscopic examinations of shales and mudstones in thin section are of great value if only because they provide a descriptive picture that can incorporate the results of all other types of analysis.

In shales, flakes of the clay minerals tend to lie nearly parallel to the bedding (Figure 12-2B), and this is a principal cause of fissility in shale. In thin sections cut normal to the stratification one sees the flakes on edge, and, because the extinction position and the slow vibration direction are parallel to the apparent elongation of the flakes, the thin sections as a whole tend to show aggregate positive elongation. When rotated between crossed polarizers, an entire thin section may appear to extinguish in two principal directions, as though it were cut from a single crystal. In many shales the particles are so small that general changes in polarization color throughout the slide are all that can be recognized. If, on the contrary, the clay particles are oriented more at random, as they generally are in the more massive claystones and mudstones, the appearance of a thin section does not vary appreciably when rotated between crossed polarizers, and it does not tend to extinguish preferentially in any two directions.

The tendency of clay particles to assume parallel orientation, and the fissile structure of shales, is probably controlled by a number of different factors but is not fully understood. Dispersed clay flakes when settled in water probably tend to lie parallel to bedding surfaces. But most detrital clays are laid down in a partly flocculated condition and consist of clusters of minute flakes. The clusters, or floccules, may be large enough to settle rather quickly in water and the individual clay flakes within them are likely to have more or less random orientations. Compaction after deposition may force the flakes into more parallel positions, but numerous

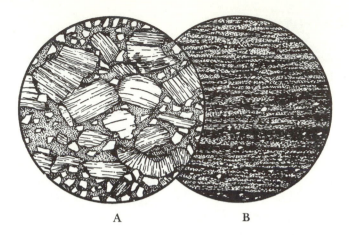

Figure 12–2. Argillaceous Sediments

A. Sandy fire clay (Eocene), Ione, California. Diam. 3 mm. Angular detrital quartz grains in a kaolinite matrix (stippled). Abundant vermicular kaolinite crystals developed authigenically after deposition, as shown by their elongation normal to the cleavage (lower right and top). Note displacements along cleavage in some crystals.

B. Lias shale, Württemberg, Germany. Diam. 0.5 mm. Tiny flakes of clay minerals are oriented roughly parallel to the bedding. Some angular silt particles composed of quartz are recognizable. Very finely laminated structure is caused by streaks and minute lenticles of a dark-brown organic substance which clouds and masks the clay (dark line pattern).

firmly compacted, massive mudstones and claystones show that compaction alone does not cause a fissile structure. Rocks composed largely of illite and smectite appear to have shaly structure more commonly than those composed of kaolinite. Also, accessory constituents in a clay deposit probably influence the orientation of the clay particles. Perhaps, for example, the presence of organic compounds may increase the tendency toward parallel orientation, for many highly organic clay rocks are very fissile. Or the presence of carbonate mud may decrease the tendency toward fissility.

Clay mineral crystals several hundred micrometers long appear in some clay rocks. Most of these have somewhat indefinite boundaries, being especially ragged on the ends. Some of them may be compound clusters of small flakes that, since deposition, have become integrated into single crystals. Some of the large clay crystals, however, undoubtedly grew within the clay deposit entirely after deposition; such, for example, are large vermicular crystals of kaolinite having a slightly curved, accordionlike aspect, which occur in some fire clays (Figure 12-2A). Fine-grained mosaics of kaolinite, probably recrystallized after deposition, are usually masses of equant crystals or feathery grains that can be identified by their uniformly low birefringence. Aggregates of smectite or illite, because of their higher birefringence, have a more colorful, twinkling appearance when rotated between crossed polarizers.

To one who is examining shales and mudstones in thin sections for the first time, the amount of silt that is generally present may be surprising. There are few such rocks in which silt is lacking, and in many it is abundant, although it may be scarcely evident in hand specimens. The most common silt particles are clear, angular fragments of quartz and feldspar and flakes of detrital mica. Differences in the relative amounts of silt and clay in adjacent laminae often give the rocks a streaked appearance, and many rocks are distinctly laminated. In thin sections, as in hand specimens, the paler laminae are usually siltier and the darker ones richer in clay.

Silt laminae in some shales and mudstones are discontinuous thin lenses and presumably indicate weak currents capable of winnowing the silt content of the deposit at the site of deposition. Other silty laminae are more regular and continuous. Some of them are distinctly graded, being sharply defined and coarser grained at the base and progressively finer grained and more clay-rich upwards. Lamination may also be caused by varying concentrations of authigenic components. Highly calcareous laminae, for example, often alternate with more argillaceous ones; similarly, siliceous laminae commonly alternate with less siliceous, more argillaceous ones.

However formed, the continuity of original lamination often is broken by bioturbation or by slumping after deposition. Traces of organic activity may appear in argillaceous deposits as irregular tubular structures cutting across or paralleling original laminae; these are the trails and borings of benthonic organisms that have been filled by sediment that differs slightly in grain size or composition from the surrounding rock. Indeed, some massive mudstones in which silt particles are scattered more or less randomly throughout the clay matrix may have been completely churned by the vigorous activity of bottom-dwelling organisms.

Most dark-colored argillaceous rocks include carbonaceous organic matter, which, like silt, tends to occur either in tiny lenses, elongated parallel to the bedding, or as scattered grains and specks (Figure 12-2B). The organic material itself is usually a dark-brown or black, opaque, structureless substance, and clays in which it is very abundant are particularly dark. The opaque matter in shales and mudstones is not all organic, however. Some of it consists of minute cubic crystals and microconcretions of pyrite. Ferric oxides, also, tend to be distributed throughout some argillaceous rocks, and if very abundant they make thin sections nearly opaque; usually they produce a reddish cast that is clearly evident.

Authigenic grains of calcite, dolomite, or siderite are to be found scattered through some clay deposits. If the crystals are very small, they are conspicuous between crossed polarizers only as sparkling specks recognizable by their high birefringence; in plain light they show pronounced change of relief as the stage is rotated. If the grains are larger, they are, of course, very easily recognized. In general, crystals of calcite tend to be less idiomorphic than those of the other carbonates. If carbonates are very abundant, they constitute a firm cement for the particles of clay and silt with which they are associated.

GENERAL REFERENCE

For general reference, see R. E. Grim, *Clay Mineralogy*. New York: McGraw-Hill, 1968; P. E. Potter, J. B. Maynard, and W. A. Pryor, *Sedimentology of Shale*. New York: Springer-Verlag, 1980.

13

Sandstones

CLASSIFICATION

Sandstones are the rocks in which detrital grains of sand predominate. The sand grains may have a single size or a range of sizes, but they pack together grain against grain to form a *framework* that provides the primary solid support in the deposit. If the sand is muddy, smaller particles of clay and fine silt lodge between the sand grains and clog, or partially clog, the pores of the sand framework, forming what is termed an *argillaceous matrix*. As the volume of matrix increases and approaches that of the sand grains, the grain-against-grain contacts in the sand framework become less extensive and fewer until muddy sandstone passes into sandy mudstone, which has mud-supported rather than grain-supported texture (p. 303). But if the sand has been washed essentially free of clay and silt, the pores between the framework grains are clear and relatively clean, although later they commonly become filled with a *cement* consisting of authigenic minerals precipitated from pore solutions.

The mineral composition of the framework grains reflects in a general way the paleogeology, paleogeography, and paleoclimate of the region from which the sand was derived. It provides the primary basis for classifying sandstones. Variations in texture, particularly the presence or absence of clay and fine silt, reflect varying conditions of deposition and are an additional basis for classification.

The most abundant framework grains, and also the most stable, are quartz (including chert). In some sandstones, quartz grains have been concentrated almost to the exclusion of any others, but most sandstones have a substantial content of less stable particles. Of these, by far the most numerous are feldspar grains and small fragments of aphanitic rock collectively called *lithic grains*; other detrital grains are subsidiary. The relative proportions of these three principal kinds of framework grains lead to a basic threefold classification of sandstones. If more than 90% of the grains are quartz, the rock is called a *quartz sandstone*. Among sandstones containing less than 90% quartz, those in which feldspar grains exceed lithic grains are called *feldspathic sandstone*, and those in which lithic grains predominate over feldspars are called *lithic sandstone*. Each of these three kinds of sandstone may be subdivided further according to texture into either *arenite* (well-washed, clean sandstone) or *wacke* (texturally immature, argillaceous sandstone). These basic sandstone subdivisions are illustrated in Figure 13-1.

The compositional and textural variations among sandstones are completely gradational, so that all subdivisions are necessarily arbitrary. The distinction between arenite and wacke, in particular, is unsettled. To differentiate them, as is customary, by some arbitrary percentage of matrix—whether 5%, 10%, or some other limit—is inherently troublesome for reasons that should not be ignored. First, matrix material itself is variously defined, and its percentage in sandstones is difficult to measure or estimate consistently. Also, grains and matrix are not independent elements, because the proportion of detrital clay and silt tends to vary with the size of the associated sand grains. For example, the coarsest portion of a graded sand bed may contain less than 5% argillaceous matrix while the finer portions of the same bed contain 15% or more. Furthermore, the volume and the character of matrix in a sandstone varies with degree of compaction and diagenetic recrystallization, and in many firmly lithified sandstones (graywackes), the matrix is largely diagenetic rather than detrital (see pp. 343–344).

The presence or absence of argillaceous matrix (clay and fine silt) in a sandstone is more meaningful than its absolute percentage and is the only criterion that can be expected to yield operationally consistent classification. Accordingly, *arenite* is defined here as a sandstone effectively free of clay and fine silt.* It represents depositional conditions in which

---

*In the first edition of this book, wacke was defined as a sandstone containing 10% or more of silt-clay matrix, and arenite as one containing less than 10%. Most authors have chosen to place the boundary at 15% matrix, or more. However, Folk (*Petrology of Sedimentary Rocks*, Austin, Texas: Hemphill's, 1965, pp. 102–103) considers a sandstone to be texturally immature if it contains as little as 5% of clay and very fine silt, and to be texturally mature if clay and fine silt are lacking. *Arenite*, as used in this book, corresponds in general to the submature and mature sandstones of Folk; *wacke*, to his immature sandstone.

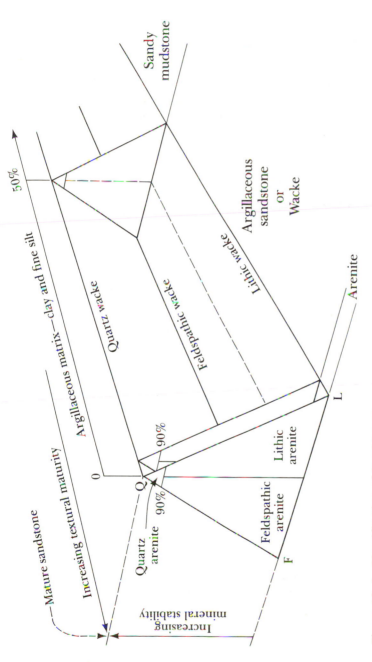

Figure 13–1. Classification of Sandstones

Figure modified after Dott, *Journal of Sedimentary Petrology*, vol. 34 (1964): p. 629. Three mineral components of sand—quartz (*Q*), feldspar (*F*), and lithic grains (*L*)—are represented by the three apices of the triangles; points within the triangles represent relative proportions of these three components. Percentage of argillaceous matrix is represented by a vector extending toward the rear of the diagram. The term *arenite* is restricted to sandstones that are essentially free of matrix material; all others are argillaceous (muddy) sandstone, or *wacke*.

the sand is winnowed by depositing currents and washed free of the finest detritus. The texturally immature argillaceous sandstones, or wackes, in contrast, are laid down in environments where some muddy suspension load is deposited with and remains with the sand; they represent depositional sites where winnowing of the sediment does not occur after the initial deposition. They cover a broad textural spectrum from sandstones in which argillaceous matrix comprises half of the rock to those in which it is a minor constituent, but in most of them the volume of matrix is less than a quarter of the rock.

A sandstone consisting of quartz grains that are rounded, well sorted, and completely free of silt and clay—a pure quartz arenite such as the St. Peter Sandstone illustrated in Figure 11-5A—represents the ultimate product of the combined chemical and physical processes that are involved in the origin of sandstones. It is therefore regarded as thoroughly *mature* both texturally and mineralogically. In theory, we may envisage the gradual maturing of detrital sediment as it progresses through one or several cycles of sedimentation.[1] For example, a sediment mixture might initially be represented by some point within the diagram in Figure 13-1, and as clay, silt, and unstable minerals are eliminated, it could be thought of as progressing gradually toward maturity at the quartz (Q) apex in the front of the diagram. But in natural sediments the trend toward textural maturity and mineral stability usually is not continuous or direct and in most bodies of sediment is not completed. Different sandstone deposits develop differing degrees and combinations of textural maturity and mineral stability as a result of differences in the sources and the processes that produce them.

Three classes of sandstone according to mineral composition—quartzose, feldspathic, and lithic—generally suffice, but a few additional subdivisions are widely used. Among feldspathic sandstones, those in which both quartz and potassic or sodic feldspars are abundant have long been known by the term *arkose*.* Lithic grains may be included, but typical arkoses have a granitic or granodioritic aspect that suggests derivation

---

*The term *arkose* is used in this book in the original sense; see S. S. Oriel, *American Journal of Science*, vol. 247 (1949): pp. 824–829. Not included among arkoses are abundantly feldspathic sandstones that contain little or no quartz. Most of these are volcanic sandstones containing abundant plagioclase together with lithic grains derived from intermediate and basic volcanic source terranes. Such feldspathic sandstones differ significantly from the general granitoid composition that was originally implicit in the term arkose; they also have a different provenance, being derived largely from supracrustal rather than basement terranes.

from quartzo-feldspathic rocks of the coarsely crystalline basement. *Arkose* and *arkosic sandstone* are names that refer specifically to feldspathic sandstones in which feldspar exceeds 25% of the framework grains; abundant quartz is implicit. Feldspathic sandstones that are more quartzose and contain less than 25% feldspar are often referred to as *subarkose*.

When the framework compositions (modes) of sandstones are determined with reasonable precision, the relative proportions of quartz (Q), feldspar (F), and lithic grains (L) may be expressed numerically to provide simple quantitative rock designations, for example, feldspathic arenite $Q_{56}F_{28}L_{16}$ or lithic wacke $Q_{70}F_{12}L_{18}$. To express the proportion of matrix (M) in a wacke, a fourth numerical term may be included, for example $Q_{55}F_{10}L_{20}M_{15}$. Alternatively, using two adjectives, the foregoing examples may be described as lithic feldspathic arenite and feldspathic lithic wacke, respectively.* Numerical designations, however, provide a more precise description of rock composition than any word descriptions or special rock names.

Many sandstones that have been buried deeply, especially older ones, have become hard, impervious, dark-gray or dark-green rocks called *graywacke*, a loosely defined term that refers primarily to the general character of the unweathered rock as seen in the field or in hand specimens. Graywackes are characterized by a dark-colored, firmly lithified matrix that pervades the rock and obscures the boundaries of individual sand grains. They all tend to be remarkably similar in overall appearance regardless of framework composition, so that different compositional varieties of graywacke commonly are indistiguishable by casual observation of hand specimens. In thin sections, however, they can be classified and named, like other sandstones, according to their framework grains (e.g., feldspathic, lithic, and quartzose graywacke).

As the grains in sandstones become smaller, fewer polycrystalline lithic grains and relatively more single-crystal mineral grains are to be expected, so that most very fine-grained sandstones are either quartzose or feldspathic rather than lithic. Also in this category are the *siltstones*, in which the majority of grains are silt size; usually they contain abundant clay, and invariably they are either feldspathic or quartzose, not lithic.

Some detrital deposits are coarser grained than sandstones. When more than 50% of the aggregate consists of pebbles and cobbles that have

---

*K. A. W. Crook (*American Journal of Science*, vol. 258 [1960] p. 425) has formalized a dual-adjective scheme of nomenclature for subquartzose ($Q<75$) sandstones, as follows: feldspathic, $F/L>3$; lithofeldspathic, $3>F/L>1$; feldspatholithic, $3>L/F>1$; lithic, $L/F>3$.

been somewhat rounded, the rock is called *conglomerate*, or *breccia* if the fragments are angular. If relatively few larger grains are present, the rocks are called *pebbly sandstone* (arenite or wacke) or *pebbly mudstone*. A general terminology for these rocks is suggested in Figure 13-2.

In most conglomerates a matrix of smaller particles is present between the pebbles and cobbles. The matrix of a poorly sorted deposit consists of sandy mud, that of a better-washed and -sorted deposit of clean sand. Pebbles and cobbles are almost invariably fragments of polycrystalline rock, so that all conglomerates and breccias are understood to be "lithic." Different kinds of conglomerate and breccia are specified by adjectives denoting the identity of the predominant lithic detritus, and they range from those containing abundant unstable pebbles, such as *granite con-glomerate* and *volcanic conglomerate*, to more stable types such as *chert con-glomerate* and *quartzite conglomerate*. Important as these rocks are, the large size of their detrital grains makes them less adapted than sandstones to examination in thin sections, and they are not treated further in this book. The sandy matrix in conglomerates can, of course, be described in the same terms as the sandstones discussed on the following pages.

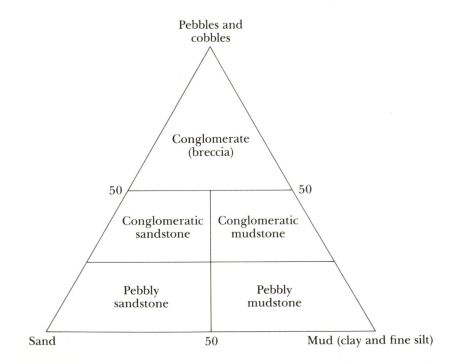

Figure 13–2. Suggested Nomenclature for Detrital Sedimentary Rocks That Contain Pebbles and Cobbles

Our petrographic descriptions of sandstones follow the scheme of classification just outlined. Graywackes and the immature feldspathic and lithic sandstones are considered first and the more mature, quartz-rich arenites last. But before beginning those descriptions, we consider aspects of mineralogy that apply to sandstones in general.

DETRITAL MINERAL GRAINS

Quartz

Quartz does not vary in composition, and it resists the various kinds of alteration that commonly affect other detrital minerals. It appears in thin sections as relatively clear colorless grains having low birefringence and refractive indices that are only slightly higher than that of the mounting medium (see footnote, p. 318). For most purposes it is sufficient to describe a sand grain simply as quartz, but some visible differences serve to distinguish several varieties of quartz.

Many detrital quartz grains are polycrystalline, composed of a number of separate crystals each large enough to be readily identified as quartz. Grains consisting of microcrystalline quartz are classified as chert and are described with lithic grains (p. 335). In the polycrystalline quartz grains derived from metamorphic rocks, individual crystals may be either elongate or subequant and are typically in roughly parallel crystallographic orientation. They are locked together along intricate boundaries and usually include small quartz crystals developed randomly between the larger ones. Polycrystalline grains derived from igneous rocks also consist of irregular anhedral crystals, but usually the crystals are more equant, less oriented, and of more uniform size. In the comblike quartz that originates in open veins and cavities, individual crystals tend to have relatively smooth planar boundaries.

Quartz phenocrysts in volcanic rocks have a distinctive bipyramidal form that is characteristic of high-temperature quartz crystals. They are bounded by pyramid faces with little or no development of prism faces, a form distinctly different from the prismatic habit of low-temperature quartz crystals or the irregular form of quartz grains in plutonic rocks (Figure 11-2). Among detrital sand grains, crystals of volcanic quartz are best identified in reflected light using a stereoscopic microscope where their euhedral or subhedral form is most evident; in thin section, they appear as limpid equant grains, some of which have roughly square or diamond-shaped outlines.

In quartz crystals that have been plastically deformed (strained), the orientation of the optic axis varies by a few degrees in different parts of a single crystal. This results in *undulatory extinction* as the crystal is rotated between crossed polarizers. Much of the quartz in schists, gneisses, and granitic rocks shows undulatory extinction, whereas the quartz phenocrysts in volcanic rocks and the quartz of cavity fillings typically does not. In sandstones, undulatory extinction in a quartz grain occasionally appears to be caused by pressure from an adjacent detrital grain, but generally it has been inherited from the source rocks. In some quartz grains derived from metamorphic rocks, sharp boundaries separate sectors each differing in extinction by a few degrees with respect to its neighbors—an effect that at first glance might be mistaken for twinning. These composite grains actually are strained grains once showing undulatory extinction but subsequently annealed in the final stages of metamorphism.

Dustlike particles and minute fluid-filled vacuoles are included in many quartz crystals. Vacuoles usually appear as tiny opaque inclusions, but under high magnification, the largest of them are seen to be transparent bubbles with strong negative relief. Usually these inclusions are concentrated in zones and along subparallel surfaces within the quartz (Figure 13-4A), especially in quartz derived from metamorphic rocks; but they may also be scattered so abundantly as to cloud what would otherwise be limpid quartz grains (Figure 13-15A,B). The milky quartz of hydrothermal veins is characteristically crowded with fluid-filled vacuoles; the high-temperature quartz of volcanic rocks, in contrast, tends to be clear and essentially free of such inclusions.

Exceedingly slender, needlelike crystals of rutile occur in the quartz of some granitic rocks, and similar needlelike crystals of sillimanite or kyanite may occur in metamorphic quartz. Small crystals of many other minerals—for example, feldspar, mica, magnetite, apatite, and zircon—also may occur as inclusions in quartz grains.

Feldspar

Like quartz, feldspars in thin section are typically clear and colorless and have low birefringence, but generally they can be distinguished from quartz by their cleavage, twinning, or refractive indices. Feldspar grains also may be partially decomposed, and then they appear cloudy or turbid in contrast to quartz grains, which are invariably unaltered and relatively clear. To identify sand grains simply as feldspar, however, is sufficient only for the most general rock classification. In petrographic studies, different varieties of feldspar must be differentiated and identified, for they vary in chemical composition and are among the most reliable indicators of provenance.

Detrital grains of perthite, microcline, and orthoclase usually indicate source areas where granitic rocks and high-grade gneisses and schists are exposed. Sanidine, $(K,Na)AlSi_3O_8$, the high-temperature counterpart of the foregoing minerals, implies a silicic or alkalic volcanic source, while detrital albite comes chiefly from low-grade regionally metamorphosed terranes. The principal sources of detrital oligoclase (An_{10-30}) and andesine (An_{30-50}) are igneous rocks of intermediate composition, either plutonic or volcanic, and also schists and gneisses of amphibolite facies. More calcic plagioclase is abundant in basic igneous rocks and in some high-grade metamorphic rocks. Zoned plagioclase phenocrysts derived from volcanic rocks are distinguishable, because the compositional zones tend to follow and accentuate the euhedral outlines of the crystals; usually they are associated with lithic particles of andesite or basalt.

In sandstones, the potassic and sodic feldspars are most numerous because they are relatively stable and are widespread and abundant in continental source terranes. Calcic plagioclase (An_{50-100}), because it is unstable in the sedimentary realm, is distinctly less widespread in sandstones. But any variety of feldspar may be encountered; in sandstones of mixed provenance, a number of different varieties are likely to be found in the same rock. Feldspars can best be differentiated in thin section by their relief with respect to the mounting medium. Pure albite and the K-feldspars have negative relief; in calcic albite and sodic oligoclase, one refractive index approximately matches that of the mounting medium; the more calcic plagioclases have positive relief, which increases with increasing calcium content. In microcline the visible trademark is cross-hatch twinning, and in perthite it is the intergrowth of two alkali feldspars. Sanidine is generally similar to orthoclase, but it is usually more limpid and it can be identified by its small optic axial angle. Also, sanidine is likely to be accompanied by high-temperature quartz crystals. Plagioclase feldspars are, of course, commonly characterized by lamellar twinning, and sometimes by compositional zoning, and the compositions of favorably oriented crystals can be determined by extinction angles measured against twinning or cleavage planes. Water-clear albite and oligoclase derived from metamorphic rocks are commonly untwinned, however, and care must be taken lest untwinned feldspars of low refractive index be incorrectly identified as orthoclase or sanidine.

For accurate and reproducible quantitative determinations of feldspar content, a representative sample of rock should be stained so as to make all the feldspar grains easily distinguishable. The techniques used to stain feldspars all involve brief etching (fuming) of a clean, smooth rock surface with hydrofluoric acid (HF) followed by immersion in solutions that produce effects of contrasting color on plagioclase and K-feldspars. Several techniques and solutions have been recommended and are in use.[2]

Alteration of feldspars may occur during surface weathering in the source area or later during diagenesis, and it may also occur during earlier hydrothermal alteration of the source rocks themselves. Under all these conditions, the calcic feldspars are more generally decomposed and extensively altered than sodic and potassic feldspars. During surface weathering, the feldspars alter to clay minerals; decomposition begins along crystal boundaries and along twin planes and cleavages within the crystals. In partially decomposed feldspars, the alteration products tend to be concentrated along these surfaces, or in seemingly random patches, but as decomposition proceeds, entire crystals may be converted to turbid, microcrystalline aggregates of clay minerals.

Alteration by diagenetic and hydrothermal processes generally produces secondary minerals that are somewhat coarser grained than the clay minerals formed during weathering. Calcite, epidote, chlorite, prehnite, and albite, in fine-grained aggregates, often partially replace plagioclase, and crystals of illite or muscovite may be seen in the K-feldspars. Calcic plagioclase, especially in sandstones that have been deeply buried, tends to be replaced by albite or by laumontite.

Lithic Grains

Most of the relatively dark-colored grains seen in sandstone hand specimens are small fragments of aphanitic rock collectively called lithic grains. They include a wide variety of rock types and are indicative of the kinds of source terranes from which the sand grains were derived. Unfortunately, the identification of aphanitic rocks is commonly difficult in fragments as small as sand grains, and inevitably some lithic grains cannot be identified with confidence by optical methods.

Small particles of volcanic rock are characterized by the various textures to be found in rocks of volcanic origin. Those of intermediate and basic composition can be broadly described as microlitic; they consist largely of very small, lath-shaped crystals of plagioclase in various textural arrangements, some of which are illustrated in Figures 2-9 and 2-10. Many of these grains include small euhedral or subhedral phenocrysts, usually of feldspar, and these are a clear sign of volcanic origin. Many volcanic grains of highly silicic composition, in contrast, are essentially microcrystalline aggregates of quartz and alkali feldspar, and their textures may be microfibrous or microgranular (felsitic). Some of them may include faint curvilinear outlines of relict glass shards. Except for scattered small phenocrysts, the crystals in most felsitic grains are too small to be discriminated individually, but the grains may show internal relief representing the contrast in refractive index between tiny crystals of quartz and alkali feldspar. The finest grains of acid volcanic rock, lacking microphenocrysts, often cannot be identified with assurance and may even be confused with chert grains.

Most lithic grains derived from sedimentary source rocks are restricted to fine-grained, firmly lithified types that are not readily disintegrated during erosion. The most stable of these is chert, which consists essentially of microcrystalline quartz. Chert grains are likely to be traversed by thin quartz veinlets and some of them include identifiable siliceous microfossils or ovoid patches of clear chalcedony representing the ghosts of microfossils (Figure 15-1A). Fine-grained terrigenous, authigenic, or pyroclastic components may occur in the less pure cherts; some of these impurities are seen simply as patchy variations in color, others are small birefringent crystals of carbonate minerals or of phyllosilicates. Chert, porcellanite, metachert, and silicified volcanic rock—all of which may have generally similar appearance in small detrital grains—are commonly designated together as chert grains.

Less stable fragments of siltstone and fine-grained sandstone are distinguished by clastic texture; bits of argillite and shale can usually be recognized by scattered silt particles embedded in the clay matrix (Figure 12-2B). Either may show traces of bedding. Fragments of fine-grained limestone and dolomite occur as grains in some sandstones and are distinguished by the high-order interference colors and pronounced variation relief that characterize the carbonate minerals.

Metamorphic rock fragments are usually recognized by schistose or semischistose fabric. Most metasedimentary grains are fine-grained quartzite or slate, phyllite, and low-grade schist consisting largely of fine-grained quartz with mica in more or less parallel orientation. Metavolcanic grains of acid composition may be texturally similar but generally contain albite. Metavolcanic grains of intermediate and basic compositions usually contain chlorite, epidote, or amphibole, typically with abundant sodic feldspar. Distinctive metamorphic minerals, such as garnet, sillimanite, or glaucophane, serve to distinguish some metamorphic grains.

Mica

Micas are widespread and often abundant in both igneous and metamorphic rocks. They are certainly to be expected as detrital grains, but the distribution of detrital mica is primarily a function of sorting and is determined largely by the hydraulic behavior of mica flakes. They tend to be transported in suspension by currents, and consequently they tend to accumulate with clay, silt, and fine sand. Flakes of mica are commonly larger than the associated detrital grains (Figure 13-7). Invariably, they are deposited with their flat surfaces approximately parallel to the bedding plane, and they tend to be concentrated in thin partings, where they appear as glistening micaceous coatings on discrete bedding surfaces. Randomly oriented mica flakes, including numerous flakes positioned at high angles to the plane of stratification, are good reason to suspect some disturbance of the unconsolidated deposit, such as bioturbation or slumping. Also, as a sand deposit is compacted by the pressure of overlying strata, mica flakes tend to be bent and squeezed between more rigid sand grains, and in hand specimens they have a crimped appearance where other grains have been pressed against them (Figure 13-7A). This does not happen, however, if the deposit is firmly cemented by authigenic minerals before compaction of the detrital grains, and it is also less likely to occur in very argillaceous deposits.

Both muscovite and biotite are common as detrital grains. Muscovite, like quartz, is one of the most stable minerals and effectively resists chemical alteration in the sedimentary environment. In thin sections cut across the stratification, muscovite appears as thin, elongate grains, colorless but with positive relief and high birefringence. Biotite, in contrast, is strongly pleochroic. Furthermore, biotite is often replaced, or partially replaced along cleavage planes, by green chlorite that can be recognized by its color and very low birefringence.

Heavy Accessory Minerals

Accessory detrital minerals are a very small percentage of most sandstones, and in thin sections only a few grains are likely to be seen. Nearly all of them have densities above those of quartz, feldspar, and most lithic grains, and they can therefore be concentrated by "heavy" liquids in which the abundant, low-density sand grains will float and the denser accessory minerals will sink. This technique is widely used, but it requires that the sand be disaggregated, and it is therefore best suited for study of deposits that have not been firmly lithified. The assemblages of heavy accessory minerals in sandstones are particularly informative in studies of provenance and dispersal of sands.

Relative stability varies greatly among the accessory minerals. Zircon, tourmaline, and rutile are exceedingly resistant to weathering and solution; and individual grains appear to survive through successive cycles of sedimentation. They are the only accessories likely to occur in mature sandstones of all ages in all environments. By contrast, olivine and pyroxene are very unstable both during surface weathering and during diagenesis. They occur in immature recent sands, but they are not common in sandstones of pre-Tertiary age. Minerals of intermediate stability, such as hornblende, epidote, garnet, sphene, magnetite, staurolite, andalusite, kyanite, sillimanite, and tremolite-actinolite are commonly seen in sandstones, but they may be eliminated from old, deeply buried deposits by gradual intrastratal solution. Stages in this solution process are to be seen in the etched character of some grains, as for example in the hypersthene illustrated in Figure 13-10A. Firmly cemented portions of some sandstone strata may contain numerous grains of unstable accessory minerals such as epidote, sphene, and hornblende, which have disappeared from uncemented portions of the same strata.

AUTHIGENIC CEMENT IN SANDSTONES

Many different minerals may crystallize in a sand after its deposition, but only a few are widespread and abundant as cementing materials. Among those that may constitute the chief cement in a sandstone, and that commonly serve alone to consolidate it, are the carbonates (calcite, dolomite, ankerite, and siderite), quartz, chalcedony, opal, various phyllosilicates (smectite, chlorite, illite, kaolinite; see Figure 12-1), anhydrite and gypsum, barite, and collophane (apatite). Of these the carbonates, quartz, and phyllosilicates are by far the most important, the others being developed more locally. Such other authigenic minerals as alkali feldspars, pyrite, chlorite, and laumontite commonly contribute to cementation, but not often do any of these constitute the principal cement.

At least two general tendencies are apparent in the occurrence of cementing materials in sandstones. First, cement is much more common in clean sandstones than in poorly sorted, muddy ones. The presence of an argillaceous matrix may inhibit formation of chemical cement, perhaps because of its relative impermeability, but, conversely, the presence of an early-formed cement, usually calcite, probably inhibits the development of authigenic phyllosilicate matrix of the graywacke type (p. 343). Second, the composition of the sand itself influences somewhat the kind of cement that is precipitated. Quartz cement, for example, is abundant only in quartz-rich arenites, where it grows as enlargements on the detrital quartz; likewise, the authigenic feldspars that occur in sandstones are found chiefly where there is also detrital feldspar, and, like quartz, they develop principally as secondary overgrowths on detrital grains. Laumontite and chlorite are generally found in sandstones rich in basic volcanic detritus and are much less abundant in others. Carbonate cements, although found in all types of sandstones, are invariably formed wherever the original sand contained primary carbonate. There may be various explanations for this tendency of particular cements to concentrate in certain sands. Some sand grains act as crystal nuclei on which cement of the same composition is precipitated from solutions barely above the point of saturation, and thus partly control the development of that cement. Cementing material may also be derived from the deposit in which it precipitates, by partial solution of some of the detrital constituents.

The habit of different cements is quite variable and of considerable interest. Calcite, for example, develops between quartz and silicate grains in some rocks as exceedingly fine-grained aggregates; in other rocks it is very much coarser. The largest crystals of calcite in sandstone may incorporate whole domains of detrital grains within a single crystal and produce what is known as *luster mottling* (Figure 13-4A). Barite and gypsum cements also develop this habit locally (Figure 13-3A). Organic fragments composed of single calcite crystals, like echinoid plates, are commonly rimmed with clear authigenic calcite grown in crystallographic continuity with the fragment. The other carbonates, dolomite, ankerite, and siderite, tend to be more uniformly fine- or medium-grained equigranular aggregates, within which the individual crystals tend to be more idiomorphic than those of calcite. Many of these crystals reveal concentric zones that differ somewhat in composition, particularly in iron content.

Quartz cement typically precipitates on the surfaces of detrital quartz grains, with which it commonly is crystallographically continuous. The

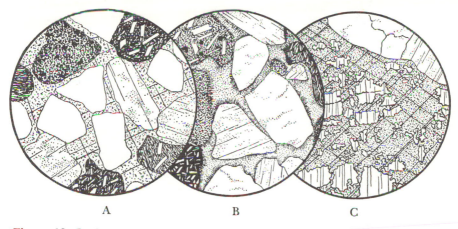

A   B   C

Figure 13–3. Cements in Sandstones

A. Lithic arenite (Miocene, Temblor Formation), 2500 m below surface, Kettle-man Hills, California. Diam. 1 mm. Lithic grains, quartz, and plagioclase enclosed in and cemented by a single barite crystal. Note uniformly oriented right-angle cleavages in barite.

B. Volcanic arenite (Miocene, Temblor formation), 1000 m below surface, Jacal-itos Field, California. Diam. 1 mm. Cement is chlorite. A microfibrous fringe rims each grain, but in the centers of pores the chlorite appears microgran-ular.

C. Arkose (Miocene, Topanga Formation), Santa Monica Mountains, California. Diam. 1 mm. Calcite replacing plagioclase, irregular patches of uniformly oriented feldspar being enclosed within a single calcite crystal. An adjacent quartz-feldspar grain (upper left) is not replaced.

authigenic enlargements or *overgrowths* on individual grains are perfectly clear and can usually be distinguished from the detrital quartz by lines of impurities that delineate the somewhat dirty surfaces of the original sand grains; or the detrital quartz may be clouded throughout by minute inclusions and thus stand in marked contrast to its limpid authigenic rim (Figures 13-4A,B and 13-15A). Wherever the overgrowths have formed in open pore spaces, they are bounded by rational crystal faces, usually faces of the prism or rhombohedron forms. But unless the faces stand nearly normal to the plane of the thin section, they may go unnoticed; however, slight tilting of the thin section mounted on a universal stage brings them into sharp focus and reveals their true identity.[3]

Phyllosilicate cements usually are finely fibrous, the fibers oriented nearly normal to the surfaces on which they develop (Figures 13-3B and 13-4C). Chalcedony is microcrystalline and commonly fibrous and has a

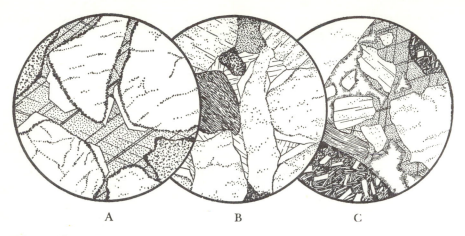

Figure 13–4. Cements in Sandstones

A. Pennsylvanian sandstone, Zuni Mountains, New Mexico. Diam. 1.5 mm. Quartz and turbid rock particles coated with ferric oxide (black), locally covered in turn by clear euhedral overgrowths of quartz, and the whole cemented by calcite (stippled). Note trains of globular opaque inclusions in quartz grains.

B. Cretaceous arkosic arenite, Gualala, California. Diam. 0.5 mm. Local clear euhedral overgrowths of authigenic quartz on detrital quartz (center, lower right, and left). Quartz overgrowths covered and remaining pores filled by the zeolite laumontite (cleavage lines but no stippling).

C. Lithic sandstone (Miocene, Temblor Formation), Reef Ridge, California. Diam. 0.75 mm. An incomplete cement of uniformly oriented calcite (stippled, with cleavage lines); voids fringed with microfibrous chlorite covering both calcite and detrital grains alike; chloritic fringe covered with opal (blank).

refractive index very slightly higher than that of the mounting medium; opal is optically isotropic and has pronounced negative relief. Collophane may be sensibly isotropic or microcrystalline, but it has distinct positive relief.

Several cement minerals may occur in a single rock. They may be associated intimately, crystallizing together in the same pore space, or they may be distributed in more or less isolated patches. Where two or more cementing minerals occur together, their relations to each other are of particular interest as possible indications of the sequence and manner in which they formed. The two most commonly observed in contact are quartz and calcite. Usually the calcite appears to have crystallized on top of euhedral quartz outgrowths (Figures 13-4A and 13-15A); but euhedral calcite against quartz has been recorded. The other carbonates, and also authigenic feldspars, are almost invariably euhedral in contact with either authigenic quartz or calcite.

Idiomorphism of one mineral relative to another is difficult to interpret, however. It may result from a sequence in which the euhedral mineral forms first within a void and is later covered by the anhedral mineral, which fills the remaining cavity. Or it may result from a process whereby the euhedral mineral has replaced the other and is therefore the later of the two. This is surely the origin of the euhedral quartz, alkali feldspar, and dolomite often found in limestones. Calcite, however, has so little tendency to be euhedral that it does not usually assume its crystal form when it replaces other minerals; in fact, the contacts it forms by replacement are irregular in detail (Figure 13-3C).

The only really reliable single criterion for the sequence of crystallization among the cementing minerals in sandstones is their disposition in the pore. If a pore is lined more or less completely with one mineral and the central parts are filled in with another, one can logically infer that the mineral along the margins was the first to crystallize (Figure 13-4C). But there is no invariable sequence of formation among cementing minerals, for the conditions of cementation are diverse. The sequence quartz-calcite is common, yet calcite sometimes forms before quartz; and of the two minerals dolomite and calcite, either may develop first.

GRAYWACKE

Typical graywacke consists of sharply angular sand and silt grains bound together in a firmly lithified, dark matrix. The larger grains may be elongate chips oriented with their longer dimensions approximately in the plane of stratification (Figure 13-5A), but more often they are approximately equant and unoriented. Most graywackes are varieties of lithic or feldspathic sandstone in which grains of quartz are abundant but are accompanied by numerous grains of feldspar and small bits of schist, slate, argillite, shale, chert, or volcanic rocks. Relatively few graywackes are sufficiently rich in quartz to be classified as quartzose, but many are quartz poor and almost wholly lithic, especially those made up of basaltic or andesitic volcanic debris (Figure 13-9C). The most characteristic feldspars are sodic plagioclase—albite, oligoclase, or andesine—and commonly they are unaltered and clear with distinct lamellar twinning. Potassic feldspars may occur but they are rarely abundant and have more limited distribution. The calcic feldspars, which are to be expected

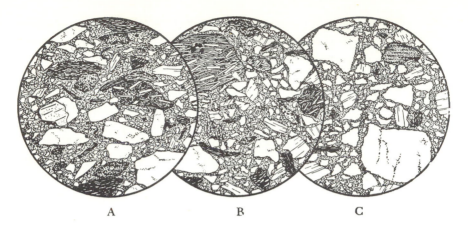

A   B   C

Figure 13–5. Graywacke

A. Ordovician lithic graywacke (Fortune Formation), Lawrence Harbor, New-foundland. Diam. 1.5 mm. An unsorted aggregate of angular grains of sand and coarse silt set in an abundant argillaceous matrix. Grains are quartz (clear or lightly stippled), feldspar (chiefly plagioclase, shown with cleavage), a few shreds of mica, and particles of phyllite, argillite, chert, and andesite or basalt. Long dimensions of most grains lie roughly parallel to bedding plane which is nearly normal to the section.

B. Franciscan graywacke, Mendocino County, California. Diam. 1.5 mm. Gen-erally similar to A, but shows less orientation of grains, slightly less matrix, and more grains of feldspar and basalt. This specimen is typical of many Franciscan sandstones that fall near the boundary between lithic and feld-spathic types.

C. Precambrian feldspathic graywacke, Hurley, Wisconsin. Diam. 1.3 mm. Tex-turally like B, except that the margins of the grains are corroded. Quartz grains are very abundant, feldspar is common, and rock chips are sparse. This is a well-known chemically analyzed graywacke (*U.S. Geological Survey Bulletin*, vol. 150 (1898): pp. 84–87).

in graywackes of andesitic or basaltic composition, have usually been replaced by turbid albite or by laumontite. In many graywackes, a few flakes of detrital micas are to be seen, but heavy accessory minerals are typically sparse.

The matrix in some graywackes comprises one-third or more of the rock, sufficient that in thin sections the sand grains appear to be sepa-rated by it (Figure 13-5), but the proportion of matrix is often much less. As seen in thin section, graywacke matrix is clearly microcrystalline. It is a mixture chiefly of phyllosilicates (commonly chlorite and mica) and quartz, but the individual crystals are difficult to discriminate. In contrast to clean, monomineralic phyllosilicate cement precipitated in open pores like that illustrated in Figure 13-3B, graywacke matrix is heterogeneous and murky and is usually ill defined. Indeed, it may appear remarkably similar to some of the finer-grained lithic fragments, and then the distinction between grains and matrix becomes particularly troublesome. What appear to be matrix-filled gaps in the grain framework of some graywackes are probably unstable lithic grains that have been diagenetically altered, and in some rocks weak lithic grains squashed and drawn out between more rigid sand grains constitute a sort of pseudomatrix.[4]

In numerous graywackes, especially the older ones, the marginal parts of quartz and feldspar grains are penetrated by minute flakes of chlorite or mica, so that the grain boundaries are hazy and indefinite. Commonly the replacing flakes are oriented at a high angle to the grain surfaces, and where numerous flakes penetrate a grain margin the boundary between grain and matrix has, in detail, a comblike appearance. If most of the grains in a rock have been partially replaced, a particularly indistinct or fuzzy texture is produced in which the grains seem to fade into the matrix. Clearly this is a result of postburial crystallization of secondary phyllosilicates.

Graywackes are among the most widespread types of sandstone, but they occur primarily in orogenic belts of early Tertiary or pre-Tertiary age. They do not appear in the sedimentary sequences characteristic of stable cratons. The great majority of them are graded sandstone beds in flysch sequences and were presumably deposited by turbidity currents in deep marine basins. They range in composition from sandstones that are rich in stable quartz or chert grains to sandstones that consist largely of unstable grains, either feldspathic or lithic. It thus appears that the genetic common denominator among graywackes is more environmental than compositional. Most of them have been deposited in tectonic settings that resulted in very deep burial and later regional deformation, and this history probably has been responsible for development of their unique characteristics, particularly the graywacke matrix.

The origin of matrix material in graywackes is uncertain and is the subject of debate. Some petrographers consider the matrix as detrital silt and clay compacted, recrystallized, and firmly lithified during diagenesis. This assumption accounts for some graywacke matrix, especially where grains in the sand framework are poorly sorted and numerous silt-size particles of quartz and feldspar appear to be integral parts of the matrix.

However, few modern sand deposits, particularly modern deep-sea turbidite sands, contain enough detrital silt and clay to provide the volumes
of matrix seen in many graywackes, especially in view of the volume
reduction to be expected as a result of compaction at depth. This fact
and the general occurrence of graywackes in older, deformed flysch formations suggest that matrix material, in part at least, is a by-product of
diagenetic destruction or alteration of unstable framework grains. Probably the modern counterparts of ancient graywackes are immature sands
containing unstable framework grains, or detrital clay matrix, or both.
Sands of this kind tend to be converted to graywacke if they are deposited
in unstable tectonic settings that lead to subsequent very deep burial with
resulting prolonged diagenesis at elevated temperature and pressure and
later deformation. Perhaps graywacke should be regarded as a product
of incipient metamorphism, rather than diagenesis, but the difference
is largely semantic.

In the tectonic settings where graywackes have developed, calcareous
sandstones also commonly occur. They occur, for example, in the type
flysch deposits of the Alps and northern Apennines, and also in those of
the Pyrenees, the Carpathians, and the Coast Ranges of California. They
occur in graded beds interstratified with mudstones, and they tend to
have the same overall appearance as noncalcareous graywackes. And like
the latter, they were presumably resedimented sands laid down in deep
water by turbidity currents but differed initially by having some content
of carbonate sand or mud. A few are so carbonate-rich as to constitute
sandy limestones. The postdepositional history of these calcareous strata
must have been essentially the same as that of associated noncalcareous
graywackes.

Some calcareous graywackes contain carbonate sand grains that show
relict organic fabrics; in others, granular patches of calcite appear to
occupy gaps between silicate grains in the sand framework and probably
represent carbonate sand grains that have been recrystallized. Calcite
may also appear in small, highly irregular patches within the interstitial
phyllosilicate matrix. In more richly calcareous sandstone strata, fine-
grained calcite partially replaces silicate framework grains and is generally clouded by small silicate inclusions; its texture appears irregular
and is generally indistinct. Such calcite probably developed by recrystallization of carbonate mud initially present in the sand deposit. If so, it is
the carbonate counterpart of the phyllosilicate matrix in less calcareous
graywackes and may be considered as a stable calcite matrix, rather than
pore-filling cement.

FELDSPATHIC SANDSTONES

Arkosic sandstones are characterized by quartz and an abundance of feldspars, including sodic plagioclase (An_{00-50}), orthoclase, microcline, and perthite. Either potassic or sodic feldspar may predominate, but calcic plagioclase is not to be expected. Fragments of slate, phyllite, low-grade schist, volcanic rocks, chert, shale, sandstone, and limestone are likely to occur as more or less subordinate constituents. For those arkosic sandstones that contain relatively few lithic grains, the name *arkose* is particularly appropriate. With increasing content of fine-grained rock fragments, arkosic deposits grade into lithic ones, and sandstones having compositions close to this boundary are numerous.

Among the most immature of all sandstones are some arkoses that are residual and lie directly upon the granitoid rock from which they were derived, or that have been transported but a short distance. These can look very much like granite, and it is not always easy to distinguish them from the partly disintegrated parent rock, whose minerals they may contain in approximately the original proportions. Their grains not only are angular but tend to be very irregular, like those in disintegrated rubble and sand on weathered granite surfaces.* Having been broken asunder, however, the grains do not interlock in arkose as they do in the parent rock, and between the larger fragments there is usually a breccialike mortar of sharply angular smaller fragments and clay (Figure 13-6A). In old residual arkoses that are firmly compacted the mica flakes are commonly distorted and even split apart by the pressure of other grains; commonly, too, curved twin lamellae give evidence of internal strain in plagioclase grains.

Sorting in most residual and slightly transported arkoses is negligible, and bedding tends to be indistinct. The deposits tend to be coarse grained and generally include pebble-sized chunks of quartz and feldspars, together with pieces of undisintegrated granitic rock. Much of the feldspar may be clouded by partial decay, and clay usually is irregularly distributed throughout the rock. Ferric-oxide pigment is common, coloring the rock red, but much arkose is gray or buff. Precipitated cements, such as calcite, are less common than in the better-sorted transported arkoses.

---

*By surface weathering, granitic rocks commonly disintegrate, with little chemical decomposition, to a fragmental rubble called *grus* that accumulates essentially in situ and corresponds to an unconsolidated residual arkose.

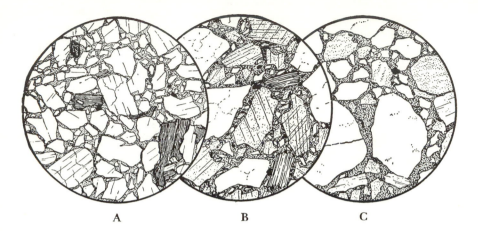

A   B   C

Figure 13–6. Arkosic Sandstones

A. Arkose (Tertiary), Lake Manapouri, New Zealand. Diam. 2.5 mm. Unsorted angular grains of orthoclase and oligoclase (with cleavage) and of quartz (clear), accompanied by large and small unoriented flakes of biotite and a grain of sphene (upper left), all bound together by a mortar of silty clay slightly stained with limonite. Essentially residual, resting on granitic rock from which it was derived.

B. Arkose (Pennsylvanian, Fountain Formation), Boulder, Colorado. Diam. 2.5 mm. Poorly sorted angular grains of quartz, turbid oligoclase, and microcline (both feldspars stippled and showing cleavage), and accessory flakes of muscovite, all bound together by a matrix of silty clay stained red by ferric oxides. The deposit has been transported but suggests a near-by granitic source.

C. Torridonian arkose (Precambrian), Loch Assynt, Scotland. Diam. 2.5 mm. Poorly sorted subangular grains of quartz (clear and very slightly stippled) and of microcline, orthoclase, and oligoclase, firmly bonded in a matrix of micaceous clay. Feldspars are in part fresh (shown with cleavage) and in part very turbid (stippled). A few rock fragments (schist) are not shown.

Arkosic detritus that has been transported for some distance accumulates in deposits having a less granitoid appearance, although the deposits have an overall composition that suggests granitic heritage (Figures 13-6 and 13-7). As a result of abrasion and sorting of the weathered material, they differ somewhat in mineral composition from their parent rocks, being generally richer in quartz and poorer in feldspars and other destructible minerals. They are frequently contaminated with detritus derived from fine-grained sedimentary, volcanic, or metamorphic rocks, and when this is so their composition may be noticeably different from

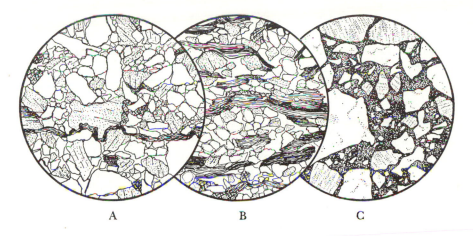

A   B   C

Figure 13–7. Arkosic Sandstones

A. Miocene arkosic arenite, or arkose, 3000 m below surface, near Simmler, California. Diam. 2 mm. Very tightly packed angular and subangular grains: not well sorted, but free from clay. Consolidated by compaction without cement. Plagioclase, orthoclase, and microcline (all lightly stippled) and quartz (blank) are about equally abundant; grains of calcite (heavily stippled) and biotite are accessory. Note pinched and contorted mica.

B. Micaceous arkosic arenite, or arkose (Triassic), Portland, Connecticut. Diam. 2 mm. Fairly well-sorted angular-to-subangular grains of feldspar (lightly stippled) and quartz (blank); abundant parallel oriented flakes of muscovite and chloritized biotite, larger than other grains, lie parallel to the bedding. The rock is lightly cemented by scattered grains of calcite (heavily stippled and showing cleavage) and secondary quartz overgrowths (separated from detrital quartz by dotted lines). Porosity high. A few schist particles, not shown in this field.

C. Red arkosic wacke, or arkose (Triassic), Mt. Tom, Massachusetts. Diam. 3 mm. Unsorted angular-to-subangular grains of quartz and turbid feldspar, in a very abundant matrix of ferruginous clay.

that of granites or granodiorites. They may also contain subangular and subrounded grains that show distinctly the effects of abrasion. Some of them have an abundant clay matrix and are *arkosic wackes*, whereas others have been sorted sufficiently to be classed as arkosic arenites (Figure 13-7).

Feldspar grains often show different degrees of decay within the same deposit. Some may be so clouded by alteration to clay as to be scarcely determinable, whereas others are limpid and unaltered (Figure 13-6C). Where this difference is recognizable in the same species of feldspar within a single rock specimen, it suggests that the climate of the source area was such as to promote chemical decay, and that materials in different stages of decay were fortuitously mixed together by rapid erosion. In most well-sorted arkosic arenites the feldspar grains are not much altered, and in some of them all the feldspar is perfectly fresh and water-clear.

Micas are rarely lacking in arkosic rocks, and they may form as much as 5% or 10% of the rock, their abundance being largely controlled by sorting (Figure 13-7B). They are most numerous in the finer arkosic sandstones and in associated argillaceous strata. Micas have often been pinched and sharply bent between tightly packed sand grains (Figure 13-7A).

In most sandstones, the proportion of quartz exceeds that of feldspar. Quartzose feldspathic sandstones, or *subarkoses*, in which feldspar comprises less than 25% of the rock, probably have had longer or more rigorous sedimentary histories than arkosic deposits, and some of them may be second-cycle deposits. Those that have been derived from crystalline basement terranes are akin to arkoses, differing only in the proportion of feldspar that survived weathering and transportation. Others derived from supracrustal terranes that included feldspathic sedimentary and metamorphic rocks usually contain numerous lithic grains and grade into lithic sandstones. As the proportion of feldspar declines, the rocks ultimately grade into quartz sandstones. In fact, the finer-grained beds and laminae in some first-cycle quartz arenites contain considerable amounts of feldspar, whereas coarser layers of the same deposit may contain none (Figure 13-14C). Most of the feldspars that occur in these rocks are relatively unaltered K-feldspars or highly sodic plagioclase, because only the most stable grains are likely to survive the extensive weathering and transportation that usually is involved in the formation of very quartz-rich sandstones. Some quartz sandstone formations that lie unconformably on old feldspathic rocks may contain moderate amounts of feldspar in their lower portions but decidedly less in their upper portions.

Minor accessory minerals in feldspathic sandstones may include hornblende, sphene, apatite, zircon, tourmaline, rutile, epidote, garnet, magnetite, and ilmenite, together with some others of more sporadic occurrence. Many of the older deposits, however, contain only the most stable of these minerals, especially zircon, rutile, and tourmaline, associated, rather surprisingly, with abundant fresh feldspars. This association, which at first seems anomalous, probably results from solution of the unstable accessory minerals during diagenesis. In contrast to these min-

erals, the potassic and sodic feldspars are generally stable during diagenesis, and may even be added to by authigenesis.

Any of the clay minerals may be included in the argillaceous material
found in feldspathic wackes or in the finer beds associated with them. As
seen in thin sections, this material is usually so very fine grained and
turbid that its exact mineral content cannot be determined by visual
means, especially where it has been stained by either iron oxides or
organic products. In some of the older rocks, however, the original clay
matrix has been recrystallized to fine-grained aggregates that can be
recognized as chiefly kaolin-type clay minerals or chiefly mica- and chlorite-type minerals. Especially when the recrystallized matrix is illitic or
chloritic, these rocks are likely to be rather dark colored.

Many of the better-sorted feldspathic arenites are porous and permeable (Figure 13-7A,B), and many of these are petroleum reservoir sands
(Figure 11-5B); others are firmly cemented by authigenic minerals crystallized around the sand grains, so that they are nonporous and
impermeable (Figure 13-8A,B). The most common cement is carbonate,
which not only fills the pores between the grains but often partly replaces
the grains themselves (Figures 13-3C and 13-8B). Other kinds of cement
are developed locally. Some arenites, however, have been consolidated by
pressure that has forced the grains into intimate contact; these contain
little of either cement or clay matrix, yet they may be very firm. An
arkosic arenite consolidated in this way is illustrated in Figure 13-7A, and
although this rock retains considerable porosity, the grains are tightly
interlocked and the rock is firm.

Authigenic glauconite is often present in feldspathic sandstones that
have been deposited in neritic marine environments. Except in very
muddy deposits, glauconite is probably moved about and sorted with the
clastic debris on the sea floor, so that in arenites it tends to occur in pellets
of about the same size as the sand grains with which it is associated, and
it is commonly concentrated in thin laminae.

Organic materials, also, form a part of many feldspathic sandstones,
their character and abundance depending largely on the environment
of deposition. In marine deposits, for example, whole or fragmented
calcareous shells may be abundant. Some of these are broken bits of
heavy molluscan shells, such as those of oysters, but small and fragile
shells, even whole tests of foraminifers, are frequently preserved. Most
arenites containing such calcareous material are cemented by additional
carbonate, probably precipitated around the shells as nuclei. Many calcareous arenites, however, show no trace of organic structures (Figure
13-8).

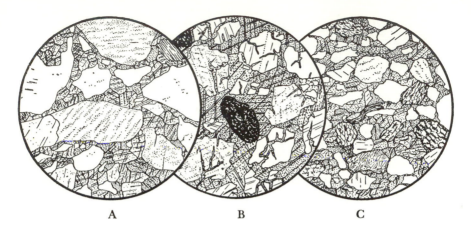

A   B   C

Figure 13–8. Arenites Cemented with Calcite

A. Calcareous arkose (Permian), Shensi Province, China. Diam. 2.5 mm. Sub-angular medium-to-very-coarse sand grains, solidly cemented with finely granular calcite (darkly stippled and showing two cleavages). Grains are perthite (lightly stippled and showing traces of cleavage), quartz, and biotite (upper right). Loosely packed, so that grains are separated by cement.

B. Calcareous arkosic arenite (Lower Cretaceous), Queen Charlotte Islands, British Columbia. Diam. 2.5 mm. Coarse sand grains, loosely packed and cemented with calcite much coarser than that in A. Grains are oligoclase and perthite (with cleavage), quartz (blank), and two fragments of metamorphic rock. Note veining, and perhaps replacement, of detrital grains by calcite cement.

C. Calcareous lithic arenite (Macigno), Bologna, Italy. Diam. 2 mm. Loosely packed subangular grains, fairly well sorted, firmly cemented with finely granular calcite. Grains are predominantly of quartz, but there are abundant grains of sodic plagioclase (with cleavage), particles of quartzose schist, and fragments of serpentine (lower center and left, dash pattern); a bent mica flake in center.

LITHIC SANDSTONES

Among lithic sandstones, both the composition and the proportion of lithic particles vary widely.[5] In some, lithic grains predominate by far over all other detrital particles, but quartz grains are commonly more abundant. Some sandstones derived from regionally metamorphosed terranes contain an abundance of slate and schist fragments; volcanic sandstones contain numerous fragments of andesite, basalt, or felsite. Other sandstones contain mixtures of many different kinds of rock par-

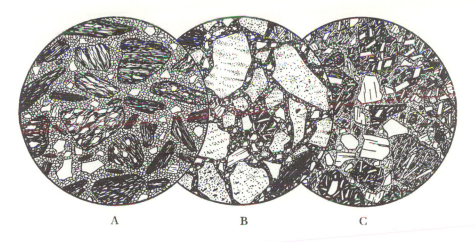

A   B   C

Figure 13–9. Lithic Arenite and Lithic Graywacke

A. Calcareous lithic arenite (Miocene Modelo Formation), Santa Monica Mountains, California. Diam. 2.5 mm. Fairly well-sorted sandstone consisting of subangular and subrounded slate and schist fragments and smaller angular grains of quartz and feldspar (trace only) cemented with fine-grained calcite.

B. Bragdon lithic graywacke (Mississippian), Trinity County, California. Diam. 2.5 mm. An unsorted aggregate of angular grains set in a dark argillaceous matrix. Less matrix than in graywackes of Figure 13–5. Grains are largely chert and devitrified rhyolites (stippled), andesite, and slate; there are fewer angular quartz grains (clear) and a trace of plagioclase (with cleavage). No preferred orientation of grains is visible.

C. Volcanic graywacke (Triassic), southern New Zealand. Diam. 2.5 mm. An unsorted aggregate of angular and subangular grains in a matrix containing much microcrystalline chlorite. Grains are chiefly fragments of andesitic or basaltic rocks; plagioclase grains (with cleavage) are common; and quartz (clear) is subordinate.

ticles, and in some, the lithic component is mainly chert (Figure 13-9B). Numerous lithic sandstones of relatively local occurrence reflect particular source rocks, as exemplified by the serpentine arenite illustrated in Figure 13-10C, or by *calclithite*, a detrital limestone or dolomite derived by erosion of older carbonate formations.*

---

*Most calclithites are conglomeratic and their lithic fragments are a mixture of different types of carbonate rock, usually with chert fragments derived from the carbonate source formations. They are of local significance and occur most commonly in deposits of molasse facies formed in relatively arid environments. They are in distinct genetic contrast to the more widespread clastic limestones discussed in Chapter 14.

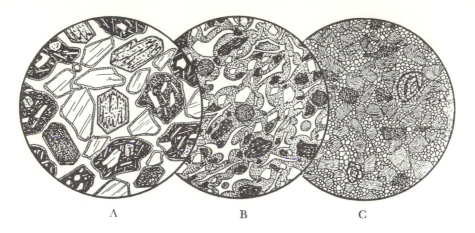

A   B   C

Figure 13–10. Miscellaneous Lithic Sandstones

A. Andesite arenite (Upper Miocene, Neroly Formation), Mount Diablo, California. Diam. 2.5 mm. Well-sorted, loosely packed, subangular grains of andesite rock, andesine (clear, with cleavage), hypersthene (center and top), and hornblende (lower left and right). Each grain enclosed in a thin fibrous rim of smectite. Hypersthene and hornblende are euhedral, but hypersthene has been etched by intrastratal solutions after development of smectite rims. This is an epiclastic arenite, not a tuff or a tuffaceous arenite.

B. Calcareous tuffaceous sandstone (Oligocene, Tunnel Point Formation), Coos Bay, Oregon. Diam. 3 mm. A mixture of pyroclastic and epiclastic material deposited in a marine environment, where it was mixed with glauconite and cemented with very fine-grained calcite (stippled). Curved glass shards and detrital quartz and feldspar are clear; turbid fragments of meta-andesite and phyllite, and spheroidal pellets of glauconite, are darkly stippled.

C. Calcareous serpentine arenite (Eocene), southeastern Monterey County, California. Diam. 3 mm. Angular and subangular grains of serpentine (line pattern), together with microcrystalline carbonate pellets (stippled), firmly cemented with finely granular calcite. Note two unbroken foraminifers.

Lithic sandstones are derived largely from supracrustal terranes where the coarsely crystalline basement is buried by sedimentary, volcanic, or regionally metamorphosed formations. Their provenance thus differs from that of arkoses. But the proportion of feldspars varies widely in lithic sandstones, and as it increases, the rocks grade into feldspathic sandstones. A great many sandstones have compositions close to this boundary and plot in the central part of the *QFL* triangle in Figure 13-1. Quartz-rich lithic sandstones also are numerous, and they grade into mature quartz arenites.

The feldspars in lithic sandstones are chiefly sodic plagioclase. Scattered grains of K-feldspar may be present, but they are rarely as abundant as plagioclase. In young volcanic sandstones of andesitic or basaltic composition, calcic plagioclase is common, but in older rocks only the more sodic plagioclases are characteristic, presumably because unstable calcic feldspars generally have been decomposed or have been replaced during diagenesis. Much of the sodic plagioclase in lithic sandstones is unaltered and has distinct lamellar twinning, but feldspar grains clouded by alteration products are not uncommon and may occur in the same sandstone with unaltered grains.

The sandstones of the Cenozoic Siwalik series in northern India, long ago described as *schist arenites*,[6] provide excellent examples of lithic sandstones derived from an area of regionally metamorphosed rocks. They include both arenites and wackes, and they contain, on the average, about 40% quartz, 15% feldspar, 35–40% schist and phyllite fragments, and 5–10% accessory detritus. Both feldspar and schist particles range from unaltered to extensively decomposed. Because of their foliated structure, the schist fragments tend to be platy and to be oriented parallel to the bedding, which makes the rocks appear highly micaceous. The Siwalik series was produced by relatively rapid erosion of a mountainous upland area and was deposited on a broad fluvial plain. The total deposit is exceedingly thick, and some of the sandstones in the lower part of the series are described as graywacke.

Volcanic sandstones of epiclastic (detrital) origin are widespread, and modern deposits are associated with all predominantly volcanic terranes. Deposits of Tertiary and Mesozoic age are particularly numerous among the strata laid down in the orogenic belts around the Pacific Ocean. Volcanic sandstones of all ages are known, however, particularly in stratal successions that have been deposited in orogenic regions. Some of them include a mixture of differing volcanic types, but most of them reflect a volcanic source of more or less uniform composition.

The most numerous and widespread volcanic sandstones consist largely of detritus derived from intermediate and basic volcanic source rocks, perhaps mixed with smaller amounts of debris from other sources (Figures 13-9C, 13-10A, and 13-11). They are characterized by abundant lithic grains of andesite or basalt, which are easily recognized by their microlitic textures. They also typically include such volcanic minerals as zoned plagioclase, augite, hypersthene, and hornblende or oxyhornblende (Figure 13-10A). These minerals occur both as phenocrysts in some of the rock fragments and as separate grains. Many of the latter,

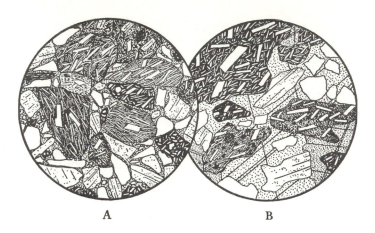

A   B

Figure 13–11. Volcanic Sandstones

A. Volcanic wacke (Eocene), Tyee Formation, Umpqua River, Oregon. Diam. 1.2
mm. Poorly sorted angular and subangular grains of coarse silt and sand
tightly packed in an argillaceous matrix colored green by chloritic material.
About half of the grains are particles of volcanic rocks, chiefly andesite; about
30% are plagioclase, chiefly andesine (lightly stippled, with cleavage); and
about 20% are quartz (clear).

B. Miocene arenite, 3700 m below surface, south of Lost Hills, California. Diam.
1.2 mm. Loosely packed, subangular grains of andesite, plagioclase (lightly
stippled, with cleavage), and quartz firmly cemented by coarse calcite (stippled, with two cleavages). Single calcite crystal in center encloses many sand
grains.

as well as the phenocrysts embedded in rock fragments, are partly
bounded by crystal faces, which can be recognized by their habit or by
their relation to compositional zones within the crystals. Mineral grains
of this sort are almost as clear an indication of volcanic source rocks as
the rock fragments themselves, and in the finer-grained volcanic sandstones they are typically more abundant. A few fine-grained volcanic
sandstones consist largely of plagioclase. Volcanic sandstones of andesitic
and basaltic composition are, of course, deficient in quartz, and the percentage of quartz they contain provides a rough approximation of the
degree of dilution by other source materials.

Basic volcanic detritus is particularly susceptible to alteration during
diagenesis. Glassy particles are very unstable and have commonly been
altered to smectite, chlorite, or zeolites; these minerals are also the principal constituents in the matrix of most volcanic wackes. Epidote and

chlorite are commonly formed by alteration of pyroxenes and horn-blende, and calcic plagioclase commonly has been replaced by albite or laumontite.

Volcanic sandstones of silicic composition consist principally of felsitic grains and are likely to include subhedral grains of sanidine and high-temperature volcanic quartz. Fragments of silicic volcanic glass are abundant in general but are unstable, and in pre-Tertiary sandstones they usually have been replaced by microcrystalline aggregates of clay minerals, zeolites, albite, and quartz.

The volcanic particles in many sandstones have, of course, been blown directly from volcano to basin of deposition and incorporated directly into the accumulating sediment; in other words, they are pyroclastic rather than epiclastic (Figure 13-10B). A rock that contains pyroclastic grains is properly called *tuffaceous sandstone*, or *tuff* if the pyroclastic material predominates. It is often difficult to distinguish between pyro-clastic deposits and volcanic deposits of epiclastic origin, and the term *volcaniclastic* includes both. However, recognition of pyroclastic fragments is important, because they indicate an active volcanic source in contrast to an older volcanic terrane that has been eroded by surface processes.

Among the commonest of all sandstones, including modern sands, are the lithic sandstones of mixed and largely nonvolcanic heritage (Figures 13-12 and 13-13). They contain abundant grains of quartz together with feldspars and numerous fragments of chert, argillite, siltstone, and slate, phyllite, or schist, commonly with subsidiary volcanic or metavol-canic grains and perhaps with a few grains of limestone or dolomite. The combination of numerous dark lithic grains with light-colored grains of quartz and feldspar tends to produce a pepper-and-salt aspect and an overall gray color in the unweathered rocks. Such sandstones are common among the Tertiary and Cretaceous strata in the Coast Ranges of the western United States and in the stratigraphic successions along the Gulf Coast, where they may be important petroleum reservoir sands (Figure 13-12C). They also comprise much of the Swiss molasse (Figure 13-12B).

A special significance attaches to the lithic grains classified as chert, for they are essentially composed of quartz and are thus more stable than other lithic grains. Detrital chert grains commonly include several different kinds of rock, all microcrystalline and quartzose but of variable origin and stability. Some, often of uncertain provenance, are made up

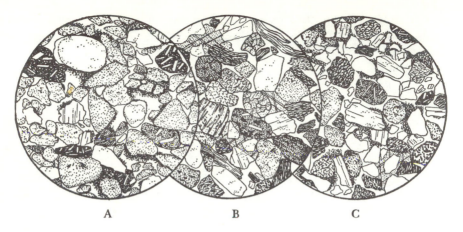

A   B   C

Figure 13–12. Lithic Arenites

A. Belly River Sandstone (Cretaceous), near Calgary, Alberta, Canada. Diam. 1.8 mm. Fragments of chert and siliceous volcanics (heavily stippled), andesite, quartz (clear or lightly stippled), and a little feldspar (with cleavage). Rock fragments greatly predominate. The grains are subangular and subrounded and firmly packed, though the texture remains porous. Note authigenic overgrowths on two quartz grains (upper left and center) and recrystallized ferruginous clay in some pores.

B. Molasse, Bach, Switzerland. Diam. 1.2 mm. A well-sorted arenite containing rock fragments and mineral grains in about equal amounts; fragments of metamorphic and volcanic rock, limestone (center), chert, quartz (nearly clear), feldspar (with cleavage), and mica (upper right and lower left). A grain of glauconite (heavily stippled) occurs in upper left; there are two local patches of calcite cement.

C. Eocene sandstone (Wilcox), 1000 m below surface, central Louisiana. Diam. 1 mm. Fine-grained friable sandstone consisting of angular and subangular grains and local patches of clay and calcite cement (lower right). Loosely packed and porous. The grains are rock particles (chert, slate, phyllite, and volcanic rocks), together with abundant quartz (clear) and fresh feldspar (with cleavage), largely plagioclase.

of microgranular or microfibrous quartz with abundant impurities; some show the shadowy outlines of organic textures and may contain minute calcite or dolomite crystals, suggesting derivation from bedded chert formations or from silicified limestones; others are metacherts or silicified volcanic rocks. Although chert is a very widespread type of rock, it is invariably less abundant than the less stable rocks that characterize the source terranes of most sandstones. A concentration of chert grains can develop only as less stable rock fragments are progressively eliminated

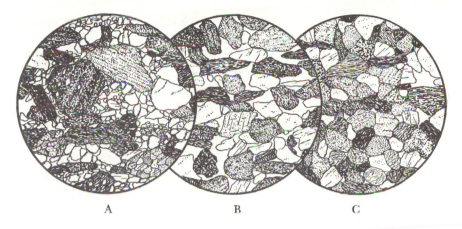

A   B   C

Figure 13–13. Lithic Arenites

A. Triassic sandstone, Boonton, New Jersey. Diam. 2 mm. Not well sorted, but contains little or no clay. Composed of angular and subangular grains derived from sedimentary and low-grade metamorphic rocks. Rock fragments of shale, slate, argillite, and limestone (lower left and right); also ragged grains of quartz and very few of feldspar.

B. Chico Sandstone (Cretaceous), near Chico, California. Diam. 1 mm. Fine-grained, well-sorted arenite consisting of subangular grains; poorly consolidated and very porous. Rock fragments are slate and fine schist, with a little chert; quartz (clear or slightly stippled) is abundant, and feldspar (with cleavage), both fresh and cloudy, is common; hornblende and epidote (darkly stippled, with cleavage, in upper left and at bottom) are present in every thin section; a bent flake of biotite in upper left.

C. Triassic sandstone (Keuper), Stuttgart, Germany. Diam. 1 mm. Tightly packed subangular grains; porosity relatively low. Abundant schist and microgranular rock particles (lined and stippled); abundant quartz and feldspar (lightly stippled with cleavage), both orthoclase and plagioclase; some mica flakes. Grains of mica schist are commonly oriented parallel to bedding and give the rock a very micaceous aspect in hand specimen.

by sedimentary processes. The proportion of chert grains relative to other lithic grains in a sandstone is, therefore, a rough index of the compositional maturity of the deposit; it can be expressed numerically as the ratio of chert grains (C) to total lithic grains (L). Chert arenites, in which C/L approaches unity, usually consist of both chert and quartz grains with little or no feldspar, and, regardless of the percentage of quartz grains, they are relatively mature sandstones.

Subordinate constituents, in addition to those that define the rock, are to be expected in most lithic sandstones. Both muscovite and biotite are common accessory minerals, especially in sandstones derived from metamorphic source areas, and they are abundant in some of the finer and more argillaceous strata. Heavy minerals such as zircon, tourmaline amphiboles, epidote, sphene, and garnet are common minor accessory minerals in lithic sandstones, because they are widespread in a variety of source rocks. In sandstones derived from metamorphic terranes, accessory heavy minerals may also include such minerals as staurolite, andalusite, sillimanite, and kyanite, and pyroxenes may occur in sandstones derived from basic igneous rocks. But although any of the numerous heavy minerals may be encountered, they rarely occur in more than very minor amounts, and only the most stable varieties are likely to survive in very old lithic sandstones.

Pellets of authigenic glauconite often appear in sandstones that have been deposited in marine environments. They usually are sorted with the detrital sand grains and are a part of the sand framework, though they may be concentrated in certain laminae. Organic remains also are commonly seen in marine strata, either as whole fossils or as abraded and sorted fragments.

The most common authigenic cements in lithic arenites, as in other types of arenite, are either calcite or quartz. But the composition of the lithic grains, as well as the diagenetic conditions, have determined the cements in many rocks. Smectite and chlorite, for example, are most often seen as fibrous or microgranular cements in the pores of andesitic and basaltic arenites (Figures 13-3B and 13-10A), and chlorite is responsible for the greenish color of the matrix in many volcanic wackes. Lithic arenites rich in chert particles are often cemented by chalcedony rather than by the quartz overgrowths that form typically on detrital quartz grains. Many lithic arenites are uncemented but have been firmly lithified by compaction and close packing of the sand grains (Figure 13-13C). The compaction may be so extreme that the grains effectively interlock, and in some sandstones microstylolitic contacts between grains are indicative of pressure solution along grain boundaries. This is particularly noticeable between closely packed chert grains in chert-rich arenites, but it may also occur between quartz grains (Figure 13-15B).

QUARTZ SANDSTONES

In quartz sandstones, grains of quartz and quartzite make up 90% or more of the detrital silicate material (Figures 11-5A and 13-15). Most of the finer-grained ones also contain at least a little feldspar, chiefly microcline, orthoclase, or albite. The typical accessory minerals include only the stable species such as zircon, tourmaline, and rutile, and these are invariably so scarce that only a few grains are found in any one thin section. Muscovite may be present in small amounts, for it is a widespread stable mineral, but its occurrence is chiefly a result of local sorting.

Texturally, most quartz sandstones are characterized by thorough sorting. They are usually clean, well-washed arenites devoid of argillaceous material. Presumably they are deposited in stable environments, either continental or marine, where deposition was relatively slow and the particles were so winnowed by currents before final burial that any argillaceous materials were almost completely washed away. At least a few of them are the result of more than one cycle of erosion and deposition. Under these conditions, the sand grains tend to become subrounded or well rounded (Figures 11-5A and 13-14B). But some first-cycle quartz sandstones consist of angular and subangular grains, and a few of them contain enough clay material to be classed as quartz wacke.

Chert arenites are classed in this book with lithic sandstones (p. 357) because of the relatively dark color and lithic appearance of microcrystalline chert particles. Like quartz arenites, they are mature sandstones in which the grains are usually both well sorted and subrounded. In fact, they grade into quartz arenites, many of which include a small percentage of chert grains.

Many of the widespread quartz arenites were deposited in warm, clear, shallow seas, where calcareous organisms are likely to abound, and it is therefore not surprising that many quartz arenites contain numerous organic fragments. Indeed, carbonate particles may predominate over quartz, and then the rocks become varieties of *calcarenite* (Figure 13-14A). In some early Paleozoic rocks, phosphatic shell fragments are more common than calcareous ones. Many but not all of the organic fragments are rounded and sorted so as to be of about the same size as the accompanying grains of quartz. Calcareous material, however, is readily recrystallized by diagenesis, so that original calcareous fragments in an arenite may be represented simply by patches of recrystallized calcite having the aspect of a local cement that constitutes a calcite-filled gap in the framework of quartz grains.

Another authigenic mineral in many marine quartz arenites is glauconite, which occurs as clastic green pellets sorted with the quartz grains (Figure 13-14B). Ferric oxides, also, as coatings on quartz grains, are common in quartz arenites laid down in environments where oxidizing conditions prevailed.

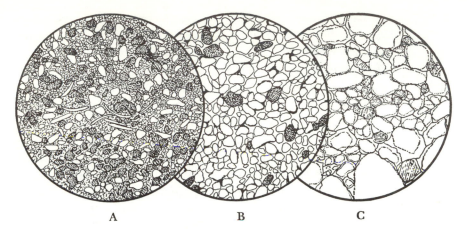

A   B   C

Figure 13–14. Quartz Arenites Containing Impurities

A. Aux Vases Sandstone (Mississippian), southern Indiana. Diam. 2.5 mm. A well-sorted mixture of subangular and subrounded quartz grains and pellets of microcrystalline calcite (dark stippling), together with a few elongate curved shell fragments (center). Firmly cemented by very fine-grained calcite. This specimen is a sandy clastic limestone, a typical associate of quartz arenites.

B. Flathead Quartzite (Cambrian), near Jackson, Wyoming. Diam. 2.5 mm. Well-sorted subrounded quartz grains and glauconite pellets (stippled) of about the same size; patches of ferric oxide (black). The rock is firmly cemented with authigenic quartz outgrowths on every detrital grain, like those in Figures 13–14C and 13–15C; here the authigenic quartz has been omitted to make clear the open texture and the shapes of the detrital grains.

C. Potsdam Sandstone (Cambrian), Lake Champlain, New York. Diam. 2 mm. This specimen is a feldspathic arenite consisting of 10–15% orthoclase and microcline and 85–90% quartz. Note that feldspar (lined) is found only among the smaller grains; coarser layers of the same deposit contain very little. Grains subrounded but not so well sorted as in B. Both quartz and feldspar grains have been enlarged by clear authigenic outgrowths (quartz on quartz and feldspar on feldspar) constituting a firm cement. Porosity is negligible, although the detrital grains are loosely packed.

Some quartz arenites have been consolidated solely by cementation, so that their detrital grains are tightly locked together by authigenic minerals. Most commonly the cement is secondary quartz or carbonates or both together (Figures 13-14A,C and 13-15A,C). Consolidation may also be accomplished by tight packing of the detrital grains under pressure, usually coupled with relatively minor cementation (Figure 13-15B). Not uncommonly, however, even some of the older quartz arenites remain

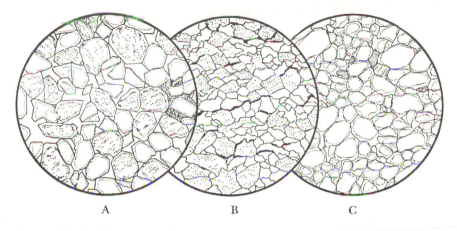

A   B   C

Figure 13–15. Quartz Arenites

A. Woodbine Sandstone (Cretaceous), 2000 m below surface, Houston County, Texas. Diam. 2 mm. Well-sorted subrounded-to-rounded, loosely packed, detrital quartz grains (outlined by dotted lines). Clear euhedral overgrowths of authigenic quartz partly cement the rock; a patch of calcite cement (right) covers the quartz overgrowths and probably formed later. This is a petroleum reservoir rock in which pores are sharply angular, being bounded by crystal faces of quartz.

B. Another bed in the same drill core as A. Tightly interlocking mosaic of detrital quartz grains. No authigenic cement, but a concentration of clay along some grain boundaries. Grains tend to be elongated parallel to bedding plane. Compare texture with that in A. Consolidation probably produced by pressure-solution of grains—that is, by deformation of grains under load, coupled with solution along grain boundaries at points of maximum strain.

C. Coconino Sandstone (Permian), Grand Canyon, Arizona. Diam. 2 mm. Well-sorted subrounded and rounded quartz grains, with sporadic microcline (upper center), tightly cemented by authigenic quartz overgrowths. Dark stippling represents remaining voids.

loosely consolidated, as is the case in parts of some of the Jurassic sandstones of the Colorado Plateau and in parts of the Ordovician St. Peter Sandstone of the northern Mississippi Valley (Figure 11-5A).

Cementation by quartz invariably takes the form of secondary overgrowths on the detrital quartz grains, from whose surfaces the authigenic quartz grows outward in crystallographic continuity with that of the grains (p. 339). Many secondary quartz overgrowths develop until they completely fill the original pore spaces in the sand; the quartz crystals are thus tightly locked together in aggregates that have been termed *orthoquartzite*. Very commonly, however, the quartz overgrowths are not so large and do not completely fill the pores of the sand. They are then terminated, along the sides of the cavities, by crystal faces of quartz, and the pore spaces bounded by them are then smooth-sided and sharply angular (Figure 13-15A,C). Where the cement is carbonate the pore spaces of the sand are usually filled completely with a crystalline granular mortar of calcite or dolomite or ankerite.

GENERAL REFERENCE

See F. J. Pettijohn, P. E. Potter, and R. Siever, *Sand and Sandstone*. New York: Springer-Verlag, 1972.

REFERENCE NOTES

[1] See, for example, R. L. Folk, *Journal of Sedimentary Petrology*, vol. 21 (1951): pp. 127–130.

[2] M. D. Wilson, S. S. Sedeora, *Journal of Sedimentary Petrology*, vol. 49 (1979): pp. 637–638; R. V. Laniz, R. E. Stevens, and M. B. Norman, *U.S. Geological Survey Professional Paper*, no. 501-B (1964): pp. B152–B153; E. H. Bailey and R. E. Stevens, *American Mineralogist*, vol. 45 (1960): pp. 1020–1027.

[3] C. M. Gilbert and F. J. Turner, *American Journal of Science*, vol. 247 (1949): pp. 1–26.

[4] The character and interpretation of lithic grains and of detrital and diagenetic elements in graywacke matrix is discussed by W. R. Dickinson, *Journal of Sedimentary Petrology*, vol. 40 (1979): pp. 501–516.

[5] A good model for the interpretation of provenance of lithic sandstones is offered by W. R. Dickinson, K. P. Helmold, and J. A. Stein, *Journal of Sedimentary Petrology*, vol. 49 (1979): pp. 501–516.

[6] P. D. Krynine, *American Journal of Science*, 5th ser., vol. 34 (1937): pp. 422–446.

14

Limestones and Dolomites

MINERAL COMPOSITION

The predominant minerals in limestones and dolomites are aragonite (orthorhombic $CaCO_3$), calcite (rhombohedral $CaCO_3$), and dolomite [rhombohedral $CaMg(CO_3)_2$]. Aragonite is invariably almost pure calcium carbonate. Calcite tends to be less pure and commonly contains significant amounts of Mg^{2+} or Fe^{2+} substituting for Ca^{2+}. In the dolomite crystal structure, Fe^{2+} substitutes readily for Mg^{2+}, producing an isomorphous series that includes crystals in which Fe^{2+} equals or exceeds Mg^{2+}; the latter are called ankerite. Most sedimentary dolomite is slightly ferrous. Magnesite ($MgCO_3$) and siderite ($FeCO_3$) also occur in sedimentary rocks, but in limestones and dolomites (dolostones), they are minor constituents, if they are present at all.

Aragonite and magnesian calcite (commonly containing up to about 20 mole % Mg) are abundant primary minerals in recent marine carbonate deposits and young limestones. Both are metastable, tending to alter to stable low-magnesian calcite, which is the dominant mineral in ancient limestones. This alteration is one of the principal diagenetic changes to occur in carbonate rocks. Another is the crystallization of dolomite, which occurs principally as a replacement of aragonite or calcite. Dolomite occurs only rarely as a strictly primary precipitate in the carbonate deposits forming today, but in some modern deposits it formed shortly after the initial sediment was deposited.

Aragonite, calcite, and dolomite are usually difficult to distinguish in thin section because their optical properties are much alike and twinning (which in metamorphic rocks provides a ready diagnostic criterion) is typically absent in sedimentary carbonate grains. Simple chemical tests may therefore be necessary to identify them.* In exceedingly fine-grained carbonate rocks, the x-ray diffractometer and scanning electron microscope (SEM) are essential to determine mineral composition and textural details, and widespread use of these instruments in recent years has provided a much improved understanding of carbonate sedimentology, particularly for the finer-grained deposits.

Many minerals other than carbonates occur in limestones and dolomites. Some are simply epiclastic or pyroclastic grains washed or blown into the depositional basin and mechanically mixed with carbonate material (Figure 14–1A,B). There is almost no limit to the variety of rock particles and minerals that may be included in this manner, but, as might be expected, quartz and clay are the most common. Other noncarbonate constituents are organic remains, such as the opal of diatom and radiolarian tests and sponge spicules; the collophane of bones, teeth, and some brachiopod shells; and the organic pigment that darkens many limestones. Authigenic minerals may also be present. These may be formed either almost contemporaneously with the calcareous deposits or later, during and after lithification. Among the commonest of these are chalcedony, quartz, glauconite, pyrite, gypsum, anhydrite, and alkali feldspars (Figures 14–1C; 14–12 B,C; and 14–13).

Rocks consisting largely of calcite or aragonite are termed *limestone*; those mainly composed of dolomite are called *dolomite* or *dolostone;* and mixtures of calcite and dolomite are referred to as *dolomitic limestone* or *calc-dolomite*. If accessory constituents are abundant, the rock name is modified adjectively to suit the particular composition. *Glauconitic, cherty,*

---

*The most common chemical tests involve etching and staining with various solutions. The stains most useful for distinguishing aragonite, calcite, and dolomite are described by J. Rodgers, *American Journal of Science* 238 (1940): 788–798 and also by G. M. Friedman, *Journal of Sedimentary Petrology* 29 (1959): 87–97.

Because calcite is more soluble than other rhombohedral carbonates, a surface of limestone is etched when treated with weak acid, leaving dolomite and other less soluble minerals standing in relief as the calcite dissolves. Carefully etched, smooth surfaces can be reproduced on acetate sheets, and the reproductions, called peels, record textural details often more clearly than do thin sections. Methods of etching and preparing peels are described by A. W. McCrone, *Journal of Sedimentary Petrology*, vol. 33 (1963): pp. 228–230 and also by P. J. Davies and R. Till, *Journal of Sedimentary Petrology*, vol. 38 (1968): pp. 234–237.

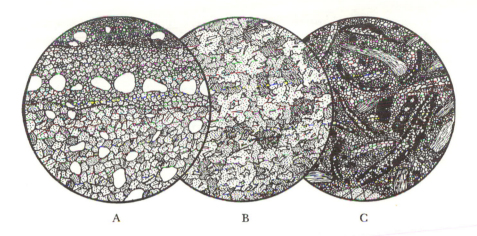

A	B	C

Figure 14–1. Impure Limestone and Dolomite

A. Sandy dolomite, Jefferson Formation (Devonian), Bear River Range, northern Utah. Diam. 2.5 mm. Subrounded detrital quartz grains in fine-grained, laminated dolomite mosaic. Laminae distinguished by abrupt change in grain size of dolomite and by difference in size and distribution of quartz grains. Rounded quartz grains "float" in dolomite; the rock was probably a sandy limestone in which original calcite has been replaced by dolomite.

B. Tuffaceous limestone (Upper Eocene, Markley Formation), Mount Diablo, California. Diam. 1 mm. Sharply angular clear shards of glass "floating" in a fine-grained calcite mosaic. Note the irregular boundaries of calcite grains and the very even distribution of glass particles. Shards rarely cross boundaries between calcite crystals. Texture of calcite mosaic indicates it is neomorphic (recrystallized).

C. Pyritic limestone (Lias), Württemberg, Germany. Diam. 2.5 mm. Limestone containing shell fragments cemented with microcrystalline calcite. The shells that consist of granular calcite have been recrystallized; fibrous ones are probably original calcite; coarse calcite grain at bottom is probably an echinoid fragment. Authigenic pyrite (black) replaces parts of the recrystallized shells and forms microconcretions throughout the calcite matrix.

sandy and *argillaceous* limestones are particularly common. Similar terms are applicable to some dolomitic limestones and dolomites. As the content of noncarbonate materials increases, calcareous rocks grade into other rock types; thus sandy limestone grades into calcareous sandstone, and argillaceous limestone into marl and calcareous shale.

This simple classification by mineral composition belies the petrographic complexity of limestones and dolomites. The two rocks, although composed essentially of carbonates, are produced by any one or more of several modes of deposition—mechanical, chemical, organic, and meta-somatic. Their textures, therefore, are exceedingly diverse. Some have clastic textures, as varied as those of detrital silicate rocks. Other textures result from processes of organic secretion and are but slightly less varied than the organisms producing them. Textures, also, that are formed by diagenetic recrystallization and replacement differ according to the size, uniformity, form, and orientation of the component crystals. Further complication arises from the fact that few carbonate rocks result from a single mode of deposition, and textures of multiple origin may be seen in a single thin section.

GENETIC COMPONENTS AND CLASSIFICATION OF LIMESTONES

A salient fact about limestones is that most of them are particulate and bimodal; that is, they consist initially of relatively large carbonate particles mixed in various proportions with carbonate mud. Inorganic precipitation is responsible for some of the carbonate mud that is deposited in lakes and seas, and it also produces larger grains by chemical accretion around nucleus particles and by cementation of clusters of small particles. However, most of the carbonate sediment in modern marine environments, and also in most ancient limestones, has been produced by organisms. A great variety of organisms secrete calcium carbonate to form their skeletons or shells, some of which are large and others minute. This skeletal material may accumulate either intact or more or less fragmented. Some of the larger organic structures are weak or poorly integrated, and they disintegrate readily into smaller carbonate particles. Skeletons of crinoids, for example, break down into sand-size fragments, whereas minute aragonite crystals secreted by some algae accumulate as aragonite mud when the organism dies. Some organisms bore into and weaken large shells and other skeletal structures, which then tend to be more readily disintegrated. Predators and scavengers, which abound in the marine environment, break up much skeletal material; some of these creatures crush shells and break bits from reef structures, while others, in their search for edible organic matter, ingest the sediments accumulated on the bottom and excrete the indigestible carbonate material. The combined product of all these processes is particulate carbonate sediment, including much clay-size material and a variety of larger fragments, and all of it originates within the depositional basin.

Much less important is terrigenous carbonate detritus derived by subaerial erosion from older formations outside the basin of deposition. In fact, terrigenous carbonate is significant only where erosion has been rapid and the detritus has not been transported far, especially where a relatively arid climate has impeded solution of the carbonate detritus and favored its preservation.

Some limestones consist essentially of unbroken fossil material that has not been moved from the habitat in which the organisms lived; indeed, in some the fossils remain in the position of growth. Long ago, A. W. Grabau classified such rocks as *organic* or *biogenic limestones* and distinguished them from *clastic limestones* composed of transported calcareous particles that have been more or less abraded and sorted by currents.[1] Clastic limestones were further subdivided according to average grain size into *calcilutite* (carbonate mudstone), *calcarenite* (carbonate sandstone), and *calcirudite* (carbonate conglomerate or breccia). In principle, biogenic limestones stand in contrast to clastic limestones, but in practice the distinction is less clear, for most clastic limestones consist largely of organic material. To be considered clastic, how much, or how little, must skeletal material be broken or disintegrated, and how much, or how little, must it be moved from its place of origin? Classification by such genetic criteria is likely to be variable and inconsistent. Nevertheless, Grabau's concept is valid, and his terms are useful.

The descriptive classifications of limestones in use today are based primarily on depositional texture as determined by the relative proportions of carbonate mud (microcrystalline carbonate) and larger particles, called grains or allochems. The specific character of the larger particles provides the basis for further subdivision. Components to be recognized and distinguished in limestones are the following.

1. *Grains* or *Allochems* are discrete particles of calcium carbonate the size of coarse silt or larger; they include the following principal types.
   a. *Skeletal grains* (fossils) are fragments of the hard parts of any calcareous organisms and also the unbroken shells of such individual organisms as molluscs, echinoids, ostracods, and foraminifers.
   b. *Ooids* are more or less spheroidal, sand-size particles consisting of an aragonite or calcite cortex formed by chemical accretion around a nuclear particle. The cortex includes a number of concentric layers and commonly has radial fibrous texture.
   c. *Pellets* are spheroidal or ellipsoidal grains, usually of sand size, that consist of microcrystalline carbonate throughout. They have no

apparent internal structure, and this distinguishes them from ooids. No specific origin is implied by the term pellet, but most of them are tiny mudballs produced by bottom-dwelling, deposit-feeding organisms that ingest the bottom sediment, digest the contained organic matter, and excrete the undigested carbonate sediment as more or less coherent fecal pellets.

   d. *Intraclasts* are fragments of carbonate rock that, as the name signifies, originated within the basin of deposition and have been moved somewhat before being redeposited. They are produced by rupture of coherent sediment in the depositional environment and are to be clearly distinguished from terrigenous fragments eroded from much older carbonate formations outside the basin of deposition. Intraclasts may be as small as sand grains but typically are pebble size or larger. They have environmental significance, because they usually are formed either in areas of lowered wave base where waves and tides accompanying severe storms erode the bottom deposits to depths of a few tens of centimeters or on tidal flats where muddy deposits are ruptured by periodic drying and cracking of surface layers.

2. *Microcrystalline calcite* refers to microgranular calcite aggregates. In firmly lithified limestones, such aggregates consist of interlocking anhedral calcite crystals typically less than 20 mm in diameter, and they are generally considered to represent original carbonate mud. Presumably the carbonate material was deposited initially as an impalpable mud composed of minute crystals of aragonite or calcite, or both, which by cementation and partial recrystallization was converted to microcrystalline calcite—usually somewhat coarser grained than the original mud—in the lithified rock.

3. *Sparry cement* or *Spar* is relatively clear, granular calcite that has crystallized within void spaces in a carbonate deposit or a limestone, particularly in the void spaces between grains and in the body cavities of fossils.

Limestones in which grains (allochems) predominate over mud have *grain-supported texture*; that is, the grains are in contact and the grain-against-grain contacts provide the structural support for the mass. Mud, if it is present in these rocks, occurs in interstices between grains. Other limestones have *mud-supported textures*, in which mud provides the structural support and grains (allochems), if present, are surrounded by mud and are suspended in it. The change from one to the other of these two textural types does not correspond to a fixed proportion of grains and mud, because of wide variations in the shape and packing of carbonate grains, but the usual proportion approximates equal volumes of the two components. The limestone classification suggested by R. J. Dunham is based primarily on this dual subdivision according to depositional texture.[2]

A more elaborate scheme of classification, suggested by R. L. Folk and illustrated in Figure 14–2, subdivides limestones according to the occurrence of microcrystalline calcite in the rock and further according to the kinds of allochems (grains) included in it.[3] Rocks wholly composed of microcrystalline calcite are called *micrite*; those containing allochems in a micrite matrix are various allochemical micrites and are subdivided

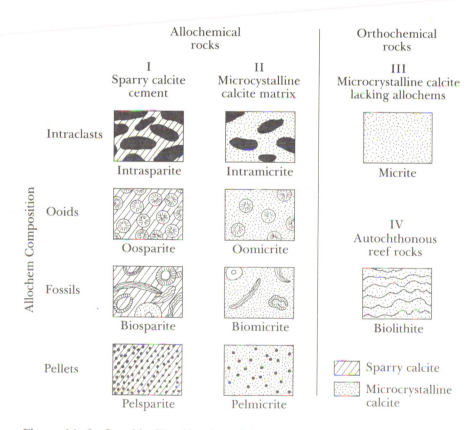

Figure 14–2. Graphic Classification of Limestones

(After R. L. Folk, *Bulletin of the American Association of Petroleum Geologists*, vol. 43 1959: p. 18.)

according to type of allochem. Limestone consisting of allochems alone, which are lithified by sparry cement, are called *sparite*, and varieties of sparite are distinguished according to type of allochem.

Figure 14–3 outlines the classifications of Dunham, Folk, and Grabau and indicates the general correlation of terms.

TEXTURE AND COMPOSITION OF SKELETAL MATERIAL

Fossils are recognized and identified generally by the form of their shells or skeletons. However, the internal fabric and mineral composition of skeletal material are also distinctive, and they may be the only clues to the organic origin of skeletal grains that have been broken into unidentifiable shapes or have been very much abraded. Furthermore, the mineral composition and texture of skeletal material have a considerable bearing on a fossil's preservation, on the way in which it disintegrates into smaller particles, and on the texture of the cement that binds it together.

Aragonite and calcite are the only carbonate minerals secreted by organisms. Skeletal material composed of aragonite is invariably almost pure calcium carbonate, but the calcite secreted by organisms contains varying amounts of magnesium. Some organisms secrete high-magnesian calcite, notably echinoids, red algae, and some corals and benthonic foraminifers.

Most skeletal carbonate consists of either calcite or aragonite alone, although some shells contain both minerals. The carbonate crystals tend to be either prismatic or finely fibrous and to be oriented either at a high angle to the shell surfaces or nearly parallel to them. The calcareous brachiopods, for example, produce shells consisting wholly of finely fibrous calcite so oriented that the fibers are parallel or nearly parallel to the shell surface.[*] The complexly layered shells of oysters also consist entirely of finely fibrous calcite; but Figure 14–4A illustrates the complex structure of a gastropod shell composed wholly of fibrous aragonite. Coral, too, whether composed of aragonite or calcite, is characterized by a fabric made up of fibrous or prismatic crystals.

---

[*]Other brachiopods secrete chitin or calcium phosphate.

A (After R. J. Dunham)

Original components were discrete loose particles not bound together during deposition				Original components bound together during deposition	Depositional texture not evident
Mud-supported texture		Grain-supported texture			
Less than 10% grains	More than 10% grains	Mud present between grains	Lacks mud; clean grains	Boundstone	Crystalline granular limestone
Mudstone	*Wackestone*	*Packstone*	*Grainstone*		

B (After R. L. Folk)

Microcrystalline limestone	Allochemical Limestones			
	Allochemical micrite		Sparite	
	Sparse	Packed		Biolithite (autochthonous reef rocks)
Micrite	*Sparse* biomicrite pelmicrite intramicrite oomicrite	*Packed* biomicrite pelmicrite intramicrite oomicrite	*Biosparite* *Oosparite* *Pelsparite* *Intrasparite*	

C (After A. W. Grabau)

Biogenic (Organic) Limestone →

Shell oozes	Shell beds	Reefs

Clastic Limestone →

Calcilutite	*Calcarenite* *Calcirudite*	

Terrigenous analogues	Mudstone and shale	Sandy or silty mudstone and shale	Argillaceous sandstone or wacke	Clean sandstone or arenite

Figure 14–3. Classification and Nomenclature of Limestones
(A, After R. J. Dunham; B, after R. L. Folk; C, after A. W. Grabau.)

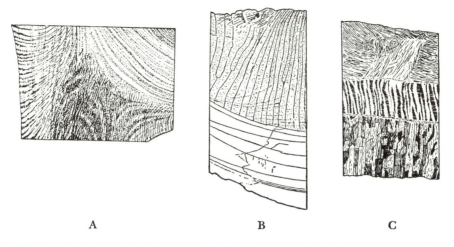

A   B   C

Figure 14–4. Organic Carbonate Fabrics (Drawn by R. S. Creely.)

A. Segment about 2 mm wide of the columella of Eocene gastropod *Potamides*, consisting of fibrous aragonite (lines and short dashes) traversed by curving growth lines (stippled). Aragonite fibers commonly cross growth lines and intersect shell edge (lower left and right) at a high angle.

B. Segment of *Inoceramus* shell, consisting of two distinct layers. Upper layer is about 1 mm thick and consists of prismatic calcite; lower layer is lamellar aragonite. The segment illustrated is near the hinge area, which accounts for irregularities in the upper part of the prismatic layer.

C. Segment of *Pecten* shell, consisting of three distinct layers. Upper layer is about 0.75 mm thick. It is the outer layer of the shell, and consists of fibrous calcite having irregular orientation; middle layer is aragonite having a somewhat irregular and hazy columnar structure; lower (and inner) layer consists of stumpy aragonite prisms standing normal to the shell surface.

In shells consisting of both aragonite and calcite, such as are secreted by many marine molluscs, the two minerals are not intimately mixed. Instead, they occur separately in different layers of the same shell. Generally the outer layer is composed of prismatic or fibrous calcite, while the inner layers, which are often pearly, are formed of fibrous or lamellar aragonite (Figure 14–4B,C). Perhaps the best-known example is the shell of the mollusc *Inoceramus*, which is made up of a thick layer of cross-oriented calcite prisms and an adjacent nacreous layer of aragonite (Figure 14–4B). Bits of the calcite layer commonly break from lifeless *Inoceramus* shells and are included in sediments as thin plates with distinctive cross-oriented prismatic structure. These have been called *Inoceramus prisms*.

A notably different fabric is found in the echinoderms, the skeletal structures of which invariably consist of coarse calcite crystals. Each calyx plate or stem segment of a crinoid is a single calcite crystal. The same is true of each plate of a sea urchin or other echinoid. Each plate is perforated, to be sure, and has an organic pattern or shape, but the calcite has the same crystallographic orientation throughout. The spine of a modern sea urchin, for example, has a very intricate pattern radial about the spine axis, but the entire structure is a single calcite crystal so oriented that its $z(c)$ axis coincides with the spine axis, and in each plate of the same organism the calcite is oriented with its $z(c)$ axis normal to the surface of the plate. The pits and pores of these organic structures, when entombed in sediments, are filled with secondary calcite that grows in continuity with the original calcite, producing large and apparently structureless crystals and at the same time cementing them into coarse-grained aggregates. It is for this reason that most crinoidal limestones have a coarse texture. Large sparkling cleavage surfaces may be evident even in hand specimens, their derivation from crinoidal remains being indicated by the circular outline of many grains and occasionally by some remnant of internal structure within the crystals.

The composition and structure of organic material have a controlling influence on the state of its preservation in sedimentary rocks. Fossils consisting of aragonite are, as a rule, poorly preserved. Aragonite shells, or shells containing aragonite layers, are weaker and more easily broken than those composed wholly of calcite, and they may be pulverized into aragonite mud in places where calcitic shells tend to form larger fragments. And, because aragonite is unstable at pressures below 4 kb and hence more soluble than its stable polymorph calcite, fossils composed of it are commonly represented simply by molds. Calcitic secretions of coarse grain are least soluble, as shown by the good preservation of many echinoids in some rocks in which other fossils are scarcely determinable.

Given sufficient time in the presence of aqueous pore fluids, aragonite inverts to the stable polymorph calcite. This inversion involves recrystallization, which obliterates the original internal structure of aragonitic fossils and replaces it with a granular calcite mosaic; only the external form of the fossils then remains to indicate their organic origin. Generally similar calcite mosaics may be formed when calcite precipitated from pore fluid fills the fossil molds formed by prior solution of aragonite. Calcitic organic remains in which detailed internal structure is distinct probably have not been altered and represent skeletal material composed originally of calcite, but even these have sometimes been recrystallized, especially if the original calcite was magnesian. The greater susceptibility of aragonite to solution and to recrystallization has been well documented, however, and most shells composed of granular calcite were aragonite originally. In some thin sections, fossils in which the original calcite fabric is preserved occur together with others that were aragonite originally and are now composed of granular calcite mosaics (Figure 14–9A).

DIAGENETIC TEXTURES IN LIMESTONES

The earliest diagenetic effects in carbonate deposits occur in marine environments while the sediment remains exposed to the marine waters in which it formed and also to the organisms inhabiting the uppermost parts of the deposit. Studies of modern carbonate tracts have shown that deposits may become at least partially cemented on the sea floor by precipitation of fine-grained aragonite or magnesian calcite in void spaces. This appears to occur most effectively where normal deposition has either been interrupted or is extremely slow. Also, on the sea floor the original fabrics of skeletal grains and of ooids may be largely obliterated and replaced by microcrystalline carbonate through the action of minute boring algae and fungi, which are abundant in many shallow marine environments. This process, called micritization, proceeds inward from grain surfaces. Where entire grains have been micritized, they take on the characteristics of pellets and undoubtedly have been identified as such in some limestones. Partial micritization, however, results in thin rims, called *micrite envelopes*, which surround a grain and truncate its primary internal fabric. Micrite envelopes in ancient limestones appear in thin section as dark, microcrystalline rims around grains (Figure 14–9A); thin chemical incrustations on grain surfaces may have a similar general appearance.

Later diagenetic changes take place under physicochemical conditions that differ from those of the original deposits, both during burial to increasing depths beneath younger deposits and also after uplift and exposure to circulating groundwaters or vadose solutions. An increase in sparry calcite and an increase in grain size generally occurs. Initially fine-grained, relatively soluble minerals such as calcite inevitably tend to coarsen with time, thereby reducing total surface energy (and hence free energy) of the aggregates. Sparry calcite may be formed both by *cemen-*

tation and by a general process called *neomorphism*, whereby a fine-grained carbonate aggregate is converted to a coarser-grained calcite mosaic. Neomorphism includes both polymorphic inversion of aragonite to calcite and recrystallization of calcite; cementation refers to precipitation of calcite from pore fluids in void spaces (cavities) in the aggregate. Cavities thus cemented include interstices between grains, body cavities in fossil shells, openings and perforations in skeletal material, and also fractures and solution cavities that commonly develop after a deposit has become somewhat lithified.

The crystals in cavity walls serve as nuclei for those that form the cavity-filling cement. On large calcite crystals, such as echinoid grains, the cement typically occurs as overgrowths of clear calcite in structural continuity with the calcite of the grains.* In contrast, on finely crystalline cavity walls, cement crystals grow at many different centers and develop fine-grained cement mosaics instead of large overgrowths. In general, as a cement mosaic develops, the more favorably oriented crystals survive and bury those that are less favorable with the result that crystal size tends to increase toward the center of the cavity filling.

Perhaps the most clear-cut example of neomorphism is alteration of original fibrous aragonite in fossil shells to mosaics of sparry calcite. Yet generally similar calcite mosaics can be formed by precipitation of calcite cement in open molds formed by the solution of aragonite fossils (Figure 14–6C). Likewise, sparry cement in a grainstone, or sparite, may resemble neomorphic spar, which replaces original mud matrix in a packstone, or packed micrite. To distinguish cement spar from neomorphic spar requires close attention to textural details, but distinctions are often uncertain.

Increase of crystal size inward from the walls toward the centers of original cavities is a common characteristic of cement mosaics, whereas in those formed by recrystallization, crystal size tends to vary irregularly. Also, cement mosaics are typically better defined and often are outlined by comblike marginal fringes of smaller crystals crudely elongated normal to the original cavity walls. Neomorphic spar is likely to include small relics inherited from the primary fabric, and the arrangement of these

---

*The calcite overgrowths are strictly analogous to the authigenic quartz overgrowths that have formed on quartz grains in many sandstones. Some echinoid grains that are enclosed by micrite matrix have similar, but usually smaller, calcite overgrowths that appear to replace the micrite along an irregular boundary.

inclusions sometimes reveals vague outlines of the original structure; cement that fills void spaces is characteristically clear and limpid. Furthermore, within cement mosaics, the boundaries of individual crystals tend to be smooth and are commonly planar; in recrystallized mosaics, crystal boundaries are typically irregular and planar boundaries are rare.*

Large void spaces in carbonate rocks, such as body cavities in megafossils and the larger openings in reef structures, have sometimes been filled during several sequential episodes. A common example of this consists of a lower layer of mud (micrite) that infiltrated and settled on the floor of the original cavity and an overlying layer of sparry calcite cement precipitated later in the unfilled upper portion of the cavity. Figure 14–5B is typical. Structures of this kind, which serve to distinguish stratigraphic top and bottom, are often called *geopetal* structures.

AUTOCHTHONOUS BIOGENIC LIMESTONES

The rocks called boundstone and biolithite by Dunham and Folk, respectively, are good examples of biogenic limestone. One tends to think first of those constructed fundamentally by sedentary colonial organisms like coral, coralline algae, and bryozoa living and growing in reefs on shallow marine banks. The calcareous skeletons of these reef-forming organisms are relatively large and firmly anchored to the bottom, and they provide not only a wave-resistant framework upon which the reef can be built but also an organic binding for the reef mass. Every living reef teems with many kinds of marine life, and fossil reefs contain the remains of many different kinds of organisms. Molluscs and algae, for example, are more abundant in some modern "coral reefs" than coral itself; some Paleozoic reefs, likewise, consist largely of bryozoa, brachiopods, and crinoids. Furthermore, every reef traps fragmented skeletal material of all sizes, and this accumulates in pockets and cavities between and within the larger organic structures. All this material becomes cemented together, but the skeletal structure of the reef-forming organisms is the integrating element. Thus it is appropriate to speak of coral or algal

---

*In monomineralic crystal mosaics, planar crystal boundaries usually are irrational planes rather than rational crystal faces. In such mosaics, euhedral and subhedral crystals are to be inferred only where typical crystal form and interfacial angles can be recognized. The distinction is readily made by simple universal-stage measurements.

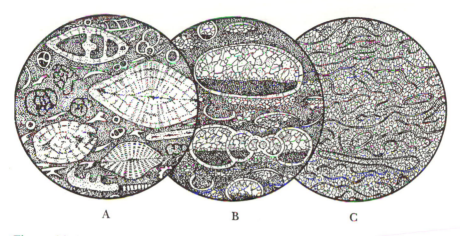

Figure 14–5. Allochemical Limestones

A. Foraminiferal biomicrite (Eocene), Italy. Diam. 3 mm. Abundant foraminifers in a matrix of microcrystalline calcite (stippled). Orbitoids predominate, but a variety of other forms is included.

B. Gastropod biomicrite (Miocene), Ulm, Germany. Diam. 3 mm. Fresh-water limestone containing abundant whole and broken *Planorbis* shells. Matrix is turbid microcrystalline calcite (dark stippling) containing patches of clear coarser calcite. Larger shells were partly filled with carbonate mud at the time of deposition. Voids remaining within shells, and also cavities under shell fragments, were later filled with coarser spar as a result of authigenic precipitation. The filling within several shells is an example of *geopetal structure*; contact between microcrystalline calcite and sparry calcite within shells is the bedding surface and is shown right side up.

C. Trilobite sparite (Silurian), Asker, Norway. Diam. 3 mm. Very abundant carapaces of the trilobite *Olenus* enclosed in sparry calcite cement in which crudely columnar crystals stand approximately normal to the shell surfaces.

biolithite, or boundstone, even though other organisms and fragmented material may predominate. These names are proper, however, only if the reef-forming skeletal structure is intact in growth position; the names should not be used for broken chunks of reef rock, however large, that occur as fragments in many calcirudites.

Laminated boundstones, or biolithites, produced by filamentous algae are called *stromatolites*. They retain no recognizable organic skeleton and are characterized instead by millimeter-thin carbonate laminae, some of which clearly formed at angles that would be impossible under the influence of gravity alone. Algal stromatolites have a variety of configurations, and they have been classified by the geometry of the laminated structure, which is probably related to the particular conditions under which they were formed. Some are crinkled, matlike layers or lenses of laminated micrite; others consist of cabbagelike heads or stacks of laminated micrite more or less linked together; some are discrete laminated spheroids, called *oncolites* or *algal pisolites*. These deposits are presumed to be analogous to modern deposits formed by filamentous green and blue-green algae, which produce sticky organic mats that, as they continue to grow, trap successive films of fine sediment and thus produce a succession of very thin laminae. At the time of deposition, the sediment is bound by the algal filaments, but when the organic material has decomposed, only laminated sediment remains. Algal deposits of this type form today in a wide range of environments, and are particularly characteristic on tidal flats and on very shallow bottoms near shore.

Beds of unbroken benthonic fossils essentially in growth position are exemplified by layers largely composed of the articulated shells of brachiopods or pelecypods or of the unbroken skeletons of echinoids. These are autochthonous biogenic limestones in the sense of Grabau, but they are not boundstone or biolithite and are included among the allochemical limestones.

MICROCRYSTALLINE LIMESTONES

Most exceedingly fine-grained limestones originated as carbonate mud or ooze and have since been firmly lithified. In hand specimen, they are dull and opaque; in thin section, they are subtranslucent and dark, usually brownish. Limestones of this kind are called *micrite, carbonate mudstone*, or *calcilutite*; those of ultrafine grain and very uniform texture, because of their use in the technology of printmaking, are called *lithographic limestone*.

Carbonate muds consist of particles that originate both by inorganic and organic precipitation and also by physical disintegration and abrasion of skeletal material. In the marine environment, they appear to be largely of biogenic origin. Regardless of how they were produced, tiny carbonate particles have been produced in great abundance in warm seas and in many lakes, and they have accumulated on the bottom wherever turbulence was insufficient to put them in suspension and currents were too weak to carry them away. Areas of this kind include much of the open sea floor as well as low-energy areas in shallow shelf seas, but there are compositional differences in the muds that accumulate in these two environments. The widespread foraminiferal and coccolith oozes deposited in the open sea consist largely of low-magnesian calcite, whereas the muds deposited on shallow shelves and on tidal flats are largely aragonite and magnesian calcite. Ancient micrites presumably originated in similar general environments and many were composed of similar constituents, although there surely were differences in the specific organisms involved and probably in the contribution made by inorganic chemical precipitation.

Micrites made of the remains of calcareous plankton have been deposited in extensive parts of the open sea since mid-Mesozoic time. During recent years, they have been encountered repeatedly in deep-sea drill cores, and similar limestones have been recognized in the deformed sequences exposed in the Alps and Apennines. The most widely recognized planktonic limestone is *chalk*, which accumulated not on the deep-sea floor but at lesser depths in marginal parts of the open ocean. The most widespread chalks are Cretaceous in age. Studies using the electron microscope show that chalk is essentially coccolith ooze composed of the skeletal debris of coccoliths together with scattered foraminifers and bits of molluscs and a few other organisms. Coccoliths and most foraminifers secrete low-magnesian calcite, the stable form of calcium carbonate, and mud deposits composed of their skeletal remains are certainly less susceptible to diagenetic recrystallization than muds containing more aragonite and magnesian calcite. The fact that chalk is so typically friable and soft may perhaps be attributed to this.

Local concentrations of impurities and differences in grain size in micrites are often discernible as subtle variations within a single thin section or hand specimen. Portions of a rock in which the material is exceedingly fine grained or impure appear darker and duller than those portions in which it is somewhat coarser or purer. The variations may represent differences in the texture of neomorphic microspar, which are to be expected in ancient micrites. Or they may represent variations that existed in the original deposit before it was lithified. For example, the vague relics of fecal pellets are likely to appear in thin section as indistinct, slightly darker patches. Churning of a mud bottom by burrowing and scavenging organisms (bioturbation) and soft-sediment deformation caused by slumping or settling of unstable deposits produce other more irregular structures that may be discernible in the lithified micrite. Tiny lenses and strings of clear calcite spar in the darker micrite matrix are referred to as *birds-eye* or *fenestral* structure. They are particularly noteworthy in some laminated micrites, probably of algal origin in a tidal flat environment. Occasionally, too, small openings later filled with sparry calcite may develop in muds during bioturbation or other soft-sediment deformation.

Veinlets of fine- to coarse-grained calcite represent fractures in lithi-fied rock that have been filled and healed by later cementation. They are characteristic of old limestones of all types, but they are particularly striking in micrites, because the coarser, lighter-colored vein calcite con-trasts strongly with the darker microcrystalline rock. The veinlets may be so numerous as to resemble small-mesh networks.

Detrital silicate impurities in micrite usually consist of clay and the finest silt, which are likely to find their way into the quiet waters where carbonate muds tend to accumulate. These argillaceous impurities may be intimately mixed with the carbonate mud or they may be concentrated in separate laminae (Figure 14–7C); by increase in the total clay content, micrite grades into *marl*, which is approximately half clay (Figure 11–1). Larger silicate sand grains are sparse in micrites, if they are present at all, and they probably are grains that have been blown or rafted into the accumulating carbonate mud deposit.

ALLOCHEMICAL LIMESTONES

A few allochems (grains) occur in most micrites. As the proportion of allochems increases, micrite grades into the various allochemical micrites, among which biomicrite, pelmicrite, and intramicrite are the predomi-nant types. A predominance of carbonate mud in allochemical micrites indicates deposition in areas of relatively low turbulence and weak cur-rents. In these rocks, allochems typically make up less than half of the rock and probably have been deposited approximately where they orig-inated with minimal transportation by currents.

Biomicrites consist of skeletal grains embedded in mud. Some skeletal grains have been transported and deposited in areas of mud accumula-tion during storms and other disturbances of loose bottom sediment, and some biomicrites that have grain-supported texture (packed biomicrites) have undoubtedly been infiltrated by mud after the allochems had accu-mulated. In general, however, the skeletal grains in biomicrites have been deposited with little or no transportation. Often they are unbroken or very slightly broken benthonic fossils and are the remains of organisms that inhabited the muddy bottoms where they became entombed (Figure 14–5A,B). In other biomicrites, the allochems are bits and pieces of skeletal material. Where angular and unsorted shell fragments are randomly oriented and distributed in the micrite matrix, the fragmentation of the shells and mixing with micrite probably has resulted from bioturbation (Figure 14–6A). Crinoidal biomicrites differ from shelly ones, because the crinoidal skeletons tend to break up into stubby, cylindrical stem segments and calyx fragments of more or less uniform size, each of which consists of a single large calcite crystal. Most of them have formed in situ on shallow sea floors where the sessile crinoids lived in profusion and provided an organic baffle to trap the mud in which their remains accumulated.

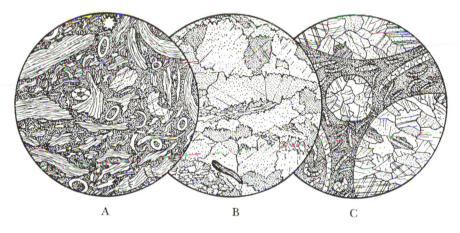

| A | B | C |

Figure 14–6. Allochemical Limestones

A. Biomicrite, Twin Creek Limestone (Jurassic), near Jackson, Wyoming. Diam. 2.7 mm. Poorly sorted, ragged organic fragments enclosed in a matrix of calcite mud (stippled). Most larger fragments are fibrous calcite and may be bits of brachiopod or of certain molluscan shells; two coarse calcite fragments are bits of echinoids. Ragged, disoriented character of the organic fragments suggests bioturbation.

B. Crinoidal limestone, Trenton Limestone (Ordovician), Trenton Falls, New York. Diam. 3 mm. Medium-grained limestone composed of tightly interlocking crinoid fragments. Pressure solution along grain boundaries has produced microstylolites between the grains. One phosphate shell fragment in lower part of diagram.

C. Cephalopod biomicrite (Silurian), Chuohle, Bohemia. Diam. 4 mm. Casts of the nautiloid cephalopod *Orthoceras* (circular cross-sections) composed of medium-grained sparry calcite are embedded in a matrix of microcrystalline calcite and small shell fragments. Absence of any trace of shell in the large casts suggests that the original shells were removed by solution and the resulting molds later filled with calcite spar.

Fossils and fossil fragments, when seen in thin section, provide a striking contrast to the surrounding micrite. Both their forms and their textures are distinctive, and they tend to be distinctly more transparent and therefore lighter colored than the micrite matrix. Body cavities in fossils may be infiltrated by mud at the time of deposition (Figure 14–5A) or they may remain empty, or partly empty, until later filled with sparry calcite cement (Figures 14–5B, 14–6C, and 14–7C). In hand specimens of crinoidal biomicrite, reflections from the cleavage in calcite crystals composing the individual allochems stand out in the dull micrite, and if the crinoidal grains are evenly distributed through the matrix, they give the general impression of porphyritic texture.

Allochemical limestones that have been deposited in higher-energy environments consist largely of allochems that have been transported and deposited by relatively strong currents. They have been washed more or less free of mud, and the allochems in many of them have been well sorted and rounded by abrasion. Silicate sand grains may be present, either mixed and sorted with the carbonate grains or concentrated in distinct beds or laminae. Current bedding is typical. The general term *calcarenite* is a fitting designation for these rocks; more specifically, most of them are biosparite, oosparite, or pelsparite.[4]

Different kinds of skeletal grains and different degrees of rounding and sorting of the grains are responsible for considerable variety among the rocks called biosparite. Two noteworthy examples are shelly biosparite and crinoidal biosparite, but there are many others. Crinoidal grains typically are more or less equidimensional and frequently round and each consists of a single calcite crystal. They are cemented by overgrowths of calcite producing crystalline granular rocks in which the original allochems are incorporated into the enlarged calcite crystals. In hand specimens, the allochems and cement may not be distinguishable, especially on freshly broken surfaces where the cleavage of the granular calcite is most evident.* In thin sections, however, the relatively clear cement overgrowths can be distinguished easily from the more turbid allochems, which commonly show some organic structure.

A different texture is to be seen in biosparites composed largely of shells or shell fragments, which typically have curved and platelike shapes and often include portions of the original body cavities. Because the shell material is finely crystalline, cement in these rocks consists of granular

---

*In the ornamental stone industry, these rocks are frequently classified as marble, because their crystalline granular texture superficially resembles that of the metamorphic rock.

crystalline mosaics rather than large overgrowths. Furthermore, the grain packing is distinctive, for shells and shell debris do not tend to pack closely, and the aggregates usually have more than 50% porosity. In thin slices through rocks of this kind, the platy allochems appear to be suspended in granular cement, and the volume of cement appears excessive (Figures 14–7A,B and 14–9A). If the allochems are mostly whole or

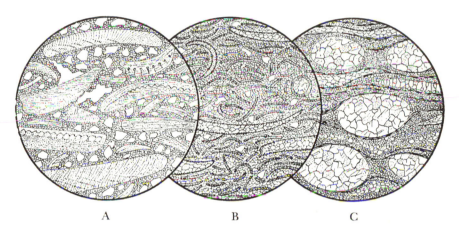

A   B   C

Figure 14–7. Biosparite and Fossiliferous Marl

A. Pelecypod biosparite, Anastasia Formation (Pleistocene), Titusville, Florida. Diam. 3 mm. This diagram illustrates the finer part of a coarse coquina composed of subangular molluscan shell fragments and detrital quartz grains; it is incompletely cemented by fine-grained calcite. Shells consist of original fibrous calcite and aragonite. Two unfilled pores in upper left. Note that the quartz grains are all about the same size but are much smaller than the shell fragments.

B. Pelecypod biosparite (Jurassic), Hannover, Germany. Diam. 3 mm. Fragments of small molluscan shells and scattered angular and subangular quartz grains, completely cemented by fine-grained calcite. Most smaller shell fragments consist of original microfibrous calcite (stippled); larger shells recrystallized to granular calcite were probably aragonite originally. Micrite envelopes occur on all shell fragments and were formed before cementation by sparry calcite.

C. Ostracod marl, Green River Formation (nonmarine, Eocene) near Kemmerer, Wyoming. Diam. 3 mm. Ostracod shells filled with medium-grained sparry calcite. Mud matrix consists of microcrystalline calcite and clay; numerous laminae composed largely of clay (line pattern) accentuate the bedding. Laminae have been bent by compaction around the ostracods, suggesting that the fragile shells may have been filled with calcite before compaction.

slightly broken shells, the original porosity and the volume of cement are likely to be even greater. The term *coquina* refers to more or less cemented aggregates of coarse shell debris; *microcoquina* is similar but is composed of sand-size shell debris. Deposits of this kind are to be seen on many marine beaches and shoals exposed to large waves and strong currents. Some of them include well-sorted silicate sand grains which, because they are more spheroidal, tend to be distinctly smaller than the platy shell fragments with which they are associated (Figure 14–7A,B).

Limestones rich in ooids have commonly been called *oolite* or *oolitic limestone*; most of them are well-sorted oosparites. Individual ooids are smooth spheroids or ellipsoids of sand size in which a nucleus is encased by an aragonite or calcite cortex of approximately uniform thickness. The nuclei in ooids may be fragments of skeletal material, fecal pellets, silicate sand or silt grains, or even bits of fractured ooid (Figures 14–8 and 14–9). In recent ooids, the cortex consists of extremely fine aragonite fibers oriented more or less tangentially in a succession of thin concentric sheaths or layers. Each layer probably represents an episode of chemical accretion, the boundaries between layers representing discontinuities in the growth of the ooid. In many ooids, for example those from the Great Salt lake (Figure 14–8A), some cortex layers are thicker and slightly coarser than others and have a radial fibrous structure. The thinnest and finest layers appear brownish and relatively dark in transmitted light, whereas the thicker radial fibrous layers are somewhat clearer, even though individual crystals cannot be discriminated with the optical microscope.

Ooids in ancient limestones are calcite rather than aragonite, but their cortexes commonly show concentric and radial structures similar to those in recent aragonite ooids. The older calcite ooids have generally been interpreted as original aragonite ooids converted to calcite during diagenesis, but such an interpretation may not be consistent with well-preserved concentric and radial structure. Perhaps, as Sorby suggested long ago, they were calcite originally.[5] Diagenetic alteration of fibrous aragonite to calcite in fossil shells has invariably resulted in granular calcite mosaics that obliterate the original aragonite fabric, and it seems doubtful that ooids would alter differently, especially where the two occur in the same rock (Figure 14–9A).

The good sorting typical of oolitic limestones is consistent with the occurrence of modern ooids, most of which accumulate in those areas of lakes and shallow seas where the grains are more or less continually rolled by waves or tidal currents and are washed free of mud. Continual agi-

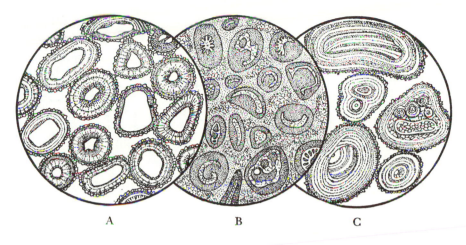

A   B   C

Figure 14–8. Oolitic Limestones

A. Pleistocene ooids, Great Salt Lake, Utah. Diam. 3 mm. Ooids consist of sub-angular detrital quartz grains enclosed by aragonite having both concentric and radial fibrous structure. Incipient cement.

B. Oomicrite, Volksen, Deister Mountains, Germany. Diam. 3 mm. Loosely packed ooids consist of nuclei encased by microcrystalline calcite (dark stippling); nuclei are shell fragments, some of which have been recrystallized to calcite mosaics. Ooids occur in a micrite matrix that has been partially recrystallized; note patches of neomorphic microspar and fine-grained spar. The allochems are called ooids, because nuclei are visible and also because vague relics of concentric structure are visible in some (not illustrated); they have probably been micritized.

C. Composite ooids (Pleistocene), Pyramid Lake, Nevada. Diam. 6 mm. Large ooids consisting of microcrystalline (stippled) and radial fibrous (clear) concentric layers. Nuclei are fragments of broken ooids, clusters of tiny ooids (right and center), and bits of granular carbonate (lower right). Incipient cementation as in A.

tation during the growth of marine and lacustrine ooids may have been responsible, as is generally assumed, for the typical uniformity of the cortex around the nucleus, but it can scarcely be considered essential for the formation of all ooids in view of those that have been formed by pedogenic processes in some calcareous soils.[6]

Fecal pellets are among the most abundant allochems in marine carbonate deposits today and probably have been equally abundant in past deposits. They are generally fragile grains, however, easily deformed and not sufficiently cemented to survive much transport or abrasion. They tend to accumulate in the muddy deposits where they are formed,

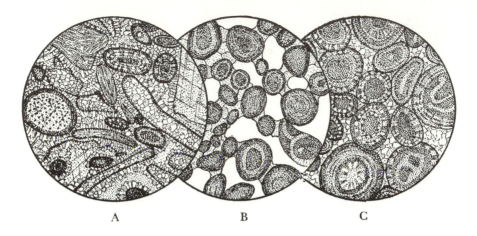

A   B   C

Figure 14–9. Oolitic Limestones

A. Oolitic biosparite (Jurassic), Bath, England. Diam. 2.5 mm. Radial fibrous calcite ooids (upper right), microgranular calcite pellets (heavily stippled, at bottom), and abraded shell fragments, all cemented with fine-grained calcite. Cement fabric consists of bladed calcite crystals rimming each carbonate fragment, with coarse calcite crystals (lightly stippled, near bottom) occupying the centers of original pores. Some shell fragments are original fibrous calcite; some are abraded single crystals, probably from echinoids (right and left); some are recrystallized granular calcite and were probably aragonite originally. Micrite envelopes on most allochems.

B. Recent ooids, coast of southern Florida. Diam. 2.5 mm. Dark microcrystalline ooids having distinct concentric structure. Nuclei are microcrystalline pellets; concentric carbonate is aragonite. Partly cemented with fine-grained calcite, which probably formed in the vadose environment. Remaining pores are blank.

C. Oosparite, St. Louis Limestone (Mississippian), Bowling Green, Kentucky. Diam. 2.5 mm. Ooids consisting of radial fibrous calcite, but with distinct concentric banding, tightly packed and firmly cemented by fine-grained clear calcite. Nuclei in ooids are mostly microcrystalline calcite pellets, but a few appear organic (right edge and lower right). Compare the looser packing in B.

but they also have been winnowed and sorted, along with small skeletal grains, by weak waves and bottom currents. Because their mineral composition and texture are similar to those of the muds in which they are often embedded, their boundaries tend to become vague. In ancient micrites, therefore, pellets are likely to be overlooked; in pelsparites they are more easily distinguished even though the sparry cement generally is very fine grained. Not all microcrystalline pelletlike allochems are fecal pellets; some are small rounded intraclasts composed of micrite, and some are skeletal grains or ooids that have been micritized on the sea floor. These distinctions may be uncertain in older limestones, but abundant pellets having essentially uniform size and shape throughout a rock are best interpreted as fecal pellets (Figure 14–13B).

Intraclasts consisting of chunks, slabs, and chips of micritic limestone are to be seen in numerous limestones and are significant because they signify a depositional environment where coherent, generally weakly consolidated sediment has been disrupted and redeposited. They may be as small as sand grains or as large as boulders, and they are particularly striking if they include portions of several strata and thus show internal stratification distinct from that of the enclosing limestone. When seen in the field, intraclasts are often similar in general character to the underlying strata, and they probably have not been moved far before being redeposited. Most of them have been somewhat rounded, but rounding of weakly consolidated fragments is not indicative of extensive transport; some have been plastically deformed, indicating that they were not rigidly lithified when they were redeposited.

Intraclasts are distinguished from the enclosing limestone by difference in composition and texture, but the differences are often slight and may be scarcely discernible. Small intraclasts can be recognized in thin sections, but the larger ones are best seen on surfaces that have been delicately etched by natural weathering, or that have been etched or polished in the laboratory. Local concentrations of pebble- or cobble-size intraclasts produce limestone conglomerates or breccias, the most distinctive of which is *flat-pebble conglomerate* characterized by thin slablike fragments. By comparison, detrital limestone conglomerates of terrigenous origin (calclithites) usually contain a much greater variety of carbonate rock fragments and also angular bits of the chert that typically occurs in older carbonate formations.

Deposits of carbonate mud and grains are initially very porous and are subject to compaction so long as open pore space exists and the carbonate particles can readjust into closer packing arrangements.* Significant compaction ceases, however, once the particles have been cemented together so that relative movement between them is restricted; if all pore space is completely filled by cement, intergranular compaction

---

*The 8% expansion when aragonite inverts to calcite may be a factor in the compaction of some limestones.

ceases altogether. The amount of compaction that has occurred in cal-carenites prior to their cementation can be roughly estimated from comparisons of grain packing in different rocks. For example, the recent oosparites illustrated in Figures 14–8A and 14–9B have not been buried and have not been appreciably compacted; the oosparite of Figure 14–9C, which is much older and has been buried, is more tightly packed and was effectively compacted before it was cemented. Even so, in the older rock, a large volume of cement has been required to fill the pore space remaining between the compacted grains. More or less open packing of this kind is characteristic of a great many calcarenites, and they contain correspondingly large volumes of cement. Furthermore, fragile shells and fecal pellets, generally, have not been deformed by compaction. It appears that most limestones have been cemented relatively early in their diagenetic history before major compaction could occur.

Thorough compaction has occurred in some limestones, however. In the calcarenite of Figure 14–6B, for example, the grains are fitted closely together and pore space available for cement has been effectively eliminated. The boundaries between grains are microstylolitic, indicating enough pressure along grain boundaries to cause differential solution as an adjunct to close physical packing. If silicate sand grains occur in rocks of this kind, they tend to penetrate into adjacent carbonate grains as though pressed into them. Textures of this kind presumably mean that effective burial, to induce pressure sufficient to cause the compaction, occurred prior to effective cementation. More commonly, however, cementation appears to have occurred earlier.

The general observation that cementation in most limestones has taken place before much compaction has far-reaching implications. Porosities in the original carbonate deposits probably averaged about 50%, whereas porosities in old limestones average less than 5%. If there has been little compaction prior to cementation, almost half the total volume of the limestones is cement. The source and the transfer of this much calcium carbonate pose important geological problems.

DOLOMITIC LIMESTONE AND DOLOMITE

The volume of dolomite in recent carbonate deposits is exceedingly small, yet many ancient carbonate formations consist largely of dolomite. This observation and the evidence of numerous structures in which calcite and aragonite have been replaced by dolomite lead to the general conclusion that most dolomitic rocks have been produced by alteration of limestones. Organic secretions, for example, are never dolomite originally, but much skeletal material in limestones has been dolomitized. No one contends, however, that all dolomites and dolomitic rocks have formed in one way.

Studies of dolomite crystallization indicate that at low temperature well-ordered, stoichiometric dolomite is unlikely to precipitate directly in the sedimentary environment, for it crystallizes with difficulty and only by exceedingly slow crystal growth. The early-formed dolomite in recent deposits is *protodolomite*, which contains excess Ca^{2+} and has a crystal structure that is imperfectly ordered. In contrast, the dolomite in ancient carbonate rocks is typically well-ordered and stoichiometric and is presumably a product of diagenesis.

Recent dolomite deposits occur in limited environments where evaporation has produced saline brines, either in shallow lakes or lagoons where water circulation is restricted or in the pore waters of deposits on broad supratidal mudflats. Although the dolomite may precipitate directly from the brine in a few lagoons or saline lakes, it forms more often by penecontemporaneous alteration of initial aragonite or calcite. The dolomite crystals that form in these environments are minute, no more than several micrometers in diameter, and the deposits are dolomitic mud or *dolomicrite*. Ancient dolomite rocks are significantly coarser grained than this, although many of them are microcrystalline. Doubtless, most of them have formed by replacement of aragonite or calcite and are products of diagenesis, but some are essentially primary or syngenetic.

Among ancient dolomite formations, those that are demonstrably onshore facies in marine carbonate complexes and also those that are interbedded with evaporite deposits are most likely to be syngenetic. However, dolomitization of fossil reefs and shell beds can scarcely be regarded as syngenetic, for dolomite does not appear to form in the subtidal zones that support flourishing organic habitation and growth. The dolomitization of these rocks is often demonstrably related to paleo-geography over large regions and through thick sequences of limestone strata, and it has been attributed to mixing of fresh meteoric water with interstitial marine water in the subsurface phreatic zone adjacent to ancient shorelines.[7]

Masses of dolomite rock having irregular or veinlike form that cut across the stratification of associated limestones, particularly those related to faults and joints, undoubtedly replaced the limestones after they had been buried and lithified. Some of these dolomites are associated with deposits of metallic sulfides and can be regarded as products of late metasomatism.

A few dolomites and dolomitic limestones are clastic rocks. Some of these are terrigenous, derived from older dolomite formations by subaerial erosion. Others are made up of grains that have been moved about and redeposited by currents after their initial development either in syngenetic deposits or during early diagenesis. A classic example of the latter occurs in the Triassic carbonate rocks of the Austrian Alps where B. Sander described dolomite laminae rhythmically interlayered with calcitic laminae, commonly on a millimeter scale, and interpreted them as the result of reworking of early-formed dolomitic sediment on the sea floor.[8] Intraclasts of dolomicrite on tidal flats are another example.

Rhombohedra of dolomite are distributed more or less regularly throughout many limestones. Here the tendency of dolomite to form euhedral crystals in contact with calcite is clearly evident, and this usually serves to distinguish the two minerals, for calcite in limestones is invariably anhedral. Commonly the dolomite crystals contain a central, somewhat turbid zone surrounded by a clearer rim, and some crystals include several zones having variable iron content. Dolomite rhombohedra are particularly striking when they occur in a micrite matrix where the larger and clearer euhedral crystals of dolomite contrast strongly with the more turbid microcrystalline calcite (Figure 14–10A). Similar dolomite rhombohedra scattered through coarser-grained calcite mosaics are not so striking but are nevertheless recognizable by their characteristic crystal outlines. Undoubtedly the dolomite crystals have replaced calcite in the limestones, for they typically contain minute calcite inclusions, often zonally distributed within the rhombohedra.

In other dolomitic limestones, crystal mosaics composed chiefly of dolomite are in contact with other mosaics composed largely of calcite. For example, dolomite may occur in certain beds or laminae that alternate with others composed of calcite (Figure 14–10B), or irregular aggregates of dolomite may be randomly distributed through the limestone. Contacts between the dolomite and calcite mosaics are irregular on a microscopic scale, but on a larger scale they may appear to be smooth. Alternating laminae of calcite and dolomite perhaps are caused by changing physicochemical conditions in the depositional environment favoring early dolomitization of some layers and not of others. Some laminated dolomitic rocks are probably of algal origin, in which thin magnesium-rich algal mats alternated with laminae of more calcium-rich

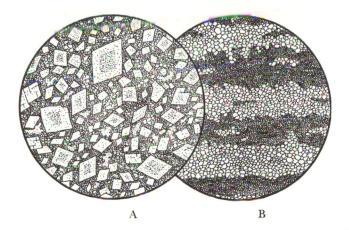

A   B

Figure 14−10. Dolomitic Limestones

A. Chazy Limestone (Ordovician), Chazy, New York. Diam. 1 mm. Microcrystalline calcite of uneven grain size containing euhedral rhombohedra of dolomite. Darker central zones are distinct in many of the dolomite crystals.

B. Triassic dolomitic limestone (Marble Bay Formation), Texada Island, Georgia Strait, British Columbia. Diam. 1.8 mm. Alternation of calcite and dolomite laminae. Dolomite is finer-grained than calcite but otherwise similar; the two are distinguishable only by chemical tests. Laminae at bottom have relatively smooth contacts; others are irregular, suggesting replacement of one carbonate by the other, probably during early diagenesis. The laminae are seen in the field to be extensive.

mud and have been preferentially dolomitized. Irregular, branching dolomite structures are common in dolomitic limestones in some places; they may be the borings or tracks of marine organisms in which dolomite has formed preferentially, or they may perhaps represent dessication structures formed on ancient tidal flats. Such structures usually are concentrated in certain beds and are disposed more or less parallel to the bedding planes.

Rocks consisting wholly of dolomite are mosaics of anhedral or subhedral grains. If the crystals are strictly anhedral and tightly interlocking, the texture is similar to that of sparry calcite mosaics and the two may be optically indistinguishable in thin section (Figures 14−1 and 14−12A). The crystals in many dolomite aggregates, however, clearly display a tendency to assume their characteristic crystal forms, and in these there is likely to be some intergranular porosity where adjacent subhedral crystals fail to interlock (Figures 14−11B and 14−12C). The texture of such rocks is often described as sugary.

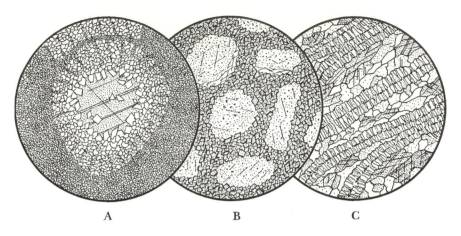

A   B   C

Figure 14–11. Dolomitized Limestones

A. Dolomitized Devonian coral, Bear River Range, northern Utah. Diam. 8 mm. Limestone matrix and septa of coral replaced by very fine-grained dolomite; coarser dolomite has filled in between septa in coral; dolomite euhedra near the center are enclosed in a single large calcite crystal.

B. Dolomitized crinoidal limestone (Silurian), Niagara River, New York. Diam. 6 mm. Coarse calcite crystals (stippled) are remnants of crinoid plates and stem segments enclosed and marginally replaced by a fine-grained mosaic of subhedral dolomite crystals.

C. Dolomitized Devonian coral (*Cyathophyllum*), Eifel, Germany. Diam. 3 mm. Coral structure cut longitudinally. Septa consist of cross-oriented prismatic dolomite; dolomite mosaic between septa is composed of interlocking larger anhedral grains, generally elongated parallel to septa.

Dolomites tend to be more evenly grained than limestones, and they are rarely as fine grained as the finest or as coarse grained as the coarsest limestones. Moreover, they have fewer recognizable microstructures, and most of those that do appear are shadowy features whose outlines suggest that they have been inherited from original limestone structures. The clastic texture of calcarenites is not to be expected in dolomites, yet many dolomites contain rounded quartz sand grains presumably inherited from original calcarenites whose calcareous textures have been obliterated by dolomitization (Figures 14–1A and 14–12C); in some of these rocks the quartz grains are disposed in cross-laminae, indicating that the original limestones were indeed deposited by currents. Fossils, too, are relatively rare in dolomites. Few organisms inhabit environments that favor syngenetic development of dolomite; furthermore, later dolomitization usually destroys the internal structure and even the outlines of fossils, and for this reason richly fossiliferous limestone and apparently

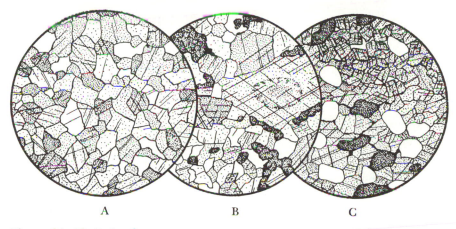

A   B   C

Figure 14–12. Dolomites

A. Lone Mountain Dolomite (Silurian), 3000 m below surface, near Eureka, Nevada. Diam. 2.5 mm. Mosaic of dolomite anhedra, not visibly different from some recrystallized calcite mosaics.

B. Glauconitic Bonneterre Dolomite (Cambrian), near St. Louis, Missouri. Diam. 2.5 mm. Inequigranular dolomite mosaic, with patches of microcrystalline glauconite between dolomite grains. Local ferric oxide (black). Compare pellet form of glauconite (stippled) in C. Relict ovoid in large dolomite grain at right may be organic. The rock contains some detrital quartz grains (not shown in this field) and is perhaps a dolomitized glauconitic calcarenite.

C. Sandy glauconitic dolomite (Cambrian, Sawatch Formation), Ute Pass, El Paso County, Colorado. Subrounded quartz grains and glauconite pellets floating in a dolomite mosaic; probably a dolomitized calcarenite. Compare the nonporous mosaic of anhedral dolomite grains at the bottom with porous aggregate of dolomite rhombs in upper part of figure. Local ferric oxide stain (black).

barren dolomite may form adjacent portions of a single stratum. Dolomitized fossils have been faithfully preserved in some rocks, however; in some others, they are recognizable but scarcely determinable.

Dolomitization is often selective. In a few limestones, only the organic structures have been replaced; in others, the rock matrix has been replaced and calcite fossils are surrounded by fine-grained dolomite. Fossil molds may occur where preferential solution has removed calcitic or aragonitic shells and left the dolomite surrounding them. In general, aragonite and magnesian calcite are more susceptible to dolomitization than is low-magnesian calcite because they are metastable, and fine-grained aggregates appear to be more susceptible than those of coarser grain (Figure 14–11B), perhaps because of relative instability due to their higher surface energy.

Ooids, like skeletal materials, are never composed of dolomite originally, but many oolitic limestones have been dolomitized. Partial dolomitization of ooids is common. The dolomite may selectively replace either the ooids or the surrounding matrix or cement, or it may partially replace both; it is not uncommon to see euhedral dolomite crystals within the ooids and also crossing their margins. Where the process has gone to completion, the oolitic structure is either entirely obliterated or the ooids are represented only by shadowy ovoid shapes that are indistinguishable from the forms of original pellets.

If complete conversion of calcite to dolomite occurred molecule for molecule, by substitution of magnesium carbonate for part of the calcium carbonate, it would cause a reduction in volume of about 12%, and hence an increase in porosity. There is no reason to suppose, however, that any natural metasomatic reaction need be represented by a precise exchange of Mg^{2+} for Ca^{2+}. Petrographic examination commonly indicates that little if any change in volume accompanies dolomitization. Dolomite euhedra in limestone are usually in close contact with surrounding calcite, and many microstructures, such as ooids and fossils, have been replaced by dolomite without overall distortion or production of porosity. This suggests that dolomitization is usually a volume-for-volume replacement. Nevertheless, some dolomite mosaics do contain pores possibly produced by dolomitization (Figure 14–12C), and the experience of most petroleum geologists dealing with carbonate rocks shows that dolomites tend to be porous reservoirs more often than limestones.

Replacement of calcite and aragonite by dolomite is so evident and widespread that most geologists overlook the possibility of the opposite reaction. Dolomite is not always the last of the authigenic carbonate minerals to form. Not only do veinlets of calcite cut through many dolomites, but late calcite replacing dolomite is irregularly distributed through some carbonate rocks, the calcite embaying and transecting individual dolomite crystals. The evidence of this process is easily overlooked, however, and is likely to be ambiguous.

AUTHIGENIC SILICATES IN LIMESTONES AND DOLOMITES

Various accessory silicate minerals often found in limestones and dolomites have been referred to in the foregoing descriptions. Opal, chalcedony, quartz, feldspar, glauconite, and illite are the most common.

Some of these are detrital; others, of greater interest, are authigenic (Figure 14–13).

First consider authigenic quartz. In some limestones the quartz forms euhedral crystals replacing calcite. The crystals may have detrital cores, to which have been added euhedral rims crowded with minute carbonate inclusions, or they may be entirely authigenic and have the form of slender doubly terminated prisms. This elongate prismatic habit generally develops in quartz crystals formed at low temperatures, and it is especially typical of those formed authigenically in limestones. The quartz prisms are usually filled with tiny calcite inclusions, often zonally arranged. Authigenic quartz is never abundant, but a few crystals are to

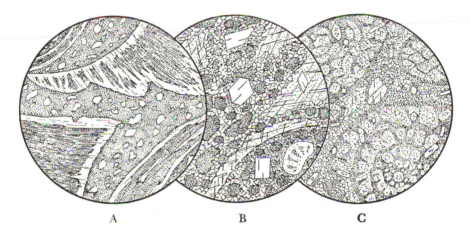

| A | B | C |

Figure 14–13. Authigenic Silica and Silicates in Limestone

A. Chalcedony in Cretaceous limestone, Guadalupe Mountains, Cochise County, Arizona. Diam. 3 mm. Fragments of oyster shells embedded in a matrix of mixed sand, clay, and microcrystalline calcite. Fibrous calcite of shells partially replaced along the margins by chalcedony (clear, with fine lines).

B. Authigenic albite in Ordovician limestone, Glens Falls, New York. Diam. 2 mm. Clastic limestone, consisting of cryptocrystalline calcite pellets (dark stippling) and coarse calcite crinoid fragments (upper half), cemented with very fine-grained clear calcite. Four authigenic albite euhedra (blank except for cleavage) replace the calcite. For published description see *Bulletin of the Geological Society of America*, vol. 40 (1929): pp. 463–468.

C. Authigenic quartz in fusulinal limestone (Upper Pennsylvanian, Naco Formation), Portal, Cochise County, Arizona. Diam. 2.5 mm. Dark microcrystalline tests of *Triticites* embedded in clearer fine-grained calcite. Euhedral quartz prisms containing minute calcite inclusions replace both tests and matrix.

be found in many limestones. They develop without any preferred orientation, and mostly in fine-grained rather than coarse-grained calcite. Their origin by replacement is indicated not only by calcite inclusions but also by the growth of many of them across carbonate structures such as oolites and across bits of organic material (Figure 14–13C). In some carbonate rocks, however, detrital grains of quartz may be corroded and marginally replaced. This reaction is particularly common in dolomites, and quartz euhedra, too, are less common in dolomites than in limestones, though whether they fail to develop there as commonly as they do in limestones or are destroyed during dolomitization is uncertain.

Authigenic feldspars have been found in many limestones, but they are much less common than authigenic quartz (Figure 14–13B). They invariably consist of K-feldspar (adularia) or pure albite, the latter being the most abundant. The crystals are euhedral or nearly so and perfectly clear, and usually they develop a thick tabular habit in which only the simpler crystal forms are present—$\{001\}$, $\{010\}$, $\{\overline{1}01\}$, $\{110\}$, $\{1\overline{1}0\}$. Albite crystals are generally twinned on the albite law, although because of their very small size they may appear to be untwinned or simply twinned. Unlike quartz, all the authigenic feldspars are free from inclusions. They as well as quartz appear to develop most readily in the fine-grained portions of limestones.

The argillaceous material in limestones and dolomites is presumed to be a detrital admixture. But the clay mineral found in limestones is so often illite that one must assume a general tendency for this mineral to develop from other detrital clays during diagenesis.

GENERAL REFERENCE

For general reference, see R. G. C. Bathurst, *Carbonate Sediments and Their Diagenesis,* 2nd ed., Developments in Sedimentology, no. 12 (New York: Elsevier, 1975); H. Blatt, G. Middleton, and R. Murray, *Origin of Sedimentary Rocks,* 2nd ed. (New York: Prentice-Hall, 1980), chaps. 13 and 14. Also useful is a symposium on the classification of carbonate rocks: W. E. Ham, ed., *American Association of Petroleum Geologists Memoir 1* (1962). Finally, instructive photomicrographs covering carbonate rock constituents and textures are presented by P. A. Schole, *American Association of Petroleum Geologists Memoir 27* (1978).

On dolomitic limestone and dolomite, see G. Larsen and G. V. Chilingar, *Diagenesis in Sediments and Sedimentary Rocks,* Developments in Sedimentology, no. 25A (New York: Elsevier, 1979); the chapter on dolomites and dolomitization presents a good summary discussion of the character and origin of dolomite. L. C. Pray and R. Murray, *Society of Economic Paleontologists and Mineralogists,* special pub. 13 (1965) is a symposium that includes three articles describing recent dolomite formation in the Netherlands Antilles, the Persian Gulf, and the Bahamas.

REFERENCE NOTES

[1]A. W. Grabau, *American Geologist,* vol. 33 (1904): pp. 228–247.

[2]R. J. Dunham, *American Association of Petroleum Geologists Memoir 1* (1962): pp. 108–121.

[3]R. L. Folk, *American Association of Petroleum Geologists Bulletin,* vol. 43 (1959): pp. 1–38.

[4]A good account of the variety of carbonate sands formed in a representative part of a modern reef complex, including their character, origin, and possible ancient analogues, is given by R. L. Folk and R. Robles, *Journal of Geology,* vol. 72 (1964): pp. 255–292.

[5]H. C. Sorby, *Quarterly Journal of the Geological Society of London,* vol. 35 (1879), p. 75. This problem has been considered recently by P. A. Sandberg, *Sedimentology,* vol. 22 (1975): pp. 497–537; see also, B. H. Wilkinson, *Geology,* vol. 7 (1979): pp. 524–527.

[6]R. L. Hay and B. Wiggins, *Sedimentology,* vol. 27 (1980): pp. 559–576.

[7]For example, see J. B. Dunham and E. R. Olson, *Geology,* vol. 6 (1978): pp. 556–559.

[8]B. Sander, *Contributions to the Study of Depositional Fabrics,* trans. E. B. Knopf. Tulsa, Oklahoma: American Association of Petroleum Geologists, 1951.

15

Miscellaneous Sedimentary Rocks

The majority of sedimentary rocks are predominantly conglomeratic, sandy, argillaceous, or calcareous, and understanding of these alone is background enough for examination of many sedimentary terranes. But some important sedimentary components, usually found as minor constituents in these rocks, become concentrated locally to constitute the chief ingredients of such deposits as cherts, ironstones, phosphorites, and evaporites, and of such bituminous and carbonaceous sediments as oil shales and coal. Some of these rocks are economically important; all are significant as lithofacies originating within particular environments. This chapter deals with cherts, iron-rich rocks, phosphorites, and anhydrite and gypsum. The more soluble saline deposits and the carbonaceous and bituminous sediments are omitted, mainly because their study generally involves techniques and training that are beyond the scope of this book.

CHERT AND PORCELLANITE

Next to mudstones, sandstones, limestones and dolomites, the most widespread sedimentary rocks are cherts and porcellanites. These are rocks composed largely of organic or authigenic SiO_2 in the form of opal, chalcedony, or microgranular quartz.

*Opal** contains varying amounts of water up to about 10% and is sensibly isotropic. Its refractive index varies but is always low enough to produce strong negative relief in thin sections. It is metastable and, in time, it alters to microcrystalline quartz. Opal therefore is most abundant in the younger rocks and is almost limited to those of Cenozoic age.

Chalcedony is a microfibrous or feathery variety of quartz with refractive indices (1.53–1.54) somewhat lower than those of normal quartz. Electron microscopy shows that swarms of submicroscopic aqueous bubbles occur in chalcedony, and these presumably are the cause of its lower refractive indices and of the brownish tinge that it often shows. The fibrous structure tends to be oriented approximately normal to any surface on which chalcedony crystallizes, but it usually develops from separate centers to produce a sheaflike or radial appearance. Generally it is length-fast, indicating elongation normal to the optic, or z, axis of the crystals, but some chalcedony (quartzine) is elongated parallel to the z axis and is length-slow. This latter variety is noteworthy because it is commonly associated with anhydrite, gypsum, and other evaporite minerals and appears to typify environments that are highly saline or alkaline; late Pleistocene cherts (10,000 to 20,000 years BP) formed in saline, highly-alkaline Lake Magadi, East Africa, are characterized by length-slow chalcedony.[1]

A form of chalcedony sometimes seen as a replacement of carbonate rock looks, under low magnification, like clouded quartz that extinguishes uniformly over relatively large areas. Under higher magnification, however, it appears to have a very fine parallel or mesh structure, somewhat like that of microperthitic feldspar, with negative elongation parallel to the predominant internal structure.

Microgranular quartz occurs as mosaics of minute, subequant crystals that show pinpoint extinction in thin section. The individual quartz crystals are usually less than 5 μm in diameter, and may be of submicroscopic size. In the coarser mosaics, the grains range up to about 20 μm in diameter and usually show wavy, incomplete extinction caused by superposition and overlapping of separate crystals.

Siliceous pelagic muds consisting largely of the opaline tests of diatoms or radiolarians have accumulated on many parts of the ocean floor, both

---

*Opal is used here in a broad petrographic sense and includes the following structural forms that can be differentiated by use of x rays and the electron microscope, namely, opal A (x-ray amorphous biogenic and volcanic SiO_2), opal-CT or *lussatite* (disordered, low-temperature cristobalite-tridymite), and opal C (more ordered α-cristobalite).

past and present. Where these organic remains accumulate uncontaminated by other sediment, deposits of nearly pure biogenic silica are
formed; more often, however, the deposits are impure, the biogenic
material being mixed with varying amounts of clay or carbonate mud.
Incoherent deposits of diatom frustules are called *diatomaceous ooze*, or
earth, and the somewhat lithified, coherent equivalent is called *diatomite*.
Similarly, *radiolarian ooze* and *radiolarite* refer to incoherent and lithified
accumulations of radiolarian tests. Opaline sponge spicules also are abundant in some of these deposits. Radiolarite occurs throughout the stratigraphic column from the Cambrian onwards, but diatomite appears
first during the Cretaceous and is confined to younger parts of the column.

Tough, nonporous, sedimentary rocks of somewhat vitreous luster
composed largely of authigenic silica are called *chert*. They consist largely
of opal, chalcedony, or microcrystalline quartz, but they commonly contain numerous impurities and occur in various colors and forms.* Rocks
called *porcellanite* (porcelanite) are not as tough, dense, or vitreous as
typical chert and are composed largely of opal. They are microporous
and have a dull luster resembling that of unglazed porcelain. Most porcellanites contain substantial argillaceous or carbonate impurities, and
they typically grade into siliceous mudstone or marl; porcellanite also
grades into impure chert.†

The origin of siliceous sediments is a subject of ongoing discussion
and debate. Silica, although notable for its very slight solubility in water
at low temperatures, is nevertheless dissolved in appreciable quantities
in most rivers.‡ Sea water contains less. Much of the silica carried to the
sea by streams has been precipitated in the siliceous tests of organisms
such as diatoms and radiolarians, and some may have been precipitated
inorganically. It seems likely that silica, precipitated in one form or

---

*Many varieties of chert have been distinguished, the following being most common. *Flint*
is a tough, gray-to-black chert that usually has distinct conchoidal fracture; it is composed
of microgranular quartz, perhaps with some chalcedony, and its typical occurrence is as
nodules in chalk. *Jasper* is red or brown chert made up of microgranular quartz colored by
ferric oxides. *Novaculite* is a term applied, in America, to certain white cherts found in
Arkansas and Texas, but more generally it refers to very tough, nearly pure, whitish chert
composed of microgranular quartz.

†Use of the terms chert and porcellanite varies. As used here, the two rocks are distinguished by general physical characteristics evident in the field or in hand specimen. When
they are defined in a mineralogical sense, cherts are quartzose and porcellanites are cris-
tobalitic opal or lussatite. Pre-Tertiary examples are almost invariably chert by either definition.

‡"Dissolved" as used here includes SiO_2 in colloidal suspension as well as in true solution.

another, has been deposited more or less continuously with the finer detrital sediments accumulating in the oceans. Two basic questions concerning the origin of the siliceous rocks, however, are these: What are the sources of the vast amount of soluble silica that has been deposited, and how has it become as concentrated as it is in the purest of these deposits?

Much dissolved silica results from decomposition of silicates by weathering. The amount produced in this way is greatest if decomposition takes place in regions of low relief and tropical or subtropical climate, as, for example, during the formation of laterites; and it has been suggested that very siliceous marine sediments have been laid down when and where such conditions prevailed on neighboring landmasses. However, the distribution of silica in modern marine deposits seems generally related to the patterns of sea-water circulation, with silica concentrations in areas where there is upwelling of deeper water. Doubtless some of the dissolved silica in surface waters is also of volcanic origin. Silica may be added to the water directly by volcanic emanations, and it may also be freed and put into solution more slowly by alteration of vitric ash. Numerous bedded cherts and siliceous shales, and also diatomites and radiolarites, have been formed during periods of volcanism, and their origin has often been attributed to availability of volcanic silica. Yet, equally numerous siliceous formations have developed without any apparent volcanic association.

The concentrations of silica in deposits of the modern oceans appear to be largely biogenic, although the ultimate source of the silica used in the organic skeletons remains uncertain. Numerous cores recovered during drilling through sediments underlying the present sea floor indicate that deposits of early Tertiary and Cretaceous age that originally contained abundant biogenic opal have been converted into cherts composed essentially of microgranular quartz and chalcedony. This transformation involves a seemingly sequential alteration of biogenic opal to lussatite (opal-CT) followed by alteration to microgranular quartz and chalcedony. The transformation of biogenic opal to quartz appears to be a function of time, and the rate of transformation to be a function of depth of burial, temperature, and the composition of associated sediment phases. The lussatite phase is virtually limited to post-Cretaceous strata. A similar transformation related to depth of burial and concomitant temperature increase has been well documented for the Miocene diatomites, diatomaceous mudstones, and cherts of the Monterey Formation in California.[2]

For the cherts and porcellanites that occur as extensive thin beds interstratified with mudstone (shale) or marl, it seems warranted to conclude that most of them have been developed by diagenetic modification of original deposits in which biogenic silica was abundant.[3] The diagenetic processes involved in this transformation appear to include solution of biogenic opal and reprecipitation of the silica as lussatite and finally quartz, with some local redistribution of the silica. The processes involved in the origin of nodular cherts are also diagenetic but are probably somewhat different. Most nodular cherts are replacements of limestone, and they usually involve substantial addition of silica to the replacement site with concomitant removal of calcium carbonate.

Bedded Cherts and Porcellanites

Bedded chert formations are characterized by thin rhythmic interbedding of chert (or porcellanite) with mudstone or shale. Individual strata are typically a few centimeters thick. Some are extensive even layers, but some pinch and swell dramatically; some are internally laminated and others are not. The chert beds usually have sharp contacts with interbedded mudstones, and, in outcrops, the alternating hard and relatively soft layers have usually been etched into sharp relief by weathering and erosion.

The older bedded cherts consist essentially of even-grained mosaics of microgranular quartz, often transected by clear veinlets of coarser quartz or chalcedony deposited in tectonic fractures. Some chert beds are nearly pure silica, but most contain scattered iron oxides, clay minerals, or minute carbonate crystals, which effectively darken the microgranular quartz matrix and give it an overall turbid appearance in thin section. In some chert beds, the impurities are concentrated in very thin argillaceous laminae, and some contain small stylolites with clay and iron oxides concentrated along them. Pre-Tertiary chert formations are generally radiolarian. Sometimes the cherts are crowded with the tests of radiolarians, often well preserved, especially where the chert matrix is very fine grained and impure. In most radiolarian cherts, however, the organic remains are scattered and have lost their detail, and many are no more than round or oval forms that stand out in the dusky microcrystalline matrix as patches of limpid quartz or chalcedony (Figure 15–1A). The best-preserved organic remains usually appear in siliceous mudstone layers interbedded with the cherts.

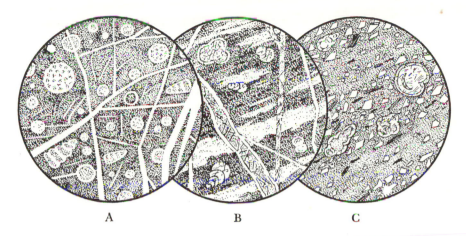

A   B   C

Figure 15–1. Bedded Chert and Porcellanite

A. Radiolarian chert (Franciscan Formation), Marin County, California. Diam. 1 mm. Microcrystalline quartz reddened with hematite dust (stippled) encloses numerous poorly preserved radiolarian tests and elongate spicules. Tests and spicules are replaced by coarser clear chalcedony (blank). Criss-crossing veinlets of clear chalcedony and microcrystalline quartz.

B. Calcareous porcellanite (Miocene, Monterey Formation), Berkeley Hills, California. Diam. 1 mm. Bulk of the specimen is dark opal, shot through with minute carbonate grains (heavily stippled); this is streaked with tiny lenses and laminae of clear microcrystalline or microfibrous quartz containing but few granules of carbonate (lightly stippled). Foraminifers infilled with coarse calcite are found in opaline portions only. The rock is transected by veinlets of coarse calcite or chalcedony (blank); one larger veinlet (at left) has rims of opal (clear) and a center of calcite.

C. Silty porcellanite (Miocene, Monterey Formation), Monterey, California. Diam. 1 mm. A mixture of silt, clay, foraminiferal tests, and opal. The matrix is opal darkened with argillaceous material (stippled). Scattered throughout are small fragments of quartz and feldspar (clear), flakes of mica (lined), and carbonaceous streaks (black). Silt particles are crudely aligned along bedding planes. Calcite tests of foraminifers are infilled with either calcite or chalce-dony, the latter shown in the large test at upper right.

Bedded radiolarian cherts occur in many eugeosynclinal assemblages, and their frequent stratigraphic association with pillow basalts has long been recognized, especially as one of the three units (serpentinite, spilitic basalt, bedded chert) that collectively make up the compound lithologic assemblage known as ophiolite. Today ophiolites are generally inter-preted as segments of oceanic crust. Radiolarites typically accumulate on the deep-ocean floor, and their origin may perhaps be connected with concurrent volcanic activity on the sea floor. However, some bedded radiolarian cherts occur as parts of deep-marine sedimentary complexes, apparently without associated igneous rocks.

Bedded opaline cherts and porcellanites are generally diatomaceous and Tertiary in age; radiolarians and sponge spicules also may occur in them. In thin section, the opaline portions of these rocks are usually rather dark and brownish, because of the pronounced negative relief of the opal and the inclusion of argillaceous, carbonate, and organic impurities. Small angular bits of detrital quartz and mica are common in many of them, especially in porcellanites. (Figure 15–1C). The opaline matrix is sensibly isotropic, but it incorporates birefringent specks and patches that represent the argillaceous and carbonate impurities. In many of these cherts, opal has been partially altered to microcrystalline quartz, which occurs in thin lenses and laminae that are irregular in detail but are generally parallel to the stratification of the deposit. Quartzose laminae are clear and colorless in contrast to the darker opaline portions of the rock, as though conversion of opal to quartz had somehow cleansed the rock of impurities. (Figures 15–1B and 15–2B).

During conversion of the original diatomaceous deposits into chert, the siliceous tests are generally destroyed or obscured, and in thin sections of opaline chert, diatom tests are rarely visible, although telltale remains frequently appear when the rocks are scanned with an electron microscope. Calcareous foraminifers are easily recognized, however, and are often abundant. Cavities in their tests may be filled either with clear chalcedony or with coarse calcite. Where these fossils are embedded in opal, they tend to be well preserved, but where they are embedded in microgranular quartz, the carbonate shells generally have been replaced by quartz and much of the organic detail has been lost (Figure 15–2B).

Interbedding of highly siliceous strata with more argillaceous or calcareous strata in bedded chert formations, and also the internal lamination within some strata, represents original layering in the deposit. Original features of stratification are commonly preserved, despite some redistribution of the biogenic silica during chertification. Some radiolarian chert layers, reported for example from late Mesozoic radiolarites in Greece and the northern Apennines, show features typical of graded turbidite layers, including sole markings and upward decrease in size of radiolarian tests along with increase in clay content. Such beds are interpreted to represent redeposition of radiolarian ooze by low-velocity turbidity currents generated on the marginal slopes of sea-floor basins.[4] And in the Monterey Formation of California, laminations in some dia-

tomites have been traced continuously and without distortion into laminated porcellanite and chert. This latter observation also indicates that
the chert was formed after compaction of the original diatomite, as might
be expected since conversion to chert presumably was induced by deep
burial and elevated temperature.

Chert in Calcareous Rocks

Chert nodules in limestones and dolomites are exceedingly widespread.
Most of them are elongated parallel to the stratification planes and are
concentrated within certain zones or along particular bedding surfaces.
Where they are especially numerous, they may coalesce to form thin
irregular beds. In some limestones and dolomites, however, the chert is
more definitely bedded, and in others it is much more irregular in form
and distribution.

Most cherts in limestones and dolomites are largely composed of chalcedony and microcrystalline quartz. Even the purest of them usually
contains a little calcite or dolomite, either as scattered crystals or as small
irregular aggregates. Less pure types consist of intimate mixtures of
spongy chert and carbonate.

Contacts between chert and carbonate rock are irregular in detail
(Figure 15–2A,C). The margins of chert nodules and masses may appear
very smooth in hand specimens; in thin sections, however, one sees intimate penetrations of chert into carbonate rock, and the marginal parts
of the chert are commonly clouded by inclusions of carbonate. These
relationships suggest a replacement origin for at least the marginal chert,
and still more convincing evidence is afforded by the presence of typically
calcareous textures preserved in the chert. Many of these are organic,
consisting of shells replaced by silica. Others are inorganic, such as oolitic
texture or the clastic textures of calcarenites. Relatively large grains of
quartz may occur in some cherts, apparently suspended in the chert
matrix. Sometimes these are distinguished only by sizes that are distinctly
larger than those of quartz crystals composing the chert (Figure 15–2A),
but often they show the rounded forms of abraded detrital grains, indicating clearly that they are residual detrital grains inherited from an
original sandy limestone.

Cavities present in rocks at the time of silicification are normally lined
with chalcedony in which the fibrous structure stands at a high angle to
the cavity walls and serves to outline the original voids. The fillings may

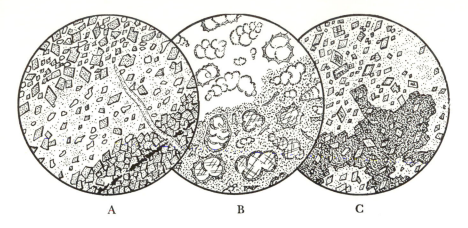

Figure 15–2. Cherts

A. Cherty portion of Madison Limestone (Mississippian), Bear River Range, northern Utah. Diam. 2.5 mm. Dolomite rhombohedra and detrital quartz sporadic grains (blank and irregular) set in a matrix of microcrystalline quartz. Chert bands like that in center parallel the bedding and alternate with others, like that at bottom, composed almost entirely of dolomite. Opaque lamina in dolomite is probably organic material. Secondary veinlet of chalcedony.

B. Foraminiferal chert (Upper Miocene, McLure Formation), Reef Ridge, California. Diam. 2 mm. In lower half, well-preserved calcite tests, infilled partly with coarse calcite (two cleavages) and partly with chalcedony (blank), are set in a matrix of opal (stippled). In upper half, matrix is clear chalcedony (blank), and calcite tests (without distinct outlines) have been largely replaced by chalcedony.

C. Chert in Helderberg Limestone (Devonian), Genesee County, New York. Diam. 2.5 mm. An irregular patch of uniformly oriented calcite (dark stippling plus cleavage) is enclosed and seemingly replaced by microcrystalline quartz (light stippling). Dolomite euhedra, some of which are zoned, are scattered through both chert and calcite.

be entirely chalcedony, but more commonly the central portions are quartz that is finely granular but coarser than that usually characterizing chert that has formed by replacement of limestones.

Where chert nodules occur in dolomite rock, a question arises concerning the relationship of dolomitization to chertification in the original limestone. In some cases, replacement by chert preceded dolomitization, because original limestone textures (organic or oolitic) are well preserved in the chert, but have been completely destroyed in the dolomite. Elsewhere the relationship is uncertain. Idiomorphic rhombohedra of dolomite are common in the cherts in calcareous rocks (Figure 15–2A,C). If scattered dolomite rhombohedra are present in a limestone, later selective replacement of calcite by chert might leave the dolomite crystals "floating" in the chert, but such an explanation cannot apply where no dolomite is present in adjacent unsilicified limestone. Alternatives would be contemporaneous formation of chert and dolomite or replacement of chert by dolomite. Dolomite euhedra and chert are so often associated in limestones that a genetic relation between the two seems probable. Perhaps a similar diagenetic environment is favorable to the formation of both chert and dolomite and has resulted in an overlapping of silicification and dolomitization.[5]

PYRITIC AND GLAUCONITIC ROCKS, IRONSTONES, AND BANDED IRON-FORMATIONS

A small amount of iron is present in most rocks and is particularly evident if the rocks are colored brown or red by iron oxides. Because of their color, reddish rocks are commonly referred to as ferruginous, but most of them contain no more iron than other rocks of less striking hue; the color reflects the state of oxidation of the iron rather than its abundance. In some sedimentary rocks, however, concentrations of iron-bearing minerals and increased iron content have resulted from some particular combination of circumstances in the sedimentary environment. Such rocks constitute significant sedimentary facies. They may be brown or reddish, but many of them are black, gray, green, or buff-colored, depending on the particular iron-bearing minerals that predominate in them.

Iron-bearing minerals are sensitive to conditions fostering oxidation or reduction of iron, so that variations in oxidation-reduction potential (Eh) and alkalinity (pH) in the depositional environments are responsible for a number of strikingly different mineralogical facies in iron-rich deposits. Pyrite, FeS_2, is the principal iron-bearing mineral formed under strongly reducing conditions that may occur in restricted environments where organic matter accumulates and oxygen supply becomes depleted. If there is a sufficient supply of iron and of H_2S from bacterial decay of organic matter, pyrite may be formed in great abundance. By contrast, under oxidizing conditions, the characteristic iron-bearing minerals are the ferric oxides, goethite (including "limonite") and hematite. Overlap between the ferric oxide facies and the pyrite facies is rarely, if ever, encountered. The principal iron-bearing minerals that tend to form under intermediate conditions are the ferrous carbonate siderite, $FeCO_3$, usually with minor substitution of Ca^{2+}, Mg^{2+}, and Mn^{2+} for Fe^{2+}; the silicates glauconite, chamosite, or greenalite; and the ferro-ferric oxide magnetite.

Two general kinds of sedimentary iron accumulations represent extraordinary concentrations of iron, namely, *banded iron-formations* and *ironstones*. The content of iron in these rocks commonly reaches levels sufficient to provide ores of commercial value. Banded iron-formations are characterized by thin layers of chert interbedded alternately with thin iron-rich layers. They are by far the thickest and most extensive sedimentary iron accumulations and are of great economic importance. Exclusively Precambrian in age, they are particularly characteristic of early Proterozoic epicontinental basins of deposition occurring on all continents. Ironstones are thinner, less extensive deposits typically interbedded with terrigenous sediments or limestones and containing little if any chert. They are almost exclusively Paleozoic, Mesozoic, or Tertiary in age.

Most iron-formations and ironstones undoubtedly have been deposited in marine waters, but in environments that varied from shallow marine with unrestricted circulation to restricted basins with impeded circulation. The sources of the iron, the causes of its concentration, and the mode of deposition are somewhat speculative but undoubtedly involved special combinations of tectonic, physiographic, climatic, and biological factors. However, heavy precipitation of iron from Proterozoic seas on a worldwide scale is one line of evidence suggesting unique physical and chemical properties of ancient ocean waters.

Pyritic Strata

Pyrite generally forms in sedimentary deposits laid down in relatively stagnant waters deoxygenated by bacterial decay of organic matter. It is a common authigenic mineral in *sapropelic* deposits, the highly organic black limestones and shales. Deposits in which pyrite is very abundant are relatively few, but the fact that some do occur establishes a pyritic facies among the iron-rich sedimentary rocks. Some of the precambrian black graphitic slates associated with graywackes in the Lake Superior region are abundantly pyritic; in the Iron River district in Michigan, for example, some black slates contain as much as 40% pyrite. The mineral here may be an original constituent of the deposits, but it has undoubtedly been somewhat recrystallized during late diagenesis or weak metamorphism. It is disseminated throughout the rocks in very small subhedral crystals that are clearly evident only on polished surfaces. Among the Wabana (Ordovician) iron-rich strata in Newfoundland, some graptolitic shales contain exceedingly abundant oolitic pyrite.

Glauconitic Rocks

Glauconite is a hydrous potassium aluminum silicate containing both ferrous and ferric iron and some magnesium. It is a typically sedimentary mineral, formed by marine authigenesis, and is usually found as green sand-size pellets that are microcrystalline and have low aggregate birefringence (Figures 15–3 and 15–4). Fresh glauconite pellets are bright green, but oxidation to brown limonitic aggregates and leaching to gray kaolinlike aggregates proceed rapidly where the mineral is exposed to weathering. As a detrital mineral exposed to subaerial weathering, therefore, it is unstable.

The green color of the mineral is one of its more distinctive properties, although color alone does not distinguish it from some other iron-bearing silicates, such as chamosite, greenalite, and some illite. Deposits containing abundant glauconite are green throughout. The most typical glauconitic rock is *greensand,* which consists largely of glauconite pellets. There are all gradations from greensand to sandstones, mudstones, shales, and limestones in which glauconite is but a minor constituent.

Although the pellets are highly characteristic, glauconite is structurally a mica. When very fine it often forms the green pigment in some marine muds, but its micaceous character is most often seen where large authigenic crystals have been formed by recrystallization of microgranular pellets. These crystals resemble green mica, though they have only moderate birefringence and faint pleochroism, but their outlines remain more or less ovoid. Some are vermicular accordionlike grains elongated normal to the cleavage, a form that would scarcely survive appreciable transportation and must therefore develop in place (Figures 15–3B and 15–4C). An intermediate stage in the recrystallization of microgranular pellets shows patches and ragged flakes of micaceous glauconite within the microgranular material.

Glauconite is forming at the present time on some parts of the sea floor where other sediments are accumulating very slowly or not at all.

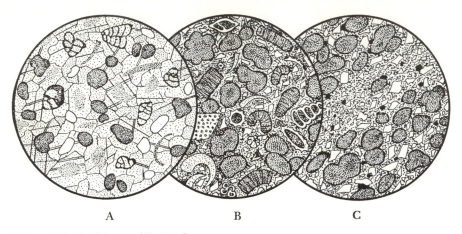

A   B   C

Figure 15–3. Glauconitic Rocks

A. Glauconitic foraminiferal limestone, Hollybush Hill, Malvern, England. Diam. 2 mm. Pellets of glauconite (dark stippling), subrounded quartz grains (blank), and microcrystalline foraminiferal tests float in a mosaic of granular calcite (light stippling). In the matrix, numerous dark calcite pellets (medium stippling) are kernels within calcite crystals, with which they are optically continuous. The rock is probably a recrystallized calcarenite.

B. Calcareous greensand (Oligocene), Duntroon, southern New Zealand. Diam. 2.5 mm. Very abundant glauconite in the form of microcrystalline pellets (heavy stippling) and vermicular crystals (heavy stippling plus cleavage), cemented with microcrystalline calcite (light stippling). Numerous fragments of coarse organic calcite (lower left), foraminiferal tests, and angular quartz grains (blank). Note glauconite within tests of foraminifers (center and right). Two zoned triangular grains in lower right are phosphatic fish teeth.

C. Pliocene greensand (Wildcat Group), Humboldt County, California. Diam. 2.5 mm. Very glauconitic sandy mudstone. Pellets of glauconite and angular fine sand and silt grains composed of quartz and feldspar (blank) enclosed in an argillaceous matrix. Glauconite concentrated in certain laminae (top and bottom). Scattered opaque microconcretions of pyrite.

Probably it develops by reconstitution of marine mica-type clay minerals on the sea floor. Its common occurrence as internal casts in the shells of foraminifers (Figures 15–3B and 15–4B) suggests that organic agents may be involved in its production, and some glauconite has been thought to develop from the fecal pellets of scavengers. In any case, it seems to form in rather restricted marine environments, which nevertheless have existed often and in many places, chiefly if not exclusively in shallow water where slightly reducing conditions prevail.

Like any other authigenic mineral, glauconite may be transported from its place of origin and incorporated in sediments elsewhere. Much

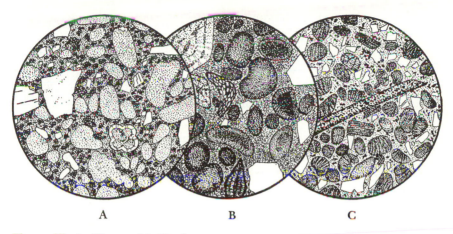

A   B   C

Figure 15–4. Glauconitic Rocks

A. Eocene greensand, 1600 m below surface, west of Stockton, California. Diam. 2.5 mm. Numerous pellets of yellowish-green glauconite in a green argilla-ceous matrix (stippled), which is probably also glauconitic. Dark granules of siderite or ankerite dispersed through the matrix; also numerous larger angular grains of quartz (blank) and a few of feldspar. One foraminiferal test filled with glauconite.

B. Glauconitic nummulitic limestone, Kressenberg, Bavaria. Diam. 2.5 mm. Dark-brownish pellets of glauconite partly altered to ferric oxides, angular quartz grains, and tests of Foraminifera, all enclosed in fine-grained calcite. Note glauconite in pores of two foraminifers (center and bottom).

C. Cretaceous greensand, Hazlet, New Jersey. Diam. 2.5 mm. Vermicular crys-tals and microcrystalline pellets of glauconite, in a green argillaceous matrix that is probably also glauconitic (stippled). Abundant angular silt and fine sand particles (blank) of quartz and feldspar, a phosphate shell fragment (center), and scattered small patches of pyrite and limonite (black).

pellet glauconite has probably been moved about on the sea floor, espe-cially that which is mixed and sorted with detrital sand. Because the mineral is readily weathered, it is not likely to survive much subaerial erosion, and it seems safe to assume that its occurrence indicates a marine environment of deposition.

Sideritic and Chamositic Ironstones

Siderite and chamosite, two minerals rich in ferrous iron, are the chief constituents of many ironstones. In some ironstones siderite alone is the iron-bearing mineral; in others the two minerals are to be found together; and in a few chamosite alone contains the iron. Presumably they have usually been formed in shallow marine waters where conditions were not strongly oxidizing. Siderite is ferrous carbonate ($FeCO_3$). Crystals of it are either colorless or light yellow, and may be clear when fresh and large enough to be examined individually. They tend to be zoned in shades of yellow and brown because of variations in composition. Chamosite is a hydrous aluminum silicate of ferrous iron. In individual crystals large enough to be distinguished the mineral has a micaceous aspect, but it is usually micro- or cryptocrystalline. It is bluish green when fresh, but yellowish green when somewhat oxidized.

Some ironstones contain little besides carbonate minerals. The most ferruginous of these are *siderite mudstones,* consisting essentially of microgranular or finely granular siderite. They are marine and were probably laid down as fine siderite muds, later recrystallized to more compact aggregates. *Sideritic limestones* contain mixtures of calcite, dolomite, and ankerite with siderite. In these, as with dolomite in limestones, siderite may occur as rhombohedra scattered throughout the rock or as more or less irregular aggregates. Replacement of calcite by siderite is common and is especially evident where organic remains have been sideritized; yet unaltered calcareous fossils are sometimes enclosed in a siderite matrix. There are all gradations between pure siderite rock and sideritic limestones of low iron content.

Green microcrystalline rocks in which the chief constituent is chamosite are called *chamosite mudstones.* These are usually associated with siderite mudstones; the two minerals, in fact, are often intimately intermingled. Most striking of the chamositic rocks, however, are those having oolitic structure. Chamosite ooids invariably have a pronounced concentric, or onionlike, structure. The mineral is so oriented that the direction of the slow ray is tangential to the ooid structure. The aggregate birefringence is low, but each ooid produces a well-marked extinction cross under crossed polarizers. In many ooids there appears to be no nucleus; others have developed around large flakes of chamosite or around particles of sand, silt, or carbonate. Where ooids are particulary abundant, they may be flattened and distorted as though they had been plastic.

Ironstones containing chamosite ooids vary because of differences in the matrix. In some the matrix is chamositic mud, usually containing some siderite and scattered grains of sand, and the rock has an overall greenish color. In others the matrix is finely granular carbonate (Figure 15–5B). The textural relationship of siderite to chamosite often suggests that the former has replaced the latter. Ooids of chamosite in a siderite

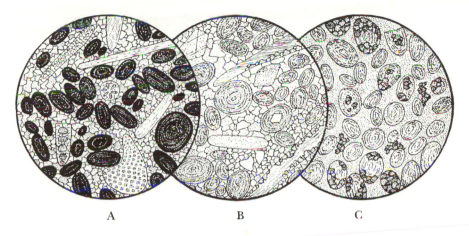

A   B   C

Figure 15–5. Ironstones

A. Frodingham Ironstone (Lias), Scunthrope, Lincolnshire, England. Diam. 2 mm. Ovoid limonite ooids in a shelly limestone. Ooids are brown, concentrically banded, and translucent in thin section. The matrix is finely granular calcite, containing a variety of abraded shell fragments, some of which are granular and some fibrous. Cavities in three shell fragments (center and lower part) are filled with green chamosite (stippled).

B. Northampton Sand Ironstone (Lias), Corby, Northamptonshire, England. Diam. 2 mm. Sideritic limestone containing numerous chamosite ooids (stippled lightly) and also shell fragments and grains of detrital quartz (blank). One ooid has quartz nucleus. An abraded phosphate shell fragment (stippled) in lower center, two fibrous shell fragments marginally replaced by siderite.

C. Northampton Sand Ironstone (Lias), Irthlingborough, Northamptonshire, England. Diam. 2 mm. Chamosite ooids in a matrix of chamosite mud. Both matrix and ooids partly replaced by patches of granular siderite.

matrix may have irregular outlines as though penetrated by growing siderite, and some have been extensively replaced. Completely sideritized ooids are granular, and their origin is indicated only by their spheroidal outlines. In a single rock nearly pure chamosite ooids often occur side by side with ooids that are more or less completely sideritized (Figure 15–6B), as if they had formed in different environments and became mixed mechanically by being moved about on the sea floor. A similar process probably accounts for the frequent occurrence of fractured ooids. Sideritization has also etched and frayed the margins of many detrital quartz grains.

The ferrous iron of siderite and chamosite is readily oxidized by weathering, and at the surface the rocks containing them have generally

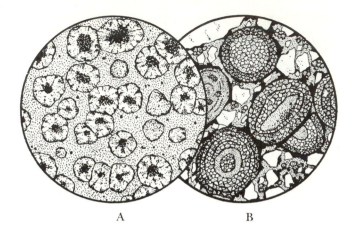

A   B

Figure 15–6. Ironstones

A. Siderite spheroids in clay matrix (Carboniferous), Lawrence County, Ohio. Diam. 2.5 mm. The central zone in most spheroids is darkened with ferric oxide. Coalescing spheroids meet along nearly plane surfaces, and where they are intergrown in aggregates each has polygonal form. Clay between closely adjacent spheroids tends to be oriented with flakes more or less tangential to spheroid surfaces, producing local aggregate polarization of the matrix.

B. Sandy ironstone (Eocene), Department of Boyaca, Colombia. Diam. 2.5 mm. This specimen is from a bed that underlies hematitic ironstones of higher iron content. Angular detrital quartz grains (blank) and siderite-chamosite ooids enclosed in a chamosite matrix (stippled). Scattered siderite granules and limonite (black) occur in matrix. One ooid with pronounced concentric structure consists almost entirely of chamosite (left), one at the top consists entirely of granular siderite, and others contain both minerals. Probably granular siderite has replaced chamosite. Note broken ooid in lower right.

been converted to limonitic aggregates. But the hydrous oxide, goethite, also occurs in them as lustrous dark-brown ooids, the so-called *ironshot*, which may be either original ferric oxide ooids or the result of oxidation of chamosite ooids during diagenesis (Figure 15–5A).

Mixtures of clay and siderite are *clay ironstones*. They are essentially mixtures of siderite mud and argillaceous material, deposited together either in shallow marine waters or in stagnant lagoons and lakes. Some of these rocks form extensive thin strata, but they occur mainly as nodular or lenticular beds in shales associated with coal seams.

Particularly striking are clay ironstones containing siderite spheroids. The spheroids are small (usually about 0.5–1 mm across) and set in a clay matrix. Where they are widely scattered, they are of only scientific interest, but in places they become so abundant as to coalesce and form iron-rich deposits, occasionally of commercial value (Figure 15–6A). Siderite spheroids in clay are perhaps produced by recrystallization of siderite originally disseminated through the deposit. In coal-bearing strata they are often found at the base of an underclay.

Siderite in spheroids either is finely fibrous or forms sectorlike crystals radially arranged. The crystals are elongated parallel to the $z(c)$ axis, for the brush of high relief is invariably oriented at right angles to the plane of the polarizer. Roughly concentric zones within the spheroids are frequently made evident by being colored in various shades of brown. They are rarely pure iron carbonate, and usually contain both magnesium and calcium in minor amounts. Most of them are without nuclei, but some have engulfed at random particles of silt that apparently were in their path as they grew. The finer clay particles appear to be excluded from most spheroids, but clay flakes in the matrix appear to have been shouldered aside and crudely oriented by growth of the spheroids.

Hematitic Ironstones

Sedimentary ironstones in which hematite is the chief iron-bearing mineral are well known, and in some places they constitute important iron ores. Among the latter are the Clinton (Silurian) iron ores of the Appalachian region. Essentially similar deposits, both young and old, are found in many other parts of the world.

One type of hematitic ironstone is oolitic. The hematite ooids are apparently built by accretion around nuclei, for they have a well-developed and delicate concentric structure. The hematite is microcrystalline and is nearly opaque in thin section, so that the internal structure of the ooids is not clearly evident unless the section is exceedingly thin. On polished surfaces or polished thin sections, however, the concentric layers are clearly visible in reflected light because they are colored in different shades of dark red and brown. Some ooids contain interlayered thin concentric bands of chamosite; indeed, this association may be more common than superficial examination suggests, for the thin layers of chamosite are usually obscured by the reddish hematite around them.

The nuclei of hematite ooids are usually grains of sand or silt, bits of shell, or pieces of fractured ooids, although some ooids have no apparent nucleus. Most strata composed of oolitic hematite are, in fact, clastic deposits, in which the ooids themselves may have been mechanically mixed and more or less sorted with other detrital and fragmented materials. Where the ooids predominate, they are close-packed and usually somewhat flattened parallel to the bedding plane. Being disk-shaped rather than spherical, such aggregates have an appearance that has caused them to be referred to as *flaxseed ore.*

Hematite may impregnate sandstones and silty mudstones as a cement. These rocks, because of their high quartz content, contain less iron than those that are abundantly oolitic, but the two types intergrade. Commonly, for example, hematite ooids containing nuclei of detrital quartz are scattered through a matrix of hematitic sand or silt (Figure 15–7C).

Some hematitic ironstones are essentially biosparites in which hematite has precipitated around each fossil fragment and has filled the pores in the skeletal matter (Figure 15–7A). Locally hematite has replaced the calcareous structures, producing deposits of higher than usual iron content. In the Appalachian region, where this type of ironstone is common, it is called *fossil ore.* Siderite occurs in some of these hematitic limestones but is not as widespread as calcite.

Such hematite-rich deposits are generally interstratified with other shallow marine deposits. They are usually thin, but are persistent and more or less uniform in character. It therefore seems probable that they were formed on the sea floor in places where favorable environmental conditions existed. The hematite may be a direct precipitate from sea water or an early diagenetic replacement of other materials. The intimate interlayering of hematite and chamosite (ferric and ferrous minerals respectively) in some ooids suggests that the hematite in them is not simply the result of oxidation of an earlier mineral such as chamosite.

Banded Iron-Formations

The banded character of the Precambrian iron-formations results from interlayering of thin chert beds with other thin beds in which iron-bearing minerals predominate. Typically the chert comprises 30% to 40% of the formations. It consists essentially of microgranular quartz, although it is generally ferruginous and contains minor amounts of the associated iron-bearing minerals. Oxide, silicate, and carbonate facies of iron-formation, characterized by differing assemblages of the iron-bearing minerals,[6] are more or less comparable to the facies of younger ironstones, but there are several noteworthy mineralogical differences. In the Pre-

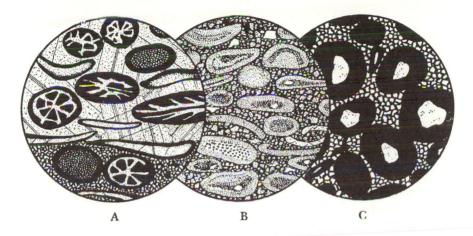

A   B   C

Figure 15–7. Clinton Hematitic Ironstones

A. Hematitic fossiliferous limestone ("fossil ore"), western New York. Diam. 3 mm. Each shell fragment is encased by dark-red hematite (black), and all are cemented by calcite. Note branching form of bryozoan (*Helopora*) solidly filled with hematite; little replacement of organic calcite by hematite appears in this specimen, but it is common in some. Matrix in upper part is coarse calcite; in lower part it is composed of fine-grained carbonate rhombohedra cemented by hematite.

B. Chamositic lamina in hematitic limestone, western New York. Diam. 3 mm. Curved and elongate ooids of chamosite (stippled), surrounded by granular calcite impregnated with hematite. Concentrically banded chamosite has usually formed around carbonate nuclei (darker stippling), either organic fragments or granular pellets. Small angular detrital quartz particles (blank) are scattered throughout; recrystallized calcite shell fragments at right edge of figure.

C. Sandy oolitic hematite, western New York. Diam. 3 mm. Dark-red hematite ooids enclosing detrital quartz grains and surrounded by fine quartz sand impregnated with bright-red hematite. The ooids have a distinct concentric structure not shown here.

cambrian iron-formations, iron oxides are invariably anhydrous hematite or magnetite, or both; the hydrous oxide goethite does not occur. Furthermore, the iron silicates in iron-formations are nonaluminous, chiefly greenalite; chamosite, the characteristic iron silicate in the younger ironstones, does not occur in the iron-formations. The characteristic iron-bearing carbonate in iron-formations is siderite, as it is in the ironstones, but among accessory carbonate minerals dolomite is more common than calcite.

Hematite and magnetite are the most abundant iron-bearing minerals in the iron-formations of the Lake Superior and Labrador regions. A purely oxide facies consists of thin beds and laminae of fine-grained, commonly steel gray hematite interbedded with chert. Some hematitic layers are oolitic, the individual ooids consisting of concentric zones of hematite and of quartz; some ooids are highly quartzose, others largely hematite. In some of the iron-formations, magnetite predominates, but, unlike hematite, it usually occurs with associated iron silicates and siderite. Where iron-bearing silicates predominate, magnetite is a typical associated phase. The carbonate facies is characterized by finely granular siderite interbedded with chert, but although the siderite may occur alone, it commonly is associated with iron silicates and magnetite. Pyrite is less widespread but is locally abundant, as in the Iron River District of Michigan (p. 408); it does not occur in association with either magnetite or hematite.

As might be expected in view of their Precambrian age, many iron-formations have been recrystallized to varying degrees by regional metamorphism. Increasing grain size in the chert beds appears to correlate with increasing degree of metamorphism. The formations characterized by chalcedonic or very fine-grained chert presumably have been little altered and include iron-bearing minerals that were present in the original deposits. Indeed, the remains of primitive microscopic plants are recognizable in some of the cherts, for example, in the Gunflint iron-formation of Ontario.[7] In these rocks, greenalite is the characteristic silicate; in metamorphosed iron-formations the characteristic silicates are minnesotaite, stilpnomelane, and iron-rich chlorite, and in some regions, notably western Australia and South Africa, fibrous riebeckite (crocidolite).

PHOSPHATIC SEDIMENTS

Phosphate in rocks occurs largely in the form of apatite, $Ca_5(PO_4)_3$ (OH,F,Cl), a widespread minor accessory mineral in igneous, sedimentary, and metamorphic rocks. In sedimentary rocks, detrital terrigenous apatite occurs in the heavy accessory fraction of many sandstones, but in richly phosphatic deposits, the calcium phosphate has been precipitated from marine waters. Sea water below the surface layers is virtually saturated with tricalcium phosphate. It is, of course, an exceedingly dilute solution, yet to maintain equilibrium about as much calcium phosphate must be deposited in marine sediments as is continuously added to the ocean by streams and springs. The total amount deposited is very large, but most of it is widely scattered as minor constituents in many sedimentary rocks.

In the rocks known as *phosphorite,* calcium phosphate has been concentrated. In these, the most common form of sedimentary phosphate is *collophane,* a cryptocrystalline phosphatic substance producing the x-ray pattern of apatite. It may consist of any of several varieties of apatite, and warrants a separate name because of its distinctive appearance. The minuteness of the crystals in collophane, combined with their low birefringence, makes it appear sensibly isotropic in thin sections, although it sometimes shows a faint birefringence and a suggestion of microfibrous structure. It is distinguished from other isotropic or nearly isotropic minerals in sedimentary rocks by its high relief (refractive index 1.59–1.63). Its color varies. In thin sections it may be clear and nearly colorless, but more commonly it is translucent and tinted in shades of brown or yellow. Deeply colored fibrous varieties may be pleochroic. When clouded by inclusions, usually of organic matter, collophane is likely to be black and may be opaque. During diagenesis, recrystallization and replacement may produce more coarsely crystalline varieties of apatite. Apatite also occurs as cavity linings, as crystalline cements in sandstones, and as disseminated crystals in various sediments.

Much collophane is biogenic. It is the chief constituent of bones and teeth and of some marine shells, especially those of certain brachiopods, and as such is widely scattered through many sediments. The fecal residues, also, of many organisms are decidedly phosphatic, and a fecal origin has been attributed to many ovoid phosphatic pellets in sediments. Like glauconite, organic phosphates become major constituents in sediments only where other sedimentary material is relatively scarce and slowly deposited. Glauconite and particles of organic collophane are, in fact, often associated in greensands (Figures 15–3B and 15–4C).

Inorganic collophane, also, is widespread. Many phosphorites, for example, consist simply of sand, silt, or clay cemented with collophane. In the most phosphatic of these, usually encountered as nodules in mudstones, detrital particles float in a cloudy collophane matrix, as if the phosphate, while being precipitated, included a little detritus by chance. Perhaps these nodules are akin to the collophane nodules now forming by slow chemical accretion on submarine banks where detrital accumulation is negligible. Some of them are small, but others are large and consist of many thin layers, which vary in content of impurities of sand, silt, clay, manganese oxide, and foraminiferal tests. Beds containing an abundance of such nodules may be interstratified with nonphosphatic shale, mudstone, sandstone, limestone, and chert.

In some sandy phosphorites there is a collophane cement in the interstices between the sand grains. Here the texture is determined by the detrital material, and the phosphate is probably precipitated during early diagenesis. It is usually colorless and slightly birefringent, and each sand grain tends to be rimmed with a thin layer of minute prisms standing normal to the grain surface and projecting into the pores. These are easily recognized if cementation is incomplete, for the pores then become tiny geodes lined with minute apatite crystals (Figure 15–8). When collophane completely fills the pore space in a sand, it may appear cryptocrystalline throughout, but close examination usually reveals a thin fibrous rim around each detrital particle.

Collophane ooids are common in many phosphorites. They are generally ovoid and have a fine concentric layering revealed by slight variations in color (Figure 15–9A). Most of them appear isotropic through-

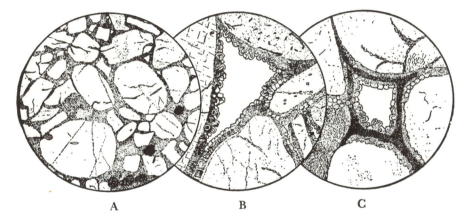

A   B   C

Figure 15–8. Phosphate Cement in Sandstone, Otago, New Zealand

Drawn by C. O. Hutton and previously published in Royal Society of New Zealand, *Transactions*, no 72, pt. 3 (1942): p. 196.

A. Grains of quartz, with some albite and glauconite (the dark pellets in lower right), set in a microcrystalline cement of francolite. Diam. 3 mm.

B. Geodelike cavity between detrital sand grains; francolite crystals radiating outward from quartz and albite. Diam. 0.7 mm.

C. Geode of francolite between quartz grains; areas adjacent to glauconite grains (lower left) are completely filled with francolite. Diam. 0.7 mm.

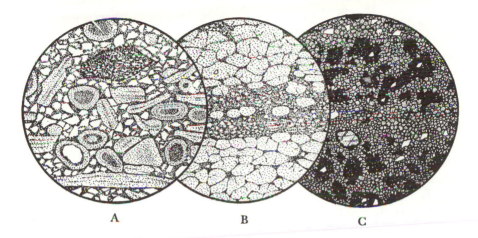

A   B   C

Figure 15–9. Phosphorites of the Permian Phosphoria Formation

A. Sandy clastic phosphorite, near Jackson, Wyoming. Diam. 3 mm. Collophane is stippled; angular and subangular detrital quartz grains are blank. Fragments of bone and phosphatic shells (some no doubt have been phosphatized); collophane ooids and ovules; a chip of silty phosphate rock (top center); and numerous quartz grains; all cemented together by collophane.

B. Phosphatic shale, Pris, Idaho. Diam. 3 mm. Collophane ovules (light stippling) in a matrix of micaceous clay (dark stippling) containing numerous clear silt grains composed of quartz. Phosphate ovules are tightly packed and somewhat molded to each other, but are separated by films of silty clay. The phosphate is concentrated in two laminae at the top and bottom of the figure.

C. Phosphatic limestone, Conda, Idaho. Diam. 3 mm. Dark-brown or opaque collophane (black) forms patches in a mosaic of granular calcite. There is a little intergranular phosphatic material in the calcite mosaic. Scattered particles of detrital quartz (blank). Collophane patches are concentrated in two horizontal laminae in upper and lower parts of figure; they are ragged, perhaps being collophane pellets partially replaced by calcite.

out, although in some a few of the concentric layers are finely fibrous and slightly birefringent. Nuclei are apparently lacking in many phosphate ooids, though a thin section might by-pass many nuclei, but at the centers of some one finds a grain of detrital mineral, a tiny phosphate pellet, or even a group of pellets. Some ovoid collophane grains have no concentric structure. These have been termed *ovules* (Figure 15–9B) in order to distinguish them from the concentrically layered ooids undoubtedly formed by accretion. These textures are well represented in strata of the Permian Phosphoria Formation in the northern Rocky Mountain region (Figure 15–9A,B).[8]

Probably much inorganic collophane is an original precipitate, but some was certainly formed by diagenetic replacement. Some calcareous shells and parts of some limestones, for example, have been replaced by collophane, and in some places phosphorites have been formed by this process.

Clastic phosphorites deposited in marine waters may contain collophane in ooids, ovules, fragments of bone and teeth, phosphatized shell fragments, and bits of older phosphorite rock. One or another of these may predominate in a particular stratum, or there may be a mixture of fragment types (Figure 15–9A). Increasing admixture of sand and argillaceous material produces all gradations from richly phosphatic rocks to those containing little phosphate. Other clastic phosphorites are concentrations of phosphatic nodules and pebbles produced by subaerial weathering and erosion of older phosphatic beds. Some of these are essentially residual, while others have been transported, including the pebble phosphate deposits of Florida.

ANHYDRITE AND GYPSUM

Deposits of anhydrite ($CaSO_4$) and gypsum ($CaSO_4 \cdot 2H_2O$) are strictly chemical deposits, for these minerals are not biogenic and very rarely form clastic deposits. They and halite (NaCl) are the most common constituents of *evaporite* deposits, so called because they are formed as a result of evaporation.

Large evaporite sequences such as occur in numerous extensive intracratonic basins—for example, the Permian Zechstein Basin in northern Europe and the Salina Basin of Silurian-Devonian age in the Michigan-Ohio area—may reach thicknesses of several hundred meters and extend laterally for many kilometers. Such evaporites have been assumed to be precipitated from standing bodies of water, generally from restricted or landlocked portions of the sea. However, extensive evaporite deposition also may occur from saline pore waters in sediments exposed on arid supratidal flats, and also in shallow nonmarine brine pools and playas. Gypsum and anhydrite are being deposited today, for example, on the sabhkas of the Trucial coast, on the southern side of the Persian Gulf, where the rate of evaporation is high and extensive supratidal flats are periodically flooded by sea water. There is good evidence to suggest that at least some of the ancient marine evaporites, rather than being precipitated from standing water bodies, are thick accumulations that originally were deposited on supratidal flats and after burial were modified extensively by diagenesis.

Evaporites in general, because they are more soluble than other rocks, are particularly susceptible to diagenetic changes. Saturated with brine, and perhaps subjected to rather high temperature and pressure as a result of deep burial, they tend to recrystallize; replacement of one mineral by another is commonplace, including in some deposits various complex and highly soluble salts such as occur in the Stassfurt deposits in Germany, where the typical reconstituted salt assemblages are, in fact—as Van't Hoff demonstrated early in the century—metamorphic systems in a state of thermodynamic equilibrium. One diagenetic change, the conversion of anhydrite to gypsum, is accompanied by such expansion of volume that intricate contortion of the beds commonly results. Deep burial would be expected to cause the reverse change, from gypsum to anhydrite. The fact is that gypsum typically occurs only at relatively shallow depths.

Many deposits of gypsum and anhydrite are intimately associated with detrital deposits of mud and sand, and in most evaporite basins the deposition of sulfate is preceded and overlapped by formation of limestones and dolomites. Gypsum and anhydrite also may replace lithified carbonate rocks, or fill cavities in them, and they may form authigenic cements in detrital accumulations. Hence, although some anhydrite and gypsum deposits are very pure, one usually finds in them admixtures of sand, silt, clay, and carbonates (especially dolomite). Common minor constituents include magnesite, hematite, carbonaceous pigment, pyrite, barite, and celestite.

Anhydrite is a colorless orthorhombic mineral possessing three good cleavages at right angles to each other. Its birefringence is moderate, generally producing second- or third-order interference tints in thin sections, and its refractive indices are high enough (1.57–1.62) to produce moderate relief. Gypsum, in contrast, has low refractive indices (1.53±), very low birefringence, and cleavage that is perfect in one direction and imperfect in two others not at right angles.

The textures of anhydrite aggregates are of three general types: (1) mosaics of interlocking anhedral grains usually having stumpy rectangular habit but sometimes shapeless; (2) aggregates of elongated or bladed crystals with irregular margins, often arranged in radiating fanlike groups but sometimes having subparallel orientation; (3) nonuniform aggregates in which rectangular porphyroblasts or radiating groups of elongated crystals are set in a mosaic of anhedral grains. The grain size of all types may vary from fine to coarse. Whether all anhydrite fabrics are to be classed as crystalloblastic is uncertain, but it seems likely that any anhydrite buried to considerable depth has been recrystallized. Figure 15–10 represents several of the commonest textural types.

Most anhydrite rock contains magnesium-bearing carbonate. This is chiefly dolomite; magnesite is common, but calcite is relatively rare. Some dolomite aggregates are enclosed and probably replaced by anhydrite (Figure 15–10A,B), the most striking of these being finely granular dolomite masses caught between radiating blades of anhydrite. In some rocks idiomorphic dolomite rhombohedra (Figure 15–10C) or small, round anhedral grains are scattered through anhydrite mosaics, and

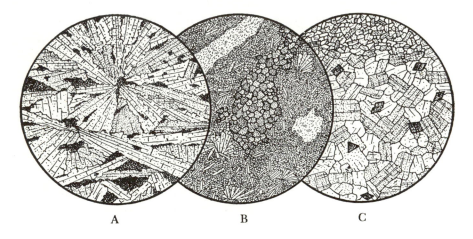

A   B   C

Figure 15–10. Anhydrite

A. Radial fabric in coarse anhydrite (Permian), Stassfurt, Germany. Diam. 3 mm. Bladed crystals of anhydrite (light stippling plus cleavage) radiating from one center in the field and several others outside it. Between anhydrite crystals are microgranular aggregates of dolomite or magnesite. Note very irregular margins of anhydrite grains.

B. Nonuniform fabric in anhydrite (Mississippian), Hants County, Nova Scotia. Diam. 3 mm. Finely granular anhydrite, enclosing larger scattered prisms and radiating groups of prisms of anhydrite. Irregular patches of granular dolomite (center) occur throughout the rock and have been partly replaced by anhydrite. Local gypsum veinlets and patches (top left and right; shown in light dashes) replace anhydrite.

C. Granoblastic anhydrite (Permian), Eisleben, Thuringia, Germany. Diam. 3 mm. Mosaic of stumpy rectangular anhydrite crystals (note right-angle cleavage), medium-grained below and fine-grained above, containing scattered euhedra of dolomite (dark stippling).

some may be enclosed within single crystals of anhydrite. Replacement of dolomite by anhydrite is frequently reported, but the dolomite occurring as scattered euhedra, at least, is probably deposited contemporaneously with the anhydrite or has replaced it.

Micaceous clay minerals, commonly accompanied by brick-red ferric oxide and sand, are dispersed throughout some anhydrite deposits. These were presumably formed by precipitation of anhydrite contemporaneously with the washing in of fine clastic debris. More commonly precipitation of anhydrite alternated with the influx of detritus, and the deposits consist of thin layers of sand, clay, or marl interlaminated with anhydrite. Such deposits grade into a type in which lenticles of anhydrite are enclosed by streaky clay. Anhydrite is also a minor constituent in many deposits of rock salt.

At moderate depths where only partial hydration by groundwaters has occurred, anhydrite and gypsum are found together. The gypsum occurs in irregular patches or veins replacing anhydrite rock (Figure 15–10B), in intricate networks of veinlets replacing single anhydrite crystals, or in large porphyroblasts enclosing bits of uniformly oriented anhydrite.

Nearer the surface, rocks composed wholly of gypsum are common. The textures of these, even within single specimens, are variable as to both size and form of grains. Some fine-grained gypsum beds are fibrous mosaics, being aggregates of interlocking anhedral grains each of which appears fibrous and ill-defined between crossed polarizers. In such fabrics it is almost impossible to trace the boundaries between the crystals. In some specimens the fibrous mosaic forms a matrix enclosing larger gypsum crystals, which also have a somewhat fibrous aspect and hazy outlines (Figure 15–11B). Other gypsum fabrics consist of intricately interlocking equidimensional anhedra that are not fibrous but extinguish sharply. These may be even-grained (a common texture in alabaster), but more commonly they vary markedly in grain size (Figure 15–11A). Gypsum that occurs in veinlets or veinlike lenses, whether parallel to the bedding or cutting across it, tends to form prisms or fibers normal to the walls of the veins (Figure 15–11C). Furthermore, gypsum is readily deformed under nonhydrostatic stress at low confining pressures, so bending and kinking of gypsum grains (as seen in sections parallel to {100}) is widespread in some gypsum rocks.

Deposits of gypsum, like those of anhydrite, are likely to be impure. Clay, sand, and carbonate admixtures are the most common; these are often interlaminated with gypsum or disseminated through it in a streaky

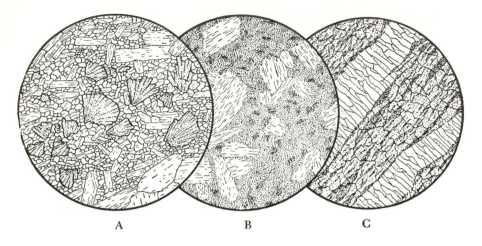

A   B   C

Figure 15–11. Gypsum

A. Triassic gypsum (Keuper), Württemberg, Germany. Diam. 3 mm. Uneven fabric, consisting of fine granular gypsum shot through with larger tabular crystals. Scattered radial groups of barite crystals, which appear to have been partly replaced by gypsum.

B. Pliocene gypsum, San Marcos Island, Gulf of California, Mexico. Diam. 3 mm. Fine-grained mosaic of indistinct fibrous gypsum grains, enclosing scattered larger gypsum crystals (porphyroblasts). Many of these are fibrous and have irregular margins, but some are nearly euhedral (top).

C. Argillaceous gypsum (Silurian), Monroe County, New York. Diam. 3 mm. Granoblastic mosaic of medium-grained gypsum, traversed by lenticular streaks of micaceous clay (short-line pattern) and by two clear gypsum veinlets. Note that the boundaries of gypsum grains have no relation to clay streaks. The two veinlets, filled by cross-oriented gypsum prisms, trend approximately parallel to the bedding. The rock is probably a recrystallized gypsiferous shale.

fashion parallel to the bedding (Figure 15–11C). Such impurities commonly represent sediments that become mixed with a primary calcium-sulfate precipitate that was being deposited in a desiccating basin. In many arid regions, however, gypsum tends to crystallize later within the soil and in shallow layers of sediment, forming fine-grained impure deposits called *gypsite*. In contrast to this material is *selenite*, which forms large clear crystals. Much selenite is deposited in organic shales by groundwaters, and it also forms in sands, where single crystals may enclose innumerable detrital grains.

GENERAL REFERENCE

The following volume contains a series of articles concerning the petrography and origin of calcareous and siliceous pelagic sediments and effectively summarizes the findings of the Deep Sea Drilling Project: K. J. Hsu and H. C. Jenkyns, *Pelagic Sediments on Land and under the Sea,* Special publication of the International Association of Sedimentologists, no. 1 (1974).

A succinct, up-to-date summary of the characteristics and probable origin of ironstones and iron-formations is given by R. L. Stanton, *Ore Petrology.* New York: McGraw-Hill, 1972. pp. 399–453.

Phosphorites formed on the continental shelves off Peru and South Africa are described in the following two papers, respectively: F. Manheim, G. T. Rowe, and D. Jipa, *Journal of Sedimentary Petrology,* vol. 45 (1975): pp. 243–251; G. F. Birch, *Journal of Sedimentary Petrology,* vol. 49 (1979): pp. 93–110.

For general reference on evaporite deposits, see D. W. Kirkland and R. Evans, eds., *Marine Evaporites: Origins, Diagenesis, and Geochemistry* (Stroudsburg, Pa.: Dowden, Hutchinson, and Ross, 1973); this volume presents reprints of important original papers.

REFERENCE NOTES

[1] See R. L. Folk and J. S. Pitman, *Journal of Sedimentary Petrology,* vol. 41 (1971): pp. 1045–1058. On late pleistocene cherts formed in Lake Magadi, see J. R. O'Neil and R. L. Hay, *Earth and Planetary Science Letters,* vol. 19 (1973): p. 259.

[2] M. N. Bramlette, *U. S. Geological Survey Professional Paper,* no. 212 (1946); K. J. Murata and R. R. Larson, *U.S. Geological Survey Journal of Research,* vol. 3 (1975): pp. 553–566.

[3] For example, see A. H. F. Robertson, *Sedimentology,* vol. 24 (1977): pp. 11–30.

[4] R. L. Folk and E. F. McBride, *Journal of Sedimentary Petrology,* vol. 48 (1978): pp. 1069–1102.

[5] See, for example, L. P. Knauth, *Geology,* vol. 7 (1979): pp. 274–277.

[6] H. L. James, *Economic Geology,* vol. 49 (1954): pp. 235–293; H. L. James, *U. S. Geological Survey Professional Paper,* no. 440-W (1966).

[7] E. S. Barghoorn and S. A. Tyler, *Science,* vol. 147 (1965): p. 563.

[8] For a description of the physical stratigraphy and lithology of the Phosphoria Formation, western Wyoming, see R. P. Sheldon, *U. S. Geological Survey Professional Paper,* no. 313-B (1963).

PART THREE
METAMORPHIC ROCKS

16

Metamorphism
and Its Lithologic Expression

METAMORPHISM: NATURE AND GENERAL SCOPE

Minerals of igneous rocks for the most part have crystallized as stable phases in contact with a silicate melt at temperatures from perhaps 650°C to 1100°C and pressures ranging from near-surface values to those appropriate to deeper levels in the crust—5 kb to 10 kb. Many igneous rocks themselves approximate polyphase systems in internal equilibrium over some finite pressure-temperature (P-T) regime within the above span. Some authigenic and organically derived minerals of sediments are stable under surface conditions (some such as adularia, aragonite, and opal are only metastable). Even the common clastic minerals—quartz, feldspars, and white micas—though originating in environments of higher temperatures, have survived long cycles of weathering, transport, deposition, and burial, thus demonstrating relative stability in the presence of water under surface conditions.

Extensive masses of sediment and volcanic rocks that have been tectonically depressed—for example, by basin subsidence or by sinking and deformation of ocean-trench fillings—thereby enter an alien environment of variably high temperature and pressure which itself tends toward a steady state only with the lapse of time. The situation may be further complicated by plutonic intrusion especially of granitic or granodioritic magma. Rising plutons generate foci of high temperature in their immediate vicinity. Where large-scale intrusion is more or less synchronous with orogeny, as is common, its influence may profoundly affect the regional thermal regime. Even the plutonic rocks themselves during protracted post-consolidation cooling and partial unroofing become exposed to physical conditions significantly different from those under which they crystallized. Thus in the deeper levels of the crust, rocks are habitually subjected over periods of millions of years to a changing *P-T* regime that departs notably from that of their origin. The spontaneous internal mineralogical and textural responses to the imposed conditions that thus become possible collectively constitute rock metamorphism—a process that takes place in the solid rock medium without intervention of a melt phase. All metamorphic changes, since they are spontaneous, must, according to thermodynamic principles, result in a decrease in free energy within the reacting system; they tend, in other words, toward a new state of equilibrium governed by the imposed conditions of pressure and temperature. Whether truly stable equilibrium is ever achieved depends on kinetic factors—lapse of sufficient time and the accelerating influence of permeating aqueous fluids, shear stress (which promotes solid-state flow and generates strain energy in individual crystals), and, above all, high temperature (for reaction rates increase exponentially with temperature).

The products of metamorphism, our immediate concern, are *metamorphic rocks*. There can be no sharp lines of demarcation between metamorphic and either sedimentary or igneous rocks. At low temperatures mineral products of incipient metamorphism are identical with some that are conventionally treated by sedimentologists as diagenetic—"chlorites" and "white micas" (both in the broad sense of the terms), zeolites, albite, quartz, and carbonates. Petrologic usage here merely reflects the attitude of the individual investigator. It is now generally realized that appearance of newly generated zeolites, prehnite, and chlorite as a matrix enveloping clastic quartz and feldspar in graywacke is the normal precursor to more complete reconstitution that can only be termed metamorphic.

At the other end of the depth scale, the respective regimes of metamorphism and partial fusion must surely overlap. Rocks that are thought to reflect these conditions are the *migmatites*: Seen in hand specimen or outcrop, these are streaky and crudely layered rocks with interlaminated dark metamorphic and light "granitic" (probably magmatic) materials. Migmatites are extensively developed over large areas of the continental shields and in the deeper levels of mobile belts.

RECOGNIZABLE CATEGORIES OF METAMORPHISM

In keeping with the general descriptive approach of this book, our principal concern is to marshal the mineralogical and textural data of metamorphic rocks—with an emphasis on microscopic observation—against the background of their field settings. Questions of petrogenesis are beyond our present scope and are discussed in texts on petrology (e.g., F. J. Turner, 1981). At this point we completely exclude reference to tectonic models that involve metamorphism. These must be framed in terms of the objective data of petrographic and field observation, not vice versa.

A nongenetic statement of recognizable categories of metamorphism adequate to accommodate our descriptive data is simple enough. It is a dual scheme based on classic definitions of long standing and it places the collective laboratory data in their appropriate field settings. The two principal categories are *contact* and *regional* metamorphism, the latter being much the more important of the two with regard to both areal extent and tectonic significance. To these we have added a minor category of localized *dislocation* metamorphism, whose products shed valuable light on processes of rock failure located along deep-seated faults. To facilitate later discussion, supplementary terms, some with a genetic flavor, are introduced: for example, "burial" and "ocean-floor" metamorphism, both within the general regional category, and "hydrothermal" as a special facet of contact metamorphism involving effects of hot aqueous fluids.

Contact Metamorphism

Contact metamorphism is localized in the immediate envelope of an intrusive pluton and directly related to the event of intrusion. Rocks showing its effects are confined to a *contact aureole*, anything from a few meters to 2–3 km wide, bordering the intrusive contact. Aureoles typically display concentric zoning; from the periphery toward the contact, enveloping rocks lose their initial character and with changing texture and grain size develop new mineral assemblages that are clearly metamorphic. The phenomenon of progressive metamorphism by which successive mineral assemblages appear with increasing intensity of metamorphism was first recorded by H. H. Rosenbusch a century ago in a contact aureole in the Vosges. An aureole clearly expresses a radial temperature gradient controlled by the central plutonic body. Take the common case of a granodioritic pluton and assume rapid emplacement of the magma already beginning to crystallize around 750°C or 800°C. Cooling models developed by J. C. Jaeger demonstrate that temperatures at the immediate contact are likely to reach a maximum (perhaps 550°C to 600°C) shortly after the intrusive event. Further out in an aureole generated in a quartzite-shale envelope, maximum temperatures would decrease and would be reached progressively later; for example, 350°C (the minimum requisite for effective reaction) about 3 km to 4 km from the margins of a cylindrical pluton 5 km in diameter, several million years after intrusion. Contact effects depend principally on the minimum dimensions of the generating pluton, less so on magmatic temperatures. It is a matter of common observation, for example, that along the borders of many diabase and gabbro sheets hundreds of meters thick (emplaced perhaps at 1000°C) contact metamorphism is obvious only within a few meters of the base; whereas below major layered gabbro intrusions several kilometers thick metamorphic effects of basic intrusions can be conspicuous.

Among localized effects commonly observed in aureoles enclosing granodioritic plutons (especially close to contacts) there may be striking hydrothermal changes effected by water-circulation systems that are set up in the envelope by the thermal influence of the invading magma body. Part of the water concerned is probably magmatic; but oxygen-isotope composition of the altered rocks suggests that much of their water content has been drawn inward from the country rock itself. This aspect of contact metamorphism, which is highly important to students of ore genesis and introduces considerable variety into the mineralogy of contact metamorphism, cannot be treated in any detail in this book.

Dislocation Metamorphism

Dislocation metamorphism is essentially a microstructural phenomenon locally associated with deep-seated faulting. It is also displayed in spectacular fashion in some Precambrian belts of intense regional penetrative deformation. Comparable textures have been produced in granitic rocks and peridotites that have been uniaxially compressed in the laboratory under confining pressures of about 5 kb, at temperatures of 300°–500°C and rapid strain rates (several percent shortening per hour). Under such conditions many crystalline rocks fail by a process of cohesive flow in which they maintain strength and cohesion although component crystals are bent, ruptured, and partially recrystallized. Dislocation metamorphism is attributed to just this type of failure, which, it must be emphasized, should not be confused with the process of brittle rupture, shattering, and loss of strength that is so familiar in the realm of ordinary experience at surface pressure.

Regional Metamorphism

The most extensive, and mineralogically the most varied, of metamorphic rocks have long been attributed by classic usage to a single very broad category—*regional metamorphism*. Most regionally metamorphosed rocks have a characteristic laminated (schistose) structure—the hallmark of intense deformation; but this is by no means a universal characteristic, especially in the initial stages of metamorphism. Certain minerals (almandine, staurolite, and kyanite come to mind) are typical products of regional metamorphism; but very few are exclusively so, and these have only a limited range of occurrence. The only universally applicable criteria of regional metamorphism are great areal extent—commonly many thousands of square kilometers—and absence of any clear detailed correlation between the total field pattern of metamorphism and synchronous plutonic intrusion. To recognize the single comprehensive category of regional metamorphism is especially suited to today's fashion of geological thinking in tectonic terms. Thus brought together are all the numerous and varied metamorphic pressure-temperature regimes that correlate with specific tectonic settings. Rocks affected by regional metamorphism indeed furnish some of the most reliable data from which to build quantitative models of plate tectonics.

Over extensive areas of eroded tectonic belts, it has been possible to map, in each case on the basis of some selected lithology (e.g., pelitic), sequences of zones of increasingly intense metamorphism. Each zone is marked by its distinctive mineral assemblage. Increasing intensity (or grade) of metamorphism is broadly correlated with increase in temperature—though other factors, notably pressure and fluid composition, may also affect the zonal pattern. Zonal boundaries, *isograds*, mark the incoming or disappearance of index minerals in the sequence of progressive (*prograde*) mineralogical transformations. A classic sequence in pelitic schists (derivatives of shale) in the Grampian Highlands of Scotland is marked by successive entries of biotite, almandine garnet, staurolite, kyanite, and sillimanite. Isograds and the zones they separate are correspondingly named: the staurolite zone, for example, lies between the staurolite and the kyanite isograds. Rather similar, but not identical, zonal patterns have been traced in shale-sandstone derivatives of other metamorphic belts, notably in Scandinavia, Japan, the Appalachians, the European Alps, and New Zealand. Zonal sequences have also been traced on the basis of other lithologies, especially in rocks of broadly basaltic parentage.

Much of the controversy that at one time was attached to causes of regional metamorphism and to the possible roles of deformation, plutonic intrusion and even of crustal depth in the general process has now dissipated. It is recognized today that regional metamorphism is a many-sided phenomenon, and that causes and effects are not always clearly separable. Metamorphism manifests itself at varied depths below the surface in a wide variety of tectonic and geographical situation—beneath the ocean floors, in the lower levels of deep sedimentary basins, in great sections of Archaean crust, and, of course, in the eroded roots of continental mobile belts. In all cases it has been governed by temperatures and pressures significantly higher than those prevailing at the surface. A general range of 3 kb to 15 kb, and 200°C to 750°C or 800°C is probably adequate for its total coverage. Every known or postulated set of metamorphic conditions within this overall spectrum is probably duplicated today at depths below 5 km in some segments of the continental crust. Possible conditions in the oceanic crust are limited to the lower values of pressure and temperature. It must be understood that regimes of heat flow in mobile sectors of the crust are, for the most part, transient when viewed on the scale of millions of years.

Thus regional metamorphism can be seen on the broadest scale as the varied mineralogical and textural response of crustal rocks to limited ranges of transient *P-T* conditions. These, in turn, reflect thermal perturbations in the underlying mantle and fluctuations in the local capacity of crustal rocks themselves to generate heat internally—for example, by radioactive decay of uranium and potassium or by exothermic metamorphic reactions (notably those of hydration or carbonation). Closely interconnected with departures from the thermal steady state are sustained and repeated dynamic perturbations and bursts of "granite" plutonism. Evidence of the former in metamorphic rocks of mobile belts is seen in folded structures and on a smaller scale in pervasive flow textures (schistosity, foliation). There can be little doubt that the stress and resulting flow (strain) to which these textures bear witness exert a powerful influence in accelerating metamorphic reaction. So it is not surprising that nonschistose rocks of low metamorphic grade dredged from the ocean floor or formed in depressed basin and geosyncline fillings by "burial metamorphism" typically are only partially reconstituted.

Interrelation of regional metamorphism and "granite" plutonism deserves further comment. Many of the great "granitic" batholiths exposed in mobile belts have no direct relation to regional metamorphism displayed in the rocks that they invade. Radiometric dating in such cases shows a lapse of many million years between the metamorphic cycle and the much later plutonic event. But granitic plutonism more or less synchronous with regional metamorphism also is a common phenomenon. Deep down in the terrane undergoing metamorphism, or wherever heat flow is locally intensified by processes still incompletely understood, streaks of granitic melt begin to develop, commonly a few centimeters in diameter and even a meter or so long, by partial fusion of the metamorphic rocks. Whenever that happens, metamorphic and igneous processes merge. If the quantity of liquid phase becomes sufficiently great, a large mass, part liquid, part crystalline, may become mobile enough to move bodily as a magma. Other granitic magmas, in a completely liquid state, may develop by squeezing away a melt phase from an unfused metamorphic residue. As such bodies ascend under gravity they may leave in their wake trails of rock heated to the point of incipient fusion. Migmatites—rocks composed of a metamorphic host material streaked and veined with "granite"—occur on a regional scale in many areas of high-grade regional metamorphism. Many such rocks, especially in Precambrian terranes, probably represent the condition described earlier, the granitic streaks having been formed by partial fusion of the metamorphic host rocks. Other migmatites, especially in the vicinity of large intrusive bodies of granite, appear to be the result of intimate injection of granitic liquid into adjoining metamorphic rocks. To such migmatites the terms *injection gneiss* or *arterite* have been applied. In both types there is extensive reaction between liquid and solid components of the migmatite as the liquid crystallizes, as a result of which the boundaries between the granitic and the metamorphic portions of the migmatite are likely to become indistinct and gradational. Such gradations once were generally cited—in most cases erroneously, in our opinion—as evidence that a solid rock has been granitized without intervention of a silicate-melt phase.

The general picture of metamorphism presented above is tentative, and it differs in detail from some of those presented by other writers. It is offered as a working basis for interpreting the petrography of metamorphic rocks.

SOME CHARACTERISTICS
OF THE FABRIC OF METAMORPHIC ROCKS

Crystal Growth in a Solid Medium

Students hitherto familiar only with the petrography of igneous rocks must realize at the outset that metamorphic textures differ completely in significance from certain igneous textures that they superficially resemble. This is so because the metamorphic fabric forms by growth of crystals, usually of several different mineral species, competing with one another for space, not in an enclosing homogeneous melt, but in a continuously solid medium. The physical properties of crystalline solids—especially those relating to velocity of growth and to stability of boundary surfaces of individual grains—vary not only from one mineral to another, but from one direction to another within an individual crystal. These differences are largely responsible for the textural details of metamorphic fabric. Similar differences play a part also in forming igneous fabrics, but there they are overshadowed by the much greater physical difference between crystalline solids in general and the melt phase within which these solids have developed.

Some of the more characteristic features of the metamorphic fabric are enumerated below, and their significance is briefly noted.

The Crystalloblastic Series

The term *crystalloblastic* has been applied to fabrics and textural relations resulting from growth of crystals during metamorphism. A grain of a metamorphic mineral bounded by its own crystal faces is termed *idioblastic*; a shapeless crystal grain is termed *xenoblastic*. Following F. Becke, the founder of modern metamorphic petrography, it is possible to list metamorphic minerals in a generalized sequence—the *crystalloblastic series* (*idioblastic order*)—such that each tends to develop idioblastic surfaces against any other mineral placed lower in the series:

Rutile, sphene, magnetite
Tourmaline, kyanite, staurolite, garnet, andalusite
Epidote, zoisite, lawsonite, forsterite
Pyroxenes, amphiboles, wollastonite
Micas, chlorites, talc, stilpnomelane, prehnite
Dolomite, calcite
Scapolite, cordierite, feldspars
Quartz

There are exceptions to this general rule; for example, the sphene of chlorite schists commonly occurs in rounded droplike grains in spite of its generally high position in the crystalloblastic series. Yet the rule is consistent enough to afford valuable evidence as to whether certain rocks are metamorphic or igneous. A hornblende-plagioclase rock in which the plagioclase crystals are euhedral must certainly be of igneous origin— that is, must be a diorite or a uralitized gabbro, not a metamorphic amphibolite. Those who might still attribute granites to metasomatism of solid rock—a kind of metamorphism—must somehow explain why well-developed outlines are common in the plagioclase of granites and are never seen in the plagioclase of typical metamorphic rocks. As yet this fact has been explained satisfactorily only by assuming that at some stage, or probably at all stages, during the evolution of granitic texture, crystals are in contact with a silicate melt.

The crystalloblastic series is obviously related in some way to the crystal structures of the minerals concerned. High in the series stand orthosilicates characterized by relatively dense packing of the component ions; then come silicates with chain, band, and sheet structures; and finally, low down in the series, come silicates having loosely packed open lattices of the three-dimensional type. The properties of the lattices of adjacent crystals control the stability of the growing interface.

Porphyroblasts

In many metamorphic rocks large crystals (porphyroblasts) of one or more minerals are associated with much smaller grains of other minerals

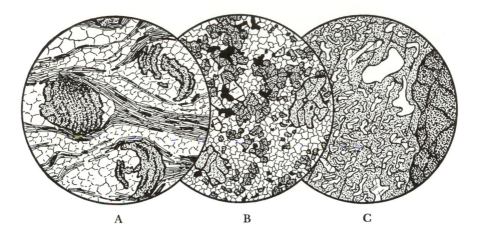

A   B   C

Figure 16–1. Metamorphic Textures

A. Porphyroblastic texture in garnet-mica-quartz schist, Perthshire, Scotland. Diam. 5 mm. Porphyroblasts of garnet enclose curved trains of graphite inclusions, the arrangement of which indicates counterclockwise rotation of the growing porphyroblasts.

B. Granoblastic texture in garnet-hypersthene-plagioclase granulite, Hartmannsdorf, Saxony. Diam. 2 mm. The two largest crystals are of almandine garnet.

C. Poikiloblastic (sieve) texture in skarn, Doubtful Sound, New Zealand. Diam. 1 mm. On the right, pink andradite garnet; on the left, part of a large crystal of epidote enclosing quartz and calcite.

(Figures 16–1A,C and 17–1A,B). Since certain minerals (e.g., garnet, kyanite, staurolite, andalusite, cordierite, albite) generally develop as porphyroblasts, and since most of these are high in the crystalloblastic series, it seems likely that the capacity of a mineral to attain large size depends at least partly on its crystal structure. But other factors also play a part. In some cases (e.g., when porphyroblasts of microcline develop in schists adjacent to granite) the determining factor is availability of certain ions (in this case K^+) in pore fluids supplied from an external source (the granite). Such an explanation does not apply where porphyroblasts of a mineral (e.g., mica) occur in a matrix that contains much smaller crystals of the same mineral. Here the porphyroblasts and the small grains possibly may have developed at different stages of metamorphism.

Porphyroblasts frequently, but by no means always, tend to have idioblastic outlines; cordierite and albite are notable exceptions. They tend also to be riddled with inclusions of other minerals that have become enveloped in the growing porphyroblast—to have *sieve structure,* or *poikiloblastic* texture (Figure 16–1C). As these inclusions have persisted from a premetamorphic or early metamorphic stage of the rock's history, they may yield interesting evidence of the course of metamorphism. Care must be taken, however, to distinguish these from inclusions of late origin—of mica, for example, formed by hydrothermal alteration of enclosing feldspar, or of chlorite developed at the expense of garnet. These secondary inclusions in a single porphyroblast will in many cases show a recognizable orientation related to the crystal structure of the porphyroblast itself; the {001} planes of muscovite flakes, for example, may be parallel to {001} or to {010} of enclosing albite.

Schistosity (Foliation) and Lineation

The term *schistosity* (or *foliation*) may be applied to any parallel structure, of metamorphic origin, that induces a more or less planar fissility in a rock. Schistosity is almost always present in rocks that have been deformed during metamorphism. The regular *slaty cleavage* of slates, the planar fissility of mica schists, and the less conspicuous and somewhat irregular foliation of quartz-feldspar schists (sometimes termed *gneissose* structure) all fall within the general category of schistosity. Some rocks have several intersecting schistosities. The term "*s*-surfaces," introduced by B. Sander, is almost synonymous with "schistosity" but has a somewhat broader connotation in that it is applied to any set of parallel surfaces, whether of metamorphic origin or not, that can be discerned in the fabric of a metamorphic rock. Relict bedding, for example, is a type of *s*-surface, though it need not impart schistosity to the rock in which it is preserved.

Schistosity is generally associated with several or all of the following structural characters:

1. A tendency for tabular crystals of micaceous minerals, and for prismatic crystals of minerals such as amphibole or epidote, to be oriented with their greatest dimensions subparallel to the plane of schistosity (Figure 16–2A,B). This is an instance of what the structural petrologist terms "preferred orientation of minerals according to external form of grains" as contrasted with *granoblastic* textures of equigranular

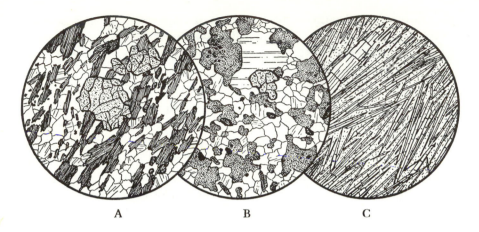

A   B   C

Figure 16–2. Metamorphic Textures: Foliation (Schistosity)

A. Parallel alignment of mica crystals in foliated andesine-quartz-biotite-garnet schist, Lake Manapouri, New Zealand. Diam. 2.5 mm. Section cut perpendicular to foliation and to lineation. The grains with high refractive index are sphene.

B. Same rock as A. Diam. 2.5 mm. The section is cut parallel to the foliation. Most of the biotite crystals have their {001} cleavage subparallel to the plane of the section.

C. Parallel alignment of amphibole needles in anthophyllite-talc-chlorite schist, Chestnut Hill, Pennsylvania.

nonschistose rocks such as hornfelses (Figure 17–3C), whose component grains are oriented at random.

2. An even stronger, though less easily recognized, tendency for most minerals to be oriented crystallographically, even though their external form may be completely irregular: xenoblastic grains of quartz may tend to be oriented with their optic axes parallel to the schistosity; shapeless grains of calcite may have their optic axes approximately normal to the schistosity, and so on. This is termed a state of "preferred orientation according to crystal structure."

3. Plane surfaces of internal slipping that define the main schistosity or more commonly (as "strain-slip cleavage") intersect an earlier set of *s*-surfaces.

4. Laminated structure resulting from segregation of simple mineral assemblages of contrasted composition into alternating layers, usually from 1 to 10 mm thick, parallel to the schistosity (e.g., quartz-albite alternating with muscovite-chlorite-epidote). This layering, a product

of "metamorphic differentiation" of initially homogeneous rocks, has sometimes been mistaken for bedding inherited from the parent rock. It is not to be confused with magmatic streaking of migmatites.

5. Parallel alignment of linear elements of the fabric in some direction within the schistosity, to give what is termed a *lineation*. The parallel elements may be prismatic crystals (e.g., of hornblende or epidote, cf. Figure 16–2C), rodlike mineral aggregates (e.g., of quartz or feldspar), axes of microfolds, or lines of intersection of different schistosities. Lineation imparts a kind of "grain," easily recognized in the field, to the schistosity surfaces. *Mullion structure* is lineation on a very coarse scale, the linear elements being macroscopically conspicuous rods (usually of quartz).

The possible dynamic significance of schistosity has been debated for more than a century. It has always been recognized as a strain phenomenon that might yield information regarding the dynamics and kinematics of regional metamorphism. But controversy has centered on the kind of inferences that might be drawn on such matters as controlling systems of stress and rheologic models of rock flow and fracture. Does schistosity develop at right angles to some simple regional compression induced by superincumbent load or by viselike motions of adjoining rigid crustal segments? Alternatively, is it the imprint of laminar slip on planes of high shear stress? Does metamorphism play an integral role in the flow process? Or again do platy and acicular crystals become aligned by passive rotation in a flowing matrix? Today most such questions, though not without some value, seem oversimplistic. Two factors especially have contributed to better understanding of the implications of what proves to be a subtle and varied textural phenomenon.

First is Sander's proposition that the soundest inferences that may be drawn concern the kinematics (flow regimes) rather than the dynamics (force systems) of rock deformation; and that the key to the kinematic pattern lies in the symmetry of the total schistose, lineated, and folded rock fabric. All-important in this connection are the data of geometric measurement on the outcrop, in the hand specimen, and (of immediate concern) beneath the microscope. Study of the second of these factors developed under the stimulus of Sander's philosophy. Initiated and vigorously pursued by D. T. Griggs, laboratory experiments under high confining pressures, differential stress, and temperature brought into the realm of petrology a new conception of strength properties, flow mechanisms, and resulting strain patterns (in normally brittle lithologic materials such as calcite and quartzite) under metamorphic conditions at different crustal levels.

Some simple generalizations appropriate to descriptive petrography have emerged from the now prolific data of fabric analysis and laboratory experiment. Planar schistosity seems commonly to approximate the plane of maximum extension (normal to maximum shortening) in the strained rock body. "Axial-plane cleavage," which actually fans perceptibly about the axis of a slip fold, expresses the imprint of pervasive laminar slip. This, too, is not far from the mean plane of maximum extension in a tightly folded rock mass. The principal lineation in such rocks usually represents the intersection of a cleavage (schistosity)—the plane of active slip*—with s-surfaces of bedding (or earlier schistosity) that have been passively rotated into their present configuration during the slip process. Such a lineation, considered alone, can have no unequivocal kinematic significance. Fabric studies suggest that slip in the s-surface of principal schistosity is generally oblique, commonly at a high angle, seldom either parallel or precisely perpendicular to any recognizable regional slip direction. Finally, strain-slip and related "cleavages" have long been recognized by petrographers as discrete surfaces of late slip induced by shear stress.

Relict Textures

Useful information about the origin and premetamorphic history of some metamorphic rocks is supplied by relict textures inherited from the parent rock. In certain metamorphic hornblende-plagioclase rocks (amphibolites) there are porphyritic or ophitic textures—termed *blastoporphyritic* and *blastophitic*, respectively—that have clearly been inherited from a parent rock of igneous origin. In metamorphosed conglomerates, also, the original pebbles can commonly be recognized, even though they have usually been stretched and flattened. In many metamorphosed sediments bedding has survived metamorphism and yields valuable evidence as to the structural evolution of the metamorphic terrane.

---

*The rheologic behavior of slaty rocks under conditions of slow strain in the presence of water appears to be rather akin to viscous flow on internal surfaces of low shear stress.

OUTLINE CLASSIFICATION OF METAMORPHIC ROCKS

Basis of Classification and Terminology

Classifications of metamorphic rocks in current use are based primarily on textural criteria, for the most part readily recognizable in hand specimens. The textural groups of fully metamorphosed rocks are thus objectively defined, unambiguous, and few in number, and convey useful broad implications as to the scope of the metamorphic regime—regional versus contact, low versus high grade, and so on. Mineralogical criteria also enter into and even dominate a few of the primary definitions (e.g., amphibolite, marble, quartzite). But, as also with igneous rocks, sharp distinctions cannot be drawn between mutually gradational types of metamorphic rock as defined on a textural, or any other, basis. What one observer terms a *phyllite* will be termed a *schist* by another. The terms *impure schistose marble* and *calc-schist* may equally well describe the same rock. Lawsuits have been fought as to whether a particular rock should be identified as *schist* or *gneiss*, each term backed by expert testimony of experienced geologists. Such controversy is trivial in that it concerns semantics rather than the rocks themselves, because there is no way to quantify the transitional natural boundary between two naturally gradational rock types.

The phenomena of metamorphism are too varied to be discussed only in terms of texturally defined categories. So other criteria enter into current terminology, supplementary to or cutting across the primary classification. Three kinds of criteria—all framed objectively in terms of descriptive data—are widely and simultaneously employed:

1. *Mineralogical:* in which dominant minerals are used as a prefix—commonly compound—to the more general term: biotite-garnet schist, hornblende-pyroxene gneiss, andalusite hornfels.
2. *Chemical:* used to convey simple features of rock chemistry: magnesian schist, ferruginous quartzite.
3. *Protolithic:* the name emphasizes the nature of the parent rock (protolith): thus pelitic hornfels (originating from shale), granodioritic gneiss, metabasalt, metagraywacke (all self-explanatory).

Within such a framework rocks can be identified in the field, subject to later petrographic confirmation. The terminology is simple and easily grasped even at the elementary level, is adequate for discussion of the broadest and most fundamental problems of metamorphic petrogenesis, and does not overstress mineralogical or geographical minutiae and rarities that are of no great significance. Yet it is elastic enough to accommodate newly recognized rock varieties that may have important implications in petrology: sapphirine granulite and quartz-jadeite metagraywacke come to mind in this latter connection. It is to be hoped that, for the general benefit of working petrologists, geochemists, and students of tectonics, classification of metamorphic rocks will continue in its present loose but practical form unfettered by any comprehensive international scheme of precise nomenclature.

Texturally-Based Primary Terminology

Objectives and Limitations

Broad terminology in current general use retains much that has been inherited from classic nomenclature developed almost a century ago. It is designed for a simple purpose—to provide the minimum that is necessary for general discussion of metamorphic phenomena among geologists, whether specialists or not. As set out below, the definitions of common metamorphic rock types are framed in descriptive terms, without regard to genetic implication. Even to the novice it must be apparent that broad, genetic inferences may indeed be drawn from certain rock types: hornfelses suggest contact aureoles; schists express metamorphism at grades beyond that of slates and phyllites; and so on. Inferences of this kind provide the basis for the overall scheme of presentation to be adopted in Chapters 17 through 20; but here, as we encounter the rocks for the first time, it seems logical to omit genetic implications, however legitimate, in the interests of objective definition.

There is nothing quantitative about the scheme of rock terminology: Texture, the primary criterion employed, is not susceptible to treatment in numerical terms. Even grain size is loosely noted without reference to any standard limits such as have proved convenient in systems of igneous and sedimentary nomenclature.

Natural transitions between rock types are the general rule in metamorphic lithology. For this reason, combined with the prevailingly loose usage of textural terms, the student of metamorphic rocks cannot expect to find in these pages precise definitions to which all common rocks can be unambiguously fitted. Instead of comprehensive, unequivocal definitions, therefore, we offer a set of type descriptions, each applicable to most, but by no means to all, of the rocks assigned to some specified group. It is possible to say how *most* schists differ from *typical* gneisses, how *most* hornfelses differ from *most* granulites. And such is our present purpose. This is not to deny, however, the existence of transitional types—some of them of great global extent—that can equally well be assigned alternative valid names. Qualifying terms are useful here, equivocal as admittedly they must be: schistose hornfels, phyllitic schist, foliated marble, and so on.

Foliation in Relation to Grain Size

The emphasis in the scheme set out below is on degree, kind, and regularity of foliation and the closely related characteristic of grain size (loosely described as fine, medium, and coarse, with no set numerical limits). A quick examination of a hand specimen is sufficient to reveal whether it is foliated or not, and in the former case the degree and pattern of visible foliations. On such features depends the way in which a specimen breaks under the hammer—preferably at the site of outcrop. Slates are highly fissile: Schists (like slates) break out of the outcrop in slabby specimens; gneisses tend to break more unevenly; the fracture of hornfelses is typically conchoidal or jagged.

Where the parental material was initially fine grained (shale, siltstone, fine volcanic ash) slaty cleavage develops early in folded rocks. The texture is dominated by planar foliation (cleavage) of maximum regularity. At the same time grain size begins to increase, but still remains in the category here termed *fine* (long dimension of phyllosilicate crystals no more perhaps than 100 μm). With progressive metamorphism through phyllite to schist the foliation tends to become more obvious but less regular as the grain size steadily increases to perhaps 1–2 mm (medium grain). Where the parent rock was itself medium or coarse grained, the onset of deformational metamorphism leads to a rather rough foliation that, with increasing strain and recrystallization, becomes progressively more regular. At the same time, however, the grain size at first becomes reduced in the recrystallizing matrix that encloses coarser relict grains surviving from the parent phase assemblage. Then, as the grade advances further, the rock, now completely metamorphic in all respects, coarsens. Again the end product is a schist of medium grain, whose foliation more often than not now becomes accentuated by segregation of light and dark components in alternating laminae a few millimeters thick.

Perfection of foliation depends not only on deformational history of the rock but even more on the crystallographic habit of its chief component minerals. Preponderance of subparallel tabular flakes of phyllosilicates—especially micas and chlorites—or of aligned prisms and fibers of amphibole is essential to the fissility that is so characteristic of slates and schists.

Conspicuously Foliated Rocks

Slates are fine-grained metamorphic rocks with perfect planar schistosity (*slaty cleavage*) that commonly is inclined at high angles to relict bedding (still recognizable in hand specimens and outcrops by color banding). Segregation banding is absent. Parent rocks for the most part were shales, mudstones, or fine volcanic ash. Principal minerals are colorless phyllosilicates and chlorites, whose precise identity can be determined only by x-ray technique. Slates spotted with knots of relatively coarse mica and chlorite or with porphyroblasts of andalusite (Figure 17–1A) are found in the outer zone of some contact aureoles.

Phyllites are rocks transitional from slates of locally similar composition. The grain size is perceptibly coarser than in slates, and broken surfaces show a lustrous sheen (due to coarsening of grain) imparted by flakes of mica and chlorite, now clearly recognizable as such, aligned along the plane of foliation. Foliation itself, while still conspicuous and regular, may already be complicated by transverse secondary *s*-surfaces (as in *strain-slip* cleavage), lineations, and incipient segregation lamination.

Schists are strongly schistose, usually lineated rocks of medium to coarse grain (perhaps 0.2 mm to 1 cm or so in maximum dimension) mostly with abundant constituent phyllosilicates—muscovite, biotite, and chlorite predominating in pelitic schists (with chloritoid in some rocks), antigorite and talc in magnesian schists. They are markedly fissile rocks, breaking readily beneath the hammer into parallel-sided slabs. Segregation lamination and lineation are usually conspicuous. The fissility of some rocks is due to parallel alignment of amphibole prisms and fibers as much as to presence of phyllosilicates; but in the group as a whole the latter predominate.

Gneisses are mostly coarser than schists, and minerals of granular habit (notably quartz, feldspar, and in some rocks garnet, hornblende, or pyroxene) greatly predominate over phyllosilicates—the latter almost exclusively biotite and muscovite. Foliation, though still conspicuous in hand specimens and outcrops, is less regular than in schists. Micas or hornblende tend to be segregated in discontinuous streaks and curving lenses that gently undulate around garnet porphyroblasts and coarse quartzo-feldspathic eyes. Thus, foliation, while still conspicuous, is uneven; and this is reflected in a tendency for specimens to break irregularly across the foliation pattern. At most, in spite of their obvious foliation, gneisses are but imperfectly fissile. All texture and mineralogical gradations exist between schistose gneisses and quartzo-feldspathic schists.

Imperfectly Foliated and Nonfoliated Rocks

Foliation, although present (even conspicuous) in individual cases, is not a diagnostic characteristic of rock types described below. And in some, notably hornfelses, quartzites, and many marbles, there is usually no trace of foliation.

Granulites typically are rather even-grained rocks of medium to coarse grain, almost or completely lacking in minerals of micaceous or acicular habit. Quartz, feldspars, almandine garnet, and, in one group, pyroxenes are the dominant minerals. In fact, granulites for the most part lack hydrous minerals of any kind apart from thin films of biotite or, in rocks transitional to amphibolites, variable amounts of hornblende. There are two principal types of granulite (with the usual transitional variants): light-colored quartzo-feldspathic rocks,° and dense, dark-colored rocks (of basic igneous parentage) composed mainly of pyroxene and plagioclase crystals. Quartzo-feldspathic granulites usually show a distinct foliation marked by subparallel, thin lenticular aggregates of medium-sized quartz grains; and in some of these rocks the foliation may be accentuated by discontinuous films of dark mica. In many pyroxene granulites, foliation is virtually lacking, and the texture is granoblastic (Figure 16–1B). There is no lithologic distinction between some of these and coarse pyroxene hornfelses; and the line between the two groups can then be drawn only on the basis of field observation.

Amphibolites are dark-colored, metamorphic rocks of medium to relatively coarse grain whose principal mineral components are hornblende

---

°Most *leptites* and *leptynites* of some European petrologists belong here.

and plagioclase (additional subordinate components may include epidote, chlorite, biotite, or garnet). Foliation is highly variable. Where the hornblende crystals are prismatic in habit, or when biotite and chlorite are present, the rock may be decidedly schistose. In other amphibolites, however, foliation is completely lacking and the rock breaks with an irregular fracture.

Quartzites are metamorphic rocks, mostly nonfoliated, composed of quartz.

Marbles are rocks composed mainly of calcite, less commonly of dolomite. Some marbles show a rude foliation due to segregation of minor silicate constituents—phlogopite, amphiboles, pyroxenes, and iron oxides—or in some cases to parallel alignment of lustrous flakes of graphite. Some marbles bear the imprint of strain involving plastic flow of individual calcite grains, which thereby assume lensoid outlines, imparting to the rock a macroscopically inconspicuous but microscopically pervasive foliation.

Eclogites are medium- to rather coarse-grained rocks of striking macroscopic aspect, in which deep-red garnet crystals are strewn in a surrounding matrix of medium-grained green pyroxene—the sodic variety omphacite. Some varieties contain green amphibole or kyanite.

Hornfelses typically lack all trace of foliation and the fine-grained types tend to break with conchoidal fracture. Granoblastic texture is characteristic; but in many hornfelses the granoblastic groundmass encloses irregularly oriented large porphyroblasts of minerals such as andalusite, cordierite, biotite, hornblende, or garnet. In some hornfelses there is a faint foliation due to a tendency for parallel alignment of mica flakes. In others relict bedding, now expressed by differences in mineral composition, imparts a visible foliation (but not accompanying fissility) to the rock. Hornfelses cover a wide range of chemical and mineral composition that reflects great diversity in protolithology. Micas, andalusite, and cordierite are typical of pelitic hornfelses. In calc-silicate hornfelses derived from impure carbonate sediments there is an abundance of such minerals as diopside, grossular-andradite, epidote, vesuvianite, wollastonite, and tremolite. Most hornblende-plagioclase and pyroxene-plagioclase hornfelses are derived from basic igneous protoliths.

Mechanically Strained Rocks

Mylonites are fine-grained rocks resulting from extreme mechanically dominated deformation of coarser, chemically stable rocks. Characteristic features are a flinty appearance and subconchoidal fracture, fine banding or streaking, and the presence of ovoid or almond-shaped eyes of undestroyed parent rock swimming in the streaked and finely crystalline matrix (see Figure 18–1). What appears as "granulation" in mylonites is now attributed to annealing of highly strained grains rather than to pulverization. Mylonites are in fact strong, highly coherent rocks.

Phyllonites are rocks resembling phyllites and commonly macroscopically indistinguishable from them, but formed, like mylonites, by "granulation" of initially coarser rocks. Chemical reconstitution is far advanced and has given rise to silky films of mica smeared out along the schistosity planes. The true identity of phyllonites is revealed only by microscopically visible relics of the parent rock—especially relict grains of high-grade minerals such as garnet.

Rock Type and Metamorphic Grade: Preliminary Notes

Classic terminology, upon which the preceding section is essentially based, was well established long before the notion of metamorphic grade was developed. For this reason and to meet what seem to be general classroom needs, we have thus far avoided the question of metamorphic grade in relation to standard rock terminology. Yet there are obvious broad, but by no means exact, correlations between rock type, field occurrence, and metamorphic regime. This topic is developed throughout ensuing chapters. One must recognize at the outset, however, that to identify and name a rock to meet classroom or museum demands is of itself an objective of only limited scope. Even the novice, therefore, must soon become aware of such generalizations, open to many exceptions though they may be, as those enumerated in the following brief notes.

Hornfelses are found almost exclusively in aureoles of contact metamorphism, and hornfels assemblages such as pyroxene-plagioclase and quartz-orthoclase-sillimanite (or andalusite), devoid as they are of hydrous minerals, represent maximum metamorphic grade in this environment.

In the realm of regional metamorphism, schists of one kind or another are the commonest rocks at all but minimum and maximum grade. Slates and phyllites are found only in zones of incipient low-grade metamorphism. At the other end of the scale, granulites typify metamorphism of maximum grade where rocks have become almost completely dehydrated. Amphibolites and gneisses are typical of medium and high (but not maximum) grades.

Certain rocks of very simple mineral composition, for example, quartzites and marbles, occur in almost any metamorphic environment. Others are more restricted—serpentinites, for example, to low-grade zones of regional and especially burial metamorphism. Eclogites, essentially bimineralic, but chemically complex, are metamorphic rocks transported from very deep sources in the lower crust or the upper mantle.

Finally, mylonites and phyllonites are restricted to zones of intense and rapidly imposed strain, such as are typically seen along sites of proven major faults. In such an environment, adjustment of the mineral assemblage to the local pressure-temperature regime cannot keep pace with purely mechanical strain involving cohesive flow without bodily distintegration or complete loss of strength of rocks so affected (as shown by laboratory experiment).

Thus, identifying rock types on a largely textural basis yields significant clues as to the parentage of the rock concerned and the physical regime and probable environment of metamorphism. But to explore and test such implications in detail we must turn to the metamorphic mineral assemblage itself and to the field environment of the rock as revealed on the map.

Nontectonites in Regional Metamorphism

In the incipient stages of regional metamorphism initially massive rocks such as graywacke and basalt begin to generate new mineral assemblages without developing either a schistosity or any recognizable sign of preferred orientation of constituent minerals. The fabric thus retains a nontectonite character. Rocks of this type—nonfoliated metagraywackes, spilites, and massive serpentinites—may crop out over areas measured in hundreds of square kilometers. Once considered to be "unmetamorphosed," these rocks have now assumed great significance among the products of low-grade metamorphism. They consist of mixtures of relict minerals inherited from the parent rock, dispersed in varying amount (depending on degrees of metamorphism) in a fine-grained matrix of truly metamorphic minerals—quartz, zeolites, albite, white micas, chlorites, prehnite, pumpellyite, epidote, and others. As rocks of this general category become increasingly deformed they tend to develop perceptible schistosity and so pass gradually through *semischists** to completely reconstituted schists.

---

*An etymologically unfortunate name introduced by one of us (Turner) but now accepted in current terminology.

Chemical Classes of Metamorphic Rocks

Many of the classes of rock defined earlier in terms of fabric can be subdivided on the basis of mineral composition, which reflects bulk chemical composition and metamorphic facies. Mineralogical details are given in the chapters that follow. In the meantime it is sufficient to note that common metamorphic rocks fall into six principal chemical classes. In each of these it may be convenient to recognize two subclasses. Rocks of the first subclass have an excess of SiO_2 and contain quartz; those of the second are deficient in SiO_2 and contain no quartz.

1. Derivatives of pelitic (aluminous) sediments (clays, shales, mudstones).
2. Derivatives of quartzo-feldspathic rocks (sandstones and acid igneous rocks).
3. Derivatives of calcareous sediments (limestones and dolomites, which may contain quartz and clay minerals as impurities).
4. Derivatives of basic and semibasic igneous rocks, including tuffs—spilites, greenschists, amphibolites, and others.
5. Hydrothermally metamorphosed peridotites—serpentinites, soap-stones and related rocks, and their derivatives such as talc-antho-phyllite schists.
6. Derivatives of ferruginous sediments (notably metacherts).

To avoid repetition these loosely defined classes will henceforth be referred to, when appropriate, as (1) pelitic, (2) quartzo-feldspathic, (3) calcareous, (4) basic, (5) magnesian, and (6) ferruginous (with manganiferous variants).

METAMORPHIC FACIES: INTRODUCTION TO PETROGENESIS

Potential and Limits of Petrography

The petrography of a single specimen of metamorphic rock reveals, through its mineralogy, its lithologic parentage (protolithology) usually in unambiguous general terms. The rock then stands identified as a metabasalt, a metagraywacke, a dolomitic marble, and so on. But there is more than a hint, too, that bears on broader problems relating to regional temperature-pressure regimes of metamorphism and the possible role of penetrative strain in the metamorphic process. These com-

plex topics are treated adequately in comprehensive and more sophisticated books covering metamorphic petrogenesis (e.g., F. J. Turner, 1981)
and structural analysis.[1] But at this point it is appropriate to open the
gateway to petrogenesis by introducing the mineralogically based concept
of metamorphic facies and its implications regarding physical conditions
of metamorphism on the regional scale.

Concept of Metamorphic Facies

During the second decade of the century the Finnish geologist P. Eskola
introduced and developed a mineralogical concept that was to become
the very foundation of modern metamorphic petrology. Its basis lies in
petrography and in the broad chemistry of metamorphic minerals and
rocks. The products of metamorphism are viewed not as single specimens
but collectively as field associations of different rock types—the wider
their chemical variety, the better. A single metamorphic facies is a unit
embracing all metamorphic rock types that habitually occur in close
mutual association in regional domains of comparable metamorphic
grade. Facies, it will be noted, have no connotations regarding age or
geographical situation. One facies includes all mineral assemblages that
occur in the chlorite and biotite zones of regional metamorphism in the
Scottish Highlands, or in the Appalachians, or in parts of the Alps, or in
southern New Zealand. To another belong the andalusite-cordierite and
hornblende-plagioclase hornfelses and associated tremolite-diopside
marbles of contact aureoles in the Sierra Nevada of California and in
any of a score of other provinces of posttectonic granodiorite plutonism.
Each facies, then, is characterized by a compound paragenesis—a set of
mineral assemblages that *collectively* is uniquely diagnostic.* Equally specific is a general, but seldom precise, correlation between mineralogy and
rock chemistry within any facies. In one facies rocks of basaltic composition are invariably represented by hornblende-plagioclase and associated metapelites by quartz-mica-garnet (with kyanite or staurolite);
chemically equivalent assemblages in another are respectively chlorite-
albite-epidote-actinolite ($\pm$ calcite) and quartz-muscovite-chlorite-albite-
epidote.

---

*Some assemblages, such as quartz-muscovite-biotite-feldspar and diopside-calcite, appear
in more than one facies. Clearly, in spite of persistent misinterpretation in some quarters,
the facies concept cannot be used as a framework for classification of individual rock types.

Although every facies is defined in nongenetic terms on the basis of petrographic data seen in a field setting, the genetic implications of facies are of the highest importance. Drawing on experience in experimental geochemistry, petrologists legitimately enter a more subjective field in which they can incorporate the data of metamorphic facies into models of metamorphic pressure-temperature regimes, of circulatory fluid systems, and even of more remotely connected tectonic systems. Through the facies concept, in short, the objective data of metamorphic petrography are channeled into the ever-changing stream of petrogenic thought.

Facies Terminology

A dozen facies are adequate to cover the total mineralogical phenomena of metamorphism. In keeping with Eskola's terminology, each facies (with one or two exceptions) is named after some typical, but not necessarily uniquely diagnostic, common rock type. To elaborate the system by proposing more precisely defined subfacies—as many of us have done from time to time—has led to confusion rather than to increased clarity. In any specific region it is indeed usually possible to recognize local subfacies; but their individuality rests merely on local variation that is inherent in geological situations and pattern. To attach to it more than local significance serves no useful purpose and introduces a note of artificiality that can indeed be misleading.

In the scheme adopted here we retain as far as possible the final terminology proposed by Eskola, modified where necessary to accommodate newer data and to conform to departures that have become entrenched in current general usage (cf. F. J. Turner, 1981, pp. 204–210).

Facies of contact metamorphism (listed in order of increasing grade)
  1. Albite-epidote-hornfels
  2. Hornblende-hornfels
  3. Pyroxene-hornfels
  4. Sanidinite
Facies of low-grade regional metamorphism
  In incipiently metamorphosed rocks:
  5. Zeolite
  6. Facies with pumpellyite (and either prehnite or actinolite)

Facies of low-grade regional metamorphism (cont.)
  In largely or completely reconstituted rocks:
  7. Lawsonite-albite-chlorite
  8. Blueschist (glaucophane-schist)
  9. Greenschist
Facies of high-grade regional metamorphism
  10. Amphibolite
  11. Granulite
  12. Eclogite

Some of the characteristics of each facies will emerge in the ensuing chapters.

Graphic Presentation of Data

Chemographic Projections

Essential relations between mineral paragenesis and rock chemistry in a facies can be brought out simply through chemographic diagrams of a type developed over sixty years ago by Eskola. Compositions of observed minerals are plotted on triangular projections of three selected influential chemical components. Each mineral plots as a point—or to represent chemical refinements now generally available, as a small circumscribed area within the projection. Ties connect pairs of coexisting minerals. We emphasize that the basis and purpose of these facies diagrams are purely descriptive; they carry no precise implications of equilibrium.*

  A single projection cannot adequately express the full spectrum of chemical and mineralogical variety in a facies or even in a local subfacies. For wider coverage several diagrams can be employed—each designed to illustrate, with varying degrees of refinement, different aspects of the total pattern. It is customary to treat in separate categories those mineral assemblages (silica-saturated class) that may include quartz and those that do not (silica-deficient class). The former is by far the more comprehen-

---

*In contrast with experimentally based triangular three-component plots widely used in petrology to depict true mineral *equilibria*.

sive; and accordingly we first consider two types of projection (*ACF, AKF*) suited to silica-saturated parageneses.

ACF and *AKF* Projections

Two oxide components, SiO_2 and Na_2O, are excluded from further consideration. The first appears "in excess" as quartz; the second is allocated to plagioclase—an approximation that is close to reality except in the blueschist and eclogite facies. Various minor components are completely allocated to ubiquitous accessory phases: TiO_2 to sphene (or rutile), P_2O_5 to apatite, CO_2 to calcite. In most metamorphic rocks K_2O is accommodated in micas, except in Al_2O_3-deficient rocks where K-feldspar is also present.

For the *ACF* diagram the three selected oxide components—recalculated from the chemical analysis to total 100(molar)—are:

$$A = (Al_2O_3 + Fe_2O_3)^* - (Na_2O + K_2O)$$
$$C = CaO$$
$$F = (FeO + MgO + MnO)$$

This projection is used in Figure 16–3, as first was done in 1915 by Eskola, to compare two chemically similar but mineralogically different parageneses—that of Orijärvi in Finland (amphibolite facies) and that of Oslo in Norway (pyroxene-hornfels facies) as had already been recorded four years earlier by V. M. Goldschmidt.†

In spite of obvious deficiencies arising from the omission of some components and the consolidation of others into compound units, the *ACF* diagram serves three useful practical purposes. It highlights contrasts between facies or between more localized parageneses within the same facies. It illustrates compatibilities and incompatibilities between minerals within the same broad paragenesis; for example, in Figure 16–3B diopside is shown to be compatible with all other phases there depicted except andalusite and cordierite. And finally, the *ACF* plot

---

*To include Fe_2O_3 in the *A* component has always seemed a dubious procedure. To suit modern analytical techniques, it may be more practical to compute total Fe as FeO.
†It was the contrast between these two parageneses that led Eskola to enunciate the metamorphic facies concept in 1915.

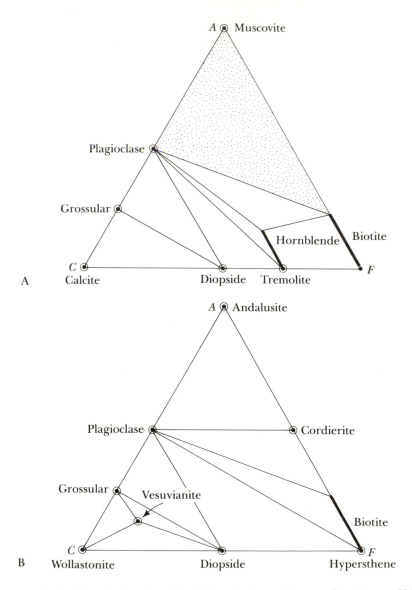

Figure 16–3. *ACF* Projections for Mineral Assemblages with Excess SiO$_2$ and K$_2$O

Quartz invariably present; K-feldspar a possible additional phase in pelitic and quartzo-feldspathic rocks.

A. Amphibolite facies, Orijärvi district, southwest Finland.

B. Pyroxene-hornfels facies, Oslo district, Norway.

enables the student of metamorphic rocks to grasp something of the gross chemical-mineralogical pattern of a paragenesis.

The *AKF* projection (Figure 16–4A) is complementary to and more restricted in coverage than the *ACF* diagram. It expresses the independent role of K_2O in parageneses that include a single calcic phase, plagioclase (stippled triangle, Figure 16–3A). This condition permits K_2O to substitute for CaO as one of the three definitive components. These are now:

$$A = (Al_2O_3 + Fe_2O_3)^* - (Na_2O + K_2O + CaO)$$
$$K = K_2O$$
$$F = (FeO + MgO + MnO)$$

Special Projections

Special projections bring out additional refinements of pattern in assemblages of limited chemical range. Such are $(Fe,Mg)\,O–Al_2O_3–SiO_2$ (Figure 16–4B) for highly aluminous rocks with silica-deficient phases such as spinel and corundum; $K_2O–Na_2O–Al_2O_3$ for clay derivatives carrying paragonite or pyrophyllite; $CaO–MgO–SiO_2$ for magnesian marbles (of dolomitic parentage) in which silicates such as tremolite, diopside, forsterite, or wollastonite may be associated with carbonates (Figure 16–5). Then there are more sophisticated projections depicting four or even more components such as the *AFM* diagram, designed by J. B. Thompson to separate the independent roles of MgO and FeO in pelitic assemblages containing quartz, muscovite, and plagioclase.[2]

All these special projections, like the more comprehensive *ACF* and *AKF* diagrams that they supplement, have a common descriptive purpose: to represent in graphic form that constant correlation between petrographically observed mineralogical patterns and rock chemistry that is one of the fundamental requirements of a metamorphic facies. But the data (and the diagrams), if themselves objective, nevertheless have genetic implications. At this point we leave petrography and enter the more speculative field of petrology.

---

*Alternatively Fe_2O_3 may be recomputed as FeO and added to the *F* component.

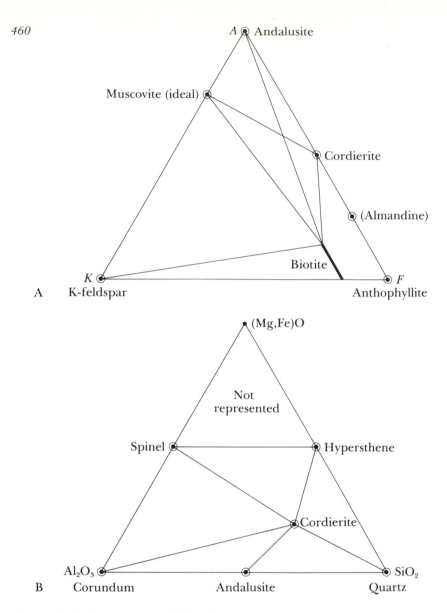

Figure 16-4. Low-Pressure Pelitic Parageneses

A. *AKF* projection for mineral assemblages with excess silica and a single calcic phase (plagioclase). Quartz and plagioclase are additional phases. P. Eskola's diagram of 1915 for rocks of the amphibolite facies, Orijärvi district, southwest Finland. Compare Figure 16–3A.

B. (Fe, Mg) O–Al_2O_3–SiO_2 projection used by C. E. Tilley in 1924 for three-phase aluminous assemblages; K-feldspar, biotite, and plagioclase are possible additional phases; silica-deficient field stippled. Pyroxene-hornfels facies, Comrie, Scotland.

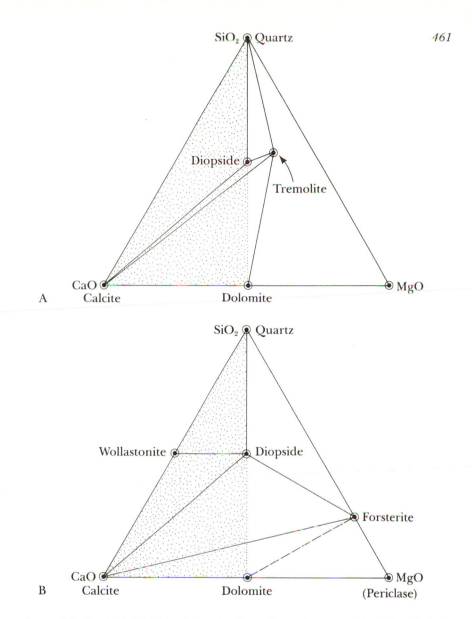

Figure 16–5. CaO–MgO–SiO₂ Projections for Siliceous Dolomitic Marbles

Rock compositions fall in stippled field.

A. Hornblende-hornfels facies (medium grades).

B. Pyroxene-hornfels facies; periclase stable (and dashed tie eliminated) only at maximum grade.

Paths of Facies Evolution

Prograde and Retrograde Metamorphism

It is generally thought that the forward (prograde) paths of metamorphic reaction, by which, for example, a shale ultimately becomes converted to high-grade quartz-mica-garnet-kyanite schist, follow the sequence of phase assemblages (a facies series) that occupy zones between successive isograds. The rationale for this concept is partly intuitive, but is familiar enough in most branches of geology: Phenomena actually observed in space are transposed into the framework of time. This particular proposition regarding the path of prograde metamorphism has found general acceptance, but nevertheless is open to question. Each specific step along the path is conventionally formulated by writing a balanced chemical equation between two chemically identical assemblages on opposite sides of the corresponding isograd; and this procedure certainly seems justified for rocks immediately adjacent to the isograd in question.

Unexpectedly perhaps, within the limits of a thin section textural evidence of prograde reaction is generally lacking. Records of chlorite-muscovite in process of replacement by biotite, or of sillimanite pseudomorphous after kyanite, are comparatively few. The possibility of imperfection in the general fund of petrographic knowledge cannot perhaps be completely discounted. But most likely there is a more fundamental explanation (which cannot be pursued here) in the theory of reaction kinetics.

Markedly in contrast is abundant and varied direct textural evidence of partial backward (retrograde) reaction—assemblages of higher grade in process of replacement by phases of a lower grade, clearly visible on the microscopic scale. Individual crystals of andalusite or cordierite, biotite, garnet, or staurolite are seen to be rimmed and streaked with chlorite; andalusite and cordierite porphyroblasts may be completely pseudomorphed by mats of fine "sericitic" mica or "chlorite." Low-grade micaceous phyllonites retain scattered, partially destroyed relics of staurolite, garnet, or kyanite, inherited from a higher grade. Most retrograde changes involve hydration or carbonation reactions; commonly they appear to have been accelerated by localized intense shearing and granulation.

Where retrograde reactions are obvious under the microscope, two distinct metamorphic facies (or subfacies) can be recognized within the field of view—the one in process of conversion to the other. The total bifacial assemblage is said to be in "textural disequilibrium." In contrast, an assemblage in which no evidence of internal reaction can be detected is said to be in "textural equilibrium." Usage of both terms is justified. But, one must ask, is "textural equilibrium" to be equated with equilibrium in the strict thermodynamic sense of the word?

Physical Implications of Facies

Basic Proposition

Once again, brought to the threshold of theoretical petrology, we conclude by noting briefly how petrographic observations may be combined with experimental data to answer questions such as that just raised, and to explore in quantitative terms the ultimate problem of metamorphic pressure-temperature regimes.

When Eskola enunciated the facies concept, and in his later writings too, he supplemented his objective definition of a facies with a statement, on which he placed equal emphasis, of its broad physical implications. Thereby he entered the field of speculation; however, with the passage of time, his interpretation has survived the test of prediction and subsequent observation to become one of the most firmly grounded propositions of metamorphic theory. Its essence is this: In the absence of textural evidence to the contrary, any metamorphic phase assemblage represents a chemical system approximating a state of equilibrium governed by some finite limited range of pressure and temperature. On a broader scale all assemblages that constitute a single facies have reached equilibrium within the same limits of pressure and temperatures. Accepting this primary proposition, we next attempt to place numerical values on specific *P-T* regimes. It was the young V. M. Goldschmidt, who, in 1912, took the first successful step along these lines, thereby launching what has mushroomed into a worldwide effort among geochemists to delineate the *P-T* stability fields of key metamorphic minerals and mineral combinations.

Equilibrium and Stability

Thermodynamic constraints apply equally to igneous and to metamorphic systems in equilibrium. But since metamorphic theory is so strongly focused on *P-T* regimes recorded by metamorphic mineral assemblages—whereas in igneous chemistry greater emphasis is placed on the bulk chemistry of evolving magmatic systems—it is now desirable to delve a little more deeply into the nature of quantitative thermodynamic constraints than was necessary in the introduction to igneous petrogenesis (as set out in Chapter 2). Once again the concepts of equilibrium and stability can be utilized only in the strictest sense of thermodynamic definition. But here we shall apply and illustrate them in a quantitative sense, rather than qualitatively as in our earlier preliminary treatment. Relevant experimentally determined data for chemically pure phases corresponding to many stoichiometrically ideal metamorphic minerals and gases are now readily accessible.*

Two chemically equivalent phase assemblages are said to be in *equilibrium* under specific conditions of P and T when there is no tendency for spontaneous change from one to the other. They are then related by a *reversible* theoretical reaction such as

$$CaMg(CO_3)_2 + 2SiO_2 \leftrightarrows CaMgSi_2O_6 + 2CO_2$$

$$\text{dolomite} \quad \text{quartz} \quad \text{diopside} \quad \text{gas}$$

The thermodynamic condition for equilibrium is that the free-energy change accompanying reaction would be zero:

$$\Delta G_r = 0$$

or

$$\Delta H_r = T\Delta S_r$$

where ΔG_r, ΔH_r, and ΔS_r are the changes in free energy, heat (enthalpy), and entropy of the system that would accompany reaction in either direction. Any real *spontaneous* reaction, such as dolomite + 2 quartz $\rightarrow$ diopside + $2CO_2$, is *irreversible*, and it must lead to decrease in free energy of the system:

$$\Delta G_r < 0$$

or

$$T\Delta S_r > \Delta H_r$$

---

*The properties in question, expressed as molar values, include volume V (cm^3) and calorimetrically determined or computed thermal quantities expressed in calories or in joules (1 cal = 4.184 joules) per mole; entropy S (calories per mole per kelvin), and heats and free energies of formation from component elements, respectively ΔH_f° and ΔG_f° (both in calories per mole). Such data, all for phases at standard room temperature (25°C = 298.15K) and atmospheric pressure (either 1.013 or 1 bar), are given in such publications as *Geological Society of America Memoir* 97 (1968): 437–458 and *U.S. Geological Survey Bulletin*, no. 1452 (1978). Simple numerical examples of their manipulation in the context of metamorphic equilibria are given by F. J. Turner, 1981, pp. 108–124.

An elementary treatment of their mutual relations is given in Appendix A.

Such is the case for the dolomite → diopside reaction at atmospheric pressure above about 200° C; but at room temperature ΔG_r has a positive value (about $+7000$ cal per mole of CO_2) and reaction is then prohibited.

Under specific conditions a phase or a multiphase system of given chemical composition is said to be *stable* when it has no capacity for internal reaction of any kind. The thermodynamic requirement for stability is minimal free energy in the system; thus for any hypothetical phase transformation that can be prescribed within the system,

$$\Delta G_r > 0$$

or

$$\Delta H_r > T\Delta S_r$$

At room temperature and pressure this condition is fulfilled for the phase assemblage calcite-dolomite-quartz ($\pm$ CO_2) found in many impure metadolomites. However, at temperatures above about 200° C, again at room pressure, this is not so; as already stated, stable derivatives of siliceous dolomitic limestones now carry diopside: for example, diopside-calcite-dolomite ($\pm$ CO_2) or diopside-calcite-quartz ($\pm$ CO_2), depending on the bulk composition of the system. Again simple computation shows that within the same compositional range the theoretical reaction

$$2CaMg(CO_3)_2 + SiO_2 \leftrightarrows Mg_2SiO_4 + 2CaCO_3 + 2CO_2$$

dolomite   quartz   forsterite   calcite   gas

becomes reversible ($\Delta G_r = 0$) at about 290°C and thus expresses equilibrium between all five phases. But this equilibrium is *unstable* (*metastable*); for at any temperature above 200°C the reactant pair dolomite-quartz itself is unstable with respect to diopside-CO_2. In other words, between 200°C and 290°C, ΔG_r is negative for both reactions:

dolomite + 2 quartz → diopside + $2CO_2$

and

forsterite + 2 calcite + 3 quartz → 2 diopside + $2CO_2$

For any prescribed set of physical conditions, the laws of thermodynamics explicitly forbid some reactions and permit others. But there is no thermodynamic argument that predicts whether a permissible reaction will actually occur. Effectiveness of any permissible reaction is determined by kinetic factors: rates of ionic diffusion and of nucleation of new phases (both exponentially related to temperature, in kelvins) and the magnitude of ΔG_r, the "driving force" of reaction. As a corollary, when an actual transformation $A \rightarrow B$ takes place there is no guarantee that B is the most stable form possible. It could be unstable (metastable) with respect to an alternative chemically equivalent assemblage C.

Limitations Imposed by the Phase Rule

The phase rule of J. Willard Gibbs is a statement, derived from first principles of thermodynamics, that defines an infallible, simple arithmetic relation between three numbers pertaining to any heterogeneous system in a state of internal equilibrium (stable or unstable): number of phases φ, variance ω, and number of components c. These numbers require individual definition.

The *phases* are the separate physically distinct entities that together comprise the system; for example, in the system calcite-wollastonite-quartz-CO_2 gas, $\varphi = 4$. So φ is determined simply by counting the phases.

There is considerable latitude in choice of actual chemical *components*; in the above case they could alternatively be designated as the three component oxides CaO, SiO_2, and CO_2; or as the four component elements Ca, Si, C, O; or as the four phases themselves; or again as various numbers of postulated ions. The *value* of c, however, is unambiguous: it is the *smallest* number of chemical components, however chosen, that collectively describes the compositions of all phases comprising the system. In the above case $c = 3$.

The *variance* (or number of degrees of freedom) ω is the smallest number of the several possible "unknowns" or "variables"—pressure, temperature, and compositional values for every phase of variable composition—that suffices to fix automatically the value of all remaining unknowns.

The phase rule is commonly written as

$$\omega = c + 2 - \varphi$$

where 2 is the number of physical variables (pressure and temperature) that influence the state of any given system. For the simple calcite-quartz-wollastonite-CO_2 gas system

$$\omega = (3 + 2 - 4) = 1$$

The system is *univariant*: If a single variable (*T* or *P*) is specified (e.g., T = 100°C), the other is automatically fixed. The four phases can coexist in equilibrium only at points that fall on a univariant curve plotted on a *P-T* diagram (cf. Figure 16–6).

Without pursuing the topic in further detail, we note several useful generalizations that follow from the phase rule:

1. Simplicity of mineral assemblages—a common condition—is consistent with, but does not prove, equilibrium. Assemblages of many phases in a system of a few components demonstrates disequilibrium ($\omega < 0$).

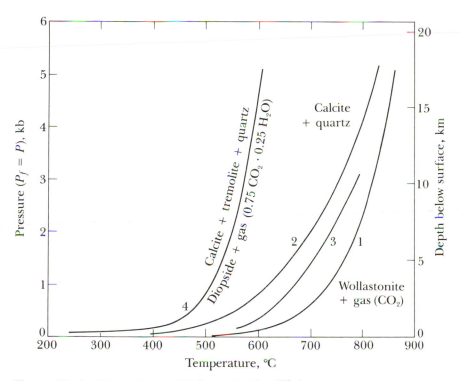

Figure 16–6. Calcite-Quartz-Wollastonite Equilibrium

1–3. Alternative curves for univariant equilibrium ($P_{CO_2} = P_f = P$):

$$CaCO_3 + SiO_2 \rightleftharpoons CaSiO_3 + CO_2$$
$$\text{calcite} \quad \text{quartz} \quad \text{wollastonite}$$

4. Computed curve for univariant equilibrium ($P_f = P$; gas composition buffered at stoichiometric value $X_{CO_2} = 0.75$; $X_{H_2O} = 0.25$):

$$Ca_2Mg_5Si_8O_{22}(OH)_2 + 3CaCO_3 + 2SiO_2 \rightleftharpoons 5CaMgSi_2O_6 + 3CO_2 + H_2O$$
$$\text{tremolite} \qquad\quad \text{calcite} \quad\; \text{quartz} \qquad \text{diopside}$$

2. Equilibrium in common assemblages is likely to have a variance of 2 or more; for repeated appearance in different geographic provinces implies general latitude in P, T, and probably in compositional variables ($X_{CO^2}^{gas}$, $X_{H_2O}^{gas}$) as well.

3. From our definition of c, two chemically identical adjacent assemblages separated by an isograd can be considered as a system in univariant equilibrium:

a. The sillimanite isograd in Scotland

$$\text{kyanite} \leftrightarrows \text{sillimanite} \ (c = 1, \varphi = 2: \omega = 1)$$

b. The second sillimanite isograd in New England

$$\text{muscovite} + \text{quartz} \leftrightarrows \text{K-feldspar} + \text{sillimanite} + \text{H}_2\text{O}$$
$$(c = 4, \varphi = 5: \omega = 1)$$

4. Laboratory investigations of phase equilibria are commonly designed to eliminate compositional unknowns. Pure phases or simple solid solutions are employed, and gas composition is buffered to some constant level. To such systems, the phase rule can be applied without ambiguity (topologies of phase-equilibrium diagrams so familiar to the student of igneous petrology in fact are rigorously constrained by phase-rule requirements). This is not the case for most natural equilibria; for here compositional variation among phases is almost universal.

5. The phase rule affirms the general validity of using experimentally evaluated numbers in geobarometry and geothermometry based on metamorphic phase equilibria, or on partitioning of common components between phases coexisting in a state of equilibrium. Consider coexisting calcite and dolomite in magnesian marble. Calcite in such a pair always contains appreciable but variable amounts of Mg; and coexisting dolomite departs slightly from the stoichiometric composition $\text{CaMg(CO}_3)_2$. The components of the system are two: CaCO_3 and MgCO_2. Unknowns are P, T, $X_{MgCO_3}^{calcite}$, and $X_{MgCO_3}^{dolomite}$. The phase rule gives $\omega = 2 + 2 - 2 = 2$. So if the two variables P and $X_{MgCO_3}^{calcite}$—are known, the other two—T and $X_{MgCO_3}^{dolomite}$—are automatically fixed. Since the compositional variables in this pair are found to be much more sensitive to T than to P, chemical analysis of the calcite phase alone is sufficient to determine T_E, the equilibrium temperature, within 25 degrees or so, within any estimated pressure range of about 5 kb (see reference note 3).

Conclusion: Equilibrium in Relation
to Scale of Observation

Let us return now to the original question on which the pressure-temperature implications of metamorphic mineral assemblages must hinge. Thermodynamic equilibrium is a condition subject to rigorous definitions and experimental test. Textural equilibrium is a subjectively judged condition that cannot be defined in precise terms. To equate the second with the first is an inference or postulate that nevertheless is open to general test in the light of accumulating experience. Some such postulate, in fact, is unavoidable if petrographic data are to be translated into precise physical terms.

From recent experience has come a general realization that what the petrographer calls textural equilibrium is a matter of scale. Minerals microscopically identified as homogeneous phases in apparent equilibrium are seen on a much smaller scale (e.g., under the electron microscope or the microprobe) to be heterogeneous. Optically homogeneous sanidine is revealed as an intergrowth of two solid solutions, $(Na, K)AlSi_3O_8$ and $(K, Na)AlSi_3O_8$. Even on the microscopic scale ilmenite and magnetite prove to be two-phase intergrowths. Here we see phases that once were homogeneous and stable at high temperature being replaced by other phase assemblages, products of exsolution as the system reequilibriated at lower temperatures during postmetamorphic cooling and decompression. Such observations in no way detract, however, from our primary postulate. Quite the reverse, in fact; for they show that thermodynamically predicted possible retrograde changes do in fact occur—but on a scale so small that the identity of the metamorphic phase assemblages is still clearly recognizable on the microscopic scale. And it is this metamorphic assemblage that we seek to evaluate in terms of pressure and temperature.

Introduction to the Petrogenic Grid

When Bowen in 1940 proposed the concept of a petrogenic grid, it was indeed hardly more than a concept.[4] Any simple univariant equilibrium between key phases with counterparts among metamorphic minerals can be represented by a curve on a *P-T* diagram (all other variables such as $X_{H_2O}^{gas}$ being held constant). This curve separates the respective stability fields of two chemically equivalent bivariant phase assemblages that can coexist in univariant equilibrium at all points on the curve. Only one such curve was initially available for Bowen's embryonic grid—that for the classic calcite-quartz$\rightleftharpoons$wollastonite-CO_2 equilibrium as computed from imperfect thermodynamic data by V. M. Goldschmidt a quarter of a century earlier. But Bowen realized the potential of new experimental techniques to develop many univariant curves of comparable or even greater significance that would one day provide the means for evaluating the whole gamut of metamorphic pressures and temperatures. It was this network, still in the conceptual stage, that he termed a *petrogenic grid*.

Today the grid has become a complicated reality possessing the full potentiality that Bowen predicted. New curves have been added in succession, and more appear every year. True, they have a transient quality. None can be regarded as immutable. Most of the earlier curves have now been modified and replaced by later versions. So petrologists find themselves in an increasingly perplexing situation in which one must choose for oneself which of several alternative curves one should accept to evaluate any particular equilibrium. The petrographer who does not wish to delve into detail must remember that the criterion of acceptability lies in neither the newness of the data nor the apparent sophistication of new experimental techniques. The true test is the implicit degree of internal consistency that exists between curves that represent equilibria with phases in common; for the topology of each curve in such a set reflects thermodynamic data drawn from a common stock (ΔH_f, S, and V values for each participating phase). Thermodynamic implications of the calcite-quartz$\rightleftharpoons$wollastonite-CO_2 equilibrium carry over to other carbonate-silicate equilibria such as tremolite-calcite-quartz$\rightleftharpoons$diopside-CO_2-H_2O.

Fortunately a recent comprehensive appraisal by H. Helgeson and coauthors has clarified what had become a confusing and extremely complex situation.[5] Their revised curves for a host of metamorphically significant equilibria are mutually consistent in the above respect. These curves are tentatively accepted by the present authors in the knowledge that from time to time the whole set will require revision in the light of new and more refined experimental data.

In conclusion, the potential of the petrogenic grid is illustrated by a few selected univariant curves that form the basis of some of the numerical pressure and temperature values that appear in the descriptive chapters that follow:

Figure 16–6 includes three alternative versions of the calcite-quartz$\rightleftharpoons$wollastonite-CO_2 equilibrium:

1. As computed from imperfect thermodynamic data by V. M. Goldschmidt in 1912.
2. As computed from more refined data by A. Danielsson in 1950.
3. As extrapolated from brackets of experimental reversal by R. I. Harker and O. F. Tuttle in 1956: currently preferred curve.

Shown for comparison is curve 4 for the equilibrium calcite + tremolite⇌diopside + gas (0.75 CO_2 + 0.25 H_2O)—univariant if the gas is held at constant stoichiometrically controlled composition.

In Figure 16–7 we show univariant curves for:

1. Jadeite-quartz⇌albite, extrapolated by thermodynamic computation from brackets of reversal at high temperature by F. Birch and P. Lecomte in 1961.
2. Aragonite⇌calcite inversion, based on experimental reversals and checked by computation, as revised by A. L. Boettcher and P. J. Wyllie in 1967.
3. Univariant equilibria and point O of invariant equilibrium ("triple point") in the Al_2SiO_5 system; data from many sources, revised by M. J. Holdaway in 1971.
4. Univariant curve for the breakdown of muscovite;* muscovite-quartz⇌K-feldspar-sillimanite-H_2O, based on experimental reversal by B. W. Evans in 1965; pressure variables are held at $P_{H_2O} = P_{fluid} + P_{solid}$.
5. Curve of univariant equilibrium for laumontite⇌quartz + lawsonite + H_2O which defines entry of lawsonite; determined by J. G. Liou in 1976 from experimental reversals.

SYNOPTIC PLAN OF TREATMENT
OF PETROGRAPHIC DATA

The present chapter provides a framework within which to review the spectrum of metamorphic petrography in a manner conducive to further application in problems of broader geological interest—for example, in

---

*Breakdown of natural muscovites is a continuous process involving preferential depletion in the sodic component. The curve is therefore a simplified version of a process that at any given pressure covers a finite, small span of temperature.

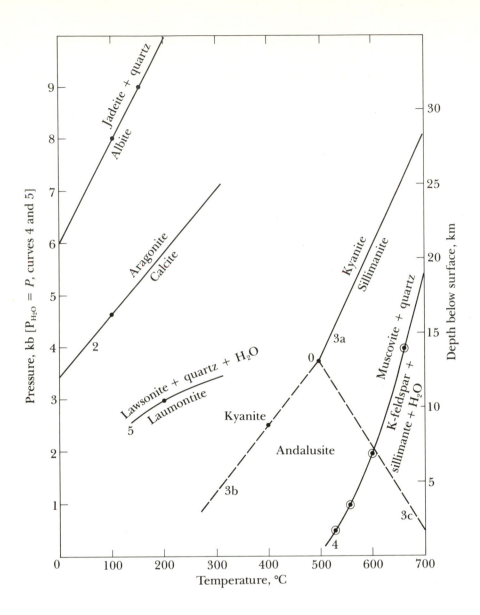

Figure 16–7. Selected Curves of Univariant Equilibrium

1. $NaAlSi_2O_6$ + SiO_2 $\rightleftharpoons$ $NaAlSi_3O_8$
   jadeite   quartz   albite

2. aragonite $\rightleftharpoons$ calcite.

3a–c. Inversions between Al_2SiO_5 polymorphs

4. $KAl_2(AlSi_3O_{10})(OH)_2$ + SiO_2 $\rightleftharpoons$ $KAlSi_3O_8$ + Al_2SiO_5 + H_2O
   muscovite   quartz K-feldspar sillimanite

5. Laumontite $\rightleftharpoons$ lawsonite + quartz + H_2O

connection with crustal evolution, tectonics, ore genesis, and plutonism. From many alternative modes of treatment we have chosen one that places first emphasis on the broad field situations in which different metamorphic patterns repeatedly occur. While individual rock types in any one of these situations are defined primarily in petrographic terms (on mineralogical and textural criteria), they have been grouped in each of the following chapters (again on petrographic evidence) into a few categories that reflect in a very broad fashion their protolithology and chemical parentage. In contrast to the treatment of igneous rocks, bulk chemical analyses have no place in the following plan; for these reflect the net combined influences of initial chemistry and later modifications (largely a matter of surmise) that for the most part cannot with any confidence be unraveled.

Our descriptive subject matter, with such ends in mind, has been marshaled under these headings:

1. Products of contact metamorphism: spotted schists, hornfelses, contact marbles, skarns, and related rocks.
2. Products of mechanical strain (as in dislocation metamorphism): mylonites, phyllonites, semischists, and similar rocks.
3. Products of low-grade regional metamorphism:
    a. Partially recrystallized (immature) rocks—slates, nontectonites lacking foliation.
    b. Completely recrystallized (mature) rocks—phyllites and schists of the greenschist, blueschist, and closely related facies.
4. Products of medium- and high-grade regional metamorphism: schists, gneisses, amphibolites, granulites, and eclogites (amphibolite, granulite, and eclogite facies).

A metamorphic facies, as stated earlier, is too broad and varied a unit to serve as a primary basis for either classifying or systematically describing metamorphic rocks. Yet it is through the facies concept that our descriptive data may become transposed into the wider and geologically significant realm of metamorphic temperature-pressure regimes in the crust in relation to tectonics and time. So in Figure 16–8 we present our tentative view of the manner in which individual facies currently appear to fit into the broad spectrum of crustal temperatures and pressures. And in most chapters we offer in tabular form the currently accepted relations of certain common metamorphic mineral assemblages to specific facies. To pursue the matter further in attempting to assign specific

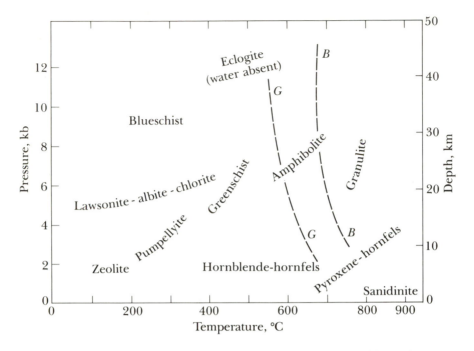

Figure 16–8. Approximate Relative Locations of Metamorphic Facies on a Pressure-Temperature Field

Diagram based on simplified assumption that $P_{H_2O} = P$—except for eclogite facies. *GG* and *BB* are approximate minimum-temperature limits for beginning of melting of granitic and basaltic rocks in presence of excess externally supplied water (simplified from Turner, 1981, figure 11–1).

metamorphic assemblages to numerically evaluated spans of pressure and temperature the reader must turn to more theoretical works (for example, F. J. Turner, 1981).

GENERAL REFERENCES

Throughout Part Three, repeated references are made to two general works:

A. Harker, *Metamorphism* (London: Methuen, 1932) is especially valuable for illustrations of rock types from classic British and European localities.

F. J. Turner, *Metamorphic Petrology,* 2d ed. (Washington, D.C.: Hemisphere, 1981) provides an expanded treatment of concepts and aspects of petrogenesis that are mentioned briefly in this book.

REFERENCE NOTES

[1]For example, F. J. Turner and L. E. Weiss, *Structural Analysis of Metamorphic Tectonites* (New York: McGraw-Hill, 1963); J. Ramsay, *Folding and Fracturing of Rocks* (New York: McGraw-Hill, 1967).

[2]J. B. Thompson, *American Mineralogist,* vol. 42 (1957): pp. 842–858.

[3]Compare J. R. Goldsmith and R. C. Newton, cited by Turner, 1981, pp. 131–133.

[4]N. L. Bowen, *Journal of Geology,* vol. 48 (1940): pp. 272–274.

[5]H. Helgeson et al., *American Journal of Science* 278A (1978).

17

Products of Contact Metamorphism

FIELD OCCURRENCE

Hornfelses and allied rocks most commonly occur in contact aureoles bordering bodies of plutonic rocks (especially granites, granodiorites, and tonalites) intrusive into rocks otherwise unaffected, or but slightly affected, by metamorphism. Large granitic batholiths, such as that of the Sierra Nevada, also enclose steeply dipping septa (separating adjacent plutons) and roof remnants composed of hornfelses; these may be several square kilometers in extent.

In the immediate vicinity of the plutonic contact, the temperatures in many aureoles were high enough to develop mineral assemblages of the pyroxene-hornfels facies. Further from the contacts, or throughout those aureoles in which the temperatures were somewhat lower, the mineral assemblages are those of the hornblende-hornfels facies. Out on the periphery of most well-developed aureoles there occur such imperfectly metamorphosed rocks as spotted slates and spotted schists, in which porphyroblasts, or coarse aggregates of new minerals, have crystallized in a matrix that still retains something of the mineralogy and texture of the parent rocks. The assemblage of newly formed minerals here may even correspond to the relatively low temperatures of the albite-epidote-hornfels facies.

Hornfelses, sometimes of unusual mineralogical composition indicating exceptionally high temperature and low pressure of metamorphism (sanidinite facies), also occur as inclusions a few centimeters in diameter in basaltic dikes and lavas, which have supplied the heat necessary for metamorphism. The tholeiitic dikes of Mull in western Scotland, and flows of contaminated basalt near Clear Lake, California, illustrate this mode of occurrence. Similar rocks also occur immediately adjacent to volcanic necks where high temperatures have been maintained by the continued flow of hot basic magma at shallow depth.

Contact metamorphism in some cases involves no appreciable change in chemical composition of the rocks affected other than addition or removal of water. Minor amounts of elements such as boron, sulfur, fluorine, and chlorine are commonly introduced from fluids of granitic origin, and these may give rise to corresponding accessory minerals in the hornfelses—tourmaline, axinite, pyrite, fluorite, scapolite, and the like. In yet other cases (e.g., in conversion of limestone to skarn consisting of silicates of Ca, Fe, and Mg) very large amounts of elements such as Si and Fe may be added during contact metamorphism. The isotopic composition of oxygen in rocks of some inner contact zones suggests that the general flux of water involved may have been inward (from the outer envelope) rather than outward (from the pluton). This finding raises doubts regarding the hitherto accepted igneous source of certain minor elements now concentrated in hornfelses close to plutonic contacts.

Hornfelsic rocks completely or almost lacking schistosity also occur on a regional scale in areas where high-grade regional metamorphism was unaccompanied by significant strain—an atypical situation.

FABRIC

Many of the principal minerals of hornfelses—quartz, feldspars, pyroxenes, grossular, calcite—usually occur in equidimensional grains. Even the micas and amphiboles are much less markedly tabular and prismatic in hornfelses than in schists and igneous rocks, and show little or no preferred orientation. Most hornfelses, then, consist of a mosaic of equidimensional, unoriented grains, and their texture (fabric) is termed *granoblastic* or *hornfelsic* (Figures 17–2B, 17–3C, and 17–5A). Several minerals formed by contact metamorphism, notably andalusite and cordierite, almost always occur as large porphyroblasts crowded with small inclusions. Other minerals, also, such as biotite, muscovite, tourmaline, grossular, vesuvianite, and scapolite, may in some circumstances be porphyroblastic. Many hornfelses are thus characterized by a combination of porphyroblastic and granoblastic fabric (Figures 17–3A,B and 17–10C).

Since deformation plays no part in the evolution of typical hornfelses, relict textures are common even in completely metamorphosed rocks of this class. Vestiges of ophitic, porphyritic, and amygdaloidal textures are readily distinguished in the fabrics of many hornfelses of volcanic origin. Stratification of parent mudstones may survive in pelitic hornfelses as delicate banding marked by variation in quantity and grain size of mica and andalusite. Curiously enough, however, metamorphism may induce reversal of relative grain size: the originally finer clay-rich bands, being highly sensitive to metamorphism, become reconstituted as coarse-grained layers crowded with large flakes of mica and spongy grains of andalusite, while the sandy bands recrystallize, without notable further coarsening, as relatively fine-grained quartzo-feldspathic layers.

Extreme contact metamorphism of slates in the inner part of an aureole usually results in obliteration of the original schistosity; but further from the contact this structure may survive, and may even be intensified, by growth of porphyroblasts of hornblende or mica with their long dimensions in the plane of schistosity. In other weakly metamorphosed derivatives of slates the initial schistosity may be retained and may even be accentuated by recrystallized fine-grained mica. Here incipient porphyroblasts of andalusite or cordierite, or scattered ovoid knots of coarser mica, may give a spotted appearance to the rock, which is then termed a *spotted schist* or *spotted slate* (Figure 17–1). Recognizable fossils may sometimes be seen in these rocks.

MINERAL PARAGENESIS

Since rocks of any initial composition may be represented among contact aureoles, and since in any particular aureole the pressure-temperature regime may be anywhere within a broad spectrum (*P* between about 100 bars and 2–3 kb, *T* between perhaps 300°C and 700°C in most cases), the mineral paragenesis of contact metamorphism is correspondingly varied. Table 17–1 shows a number of the more familiar mineral assemblages (omitting common accessories such as sphene, tourmaline, apatite, and opaque oxides) in relation to general rock composition and to facies

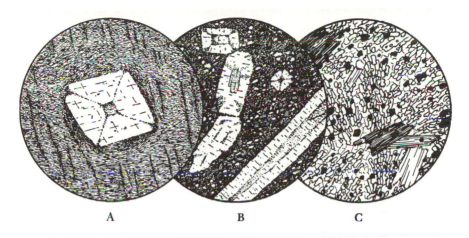

Figure 17–1. Pelitic Hornfelses and Spotted Slates

A. Chiastolite slate, Fichtelgebirge, Bavaria. Diam. 3 mm. A porphyroblast of chiastolite (now converted to a mat of indeterminate colorless micaceous minerals), cut at right angles to the z (c) axis, shows geometrically arranged graphite inclusions. The groundmass consists of finely crystalline, colorless micas, pale-brown biotite, and minor quartz and graphite. Note how the slaty cleavage (horizontal) and the cross-cutting strain-slip cleavage (steeply inclined) have been destroyed in the vicinity of the growing porphyroblast.

B. Chiastolite slate, near Mariposa, Sierra Nevada, California. Diam. 7 mm. Section cut parallel to slaty cleavage. Porphyroblasts of altered chiastolite are enclosed in a matrix of biotite, graphite, and quartz. Note the unaltered core, which has survived in the upper part of the central porphyroblast.

C. Andalusite hornfels, near Andlau, Germany. Diam. 3 mm. Spongy andalusite, biotite, muscovite, and iron oxides in a matrix of quartz.

of contact metamorphism. This table is by no means exhaustive. Some of the tabulated assemblages are elaborated in the ensuing section on petrography.

The commonest assemblages that comprise the great bulk of contact aureoles fall in the hornblende-hornfels and pyroxene-hornfels facies. Only these are illustrated in Table 17–1. At minimal grades represented in the outer fringes of some aureoles, mineral assemblages are imperfectly developed; here only an albite-epidote-hornblende assemblage in basic rocks is truly diagnostic. At the other extreme of the temperature spectrum are rare and striking mineral assemblages of the sanidinite facies. These will be mentioned in due course.

Table 17–1. Some Characteristic Mineral Assemblages (Accessory Phases Omitted) in Common Rocks on Contact Aureoles

Rock Group	Hornblende-Hornfels Facies		Pyroxene-Hornfels Facies	
Pelitic	Muscovite-biotite Andalusite[a]-muscovite-biotite Andalusite[a]-cordierite-muscovite-biotite Staurolite-biotite-andalusite[a] Staurolite-cordierite-muscovite	} Plus any or all of quartz plagioclase, K-feldspar	*With quartz* K-feldspar-sillimanite[b]-cordierite K-feldspar-sillimanite[b]	} Plus biotite (and plagioclase)
			Without quartz Cordierite-corundum-spinel Cordierite-corundum-sillimanite[b]	} Plus any or all biotite, K-feldspar, plagioclase
Calcareous				
1. Calcic marbles[c]	Calcite-tremolite (-quartz) Calcite-diopside (-quartz) Calcite-tremolite-diopside Calcite-diopside-grossular		Calcite-wollastonite (-diopside) Calcite-diopside (-forsterite) Calcite-wollastonite-diopside-grossular	
2. Magnesian marbles (metadolomites)[c]	Calcite-dolomite-tremolite-clinohumite Calcite-dolomite-forsterite Calcite-dolomite-forsterite-phlogopite		Calcite-forsterite-periclase Calcite-forsterite-monticellite Calcite-forsterite-spinel Calcite-forsterite-diopside	} Clinohumite possible additional phase
3. Calc-silicate rocks	Diopside-epidote-hornblende Diopside-grossular-epidote Diopside-vesuvianite-grossular-wollastonite Diopside and grossular, commonly with significant iron		Diopside-wollastonite-grossular-vesuvianite Diopside-grossular-anorthite (or calcic plagioclase)	

Basic	Hornblende-plagiocalse (-biotite, -almandine) Hornblende-plagioclase-diopside	Diopside-hypersthene-plagioclase Diopside-olivine-plagioclase
Magnesian		
1. Metaserpentinites	Antigorite-forsterite-tremolite Forsterite-talc-tremolite Forsterite-anthophyllite-tremolite Anthophyllite-talc	Forsterite-enstatite-spinel (-diopside)
2. Aluminous types	Cordierite-anthophyllite (-biotite) Anthophyllite-cummingtonite-biotite	Hypersthene-cordierite (-biotite)

[a] Or sillimanite.
[b] Or andalusite.
[c] K-feldspar or plagioclase, or both, possible minor phase.

[481]

PETROGRAPHY OF THE PRINCIPAL ROCK TYPES

Pelitic Hornfelses

Most pelitic hornfelses contain sufficient Al_2O_3 to cause andalusite or cordierite or both to crystallize—commonly as porphyroblasts set in a fine-grained granoblastic matrix of quartz, feldspar, mica, and graphite. The nature of the minerals in this matrix indicates the metamorphic facies to which the rock belongs. If orthoclase or microcline is present together with andalusite or sillimanite, and muscovite is absent, the assemblage belongs to the pyroxene-hornfels facies, for at lower temperatures K-feldspar is incompatible with aluminum silicates, and micas appear in its place. Consequently the corresponding assemblages in the hornblende-hornfels facies, formed at lower temperatures, are musco-vite-biotite-andalusite-quartz, muscovite-biotite-cordierite-quartz, and rarely (in highly potassic rocks), muscovite-biotite-microcline-quartz.

Andalusite has two common habits. It may form large spongy xeno-blastic crystals riddled with inclusions of quartz, graphite, and biotite (Figure 17–1C); or it may occur as the variety chiastolite in well-defined prismatic crystals, whose nearly square cross sections show symmetrical cross-shaped dark patterns resulting from concentration of included gra-phitic dust along the edges and the central axis of the growing porphy-roblast (Figure 17–1A,B). In many hornfelses crystals of andalusite are partially or completely replaced by fine-grained aggregates of white mica ("sericite"), formed by reaction between andalusite and potassic aqueous fluids with falling temperature (Figure 17–1B). Cordierite, also, occurs almost exclusively as porphyroblasts. These are irregularly bounded and crowded with inclusions of quartz, graphite, and the like. They cannot be distinguished from plagioclase except by their characteristic sector twinning,* the presence of yellow pleochroic haloes around enclosed zircons or sphenes, and a common tendency to be replaced marginally and along cracks by a colorless or yellowish member of the chlorite family (Figure 17–2B). Red-brown biotite is present in almost all pelitic horn-felses; but muscovite is restricted to rocks of the hornblende-hornfels facies. Both minerals occur as minute unoriented flakes in the grano-

---

*Clear lamellar-twinned cordierite is easily mistaken for plagioclase; but in sections of cordierite cut at high angles to intersecting lamellae these are mutually inclined at 60°—compared with almost 90° in plagioclase showing sharply intersecting twins on the albite and pericline laws.

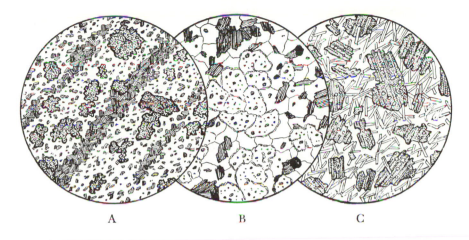

A B C

Figure 17–2. Pelitic Hornfelses

A. Silica-deficient corundum-biotite-cordierite hornfels (pyroxene-hornfels facies), Comrie, Scotland. Diam. 3 mm. The banding probably represents original bedding.

B. Cordierite-biotite-quartz hornfels, Odenwald, Germany. Diam. 3 mm. Irregular grains of cordierite are charged with small inclusions, mainly of biotite, and show incipient replacement by pale-green chlorite ("pinite") along margins and cracks.

C. Biotite-muscovite-quartz hornfels, Cascade Valley, Westland, New Zealand. Diam. 3 mm. Note the sieve texture of the biotite porphyroblasts.

blastic matrix, and occasionally as scattered porphyroblasts (Figure 17–2C) or as oval polycrystalline knots. Feldspars usually form untwinned xenoblastic grains confined to the matrix; but large porphyroblasts of microcline or albite may develop in hornfelses that have been affected by K or Na metasomatism near granite contacts. Apatite and tourmaline are almost constant accessories. If the parent rock was a marine shale, its initial content of boron would be sufficient to account for tourmaline occurring as a uniformly distributed accessory mineral. Local crystallization of abundant tourmaline, especially in the vicinity of recognizable microfractures, is usually attributable to introduction of boron from an external, probably magmatic source (Figure 17–5A,C).

In some pelitic hornfelses sillimanite appears in place of andalusite. It occurs as aggregates of slender fibers forming at the expense of biotite (Figure 17–3A) or as sheaflike segregations of much coarser prisms (Figure 17–3B). The presence of sillimanite seems to be favored by long-sustained high temperatures and possibly by the influence of magmatic

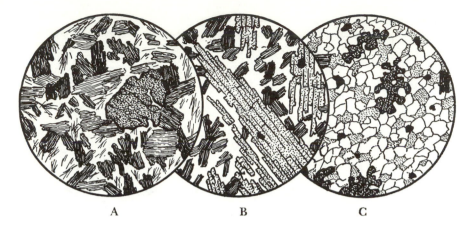

A   B   C

Figure 17–3. Pelitic and Quartzo-Feldspathic Hornfelses

A. Biotite hornfels, adjacent to contact with granite, Donegal, Ireland. Diam. 3 mm. The central porphyroblast is of yellow tourmaline. Fibers of sillimanite are enclosed in granoblastic quartz.

B. Biotite-sillimanite hornfels, adjacent to contact with granite, Donegal, Ireland. Diam. 3 mm. Coarse sillimanite, brown biotite, and granoblastic quartz.

C. Quartz-feldspar hornfels, Eberstadt, Hessen, Germany. Diam. 3 mm. The granoblastic aggregate consists of quartz, oligoclase, and minor microcline. The irregular dark grains are green hornblende. There are minor biotite and iron oxide.

fluids, for it is especially characteristic of rocks in the immediate vicinity of granite contacts. Some such rocks are so rich in sillimanite that they must be regarded as products of complex reaction between magma and hornfels, in the course of which nonaluminous components have been partly transferred to the magma and the proportion of Al_2O_3 in the metamorphic residue has been increased. Basaltic rocks sometimes contain xenolithic masses of aluminum-enriched hornfels belonging to the sanidinite facies. These may contain so much excess Al_2O_3 (as compared with SiO_2) that one of the two silica-deficient minerals corundum (sapphire) or spinel has crystallized. Where partial melting of these xenoliths has occurred, the resulting rock (transitional between igneous and metamorphic) is termed a *buchite* (Figure 17–4). Pelitic buchites may consist of cordierite, spinel, the rare high-temperature aluminum silicate mullite, and glass.

Most pelitic hornfelses contain an excess of SiO_2, so that quartz crystallizes with aluminous silicates. But some are so rich in Al_2O_3 and cor-

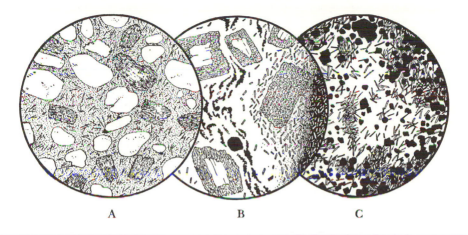

Figure 17–4. Buchites

A. Vitrified sandstone from contact with mica lamprophyre, Navajo Reservation, Arizona. Diam. 3 mm. Spongy relics of "fritted" plagioclase and of K-feldspar, partially vitrified along cleavages; corroded grains of quartz; matrix of pale-brown glass enclosing length-slow fibers of (?) mullite.

B. Partially vitrified granodiorite inclusion in basalt, Table Mountain, near Oroville, California. Diam. 3 mm. Colorless glass formed by fusion of quartz and feldspar grades through brown glass into basalt with a matrix of black glass (on right). Relict grains of plagioclase have partially vitrified "fritted" borders charged with colorless glass; the plagioclase crystal adjacent to the basalt is more completely vitrified and is charged with brown glass. Trains of magnetite dust and feldspar microlites indicate flow within the fused matrix of the quartz diorite.

C. Hornfels formed by local high-temperature metamorphism of weathered basalt, at contact with diabase neck, Tevebullaigh, Antrim, Ireland. Diam. 0.5 mm. Magnetite, deep-green spinel, and needles of mullite, in a matrix of cordierite.

respondingly poor in SiO_2 that quartz is absent, and corundum or spinel or both are present, in assemblages of the pyroxene-hornfels facies. Such is the case in the aureole of a dioritic pluton at Comrie, in Perthshire, Scotland, where silica-deficient pelitic assemblages (cf. Figure 16–4B) include corundum-cordierite-biotite, with some orthoclase and magnetite (Figure 17–2A); cordierite-corundum-spinel (biotite-orthoclase); and andalusite-cordierite-corundum.[1]

Certain aluminous metamorphic minerals—staurolite, almandine garnet, and kyanite—once thought to be alien to rocks of contact aureoles, are now known to be by no means as rare as once was thought. Thus in outer zones of the Santa Rosa aureole, Nevada, which is 1400 m in radial width, staurolite-andalusite-biotite hornfelses are associated with other pelitic rocks with cordierite-andalusite-biotite. Almandine-cordierite-andalusite assemblages are represented in the Tôno aureole of Japan. Appearance of both almandine and staurolite is correlated with high values of FeO/MgO and Al_2O_3 in the bulk composition of the rock. Kyanite is found much more rarely than either almandine or staurolite. Its presence betokens a combination of relatively high pressure and low temperature. Examples have been recorded in hornfelses from aureoles that surround some of the granodioritic plutons of Donegal; and in some of these rocks both andalusite and kyanite appear within single specimens.

Pelitic Spotted Schists

In the outer part of an aureole developed in initially slaty or phyllitic rocks, the typical products of contact metamorphism are foliated (schistose) rather than hornfelsic. There are two reasons for this: Metamorphism is not sufficiently intense to obliterate the schistosity of the parent rock, and the temperature of metamorphism (equivalent to the hornblende-hornfels facies) is low enough for muscovite and biotite to be stable minerals. The habit of these micas, though generally much less regular than in other schists, is definitely tabular; and it has distinct preferred orientation, with {001} subparallel to the original slaty cleavage, which thus may not only be preserved but even become intensified. Rocks initially high in K_2O, as many pelitic sediments are, commonly contain large porphyroblasts of biotite or muscovite, crowded with inclusions of quartz. These may completely lack regular orientation, in which case they give the rock a spotted or knotted appearance. In some rocks, however, the mica porphyroblasts are markedly tabular and tend to be oriented, with {001} in the schistosity; and bright spangles of coarse mica are then conspicuous on surfaces broken parallel to the schistosity. Andalusite or cordierite or both, developing as porphyroblasts in a schistose matrix of micas, quartz, and plagioclase, may also impart a spotted appearance to rocks of this class (Figure 17–1A,B). Note, however, that these minerals are more restricted than in rocks of the hornfels family, for at temperatures of the hornblende-hornfels facies they can crystallize only in those rocks in which the ratio Al_2O_3/K_2O is too high to allow all the Al_2O_3 to be accommodated by micas.

Though most typical of the outer parts of contact aureoles, spotted schists may also develop on a regional scale independently of exposed igneous contacts. A well-known example is afforded by porphyroblastic andalusite schists and staurolite-andalusite schists that crop out over much of Aberdeenshire, Scotland (Figure 20–3).

Quartzo-Feldspathic Hornfelses

Quartz, plagioclase, and K-feldspar, the main constituents of sandstones and of siliceous volcanic rocks (rhyolites, dacites), are stable within the temperature range embraced by the pyroxene-hornfels and hornblende-hornfels facies. Hornfelses derived from such rocks therefore consist essentially of a granoblastic mosaic of quartz and feldspars (Figure 17–3C). Blastoporphyritic textures, marked by persistence of partially recrystallized relict phenocrysts of quartz and feldspar, indicate the volcanic parentage of some rocks. In these rocks relict plagioclase commonly retains the idiomorphic outlines and complex twinning characteristic of volcanic plagioclases. Relict potassic feldspar—orthoclase in some high-temperature hornfelses, microcline in rocks of lower metamorphic grade—may show coarse perthitic structure resulting from metamorphic unmixing of sodic and potassic components that were originally combined in sodi-potassic sanidine or anorthoclase of a volcanic rock.

Biotite is almost invariably present in quartzo-feldspathic hornfelses. Small amounts of cordierite or andalusite may appear in rocks of the pyroxene-hornfels facies; but muscovite, both in porphyroblasts and in small flakes distributed through the granoblastic mosaic, is typical of hornfelses belonging to the hornblende-hornfels facies. Hornblende is less common. Tourmaline, zircon, apatite, sphene, and opaque oxides are common accessories.

Quartzo-feldspathic buchites resulting from partial fusion of inclusions in volcanic rocks are shown in Figure 17–4A,B. Tourmalinized hornfels formed by introduction of boron into sandstone (presumably via a gas phase) is illustrated in Figure 17–5A. A locally tourmalinized quartz-hornblende hornfels is shown in Figure 17–5C.

Contact Marbles

High-grade contact metamorphism of limestones and dolomites produces granoblastic rocks composed mainly of a mosaic of equant grains

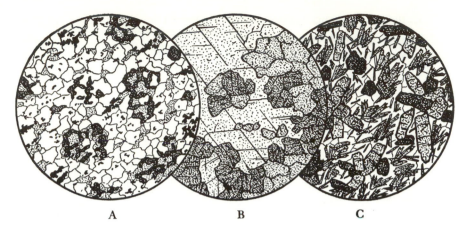

A   B   C

Figure 17–5. Pneumatolytic Hornfelses

A. Tourmalinized quartz hornfels, near Engelmine, Sierra Nevada, California. Diam. 2.5 mm. Quartz has recrystallized as a granoblastic mosaic. The original clay matrix has recrystallized to fine-grained colorless mica, and has provided an aluminous environment within which tourmaline has freely crystallized under the influence of introduced boron. Black specks are magnetite and hematite.

B. Fluorite skarn, Riddarhyttan, Sweden. Diam. 2 mm. Green hedenbergite and pale-brown andradite garnet enclosed in a single grain of fluorite.

C. Tourmalinized quartz-hornblende hornfels, Lily Lake, west of Lake Tahoe, California. Diam. 3 mm. The section, cut from rock several centimeters from a tourmaline-filled joint, shows coarse tourmaline in a quartz-hornblende matrix.

of calcite. SiO_2 and Al_2O_3 are minor impurities in most such rocks. A good many typical contact marbles are poor in silica; they contain no quartz, and are likely to contain minerals incompatible with quartz at high temperature. Among these minerals are forsterite, minerals of the humite family (particularly clinohumite), periclase, brucite, spinel, corundum, and a number of rare silicates of Ca and Mg (larnite, spurrite, tilleyite, monticellite, etc.). Where SiO_2 is more abundant, as it is in rocks transitional to the calc-silicate hornfelses, minerals compatible with quartz may appear, notably wollastonite, diopside, tremolite, and talc. The sequence of mineral assemblages that may develop in silica-bearing dolomitic limestones as the temperature rises is complex, and each metamorphic assemblage is determined partly by the relative proportions of calcite, dolomite, and quartz in the parent rock, and partly by the temperature and pressure prevailing at the time of reaction (cf. Figure

16–5). Presence of illite in the initial calcareous sediment is reflected in phlogopite in derivative marbles.

Marbles derived from what were initially pure dolomite rocks are composed of a mosaic of calcite enclosing round aggregates of finely crystalline brucite (Figure 17–6C), the component flakes of which are arranged in more or less concentric whorls that may enclose small remnant grains of periclase.* These rocks are generally interpreted as products of dedolomitization in two stages:

1. $CaMg(CO_3)_2 \longrightarrow CaCO_3 + MgO + CO_2$
   dolomite   calcite   periclase
2. $MgO + H_2O \longrightarrow Mg(OH)_2.$
   periclase   brucite

Periclase marbles with no brucite are rare. This is to be expected, for periclase is known to be stable in contact with H_2O only at temperatures approaching 900° C, and fluid H_2O at sufficient pressure to allow it to be converted to brucite must usually be present at some stage of metamorphism when the temperature is lower than this. Coarse tabular brucite occurs in some marbles (Figure 17–7C), and this points to direct conversion of dolomite to brucite and calcite without the formation of periclase as an intermediate phase, the necessary condition being high concentration of water in the fluid phase.

Forsterite-brucite and forsterite-diopside marbles are common derivatives of magnesian limestones containing chert or some other form of silica as a minor impurity. The forsterite and diopside occur as isolated, colorless, round or idioblastic grains, conspicuous in thin sections because of their high refractive index and bright polarization colors. Partial serpentinization of forsterite is common (Figure 17–7A). Other minerals likely to be present in assemblages of this kind (Figures 17–6A and 17–7C) are spinel, clinohumite or some other member of the humite family, and more rarely corundum. The spinel is usually colorless and may be recognized by its isotropic condition and octahedral habit. Most

---

*This highly characteristic texture in brucite pseudomorphs turns out to be due to regularly repeated strong kinking that has developed during radial growth of {0001} tables under stress induced by the great volume increase that attends the hydration process. Thus the general trend of visible {0001} cleavage is radial; but it is sharply kinked at intervals, first in one sense, then the other, thus imparting the concentric or tangential element to the ultimate structure.

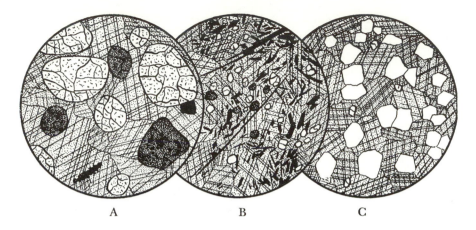

Figure 17–6. Magnesian Contact Marbles

A. Chondrodite-spinel marble, Amity, New York. Diam. 3 mm. Pale-yellow chondrodite and deep-green pleonaste in a matrix of calcite. A single crystal of pyrite (right) and a ragged flake of graphite (lower left). Addition of fluorine and sulfur is indicated by presence of chondrodite and pyrite.

B. Ludwigite-forsterite-spinel marble, Twin Lakes, Sierra Nevada, California. Diam. 2 mm. Calcite encloses round grains of forsterite and green pleonaste and slender prisms of the magnesium-iron borate ludwigite (γ = dark brown; α = dark green; refractive index 1.85–2.0; elongation parallel to γ). Presence of ludwigite indicates addition of boron and iron.

C. Brucite marble (predazzite), Predazzo, Italy. Diam. 2 mm. Colorless clear areas are of brucite, pseudomorphous after periclase; under crossed polarizers they show a complex, concentric arrangement of deformational kinks in the brucite crystals. A few round granules of forsterite are also present.

humite minerals are pleochroic in golden-yellow tints; but some are colorless and may be hard to distinguish from forsterite without recourse to a universal stage unless they show simple twinning.

The mineral assemblages noted above are characteristic of relatively high metamorphic temperatures, corresponding to the pyroxene-hornfels and the upper range of the hornblende-hornfels facies. Magnesian marbles of lower grade consist of calcite and either tremolite or talc. Tremolite typically occurs as long colorless prisms in a matrix of calcite; talc is much less common and appears to represent the lowest temperatures of reaction between dolomite and silica.

Marbles containing no MgO and with silica as sole impurity recrystallize without reaction as quartz marbles, except at or above the high temperatures corresponding to transition between the hornblende-hornfels

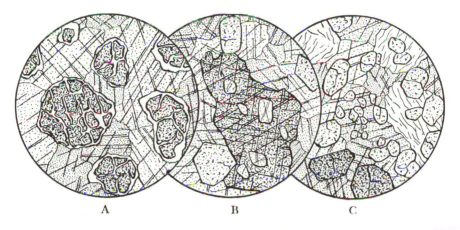

Figure 17–7. Magnesian Contact Marbles

A. Forsterite marble, Barstow, California. Diam. 3 mm. Note partial serpentinization of forsterite. The carbonate is calcite.

B. Diopside-vesuvianite marble, Crestmore, California. Diam. 3 mm. The large central crystal of diopside, showing {110} cleavage and {100} parting, encloses several grains of vesuvianite. Other grains of vesuvianite are surrounded by calcite.

C. Chondrodite-spinel-brucite marble, Crestmore, California. Diam. 3 mm. Colorless octahedra of spinel (at lower edge) and colorless and yellow grains of minerals of the chondrodite-clinohumite family are enclosed in coarsely crystalline brucite (at right) or in calcite (at center and lower edge).

and the pyroxene-hornfels facies. At such temperatures reaction occurs to give wollastonite. The assemblages wollastonite-calcite and wollastonite-diopside-calcite are therefore typical of high-temperature marbles (cf. Figure 16–5B).

There is an interesting but rare group of silica-poor metamorphic rocks formed at the low pressures and extremely high temperatures of the sanidinite facies, at contacts between flint-bearing chalks and intrusive necks of igneous rocks such as diabase. In these calcite is associated with various rare calcium silicates, most of which are colorless minerals of medium-to-high refractive index and moderately high birefringence, and commonly show lamellar twinning. They include larnite (Ca_2SiO_4; α to {100} twin plane = 13°); bredigite (Ca_2SiO_4; pseudohexagonal twins, 2V = 30°, sign +); spurrite ($CaCO_3 \cdot 2Ca_2SiO_4$; 2V = 40°, sign –); rankinite ($Ca_3Si_2O_7$; 2V = 64°, sign +, birefringence 0.009); tilleyite ($2CaCO_3 \cdot Ca_3Si_2O_7$; 2V = 90°); cuspidine [$Ca_4(F,OH)_2Si_2O_7$; relatively low refractive index, 1.60, and birefringence, 0.012]; merwinite

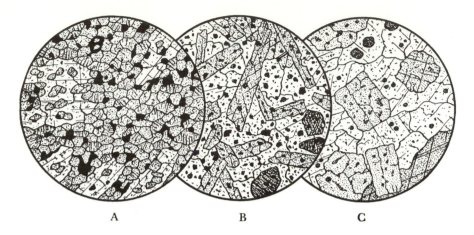

Figure 17–8. Silica-Deficient Calc-Silicate Rocks (Sanidinite Facies) at Chalk-Diabase Contacts, Ireland

A. Larnite-spurrite-magnetite, Scawt Hill, Antrim. Diam. 2.5 mm. Porphyroblasts of spurrite enclosing granular larnite (with high refractive index) and magnetite.

B. Merwinite-spurrite, Scawt Hill, Antrim. Diam. 3 mm. Large twinned tabular crystals of merwinite enclosed in spurrite. The three dark crystals at the lower right are larnite, formed by inversion of its high-temperature polymorph bredigite, the hexagonal cross section of which is still clearly retained. Grains of magnetite and of calcite and a few small granules of yellowish-brown perovskite (isotropic, with very high refractive index) are also present.

C. Tilleyite-melilite, Carlingford. Diam. 3 mm. Pale-yellow tablets of melilite are enclosed in colorless tilleyite. A few grains of green pleonaste (above) and numerous small granules of perovskite.

$(Ca_3MgSi_2O_8$; two sets of oblique twins, $2V = 70°$, sign $+$); monticellite $(CaMgSiO_4$; like forsterite, but with lower birefringence, 0.011); and melilite $[Ca_2(Al,Mg)(Si,Al)_2O_7$; uniaxial, sign $-$, untwinned]. Some of these are illustrated in Figure 17–8.

Calc-Silicate Hornfelses and Skarns

Calc-silicate hornfelses and skarns are composed entirely, or almost so, of calcium-bearing silicates. Those more properly termed hornfelses are derivatives of argillaceous limestones; "skarn" is generally reserved for petrographically similar rocks derived from nearly pure limestones and dolomites into which large amounts of Si, Al, Fe, and Mg have been

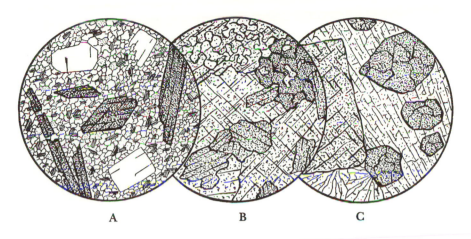

A   B   C

Figure 17–9. Skarns

A. Scapolite-actinolite-phlogopite marble, Germany. Diam. 2.5 mm. The three colorless idioblastic crystals with relatively low refractive index are of scapolite.

B. Skarn, Donegal, Ireland. Diam. 2.5 mm. Vesuvianite enveloping green diopsidic pyroxene (in lower half). Grossular (upper right) and vesuvianite (upper edge), both enclosing granular epidote-clinozoisite.

C. Skarn, Aberdeenshire, Scotland. Diam. 2 mm. Large prismatic crystal of vesuvianite (at left) and darker grains of grossular-andradite with irregular fracture, enclosed in colorless, radially prismatic prehnite.

introduced. The two modes of origin may usually be distinguished on the basis of field evidence; for skarns tend to be limited to sharply defined zones at junctions between marbles and plutonic rocks, whereas hornfelses grade peripherally into less altered rocks of similar chemical composition. With increase in calcite, skarns grade into calc-silicate marbles (Figure 17–10).

The composition of a skarn or a hornfels may vary considerably within a single hand specimen or even within a single thin section. One of the most typical mineral assemblages consists of colorless or brownish garnet of the grossular-andradite series (in some cases zoned and feebly birefringent), nonpleochroic greenish diopside-hedenbergite, and pale-pinkish or brownish vesuvianite (Figures 17–9B,C). The latter may be difficult to distinguish from birefringent garnet. In the more aluminous bands this assemblage may also contain calcic plagioclase (Figure 17–11B), a mineral of the epidote group (Figure 17–9B), or a scapolite (Figure 17–10A); more calcic assemblages are likely to include wollas-

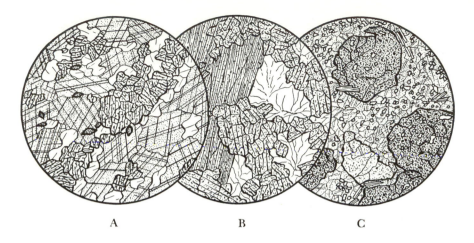

A   B   C

Figure 17–10. Calc-Silicate Marbles

A. Diopside-scapolite marble, Three Rivers, Sierra Nevada, California. Diam. 2 mm. Greenish diopside, colorless scapolite, coarse calcite, accessory granules of sphene.

B. Diopside-wollastonite hornfels, Wasatch Mountains, Utah. Diam. 2 mm. Large prismatic crystals of wollastonite (left half) and coarse granular diopside, with interstitial radial aggregates of prehnite (relatively low refractive index, colorless).

C. Grossular-vesuvianite-diopside marble, near Cisco, Sierra Nevada, California. Diam. 3 mm. The darker of the two highly refractive minerals is grossular, and the other is vesuvianite. There are small prisms of wollastonite and a granoblastic matrix of calcite.

tonite in the pyroxene-hornfels facies (Figure 17–10B) or calcite plus quartz in the hornblende-hornfels facies. Contrary to what might be expected in a mineral assemblage in internal equilibrium, epidote-bearing hornfelses frequently contain both clinozoisite and zoisite as well. The two latter may be distinguished with certainty only on the basis of optic axial angle (0–40° in zoisite, 80–90° in clinozoisite), for both are colorless minerals with high refractive index, and both tend to show anomalous blue or yellow-brown interference colors. In many rocks coarsely granular epidote contains intergrown vermicular quartz (Figure 16–1C). Wollastonite may occur in hornfelses containing abundant colorless diopside, and may then be overlooked in a cursory examination; but if its presence is suspected it is easily confirmed by measurements of extinction angle (α to [001] = 34°) and optic axial angle and sign (2V = 35–40°; sign −). Scapolites differ from all associated minerals in being

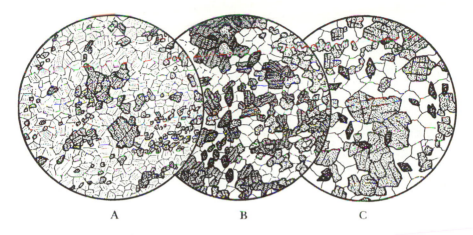

Figure 17–11. Calc-Silicate Hornfelses

A. Diopside-plagioclase hornfels with minor quartz, east of Visalia, Sierra Nevada, California. Diam. 2 mm.

B. Hedenbergite-andradite-labradorite hornfels, near Visalia, Sierra Nevada, California. Diam. 1 mm. Andradite is concentrated chiefly within the right half, hedenbergite chiefly within the left half of the slide. A string of lozenge-shaped crystals of sphene trends about vertically, left of center.

C. Diopside-quartz-sphene hornfels, Donegal, Ireland. Diam. 2.5 mm.

uniaxial and negative and in having a low refractive index and moderately high birefringence.

Sphene (Figure 17–11C), microcline, phlogopite, magnetite, and sulfides may occur in any of the above assemblages, especially in skarns. Hornblende, associated with diopside, is found only in rocks relatively poor in CaO and belonging to the hornblende-hornfels facies. Fluorite is not uncommon in skarns containing iron-rich pyroxene and garnet (Figure 17–5B).

Basic Hornfelses

High-temperature contact metamorphism of rocks of the basalt and andesite families yields dense dark hornfelses that beneath the microscope are seen to be composed of a granoblastic mosaic of labradorite, pale-green or colorless diopside, pink hypersthene (more distinctly pleochroic than the hypersthene of igneous rocks), and accessory magnetite, apatite, and sphene (cf. Figure 17–13C). In rocks of very basic parentage

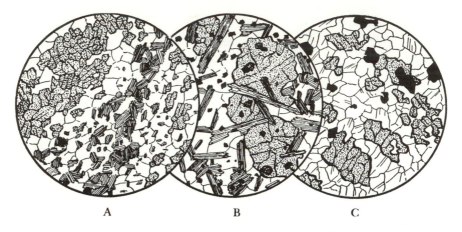

Figure 17–12. Basic Hornfelses, Comrie, Perthshire, Scotland

A. Hornblende-biotite-plagioclase hornfels. Diam. 3 mm.

B. Hypersthene-biotite-plagioclase-quartz hornfels. Diam. 3 mm.

C. Hypersthene-cordierite-plagioclase-magnetite hornfels. Diam. 3 mm.

olivine, sometimes in large poikiloblastic grains, may accompany this assemblage. Such is a European hornfels (Figure 17–13C) originally described as an unmetamorphosed basic dike rock.° Hornfelses derived from andesites commonly contain some biotite (Figures 17–12B and 17–13A), and they may show vestiges of original porphyritic texture; phenocrysts of what once was hornblende, for example, may be represented by a mixture of diopside and hypersthene.

Corresponding hornfelses of the hornblende-hornfels facies consist of the assemblage plagioclase-hornblende-diopside, often with magnetite, apatite, and sphene as minor constituents (Figure 17–13B). When biotite and quartz are also present, it is likely that the parent rock was andesitic (Figure 17–12A). Hornblende hornfelses commonly retain relict traces of porphyritic, ophitic, and other typically igneous textures, which are described by such terms as "blastoporphyritic" and "blastophitic." A typical instance of blastoporphyritic texture is seen where aggregates of hornblende, pseudomorphous after eight-sided prismatic crystals of augite, are enclosed in a granoblastic hornblende-plagioclase matrix. Plagioclase, on account of its low position in the crystalloblastic series, invariably recrystallizes as a mass of xenoblastic grains; so any

---

°*Beerbachite.*

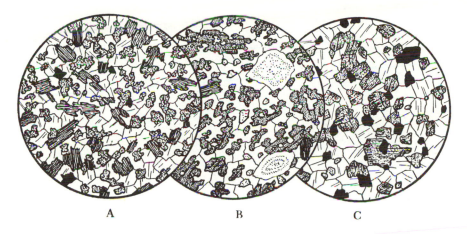

Figure 17–13. Basic Hornfelses

A. Diopside-plagioclase-biotite hornfels, near Cisco, Sierra Nevada, California. Diam. 3 mm. Diopside shown stippled; a few grains of magnetite.

B. Hornblende-plagioclase hornfels, near Cisco, Sierra Nevada, California. Diam. 3 mm. Relict phenocrysts of plagioclase retaining zonary structure indicate igneous origin.

C. "Beerbachite," Odenwald, Germany. Diam 3 mm. Hypersthene, diopside, plagioclase, and magnetite; pyroxenes show retrograde alteration to fibrous pale-green amphibole; olivine (not shown) is also present.

recognizable trace of lathy or tabular habit in the plagioclase of hornfels (e.g., in blastophitic textures) must necessarily have been inherited from the parent rock (Figure 17–13B). Relict amygdaloidal structure may be very conspicuous in hornblende hornfels of relatively low metamorphic grade, as when oval patches of recrystallized quartz, calcite, and epidote or of albite and epidote are scattered through the normal plagioclase-hornblende mosaic.

Magnesian Hornfelses and Schists

Magnesian metamorphic assemblages in rocks of contact aureoles usually develop in one of two distinct situations. In one, they are products of local magnesium metasomatism synchronous with the metamorphic imprint. In the other, bodies of serpentinite develop new mineral assemblages where they enter the aureole; here any metasomatism that may have occurred took place prior to the contact imprint.

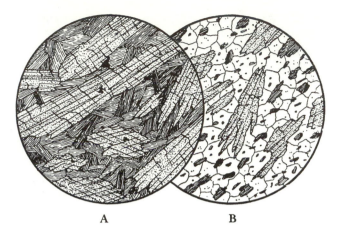

A   B

Figure 17–14. Magnesian Contact Schists

A. Anthophyllite-cummingtonite-biotite schist, Riddarhyttan, Sweden. Diam. 4
   mm. In the large central crystal and the one at the left edge, the two amphiboles are intergrown.

B. Cordierite-anthophyllite-biotite schist, Riddarhyttan, Sweden. Diam. 3 mm.
   The large crystals are pale anthophyllite.

Typical products of the first category are coarse-grained hornfelsic
rocks composed essentially of magnesian amphiboles (anthophyllite or
cummingtonite), cordierite, and in many instances biotite, almandine
garnet, or both. Two examples from Sweden are illustrated in Figure
17–14. The anthophyllite-cordierite assemblage of contact aureoles is
typical of the hornblende-hornfels facies.* At higher grades of contact
metamorphism (pyroxene-hornfels facies) chemically equivalent hornfelses consist principally of cordierite and hypersthene (cf. Figure
17–12C).

A sequence of mineral assemblages that have developed in metaserpentinite, where this enters the contact aureole of the Bergell tonalite
pluton, Italian alps, has been recorded by V. Trommsdorff and B. W.
Evans.[2] The parent rock outside the aureole is a low-grade regional
metaserpentinite consisting of antigorite-olivine-diopside. Successive
contact assemblages, in order of increasing proximity to the Bergell plu-

---

*Where the same assemblages appear locally at contacts with granodioritic plutons that
were emplaced synchronously with a regional metamorphic episode of much greater areal
extent, they can equally well be placed in the amphibolite facies. Such is the case in Eskola's
classic Orijärvi region, Finland.

ton, are: antigorite-olivine-tremolite, talc-olivine-tremolite, anthophyl-lite-olivine-tremolite. All are within the hornblende-hornfels facies. In the southern Sierra Nevada, California, assemblages of still higher grade (pyroxene-hornfels facies) appear locally at the contact between ada-mellite and serpentinite. These are hornfelses consisting of olivine, enstatite, spinel, and clinochlore—an assemblage of minerals that appears almost to duplicate that of the parent peridotite from which the serpentinite itself was originally derived.

REFERENCE NOTES

[1] C. E. Tilley, *Quarterly Journal of the Geological Society of London,* vol. 80 (1974): pp. 22–41.

[2] V. Trommsdorff and B. W. Evans, *American Journal of Science,* vol. 272 (1972): pp. 423–437.

18

Metamorphic Products
of Mechanically Dominated Strain

NATURE OF DISLOCATION METAMORPHISM

Metamorphism in general can be viewed simplistically as being compounded of two closely intertwined sets of processes: one of a "chemical," the other of a "mechanical" nature. The resulting mineral assemblage most obviously reflects the overall imprint of chemical processes—chemical reactions, recrystallization, ionic diffusion, crystal growth, and so on. Visible strain effects, in contrast, are related to some regime of internal flow within the metamorphic system under externally imposed nonhydrostatic stress; and in the newly developed fabric we see the imprint of mechanical elements in the metamorphic process. In most instances of regional metamorphism, at all but the incipient stages, both effects are clearly recognizable, though it is usually impossible to unravel in detail possible roles of chemical processes in contribution to strain and the influence of "mechanical" (especially kinematic) influences on the efficacy of chemical reconstitution.

There are certain situations, notably in contact aureoles and in domains of burial metamorphism, in which strain effects are minimal, and metamorphism is essentially a chemical response to a regime of temperature and hydrostatic pressure. But, by contrast, there are others in which the metamorphic fabric is clearly the imprint of essentially mechanical processes. Chief among these are narrow zones of profound crustal dislocation, as along the San Andreas Fault (California) and the Alpine Fault (New Zealand).

Such rocks can be called products of *dislocation metamorphism;* for their characteristic fabrics reflect penetrative internal strain locally set up by relative dislocation of adjoining crustal domains. Transcurrent fault zones obviously are ideal sites. But comparable rocks can also develop in more diffuse, less sharply defined belts of intense strain, some, as in Archaean massifs of Central Australia, even of regional dimensions. All these rocks together constitute the topic of this chapter, and their most characteristic representatives are *mylonites.* The metamorphic process itself is one of protracted cohesive flow of solid rock, with only minimal internal readjustments of a purely chemical nature. Some writers have designated it *cataclastic* flow. But this term implies an essential role for brittle fragmentation in the flow process; yet, as shown below, this now proves to be by no means universally true. Therefore, we avoid using that term.

MECHANICALLY IMPLEMENTED COHESIVE FLOW IN SOLID ROCKS

First a word must be said about the nature of the flow process itself. Classic interpretations of mylonites were based on everyday experience of the way in which different kinds of solids respond to imposed systems of nonhydrostatic stress under atmospheric pressure (i.e., at confining pressure of about 1 bar). At room temperature and pressure solids in this context are usually thought of as being either brittle or ductile (plastic). When the applied stress exceeds some limiting value ("breaking strength") *brittle* materials fail by fragmentation; the specimen as a whole loses strength, but repeated application of stress to individual fragments (as in a mortar) causes progressive fragmentation and reduction in grain size. The specimen as a whole has now completely lost internal cohesion. Most minerals are brittle at atmospheric pressure, even at temperatures not far below that of melting. *Ductile* materials, such as most metals, respond very differently to applied stress, even at low confining pressures. When the stress exceeds and is sustained above a limiting yielding strength, the solid begins to flow continuously without loss of strength and cohesion. There is, however, no sharp division between brittle and ductile behavior.

It is now known from extensive experimental research by D. T. Griggs, M. S. Paterson, J. C. Handin, and others over the past three decades, that added dimensions of high confining pressure (several kilobars) and immensely longer spans of available time (implying slow rates of strain) within the intracrustal environment permit strain regimes hitherto virtually unknown except in metallurgy. Brittle and ductile (plastic) behavior patterns can still be recognized. But ductile behavior in "normally" brittle materials such as calcite, dolomite, olivine, and even quartz can be induced or facilitated by several factors: high confining pressure and temperature, slow strain rates, and (for quartz absolutely essential) the presence of a water envelope. Moreover, under such conditions, whether failure of the specimen is mainly by brittle internal fracture or by ductile strain, it tends to retain cohesion and strength. The ensuing strain process may simply be termed *cohesive flow.* Textural responses of the rock fabric to cohesive flow under experimentally controlled conditions are many and varied; and they are precisely reproduced in the fabrics of mylonites and related rocks. The observed effects include the following:

1. During ductile flow of single crystals (or polycrystalline aggregates) of minerals such as calcite, olivine, or quartz, the individual crystal tends to retain its identity without fragmentation. But it changes shape by flattening, elongation, and bending (including sharp kinking) of one internal domain relative to another. Bending and kinking are microscopically recognizable in the long-familiar phenomenon of undulatory extinction in optically anisotropic crystals. The mechanism of intracrystalline flow involves mutual displacements of ions in crystallographically determined patterns (the concept of crystal gliding). In some minerals the glide process produces visible lamellar twinning (on $\{01\bar{1}2\}$ in calcite, on $\{02\bar{2}1\}$ in dolomite, and so on).

2. Brittle rupture yields swarms of irregular small fragments distributed within the specimen in recognizable zones of shear failure. The fragments thus produced commonly show internal undulatory extinction, an indication of permanent strain set up within each individual lattice prior to brittle rupture.

3. By virtue of stored *strain energy* resulting from myriads of newly developed lattice imperfections (*dislocations*), visible only under the electron microscope, any strained crystal has higher free energy, and so is less stable, than its unstrained counterpart. Given sufficient time, a strained crystal kept at high temperature tends to recrystallize to a mosaic of unstrained (hence more stable) smaller grains.* From the

---

*This process is analogous to those long familiar in annealing of metals.

petrographic viewpoint it is important to note that microscopically obvious "granulation," which has been recorded in minerals of so many deformed rocks, in no way implies small-scale fragmentation by brittle rupture (though this may have occurred) nor any loss of cohesion and strength in rocks that display it. The term *cataclastic*, if used, must be reserved for rocks whose textures clearly reflect failure by brittle rupture (the extreme cases being fault breccias and "gouges").

PETROGRAPHY OF REPRESENTATIVE ROCKS

Terminology of rocks affected by dislocation metamorphism need not be elaborate. Each type can be designated simply in terms of parental lithology (for traces of this can always be recognized) and fabric type. Thus two rocks with the same general type of fabric may be called *dunite mylonites* or *granite mylonites* according to their protolithology. In what follows, primary emphasis is focused on fabric patterns.

Mylonites

Mylonites are products of extreme localized deformation of chemically stable rocks, typically in fault zones, under high shear stress set up by relative motion of rocks on either side of the fault. A high rate of strain at fairly high confining pressure appears to favor the development of their characteristic fabric—at one time thought to reflect extreme grinding and milling (hence the name mylonite) of the parent material. Mylonites on the whole are fine-grained laminated rocks, individual laminae being distinguished by differences in grain size, color, and usually in mineral content. Lamination is correlated with the overall kinematic picture of flow; and the longer dimensions of the constituent small mineral grains tend to dimensional alignment subparallel to the trend of lamination. In most mylonites relics of the original rock survive as oval or elongate eyes that are aligned parallel to the trend of lamination, which swirls around them, heightening the impression of the overall kinematic picture (Figure 18–1). Component minerals of these relict eyes themselves display clear signs of internal strain: undulatory extinction, bending of the crystalline structure expressed by curved twin lamellae or cleavages, and almost universal marginal "granulation" (an effect of incipient annealing rather than fragmentation). Quartz grains, in particular,

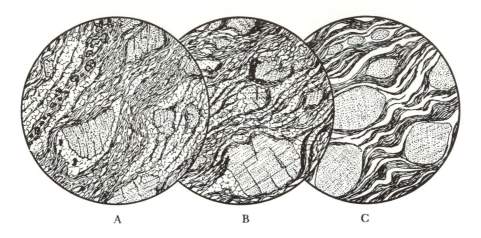

Figure 18–1. Mylonites

A. San Gabriel Mountains, California. Diam. 5 mm. Strained and broken coarse crystals ("porphyroclasts") of feldspar and a train of garnet granules set in a fine-grained schistose matrix of quartz and feldspar veined with granoblastic quartz.

B. Granite mylonite, San Gabriel Mountains, California. Diam. 5 mm. Coarse, strained, partially granulated crystals are of plagioclase, microcline, and quartz. The granular matrix is composed of quartz, feldspar, and biotite.

C. Mylonitic augen gneiss, Deadman Lake, British Columbia. Diam. 6 mm. Ovoid relict crystals of plagioclase and of K-feldspar, in a matrix of muscovite, chlorite, and quartz, traversed by swarms of stringers of later undeformed quartz.

develop undulatory extinction bands subparallel to the [0001] axis. In many quartz grains, too, there are spaced linear arrays of dusty inclusions; and swarms of thinly lenticular "Boehm lamellae" inclined to {0001} at high angles are visible between crossed polarizers by reason of slight departure in extinction from the host. Curvature of twin lamellae in plagioclase (cf. Figure 18–2B) and carbonates (cf. Figure 18–3B), or of exsolution lamellae in pyroxenes and of cleavage traces in micas testifies to strong internal deformation (structural bending) in the host crystals; for with few exceptions such visible structures, though curved at all points, preserve rational crystallographic orientation.

Mylonites, though so long pictured as intensely "pulverized" rocks, are hard, coherent, even flinty in hand specimen. The paradox thus presented by the apparently "welded" state of the "pulverized" material has been resolved by experimental experience. What we see is largely an effect of annealing (strain recovery), whereby stored strain energy is released from initially larger grains during recrystallization to a finer-grained aggregate of strain-free material. An extreme product of strain in many mylonites is a dark, optically isotropic material, *pseudotachylite*, which under high magnification typically is seen to enclose numerous relict crystal fragments. Streaks and veinlets of pseudotachylite, while partly following the fabric trend, locally trangress it and even invade fractures in coarser relics. Is this material truly a glass produced by melting at loci of high heat concentration that develop during the flow process? Or, since x-ray patterns commonly reveal the presence of crystalline material on the submicroscopic scale, does pseudotachylite represent the extreme of the grain-reduction process? Or again may it be partially devitrified glass? All three possibilities have their advocates.

Quartz and feldspars are dominant constituents of many mylonites; for both minerals are stable over a wide range of temperature and pressure even under conditions involving high shear stress and access of aqueous fluids. Very commonly, then, mylonites are derived from granites, granodiorites, quartzites, and quartzo-feldspathic metasediments. In individual fine-grained quartzose bands and lenses, the component quartz grains usually show a high degree of preferred orientation with [0001] ($= \gamma$) tending to subparallel alignment in some direction (possibly that of flow in the kinematic picture) within the lamination surfaces. This condition is readily demonstrated by inserting the gypsum plate between crossed polarizers: on rotation of the stage the aggregate tends to behave as a unit, showing dominantly blue or dominantly yellow colors in alternating position at 90° intervals. Other orientation rules also are well known in quartz of mylonites.

Mafic and ultramafic rocks whose main constituents are pyroxenes, olivine, or calcic plagioclase also give rise to mylonitic equivalents when rapidly strained in the absence of water. Mylonites are well-known local structural variants within dunites and harzburgites that make up ultramafic bodies of the "Alpine type." Large mantle-derived xenoliths of similar composition, that are so common in basanites and nephelinites (cf. p. 200), usually show clear signs of strain—presumably imprinted at their source prior to eruption—but they seldom develop a truly mylonitic fabric. In ultramafic mylonites, olivine in particular has been reduced to fine-grained, flinty aggregates streaked with trains of chromite granules. Swimming in this matrix are relict, coarse crystals of enstatite in a high state of internal strain; curved {100} exsolution lamellae in the zone of [101] are characteristic. Very rarely the highly strained domains (kinks especially) have inverted to clinoenstatite, distinguished by inclined extinction with reference to the trace of the {100} lamellae, and notably higher birefringence—$(\gamma - \alpha)$ ~0.014—than adjacent enstatite.* A classic example of olivine-rich peridotite-mylonite with intensely kinked (but not inverted) enstatite is furnished by St. Pauls Rocks, a unique island fragment of suboceanic, ultramafic mantle rock that has been squeezed up through the crust on the mid-Atlantic ridge just south of the equator.

Gabbro mylonites, too, have been recorded from many localities. They are extensively developed along with garnet-granulite mylonites across a zone over 100 km wide in the Giles complex (a deformed stratified pluton) in the Archaean basement of the Tonkinson Range, central Australia. In this region, typical gabbro mylonites show numerous large, fragmented, and strained crystals (porphyroclasts) of calcic plagioclase, diopsidic augite, and bronzite, set in a finely granular streaked-out matrix of the same minerals, cut locally by small veins of pseudotachylite. Strongly curved albite twin lamellae in plagioclase, and {100} exsolution lamellae in both pyroxenes bear witness to high degrees of internal structural strain in most porphyroclasts. Both types of pyroxene also show strong kinking at high angles to {100} (kink axes = [010]); and kink domains of some bronzite crystals display the rare phenomenon of inversion to clinoenstatite.

Augen Gneisses

The descriptive term *augen gneiss* has long been applied to quartzo-feldspathic rocks of diverse origins in which a fine-grained, foliated, often micaceous groundmass encloses megascopically conspicuous eyes of feldspar. Some augen gneisses are truly mylonitic; their large feldspars are porphyroclasts that have survived the "granulation" that was responsible for the fabric of the surrounding matrix. Others are protoclastic in that their fabric developed during flow of a plutonic mass in the last stages of crystallization. Others are essentially metasomatic, and in these the large feldspars (microcline or oligoclase) have grown under the influence of alkaline fluids. It may be difficult to distinguish in the field between augen gneisses that are purely mylonitic and those

---

*This effect was first recorded in enstatite of pyroxenite rapidly strained in the laboratory at temperatures of 300°–500°C and confining pressures of a few kilobars.

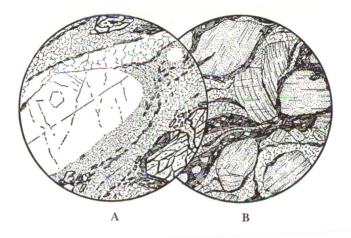

Figure 18–2. Augen Gneisses

A. Kyanite-biotite-feldspar gneiss, Nanga Parbat region, northwestern Hima-
layas. Diam. 8 mm. Kyanite at lower right and at top; at left a large porphyroblast (?) of microperthite perhaps replacing the groundmass, which is
mainly quartz and oligoclase.

B. Granodiorite gneiss deformed at high temperature, Catalina Mountains,
Arizona. Diam. 8 mm. Porphyroclasts of plagioclase (with curved lamellae,
indicating bending of the lattice, near center), in a granular matrix of biotite,
quartz, and minor feldspar.

whose large feldspars are porphyroblasts that have formed metasomatically during or after deformation. Both types are illustrated in Figure
18–2.

It may well be that many metamorphic quartzo-feldspathic gneisses—
some with large eyes of garnet—although now relatively coarse grained
and showing no trace of intracrystalline strain, have passed through a
mylonitic stage of development. Perhaps the clue to their origin is to be
sought in metallurgical experience. Annealing of strained metal can be
achieved in several ways. One is that of "recovery," already described,
whereby strain energy stored in a large crystal is released by growth of
small unstrained grains about rapidly developing nuclei. Another process
is known as "polygonization." In this case a strained crystal is transformed
to a set of a few unstrained sectors by outward migration of dislocations
which become concentrated in closely populated subplanar arrays that
constitute the sector boundaries. Sharply bounded, optically homogeneous extinction bands (differing mutually in extinction by a few
degrees) slightly inclined to [0001] in quartz probably have such an origin.

Finally there is the process of "grain growth" in the strain-free aggregate that results from "recovery." Enlargement of a few grains at the expense of the rest reduces total surface energy; but the process is excessively slow (except at increased temperature), for the ΔG driving force is weak, and extensive ionic diffusion within the crystal aggregate is essential to the process.

Many augen gneisses could well represent the end product of a sequence of processes: rapid strain and partial recovery (mylonitization) followed by slow enlargement of granules in the initially fine-grained product of the first stage. The result would be an aggregate of unstrained crystals—coarse relict eyes set in an aggregate of smaller, but not excessively small, crystals—the only reminder of the original process of solid-state flow being preferred orientation of mica tablets and the gneissic fabric itself. Augen gneisses in which such an evolutionary process can be traced in the field have been termed *blastomylonites*; but this name is petrographically inappropriate since it implies a postulated process that cannot in most cases be verified by petrographic technique.

Plastically Deformed Marbles

Calcite marbles present a special case among rocks responding to geologically imposed stress; for calcite at confining pressures of a few kilobars and temperatures above about 300°C is a weak and highly ductile material—much more so than dolomite. Marble itself is a product of metamorphism into which mechanically induced strain no doubt enters more often than not. But the effects about to be described refer to the reaction of marble to subsequently imposed (postcrystalline) stress, especially in zones of intense rock deformation.

The individual calcite grain then tends to become markedly shortened in one direction and correspondingly elongated at right angles to this. And strain is achieved by intracrystalline plastic flow, principally by a combination of twin gliding on $\{01\bar{1}2\}$ and translation gliding on $\{10\bar{1}1\}$. This latter mechanism, which turns out to be the more effective as regards degree of possible resulting strain, usually leaves no obvious microscopically visible trace. But twin gliding, the less effective mechanism, leaves a microscopically conspicuous imprint in the form of swarms of thin lamellae inclined at high angles to the principal axis of shortening of the crystalline aggregate (Figure 18–3B,C). At the same time a characteristic crystallographic orientation pattern evolves with [0001] axes clustered around the shortening axis. Curvature of twin lamellae commonly

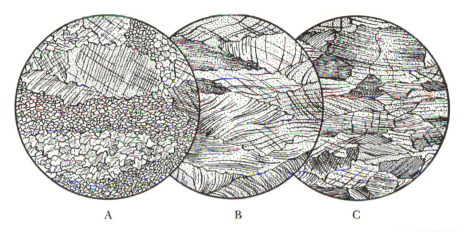

Figure 18–3. Marbles Showing Imprints of Postcrystalline Deformation

A. Calcite marble, Lans bei Innsbruck, Austria. Diam. 2.5 mm.

B. Calcite marble, Twin Lakes, Sierra Nevada, California. Diam. 2.5 mm. Note curvature of $\{01\overline{1}2\}$ lamellae reflecting internal plastic deformation of grains.

C. Experimentally deformed marble (specimen from Yule Creek, Colorado). Diam. 2.5 mm. The specimen was shortened 20% parallel to the vertical diameter of the figure by plastic flow at 300° C under a confining pressure of 5000 atmospheres. Note strong development of thin $\{01\overline{1}2\}$ lamellae, at high angles to the axis of compression and subparallel to the plane of elongation of grains.

is obvious (Figure 18–3B). In natural marbles this usually reflects intra-crystalline curvature of the lattice itself; for at all points in the grain the angle between the [0001] axis (= α of the indicatrix) and the local trace of the curving lamellae maintains its rational value, ~74°; and the grain as a whole shows strong undulatory extinction. Thus the lamellae them-selves, conspicuous though they are, record only a minor and late com-ponent of total strain imposed upon an already highly strained crystal. Perceptible biaxiality is another widespread feature in strained calcite (dolomite, by contrast, remains perfectly uniaxial).

Brittle fracture on the microscopic scale is virtually unknown in highly strained marbles, except in material rapidly deformed at low tempera-ture and confining pressure. However, postdeformational annealing to fine-grained aggregates of small crystals is rather common (Figure 18–3A).*

---

*The same result is achieved in the laboratory when calcite, first strained at room temper-ature, is heated for an hour or so above 500°C, confining pressure being maintained through-out at a few kilobars.

Phyllonites and Semischists

There are two situations in provinces of regional metamorphism where the respective imprints of closely related solid-state cohesive flow and chemically controlled growth of a new mineral assemblage may be separately identified. Two kinds of rock commonly display this phenomenon on a regionally extensive scale. They are termed *phyllonites* and *semischists*.

Phyllonites

Phyllonites ("phyllite-mylonites"), like mylonites, are fine-grained rocks formed by extreme deformation (cf. mylonitization) of originally coarse-grained rocks. Reduction in grain size during cohesive flow, however, has been followed or accompanied by recrystallization of some minerals (e.g., quartz and calcite) and growth of others appearing for the first time (e.g., mica, chlorite, albite, and epidote). An important element in the evolution of phyllonites is differential movement on closely spaced visible slip surfaces, which soon makes the rock highly schistose. Close, small-scale folding of *s*-surfaces (e.g., relict bedding or early-formed slip surfaces) is characteristic. As deformation proceeds, the limbs of these folds, which in many cases are of microscopic dimensions, shear out, and slip movements now continue on what are essentially old *s*-surfaces transposed by folding into a new direction (Figure 18–4A). The end product of these processes—the typical phyllonite—is a rock mineralogically and structurally resembling a phyllite but differing in origin and in some details of fabric and minor relict phases. Like a phyllite it consists of a mineral assemblage of low metamorphic grade, commonly belonging to the greenschist facies, and the megascopic fabric is dominated by one or more well-developed sets of schistosity surfaces, conspicuous by reason of parallel orientation of fine-grained silky mica and chlorite. The microscopic fabric is characterized by many and varied indications of differential movement: These include crests and troughs of otherwise obliterated microfolds marked by local curved trains of bent micas trending at high angles to the dominant schistosity; relict augen of partially recrystallized mineral aggregates persisting from the premetamorphic stage; flat lenses of recrystallized quartz, within many of which the component grains have a strong preferred orientation, as may be demonstrated by their similar optical reaction with a gypsum plate between crossed polarizers.

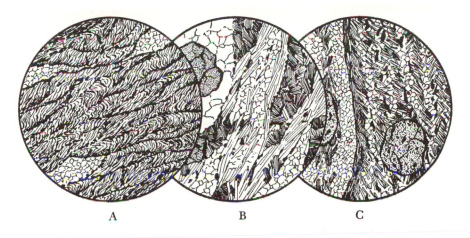

A   B   C

Figure 18–4. Phyllonites

A. Quartz-muscovite-graphite phyllonite, near Innsbruck, Austria. Diam. 3 mm. Graphite is drawn out along shear surfaces (schistosity) cutting across microfolds defined by the trend of the micas. Lenses of granular quartz, within each of which the component grains show subparallel orientation of [0001] axes, at top right and lower left.

B. Garnet-bearing phyllonite, near Innsbruck, Austria. Diam. 3 mm. Shear bands (schistosity) marked by segregation of individual minerals and by parallel alignment of muscovite flakes. Abundant chlorite shows orientation of crystals oblique to shear bands. Relict grains of almandine garnet, enclosed in the quartz lens at left, are inherited from an earlier period of high-temperature metamorphism.

C. Garnet-bearing phyllonite, Dutchess County, New York. Diam. 7 mm. Muscovite and minor chlorite in flakes mostly trending obliquely to segregation bands (shear schistosity). On the left, part of a lens of granoblastic quartz with marked preferred orientation of [0001] axes. Two rounded relict grains of almandine garnet, surviving from an earlier period of high-temperature metamorphism. Black flakes are graphite.

Phyllonites in some orogenic zones have been formed by relatively late dislocation metamorphism, at low temperature, of rocks previously metamorphosed at higher temperature—a case of retrograde metamorphism. Complex mineral assemblages that indicate chemical disequilibrium are typical of such rocks. For example, a quartz-muscovite-chlorite phyllonite may contain partially destroyed coarse relict grains of such high-temperature minerals as garnet, staurolite, biotite, and andalusite, surviving from earlier metamorphism (Figure 18–4B,C). Marginal chloritization of the first three minerals and replacement of andalusite by colorless mica supply additional evidence of prevailing disequilibrium. The larger grains of the relict minerals, moreover, commonly enclose trains of inclusions now inclined at all angles to the schistosity of the surrounding phyllonitic matrix, and representing the disrupted remnants of s-surfaces belonging to the earlier metamorphism.

The tectonic significance of thick phyllonite zones that can be mapped as formations is clearly great. Such zones, as pointed out over half a century ago by B. Sander and by E. B. Knopf, may be the only surviving relic of regional tectonic displacements in structurally complex metamorphic terranes such as the Austrian Alps (site of Sander's Innsbruck phyllonite) and parts of the Appalachians. And the identity of phyllonites that in hand specimen are indistinguishable from true phyllites (cf. p. 451) may be revealed only by microscopically recognizable scattered relict grains of high-grade minerals, notably garnet, staurolite, and kyanite.

Semischists

The term *semischist* was introduced to cover certain somewhat schistose rocks that develop in the course of transition from massive rocks such as graywacke to typical schists in the earlier stages of progressive regional metamorphism. Such rocks, in various texturally defined stages of the transition, have been mapped successfully, in New Zealand, California, and elsewhere, to delineate zones of progressive structural evolution that roughly parallel zones of mineralogical reconstitution defined by mineral isograds.

In metagraywackes of the semischist family (Figure 18–5) sedimentary grains of quartz and feldspar, showing undulatory extinction and marginal recrystallization ("granulation") are abundantly strewn through a fine-grained matrix of recrystallized quartz and feldspar. But within this matrix there are also plentiful newly generated grains of low-grade metamorphic minerals that yield valuable evidence regarding the regionally prevailing temperature-pressure regime. In some cases the new assemblage may be essentially chlorite–white mica–pumpellyite (or prehnite); in others it is chlorite–white mica–epidote; in yet others jadeite-lawsonite.

The importance of semischists was first revealed by field and petrographic observations. But modern x-ray techniques greatly enhance the value of purely petrographic data. This general topic will be pursued further in Chapter 19. Fine-grained metagraywackes in which deformation and simultaneous recrystallization have reached the point where

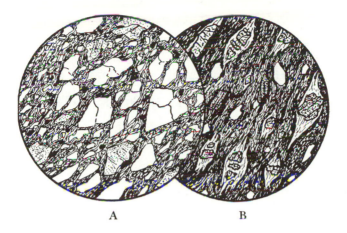

A   B

Figure 18–5. Semischists Derived from Arenaceous Sediments

A. Sheared "grit," Dunoon, Scotland. Diam. 7 mm. Relict fractured and strained sand grains—quartz (clear) and feldspar (shown stippled)—in a sheared partially recrystallized matrix of white mica, chlorite, quartz, albite, and sphene.

B. Semischist derived from tuffaceous graywacke, Lake Wakatipu, New Zealand. Diam. 7 mm. Relict grains of quartz (clear), augite, and hornblende. The augites are elongated and partially replaced by sheaths of pale fibrous tremolite. The recrystallized schistose matrix consists of chlorite, epidote, white mica, quartz, albite, and sphene.

almost all relict coarse grains inherited from the parent graywacke have been eliminated, can now be classed as true schists (Figure 19–4). From the genetic and textural standpoint they may also be viewed as a type of phyllonite.

19

Products of Regional Metamorphism, I: Rocks of Low Metamorphic Grade

PLAN OF TREATMENT

Primary Role of Metamorphic Grade

Rocks bearing the imprint of regional metamorphism will be treated primarily in terms of mineralogical criteria, with emphasis on those that most clearly reflect metamorphic grade and protolithology. Textural characteristics, while important in defining specific rock types (schist, phyllite, and so on), play only a subordinate role in the general plan of presentation.

Grade is essentially a function of temperature; for increase in grade, by definition, is accompanied by increase in total entropy of the metamorphic system—which comprises the mineral assemblage plus any gaseous phase (aqueous or rich in CO_2) that may have participated in the metamorphic process. Primary divisions in our plan of treatment are drawn in terms of relative grade, arbitrarily defined as "low," "medium," and "high." This chapter covers rocks of low grade, the upper limit of which has been chosen arbitrarily as that of the transition between the greenschist or blueschist and the amphibolite facies. While it is quite undesirable—clearly impossible, in fact—to assign precise temperature limits to low-grade metamorphism thus defined, some idea of the total temperature regime that is covered may be gained from Figures 16–7 and 16–8. The field of low-grade metamorphism itself has been subdivided by an arbitrarily drawn shadowy boundary that separates *immature* rocks, for the most part incompletely metamorphosed at very low grade, from those *mature* rocks that bear the fully fledged imprint of typical low-grade metamorphism. Certain rock types such as slates and semischists straddle the boundary.

Secondary Role of Pressure

Mineralogical indices of relative pressure are not neglected. Thus andalusite, zeolites, and prehnite are all indices of low pressure; aragonite, lawsonite, and (rarely in low-grade rocks) kyanite indicate medium pressures; jadeite-quartz and glaucophane suggest very high pressure (cf. Figure 16–7). Again the qualifying terms are used in a relative sense. The total pressure span of low-grade metamorphism covers the interval from about 1 kb to perhaps 10 or even 15 kb. In our plan of treatment only a few major divisions, cross-cutting those based on grade, are delineated with the pressure factor in mind.

Textural Constraints on Petrographic Procedure

Most products of regional metamorphism are foliated rocks, and in many of them one or more linear elements may be seen in the principal plane of foliation. Fine-scale segregation banding commonly plays a part in foliations. The resulting textural nonhomogeneity with reference to arbitrarily fixed coordinates imposes special constraints on petrographic procedures. To form a complete picture of the mineralogical and textural character of any given specimen one must examine and compare three mutually perpendicular thin sections. Most informative are a section parallel and two others normal to the principal plane of foliation; and of the latter two one should be cut normal to any conspicuous lineation. Even with the additional material thus provided for microscopic scrutiny, modal compositions can be estimated only roughly; accordingly, modes are not recorded in our treatment. Nor do we say much of the overall picture of strain kinematics that can be reconstructed from correlating geometric relations between visible structural (fabric) elements and similar data observed on the scale of an outcrop or a map area.* While the petrographer should note the existence, nature, and general relations of such textural characteristics, they play no part in the rock nomenclature or plan of treatment that we adopt here.

IMMATURE PRODUCTS OF METAMORPHISM OF VERY LOW GRADE

In two geological situations in particular, rocks that can properly be considered as metamorphic display the imprint of metamorphism (of very low grade) that has not been carried to completion. These are, respectively, regions affected by burial metamorphism and zones of incipient metamorphism on the outskirts of more extensive provinces of regional metamorphism in which strain has played a prominent role. In both situations the metamorphic process leads from an unstable to a more stable mineral assemblage. But the latter may contain metastable phases (such as chrysotile serpentine, adularia, and partially disordered albite), and there may be phases, notably certain zeolites, whose presence reflects pore-fluid compositions as much as temperature or pressure. In the fully developed mature (thoroughly recrystallized) products of low-grade metamorphism as visualized by classic petrographers, such phases tend to be eliminated in favor of others that are thought to be truly stable in the prevailing pressure-temperature regime.

Relics of parent mineral assemblages, commonly in great abundance, tend to survive in rocks of very low metamorphic grade. And inherited textures, especially in products of burial metamorphism (such as spilites) may persist almost unchanged. Such inherited features, mineralogical or textural, tend ultimately to become eliminated if metamorphism is carried on to full maturity.

In many cases of very low-grade metamorphism the full identity of the new mineral assemblage is revealed only through x-ray and related techniques beyond the scope of microscopic petrography. This, in fact, is the main reason that mineralogical aspects of metamorphism at very low grade so long remained unexplored. Today, however, enough is known for us to reconstruct with some confidence the paths of complex

---

*This is the subject matter of structural petrology, an essential element of which is based on petrographic observations.

mineral reaction that mark the onset of progressive regional metamorphism. The transitional field of overlap between diagenesis and metamorphism is known among some European petrologists as anchimetamorphism.

Systematic Petrography

The prevalence of metastable phases and the variable role of aqueous fluids in the early stages of metamorphism cause a good deal of irregularity in the overall mineralogical pattern of metamorphism at very low grade. The account that follows, therefore, is by no means comprehensive. It emphasizes regularity in paragenesis—for this is certainly perceptible—especially where this gives useful clues regarding the metamorphic temperature-pressure regimes. In particular, we have selected common rock types and omitted from consideration many less familiar though interesting variants.

Metapelites

Rocks of shale parentage are generally the first to respond to the onset of regional metamorphism. Their most conspicuous characteristics are slaty cleavage and very fine grain size. In this respect they resemble fully reconstituted slates (described below) except perhaps that the cleavage tends to be less perfect and the grain so fine that the mineral assemblage can be identified microscopically only as a mixture of indeterminate phyllosilicates (greenish "chlorites," colorless "micas"), black carbonaceous material, and colorless granular minerals (presumably quartz and feldspar). X-ray investigations have clarified the identity of most constituents, which include montmorillonites, illites, and in some rocks pyrophyllite. Coaly streaks, now of semianthracite or anthracite rank, have not yet generated crystalline graphite.

Metagraywackes

Even where there is little or no trace of schistosity massive graywackes may show perceptible signs of mineral reconstitution. A mineral paragenesis of recognizably metamorphic character begins to appear in veinlets filling cracks (that testify to brittle behavior) or dispersed as a fine-grained matrix between larger relict clastic grains (especially of quartz and feldspar). These latter usually show visible strain effects—undulatory extinction, curved twin lamellae (in feldspar grains), and marginal "granulation" (by annealing more than by fragmentation). Clastic relics of unstable minerals such as biotite, hornblende, and pyroxene show marginal replacement by minerals of the new assemblage, prominent among them chlorites and finely fibrous actinolite. The cumulative products of combined mechanically dominated strain and mineralogical reconstitution are increasingly schistose rocks transitional between graywacke and fully developed schist. These we have already considered with the emphasis on strain effects under the name *semischists* (cf. p. 512).

Successive arbitrarily defined stages in the textural transition from graywackes to fully reconstituted quartzo-feldspathic schists have been successfully used in New Zealand and California to map zones of progressive metamorphism on a regional scale. And more or less parallel changes in the metamorphic mineral assemblages—not following precisely the same pattern of correlation in widely separate areas—enhance the total metamorphic picture. The first two stages of textural development are marked respectively by nonschistose, incipiently metamorphic graywackes, and distinctly schistose semischists in which abundant coarser relict grains swim in a matrix of newly crystallized minerals of greatly reduced grain size. The third and fourth stages see a progression to almost or completely reconstituted schists of increasing grain size. These will be considered later. At present we are concerned with only the newly developed mineral assemblages of massive metagraywackes and of semischists of the second textural zone.

Zeolites are prominent in newly generated assemblages among graywackes and sandstones affected by burial metamorphism or regional metamorphism of minimal grade (zeolite facies). A generalized sequence of progressive change with increasing but still very low grade (subject to many exceptions) is:

heulandite-quartz
heulandite-prehnite-quartz
laumontite-prehnite-quartz
laumontite-prehnite-pumpellyite-quartz

Associated phases may include celadonite, montmorillonite, finely divided colorless mica, albite, and sphene. Analcime may appear in heulandite-bearing assemblages; and adularia (metastable, disordered, monoclinic K-feldspar) in others affected by metasomatism.

Metamorphic assemblages of slightly higher grade appear in massive and incipiently schistose metagraywackes of Triassic age along the north-eastern margin of the Haast Schist belt, southern New Zealand.[1] The principal ingredients of all these rocks are clastic grains of quartz (0.1 mm to 0.5 mm in diameter), partially albitized plagioclase, orthoclase, and subordinate muscovite, biotite, amphibole, rock fragments, and accessories, set in a fine-grained groudmass of newly generated meta-morphic minerals. These latter include white mica, chlorite, green pleo-chroic pumpellyite, prehnite, and sphene—the type assemblage of the prehnite-pumpellyite facies. In rather more schistose rocks (semischists) prehnite disappears, and actinolite, and in some rocks epidote, appears for the first time. The complete metamorphic assemblage then is quartz-albite-mica (colorless)-chlorite-pumpellyite (pale colored)-sphene ± ac-tinolite, epidote, calcite, and stilpnomelane. Rocks of this kind pass gradually into fine-grained schists in which clastic grains are sparse and pumpellyite is now completely absent.

All the assemblages so far described must have crystallized at pressures below 3 kb (cf. Figure 16–7). Notably higher pressures must be assumed for local development of lawsonite-bearing assemblages in other schistose graywackes (of Permian age) along parts of the northwestern edge of the Haast Schist belt. Typical of newly generated assemblages here is quartz-albite-pumpellyite-lawsonite-chlorite-muscovite-sphene (lawsonite-albite-chlorite facies).

Extreme pressures in a regime of low temperature are implied by metamorphic assemblages that appear in somewhat schistose or even massive metagraywackes that occur in belts of blueschist metamorphism in the California Coast Ranges (Franciscan Group) and elsewhere. A typical Californian specimen (near Valley Ford) consists mainly of angu-lar clasts of quartz, plagioclase, chert, and volcanic material, with relict bedding clearly traceable in streaks of finely divided colorless mica, chlor-ite, and carbonaceous matter.[2] Yet a conspicuous, newly generated phase is colorless jadeite (poorly birefringent and showing strong dispersion). This occurs as scattered single crystals and as subradial aggregates replac-ing clastic plagioclase. Associated minerals appearing sporadically in smaller amount are glaucophane and lawsonite—these too, typical of the blueschist facies.

Metabasalts

The primary mineral assemblage of basic and semibasic volcanic rocks that are so prominent in eugeosynclines—essentially plagioclase-pyroxenes-olivine—is highly unstable at low pressures in the presence of aqueous and CO_2-rich fluids. It is not surprising, therefore, to find some of them drastically altered mineralogically (while retaining their primary textures almost intact) as a result of deep burial metamorphism. Permeability to fluids greatly facilitates the process. So in thick flows metamorphic assemblages commonly are localized near the upper surfaces and in vesicular sectors. The phenomenon is perfectly displayed in large segments of the Andean Geosyncline of South America. A striking example is afforded by partially albitized porphyritic andesitic basalts in the coastal ranges west of Santiago, Chile.[3] Abundant euhedral crystals of labradorite, preserved in an unaltered state in the less permeable lower parts of lows 10 m to 40 m thick, have been completely replaced in more vesicular upper segments by turbid albite. The opaquely white aspect of these in hand specimens imparts to such rocks a most striking and characteristic appearance. While the primary pyroxenes remain largely unaltered, characteristic low-grade metamorphic assemblages appear in the groundmass and especially in vesicle fillings, where individual phases are coarse enough to be readily identified. All combinations of laumontite, prehnite, pumpellyite, and calcite are represented. Other metamorphic phases include any of albite, adularia, epidote, chlorite, white mica, and quartz. The overall metamorphic paragenesis is essentially that of spilites (see below).

Spilites, keratophyres, and serpentinites

In feebly metamorphosed or "unmetamorphosed" eugeosynclinal sequences there commonly appears an assortment of igneous rocks all of which, while showing no sign of internal deformation, have been largely converted to metamorphic mineral assemblages of very low or low grade. These rocks include spilites, keratophyres, and serpentinites—the latter commonly closely associated with spilites and cherts in ophiolite complexes. They retain much or all of their initial textures, and we have considered them already under appropriate sections of earlier chapters (pp. 113, 246). In many of them, relict minerals have been retained unaltered as essential ingredients of the total paragenesis—clear augite in spilites, olivine and chromite in some serpentinites, phenocrystic quartz in keratophyres. Here we briefly note some of the main features of the metamorphic paragenesis in relation to facies. And we emphasize that metamorphism involves extensive metasomatism, presumably through the agency of sea water and derivative fluids. These change continuously in composition (especially with respect to relative concentrations of Na^+, K^+, Ca^{2+}, and Mg^{2+}) as they circulate within the suboceanic crust, reacting all the time with the igneous assemblages they encounter.

In *spilites* and mineralogically similar metadiabases, the feldspar has been converted largely or completely to pure low-temperature albite, or less commonly to sodic oligoclase. Some degree of disorder (a sign of metastability) in albite is indicated by persistently high values of $2V_\gamma$ (85°–90°, instead of ~76°). Chlorite and either epidote or pumpellyite are also plentiful, and in some rocks actinolite is beginning to replace the original pyroxene. Parallel alternative development of epidote or pumpellyite both in spilites and in associated metagraywackes is telling evidence of the metamorphic origin of the spilite paragenesis (pyroxenes excluded). In many spilitic rocks, assemblages typical of very low-grade metamorphism appear in the vesicles—combinations of zeolites, prehnite, pumpellyite and calcite in particular (cf. Figure 3–5B,C).

Typical keratophyres differ from rhyolites mainly in profuse development of pure albite (replacing both phenocrystic oligoclase and groundmass K-feldspar). There are, however, keratophyres in which the metasomatically generated feldspar is adularia—metastable from the moment of crystallization. Adularia also appears in very rare types of spilite. Evidently the path of the metasomatic process is sensitive to the relative concentrations of Na^+, K^+, Ca^{2+}, and other ions in the agent of metasomatism—generally believed today to be ocean water or some type of derivative brine.

Metasomatism of peridotite in the upper part of the suboceanic crust through the agency of percolating sea water to yield serpentinite probably plays the primary role in the development of serpentinite bodies of ophiolites. The principal metamorphic (metasomatic) products are minerals of the serpentine family—chrysotile by far the commonest. This is a polymorph metastable from the time of formation (the truly stable phase is antigorite). Disseminated flakes of brucite (petrographically difficult to identify, but more highly birefringent than serpentine minerals) are common ingredients of serpentinites. Some rocks also contain minor talc, magnesite, or tremolite. Small white dikelets and veins commonly found in serpentinite bodies usually consist almost entirely of albite or of a varied assemblage of hydrous calcium-aluminum silicates—hydrogrossular (almost indistinguishable from grossular), clinozoisite or zoisite, vesuvianite, and others. These rocks (*rodingites*) are localized complementary by-products of the serpentinization process. Some have been shown to be metasomatized dikes and pods of gabbro or diabase.

MATURE PRODUCTS OF METAMORPHISM
AT MODERATE PRESSURES

The fully developed, thoroughly recrystallized products of low-grade metamorphism are foliated rocks of medium or even coarse grain that retain no trace of the parent mineral assemblage apart from recrystallized grains of those phases (like quartz) that remained stable under the changing metamorphic pressure-temperature regime. With some exceptions, such as relict bedding as seen in outcrops, the initial textures have also been obliterated and a completely metamorphic fabric has been substituted. We shall treat such rocks in two categories, remembering that transitional types also exist. These are rocks respectively metamorphosed at moderate pressures (perhaps up to 5 kb) and at significantly higher pressures (ranging possibly to 10–12 kb). The first category includes rocks of the greenschist facies, the second those of the blueschist facies (cf. Figure 16–8). Our immediate concern is with rocks of the first category. They are found chiefly in the zones of chlorite and biotite, on whatever basis the biotite isograd may have been mapped in pelitic rocks. In a few regions, such as southern New Zealand, rocks of slightly lower grade (in a pumpellyite-actinolite zone) also qualify. So too, at the opposite end of the spectrum of metamorphic grade, rocks of the greenschist facies may extend beyond the almandine isograd (again southwestern New Zealand is an example).

Synoptic Review of General Mineralogy

Rocks of the *chlorite zone,* where this is mapped on the basis of pelitic schists, are consistent in most respects with regard to overall mineralogical pattern. Differences between specific mineral assemblages—for example, those of pelitic compared to basic schists—depend essentially on bulk chemistry. The following notes chiefly concern essential phases found in common rocks, with typical accessory minerals. The list is by no means exhaustive.

Quartz is a principal constituent of pelitic and quartzo-feldspathic rocks (chief among them metagraywackes), a minor or possible phase in most others (metaserpentinite being the main exception). Plagioclase, the only feldspar in most rocks, is albite, commonly virtually pure (but in some rocks An_5 or An_7); it is the completely ordered low-temperature polymorph ($2V_\gamma$ ~76°). Microcline is also found in a few quartzo-feldspathic schists with high K_2O/Al_2O_3.

Muscovite (with an appreciable phengite content) is the commonest type of mica, but in some of the more aluminous metasediments there may also be fine-grained paragonite or pyrophyllite (optically indistinguishable from muscovite). Chlorites and epidotes (clinozoisite or green ferruginous epidote) are almost ubiquitous.* Stilpnomelane is widely distributed in all except magnesian schists of the chlorite zone, in most rocks as a minor but optically conspicuous constituent.

Amphiboles in the chlorite zone are tremolite (in calcareous rocks and metaserpentinites) or green pleochroic actinolite (in many basic schists). Some actinolites of greenschists are strikingly pleochroic from greenish to deep blue; though these contain appreciable (but small) concentration of the sodic crossite component, they must not be confused with lavender-blue amphiboles of the glaucophane group. Calcite and dolomite, principal components of calc-schists and marbles, are also found in smaller amount in many rocks. Common accessories in all but magnesian schists are sphene (rutile is rare except in slates), apatite (in shapeless or rounded grains), magnetite, pyrite, tourmaline (especially in metasediments), manganiferous garnet, and zircon.

Antigorite, talc, brucite, and accessory chromite and magnesite are minerals virtually confined to metaserpentinites. Manganiferous rocks (metacherts) are distinguished by abundant coexisting piedmontite and spessartine garnet.

At the low limit of the spectrum of metamorphic grade, nearly colorless pumpellyite can be a constituent of metagraywackes and metabasalts. At the opposite end of the spectrum biotite is the index mineral of the *biotite zone* and is ubiquitous in most metasediments (apart from calc-schists). Only in rare rocks does biotite coexist with stilpnomelane, which itself is absent from most assemblages of the biotite and higher-grade zones. In some regions, such as southwestern New Zealand, the greenschist facies overlaps the almandine isograd. Here at the upper limit of the grade spectrum almandine is found in quartzo-feldspathic and in pelitic schists; and in associated metabasalts (albite amphibolites) the amphibole is now aluminous green hornblende.

---

*In some rocks individual crystals of chlorite have become partially oxidized to yield a brown, pleochroic, increasingly birefringent product that has commonly been misidentified as biotite or stilpnomelane. Its secondary origin is clearly revealed when it is patchily developed in crystals adjacent to grains of limonite-stained magnetite.

Slates and Phyllites

Slates and phyllites (Figure 19–1) are fine-grained derivatives of sediments that were originally even finer-grained. The common types are therefore pelitic. Their major constituents are nearly uniaxial white mica, pale-green chlorite, quartz, and in some rocks dusty opaque graphite; common accessories are tourmaline, rutile, epidote, sphene, magnetite, and pyrite. The presence of tourmaline need not necessarily indicate permeation by fluids of magmatic origin, for most marine muds contain enough boron to account for the appearance of tourmaline in their metamorphic derivatives. The rutile of slates is chemically equivalent to the anatase and brookite of many siltstones and mudstones. Epidote is abundant only in green slates derived from tuffaceous sediments, and consequently may be accompanied by albite, which in very fine-grained rocks may be difficult to distinguish from associated quartz. Octahedra of magnetite, cubes of pyrite (Figure 19–1B), or scaly aggregates of hematite are all common.

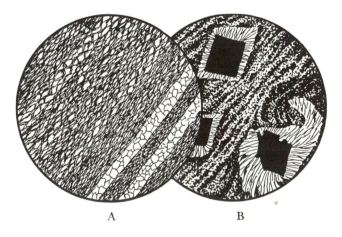

A   B

Figure 19–1. Early Stages in Development of Metapelites

A. Pelitic phyllite, Islay, Scotland. Diam. 3 mm. The section, perpendicular to schistosity, shows quartz, white mica, chlorite, and minor opaques. Quartz veins trend parallel to schistosity.

B. Pyrite-bearing graphite slate, near Carson Hill, Sierra Nevada, California. Diam. 5 mm. Fibrous quartz and a little chlorite have crystallized in "pressure shadows" around porphyroblasts of pyrite. Curvature of quartz fibers at lower right indicates rotation of the porphyroblast.

Highly characteristic of slates is the perfect fissility—slaty cleavage—imparted by a single dominant set of s-planes defined by strong preferred orientation of mica and chlorite flakes. Slaty cleavage, which is really a very regular planar schistosity, may cut relict bedding—indicated by persistent color banding—at any angle. In many slates, perhaps in most, the cleavage has developed parallel to surfaces of the differential movement (slip or shear) synchronous with recrystallization. Slaty cleavage may be accentuated by microscopic streaks of finely crystalline quartz (Figure 19–1A); but it is not until the coarser texture characteristic of the phyllite and schist stages has been achieved that segregation bands become clearly visible to the naked eye. Two or more intersecting sets of s-surfaces, referable to different stages of rock deformation, may be distinguished in many slates. To the earlier stages belong s-surfaces of *precrystalline* or *paracrystalline* deformation, marked by sharply crystalline undeformed aligned flakes of mica and chlorite, which crystallized, in part at least, later than the slip movement upon these s-surfaces. *Postcrystalline* movement on s-surfaces of later origin partially aligns favorably situated flakes of mica and chlorite, which become bent and twisted in the process. As such reorienting of mica is restricted to the immediate vicinity of postcrystalline slip surfaces, the resulting fissility is less perfect than that caused by pre- or paracrystalline movements. Some writers used to designate pre- or paracrystalline s-surfaces as *slaty* or *flow cleavage*, and to distinguish them from postcrystalline s-surfaces, which are variously termed *false, strain-slip* or *fracture cleavage*. These terms are valid only if the relation, in time, of deformation to crystallization can be determined (cf. Figures 17–1A and 19–2B).

With increasing metamorphism—caused by somewhat higher temperature, longer duration of metamorphic conditions, or greater activity of permeating fluids—slates pass into *phyllites*. These are mineralogically similar to slates, but notably coarser in grain, so that their micas form large enough flakes to impart a silky sheen to the surface of schistosity or cleavage. Occasionally the appearance of brown biotite in phyllites indicates that temperatures have risen appreciably above those that attend the transformation of mudstone into slate. The schistosity in a phyllite may be accentuated by thin but macroscopically visible segregation bands, alternately quartzo-feldspathic and micaceous, formed by the little-understood process of metamorphic differentiation—local concentration of individual minerals during chemical reconstitution, influenced by diffusion of fluids along the paths of minimum resistance offered by s-surfaces of mechanical origin.

Pelitic Mica Schists

Progressive regional metamorphism of pelitic sediments leads through natural gradational transitions, as grain size increases (to exceed perhaps 0.5 mm), from slate to phyllite to mica schist (Figure 19–2A,B). The typical mineral paragenesis of these schists is seen in chlorite zones the world over. It is dominated by muscovite-chlorite-quartz—with albite, epidote (or clinozoisite), and dolomite (or calcite) as common possible subordinate phases. The usual accessories include sphene, tourmaline, apatite, and magnetite; also not uncommon are manganese garnet, graphite, and (instead of sphene) rutile.

Muscovite and chlorite in alternating units interleaved parallel to {001} tend to build composite crystals. Muscovite of fully developed schists is a distinctly phengitic type (with appreciable Mg^{2+} and Fe^{2+} substituting for Al), it is usually coarse grained, virtually colorless and has a rather small axial angle ($2V_\alpha = 20°–40°$). Other phyllosilicates may also be present, depending on variations in bulk chemistry: chloritoid in conspicuous blue-green porphyroblasts (Figure 19–3) in rocks rich in Al_2O_3

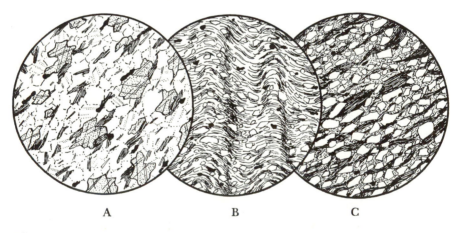

A   B   C

Figure 19–2. Low-Grade Schists, Scotland

A. Calcite-albite-chlorite schist, Dunkeld, Inverness-shire. Diam. 2.5 mm. Small amounts of muscovite and iron oxide accompany the main constituents.

B. Muscovite-chlorite-quartz schist (pelitic), Glen Esk, Angus. Diam. 2.5 mm. The main schistosity (horizontal) is crossed by a later strain-slip cleavage (vertical) due to microfolding.

C. Semischist, Falls of Leny, Perthshire. Diam. 2.5 mm. Partially destroyed relict sand grains (mostly quartz) in a schistose matrix of large bent chlorite flakes, fine-grained muscovite, and granular magnetite.

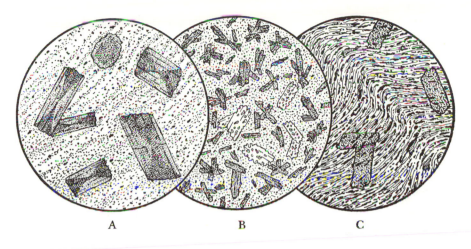

A   B   C

Figure 19–3. Pelitic Chloritoid Schists

A. Ottrelite schist, Ottrez, Belgium. Diam. 3.5 mm. Porphyroblasts of ottrelite (=chloritoid), showing hourglass structure due to local concentration of inclusions, in a matrix of quartz, muscovite, and opaque dust. Random orientation of ottrelite crystals in relation to schistosity points to late crystallization of ottrelite.

B. Chloritoid-andalusite schist, Korea. Diam. 3.5 mm. Crystals of chloritoid—many of them twinned in stellate clusters—and small spongy porphyroblasts of andalusite, in a matrix of quartz, muscovite, and minor biotite.

C. Chloritoid-muscovite-graphite-chlorite schist, Shetland Islands, Scotland. Diam. 3.5 mm. Porphyroblasts of chloritoid have crystallized late and therefore enclose trains of inclusions aligned parallel to the previously folded schistosity.

and with high Fe/Mg; paragonite (almost indistinguishable optically from muscovite) in aluminous rocks with high $Al/(K^+Na)$; small amounts of stilpnomelane only in rocks relatively low in Al_2O_3, derived from siltstones rather than typical shales. In most rocks albite and the phyllosilicates between them accommodate all the aluminum; rarely, however, an excess of Al_2O_3 gives rise, at low pressures, to a little andalusite (Figure 19–3B), or at higher pressures to kyanite (as in parts of the New England sector of the Appalachian belt).

Entry of brown biotite into comparable assemblages (excluding those that still carry paragonite or stilpnomelane) signifies a recognizably heightened metamorphic grade. However, no precise correlation can be made between biotite isograds mapped in different provinces or for different lithologies; for more than one type of interphase reaction, depending on bulk rock chemistry, can generate biotite in rocks that can be termed pelitic. In some regions, too (e.g., in the Scottish Highlands), entry of almandine garnet into pelitic assemblages marks yet another perceptible increase in grade—still, however, just within the greenschist facies (as marked by stability of albite plus clinozoisite). Throughout all such changes the composition of plagioclase remains the same—almost pure albite with no more than 5% An; it is an ordered polymorph with $2V\gamma = 76° - 78°$. Muscovite and chlorite, though both participating in most biotite- and garnet-generating reactions, remain essential constituents of typical pelitic rocks of the greenschist facies. The ultimate assemblage is essentially quartz-albite-biotite-muscovite-chlorite ($\pm$ chloritoid, clinozoisite, almandine).

Low-grade mica schists, like most fully developed schists of regional metamorphic provinces, typically show prominent schistosity (commonly more than one set of recognizable s-surfaces), lineations, and segregation lamination. The latter—marked by alternation of thin laminae of contrasted mineral composition (micaceous versus quartzo-feldspathic)—is not to be confused with bedding. Undoubted relics of bedding, apart from small-scale color banding in some rocks and lithological banding on the scale of an outcrop or a map, have been largely eliminated by the metamorphic process. The complex imprint of this latter on the fabric of most pelitic schists suggests repeated episodes of deformation accompanied and finally outlasted by interphase reaction and recrystallization. Alignment of micas in most rocks is the result of passive growth of crystals in planes and directions of least resistance to growth and of maximum permeability to fluids—both determined by strain-induced elements in the fabric. Strain-slip cleavage in some rocks (Figure 19–2B) bears witness to late minor strain imposed upon the otherwise fully developed schist. The total span of time involved in evolution of mica schists from shale may cover several million (perhaps 10 million or so) years.

Quartzo-Feldspathic Mica Schists (Metagraywackes)

The general mineralogy of mica schists derived from graywackes follows the same pattern as in pelitic schists except that quartz and feldspar (albite) are the most abundant constituents; mica and chlorite play an important but reduced role; epidote or clinozoisite are ubiquitous and more plentiful than in metapelites. Actinolite may appear (with more than usually abundant epidote and chlorite) in schists derived from graywackes that initially had a significant component of volcanic rock fragments. Pumpellyite (an almost colorless variety) is found only in finely crystalline schists of minimal grade in which segregation lamination has scarcely developed.

Whereas evolution of mica schist from slate involves continuous increase in grain size, the reverse is true in the metagraywacke-semischist-schist transition (cf. p. 447). Here, first stages of evolution lead to almost completely reconstituted fine-grained schists from which most relict grains have been eliminated; they generally show distinct lineation (cf. Figure 19–4), but segregation banding is no more than incipiently developed. In the ultimate stage of evolution the grain coarsens greatly and segregation layers become prominent (proof that these in no way are related to relict bedding). Albite of quartzo-feldspathic schists of the chlorite zone is the almost pure end member, An_0, of the plagioclase series. In many rocks it develops large porphyroblasts simply twinned on the albite law and in many cases enclosing trains of dark inclusions (remnants of early *s*-surfaces) whose S-shaped curvature indicates rotation of the porphyroblast during growth. As in most low-grade schists, most of the albite grains lack twinning; but some show sparse lamellae twinned either on the pericline or the albite law. Quartzo-feldspathic schists have

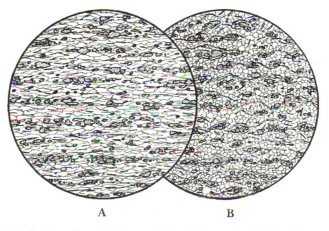

A B

Figure 19–4. Schistosity in Quartz-Albite-Muscovite Schist (Metagraywacke)

Fine-grained quartz-albite-epidote-chlorite-muscovite schist derived from graywacke, East Otago, New Zealand. Diam. 1 mm.

A. Section perpendicular to schistosity and parallel to lineation.

B. Section perpendicular to both schistosity and lineation.

Note how flakes of muscovite and chlorite maintain a more marked preferred orientation parallel to the lineation than parallel to the schistosity.

insufficient Al_2O_3 to permit development of chloritoid or paragonite. Stilpnomelane, however, is widely present within the chlorite zone; it becomes virtually eliminated at the biotite isograd. Coexistence of biotite and stilpnomelane is rare, though it has been recorded on the biotite isograd both in the Alps of Switzerland and in southern New Zealand. The fully developed metagraywacke assemblage is quartz-albite-muscovite-chlorite-epidote (or clinozoisite), with biotite an additional phase in rocks of the biotite zone, and almandine at the upper limit of grade in the greenschist facies. In rocks of arkosic parentage microcline is a possible additional phase at all grades. Sphene, apatite, and opaque oxides are common accessories. Mature schists of all grades within our present scope show all the structural complexity that is typically displayed by associated pelitic mica schists. Strain-slip cleavage, however, is much less in evidence; but there may be signs of late minor strain in quartz— undulatory extinction, "Boehm lamellae," and subparallel trains of fine dusty inclusions.

Metacherts

Metacherts of any grade can be recognized by abundance of quartz— usually coarsely crystalline—and virtual absence of feldspars. There is a good deal of variety in the mineralogical pattern depending on relative concentrations of iron and manganese, their respective oxidation states, and possible presence of thin shaly partings in the parent sediments.

Ferruginous Metacherts

Pink, slightly fissile metacherts appear here and there in the outer lower-grade zones of the Haast Schist belt, New Zealand. They are composed mainly of rather finely crystalline quartz and scaly hematite. Figure 19–5C illustrates a more complex rock containing abundant crystals of deep-red-brown stilpnomelane, minute dodecahedra of nearly colorless manganese garnet, and a few flakes of muscovite. An interesting type of low-grade metachert widely prevalent in Precambrian iron formations of South Africa and western Australia contains abundant pale-blue crocidolite both dispersed within the chert and concentrated as veinlets of blue asbestos. Crocidolite, a magnesian variant of riebeckite, $Na_2(Fe^{2+}, Mg)_3 Fe_2^{3+} Si_8O_{22}(OH)_2$, has no implication of high pressure such as is generally attributed to aluminous sodic amphiboles of the glaucophane-

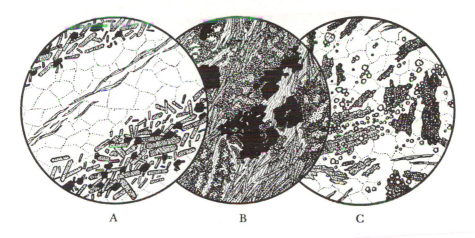

A   B   C

Figure 19–5. Low-Grade Metacherts, Lake Wakatipu, New Zealand

A. Quartz-muscovite-piedmontite schist, derived from manganiferous chert. Diam. 1 mm. The principal constituent is quartz. Foliation is marked by thin bands composed of muscovite or of piedmontite and magnetite. Piedmontite is in idioblastic prisms. Note absence of albite.

B. Magnetite-garnet-epidote-actinolite schist, derived from cherty iron-manganese ore. Diam. 2 mm. Dense aggregates of manganiferous garnet enclosing yellow epidote; acicular actinolite and minor intergrown prochlorite; porphyroblasts of magnetite.

C. Quartz-garnet-stilpnomelane schist derived from ferruginous chert. Diam. 1 mm. Deep-red-brown stilpnomelane, small idioblastic dodecahedra of manganiferous garnet, minor muscovite, and a granoblastic matrix of quartz. Note absence of albite.

crossite series $Na_2(Mg,Fe^{2+})_3(Al,Fe^{3+})_2Si_8O_{22}(OH)_2$ in rocks of the blueschist facies.

In northern Michigan and elsewhere in the iron-mining region of the Great Lakes, iron-formation (alternating cherts and carbonate sediments) has become metamorphosed to a variety of mineral assemblages in which ferriferous silicates and carbonates are prominent—chiefly combinations of stilpnomelane, minnesotaite, chlorite, and siderite, plus magnetite. Minnesotaite, the ferrous analogue of talc, is a highly birefringent uniaxial (negative) mineral, usually fibrous with positive elongation. It is almost colorless, faintly pleochroic in pale greens. The true identity of minnesotaite, while difficult to determine in thin section by optical properties alone, can be guessed with some confidence from the nature of the mineral assemblages of which it is a member.

Manganiferous Metacherts

Two contrasted types both from the chlorite zone in southern New Zealand are illustrated in Figure 19–5A,B. The first is a highly fissile lineated pink schist in which subparallel alignment of muscovite flakes induces a silvery sheen on surfaces parallel to the principal schistosity. The principal mineral is quartz, which builds a mosaic of polygonal grains. Thin bands rich in muscovite or in the manganiferous epidote piedmontite plus magnetite accentuate the foliation. The piedmontite occurs in slender idioblastic prisms elongated parallel to [010] = β; colors and pleochroism are spectacular, α = canary yellow, β = pale amethyst, γ = deep wine-red. Maximum absorption is normal to the direction of crystal elongation. Accessories in all specimens include tiny prisms of nearly colorless tourmaline concentrated in swarms within the piedmontite-magnetite bands, and small twinned prisms of rutile; apatite is common. Occasional coarse grains of barite appear in some specimens. Considered together, barite and tourmaline strongly suggest deep-sea origin for the parent rock. Similar manganiferous schists from the same region contain spessartine as well as piedmontite; and beyond the biotite isograd an additional component is phlogopite.

Figure 19–5B illustrates a nonschistose, resistant, deep-brown rock found here and there in the chlorite zone of the Haast Schist terrane. Quartz is only a minor component. Dense aggregates of pale spessartine enclose yellow epidote, acicular blue-green actinolite, and deep-green chlorite. Scattered through this matrix are irregular porphyroblasts of magnetite.

Low-Grade Calc-Schists

Limestones become converted by low-grade metamorphism into calc-schists whose main components are calcite or dolomite, or both. Minor constituents, depending on variations in initial composition, may include any combination of quartz, albite (less commonly microcline), muscovite, chlorite, clinozoisite, sphene, and graphite. Beyond the biotite isograd dolomite may begin to react in the presence of H_2O with SiO_2 to yield talc or tremolite (cf. Figure 19–6B). At this stage, too, biotite may develop by reactions involving dolomite and K-feldspar. Low-grade calcareous assemblages that develop from impure calcareous interbeds in dominantly quartzo-feldspathic schists (later Precambrian) in Scottish

Highlands consist principally of albite, zoisite, calcite, biotite, and gros-sular-andradite.

Foliation in calc-schists and schistose marbles is defined by dimensional parallelism of irregularly bounded lensoid grains of calcite and by pre-ferred orientation and local concentration of mica and chlorite flakes. It may be accentuated by variation in grain size reflecting different responses to deformation and later annealing (cf. Figure 18–3A). Lamellar twinning reflecting late postcrystalline minor strain is almost universal in calcite grains, but much less prominent in dolomite (or ankerite). Where both minerals are present they may be separately iden-tified by contrasted patterns of twinning in relation to the fast extinction direction, α'. In calcite the angle between α' and the trace of sharply defined $\{01\bar{1}2\}$ lamellae is greater than 50° (maximum value 64°). In dolomite the corresponding angle is more acute (25°–40°); for twinning in this mineral is on the acute rhombohedron $\{02\bar{2}1\}$. In grains showing two sets of sharp lamellae, α' lies in the obtuse angle of intersection in calcite, in the acute angle in dolomite.

Greenschists

Greenschists are products of regional metamorphism of basic and sem-ibasic igneous rocks at low temperatures. They are schistose green rocks, whose color is due to abundance of one or more of the green minerals chlorite, epidote, and actinolite. Segregation of chlorite-epidote-actin-olite laminae, alternating with white layers rich in albite, is characteristic.

Typical mineral assemblages (Figure 19–6) are: chlorite-epidote-albite-(calcite), chlorite-epidote-actinolite-albite, actinolite-epidote-albite. Albite (An_{0-3}) may occur as a mosaic of shapeless grains, or in large idio-blastic porphyroblasts enclosing parallel strings of epidote or actinolite needles. Where crystallization of albite has been mainly postdeforma-tional, these strings of inclusions trend roughly parallel to the schistosity of the surrounding matrix; but where, as frequently is the case, defor-mation and crystallization have proceeded simultaneously, the growing porphyroblasts of albite have slowly rotated in the yielding chloritic matrix, and the strings of inclusions have assumed an S-like pattern that reflects the direction and sense of rotation. Metamorphic albite, in con-trast with igneous plagioclase, tends to be untwinned, superficially resem-bling quartz except for its lower refractive index (less than 1.54) and

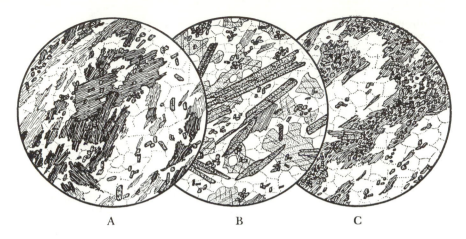

Figure 19–6. Greenschists (Low-Grade Basic Schists), Lake Wakatipu, New Zealand

A. Albite-stilpnomelane-chlorite-epidote schist. Diam. 2 mm. Deep-red-brown stilpnomelane (γ = 1.73) is associated with a smaller quantity of green prochlorite (γ = 1.62). The small, highly refractive prisms are epidote; the granoblastic colorless matrix is albite.

B. Actinolite-albite-epidote-chlorite-calcite schist. Diam. 2 mm. Slender porphyroblastic prisms of deep-blue-green actinolite, optically distinguishable from hornblende only by somewhat lower refractive index (γ = 1.65–1.66); xenoblastic nontwinned albite; small prisms of yellow, iron-rich epidote; xenoblastic calcite with widely spaced thin $\{01\bar{1}2\}$ lamellae; deep-green prochlorite; accessory sphene and apatite.

C. Albite-epidote-chlorite schist with accessory sphene, apatite, and magnetite. Diam. 2 mm.

biaxial character. Where twinning is developed, it almost invariably follows either the albite or the pericline law; simple twinning in porphyroblasts, though easily mistaken for Carlsbad twinning, in most cases follows the albite law (in a section normal to the [100] axis the {001} cleavage traces in opposite halves of a twin are mutually inclined at only 8°). The chlorite of greenschists varies somewhat in composition but most commonly is a ferruginous prochlorite with strong pleochroism, relatively high refractive index (1.61–1.63), small axial angle, positive sign, and brownish-to-purple anomalous interference colors. Some varieties have slightly higher refractive index (1.63–1.64), negative sign, and anomalous blue interference colors. The epidote ranges from yellow iron-rich types to nearly colorless clinozoisite (with anomalous blue interference tints), the latter becoming more widespread as the grade of

metamorphism advances. The epidote minerals tend to occur as idioblastic prisms elongated (parallel to β) in the directions of schistosity and
lineation. The actinolite takes the form of slender prisms and needles
bounded by {110} and imperfectly terminated. In some rocks it is of a
pale-green tremolitic variety; but much more typical is a strongly colored
mineral that is pleochroic from pale yellow (α) to deep blue-green (γ)
and optically indistinguishable from the blue-green aluminous hornblendes of amphibolites. These blue-green actinolites may contain significant though small amounts of the crossite "molecule," but in contrast
with hornblendes are low in Al_2O_3.

Among the minor constituents present in almost all greenschists are
sharply crystallized octahedra of magnetite, rounded droplike or spin-
dle-shaped grains and granular clusters of sphene, and xenoblastic crystals of apatite. Quartz is common in small quantities and may be plentiful,
along with greenish biotite, in rocks derived from tuffs. Stilpnomelane
(Figure 19–6A) is widely distributed and occurs especially as sheaves of
tabular crystals that have grown in the schistosity surfaces after deformation ceased. It was once generally misidentified as biotite, from which
it may be distinguished, however, by transverse fractures cutting across
the perfect {001} cleavage, by higher birefringence (about 0.08–0.10),
and by details of color and pleochroism.

Magnesian Schists

Most magnesian schists are formed by hydrothermal metamorphism of
peridotites, either directly or via an early stage of burial metamorphism
in which the peridotite bodies had already been converted to serpentinites (cf. p. 521) or to similar partially carbonated rocks ("ophicarbonates"
of European literature). Whether the path was direct or indirect, the
typical product of low-grade metamorphism is some form of *antigorite
schist;* for even at low temperatures the stable serpentine polymorph is
antigorite. Its platy habit and a tendency to subparallel dimensional
alignment render such rocks schistose. And in many the foliation is made
more conspicuous by subparallel irregular streaks of recrystallized
chromite. Minor amounts of CaO are accommodated by either diopside
or tremolite. Brucite is eliminated very early in the prograde sequence
by reaction with antigorite to regenerate forsteritic olivine—the first step
in progressive dehydration of serpentinite. Typical carbonate-free assemblages, recorded by V. M. Trommsdorff and B. W. Evans in the biotite

zone of the Swiss Alps, are as follows (in order of increasing SiO_2 content): antigorite-forsterite-diopside, antigorite-diopside-tremolite, antigorite-tremolite-talc. In all these the principal component is antigorite—occurring in colorless plates of micaceous habit, with good {001} cleavage; birefringence and refractive index are both low, the latter only slightly higher than in quartz. A possible additional phyllosilicate that accommodates any aluminum present is pale-green pleochroic chlorite with anomalous interference tints.

Not all low-grade magnesian schists are antigorite metaserpentinites. Some are formed by metasomatism of other rocks adjacent to granitic or granodiorite plutons. Such are chlorite-talc and chlorite-tremolite-talc schists, illustrated in Figure 19–7A, B.

Most low-grade schists derived from "ophicarbonates" in the Swiss Alps consist of antigorite, talc, and calcite. Another common type formed

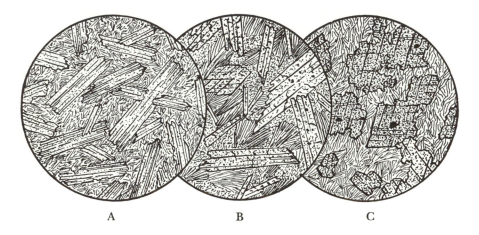

A   B   C

Figure 19–7. Low-Grade Metasomatic Magnesian Schists

A. Chlorite-talc schist, Academy Quadrangle, Sierra Nevada, California. Diam. 2.5 mm. Large idioblastic flakes of pale-green chlorite lie in a matrix of tufted platelets of talc. The rock has formed by low-temperature hydrothermal metamorphism of amphibolite, at a contact with quartz-diorite pegmatite.

B. Chlorite-talc-tremolite schist, same locality as A. Diam. 2.5 mm. The main constituents are coarse colorless prisms of tremolite and flakes of pale-green chlorite. The interstitial fine-grained material is talc.

C. Magnesite-talc schist formed locally within serpentine rock, Soapstone Hill, Bidwell Bar Quadrangle, Sierra Nevada, California. Diam. 2.5 mm. Note the subidioblastic rhombohedral form of the central grain of magnesite.

by local CO_2–SiO_2 metasomatism of serpentinite is composed largely of talc and magnesite—the latter tending to crystallize in idioblastic simple rhombohedra $\{10\bar{1}1\}$ (Figure 19–7C).

MATURE PRODUCTS OF METAMORPHISM AT HIGH PRESSURES

Distribution

A unique pressure-temperature metamorphic regime combining high pressures (up to 10–12 kb) with low temperatures (usually no higher than about 400°C) is responsible for development of glaucophane schists and allied rocks of the blueschist facies. These are virtually confined to young mobile belts, notably those that ring the Pacific Ocean and comprise the Alpine chains of Europe and Asia Minor. Their petrography has been investigated in detail especially in blueschist terranes of the Alps, Japan, the California Coast Ranges, and New Caledonia (for detailed references see F. J. Turner, 1981). In this section we describe fully evolved members of the blueschist facies, with notes on two sets of transitional rocks that respectively represent regimes of lower and of higher pressures (and grade).

General Mineralogy

Rocks of the blueschist and the greenschist facies have much in common as to mineralogical pattern. Chlorites, phengitic muscovite, stilpnomelane, albite, epidote, and sphene are widespread in both. Albite, An_{0-3}, and epidote, however, are both less common in the blueschist facies, epidote being more typical of blueschists of somewhat higher grade (beyond an epidote-garnet isograd in New Caledonia). Quartz is ubiquitous in all but some basic rocks.

The typical hydrated Ca-Al silicate of blueschists is lawsonite; and pumpellyite, found in the greenschist facies only in incompletely metamorphosed rocks of minimal grade, is much more widely developed as an essential phase of some blueschist assemblages. Both tend to diminish in quantity and even to vanish as epidote becomes more prominent. Some blueschists of relatively high grade contain an almandine garnet.

Most typical of blueschist mineralogy is the presence of special types of sodic amphiboles and pyroxenes; the former are members of the glaucophane-crossite series; the pyroxenes are either omphacites ($Jd_{40\sim50}$) or jadeites ($Jd_{>80}$) in the diopside-jadite solid-solution series. Aragonite is sporadically developed especially in vein fillings, in rocks of both the blueschist and the lawsonite-albite-chlorite facies. It is a significant index of high pressure and low temperature; but unless postmetamorphic unloading is rapid it tends to invert to calcite. Also typical of blueschists, even at highest grades within the facies, is a negative characteristic: biotite is never present (early records of biotite are due to misidentification of stilpnomelane).

A little should be said about the microscopic properties of the chief diagnostic minerals since they are almost unique to the facies, and most of them may be found as linings to open cross-cutting joints, or in monomineralic segregation bands in strongly foliated rocks, where their optic properties are readily identified.

Glaucophane is readily recognized by its pleochroism—α = pale yellow to colorless, β = violet or lavender blue, γ = deep blue—although depth of color varies with composition. It has positive elongation and a low extinction angle $\gamma \wedge [001]$. It occurs in long simple prisms, commonly zoned inward or outward to colorless tremolite. Glaucophanes form a continuous solid-solution series with crossite—the composition of which is about halfway toward riebeckite. Crossite is deeper in color, with β = dark blue and γ = violet blue; it is almost uniaxial and is optically negative.

The common pyroxene, omphacite, is pale green, scarcely pleochroic, with $\gamma \wedge [001]$ = 41° to 54°, $2V_\gamma$ = 65° to 80°. A simplified composition in terms of diopside, jadeite, and acmite end members is Di_{40-50} $Jd_{40-50}Ac_{0-10}$. Jadeite, also widespread in metagraywackes, is relatively pure, with no more than 20% combined diopside and acmite. In thin section it is an unremarkable colorless pyroxene but can be distinguished by relatively low birefringence ($\gamma-\alpha$ = 0.012–0.015), $2V_\gamma$ about 70–80°, strong dispersion, and extinction angle $\gamma \wedge [001]$ about 35°. Except in open veins, both pyroxenes tend to occur as shapeless granules, careful scrutiny of which, however, reveals the characteristic mutually perpendicular prismatic cleavages. Rarely, as in Figure 19–9C, crystals of jadeite are sufficiently large and well formed for all its diagnostic properties to be well displayed.

Lawsonite tends to occur in sharply bounded porphyroblasts, generally colorless, with perfect mutually perpendicular {001} and {100} cleavages (Figure 19−9B), lamellar twinning on {110} (two acutely intersecting sets), medium birefringence and refractive index. To the experienced petrographer aragonite is easily distinguishable from calcite, which it resembles in color, refractive index, and extreme birefringence. Most sections lack visible twinning or cleavage. Sections of minimum birefringence subparallel to {001} show all its distinctive properties—almost universal fine {110} twin lamellae in two sets intersecting at 60°, the striking interference figure in convergent light showing $2V_\alpha \sim 20°$, with the trace of the axial plane bisecting the acute intercleavage angle.

The properties of stilpnomelane and pumpellyite are nowhere better displayed than in blueschists. Stilpnomelane typically is strongly pleochroic from clear golden brown to deep reddish brown. It has perfect {001} cleavage; and a diagnostic though trivial feature is the presence of sparse interrupted cracks normal to the cleavage (a feature not seen in biotite). Like biotite it is uniaxial, negative, but its birefringence is even higher $(\gamma - \alpha) \sim 0.08-0.10$. Pumpellyite somewhat resembles epidote in refractive index, birefringence, and crystal outlines (elongated parallel to β). But its pleochroism is uniquely characteristic—α and γ are almost colorless, β is bluish-green to grass green (or pale yellow). Sections normal to an optic axis display the maximum pleochroic effect, and between crossed polarizers strongly anomalous interference tints; the convergent-light figure is unique—a strongly curved isogyre markedly red on the concave edge (dispersion very strong, $r < v$), $2V_\gamma = 25°-60°$.

Petrography of Blueschists and Related Transitional Rocks

The more striking characteristics of blueschist petrography are displayed in common rock types respectively of pelitic, graywacke, and basic (or semibasic) volcanic parentage. The following summary is based mainly on two broad regions—northern New Caledonia and the California Coast Ranges.[4] Much of the mineralogical variety among blueschists from each of these two regions reflects differences in grade. Detailed correlation with grade has been demonstrated by field mapping in New Caledonia, where an elaborate sequence of isograds has been defined. In California it is largely a matter of inference since most of the "high-grade" assemblages are represented in tectonically transported blocks, most of them smaller than 100 m^2 in surface outcrop.

All these blueschists (excluding partially reconstituted massive and roughly schistose quartz-jadeite metagraywackes, already described) are schistose medium-grained schists in which segregation banding commonly is prominent.

Metapelites

The pelitic assemblage characteristic of lower grades in the New Caledonia province is quartz, albite, muscovite (phengitic), paragonite (identified by x-rays), lawsonite, chlorite, sphene; a little spessartine or calcite appears in some rocks. At higher grades epidote displaces lawsonite, glaucophane appears and then increases in quantity; finally it is accompanied by green hornblende (rutile at this point displacing sphene).

In California, where metapelites are not so widespread they include rocks dominated by quartz, muscovite, crossite, and garnet. A pelitic glaucophane-muscovite-chlorite assemblage is illustrated in Figure 19–8A.

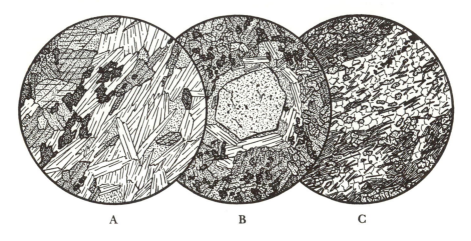

A   B   C

Figure 19–8. Glaucophane Schists

A. Glaucophane-muscovite-chlorite schist with abundant sphene, Coast Ranges, California. Diam. 3 mm.

B. Garnet-bearing glaucophane schist, Coast Ranges, California. Diam. 2 mm. Pink almandine garnet, partially replaced and enclosed by chlorite; muscovite associated with chlorite in the vicinity of garnet; abundant glaucophane enclosing granular epidote and sphene.

C. Glaucophane-epidote-quartz schist, Monument, Anglesey, North Wales. Diam. 2.5 mm. A train of hematite flakes at upper right. Elsewhere in section (not shown in figure) is abundant granular sphene.

An otherwise unusual type of aluminous glaucophane schist prominent in the Eoalpine blueschist terrane of the western Alps of Europe contains abundant chloritoid, in some cases kyanite as well, in addition to glaucophane, white micas, and garnet.

Metagraywackes

Metagraywackes are especially well developed in California. The typical assemblage consists mainly of quartz, jadeite, muscovite, chlorite, lawsonite, and sphene, ± glaucophane. Almandine or epidote, or both, may appear in rocks of presumably higher grade (cf. Figure 19–8B). A glaucophane-epidote-quartz schist from Anglesey, North Wales (Figure 19–8C) is probably of similar parentage.

Metabasalts

At lower grades New Caledonian metabasalts consist of ablite, lawsonite, pumpellyite, glaucophane, actinolite, chlorite, stilpnomelane, and sphene. At higher grades (beyond an epidote-garnet isograd) the assemblage becomes albite, epidote, glaucophane, actinolite, chlorite, muscovite, omphacite, almandine, sphene. And at maximum grade the calcic amphibole, still coexisting with glaucophane, is an aluminous green hornblende; a second white mica, paragonite, appears and rutile displaces sphene. Comparable coarse-grained Californian assemblages consist largely of: lawsonite, clinozoisite, chlorite, and sphene (Figure 19–9B); almandine garnet, muscovite, glaucophane, epidote, and sphene (Figure 19–8B); and various other combinations.

Metacherts

Some Californian metacherts duplicate those of the greenschist facies. A common assemblage is quartz-stilpnomelane-spessartine (cf. Figure 19–5C). Piedmontite-garnet metacherts have been recorded from Japanese blueschist terranes (Figure 19–10B). More diagnostic of the blueschist facies, however, are metacherts with crossite and lawsonite (Figure 19–9A) or with green aegirine from which needles of crossite sprout radially outward (Figure 19–10C). Near Laytonville, northern California, are massive blocks of coarse-grained metachert with stilpnomelane, garnet, and crossite as prominent constituents. These are almost unique in that they contain the three rare minerals deerite, howieite, and zussmanite.[5]

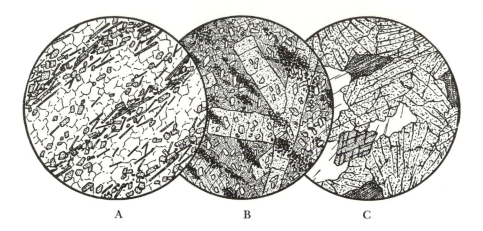

Figure 19–9. Lawsonite and Glaucophane Schists, California Coast Ranges

A. Quartz-lawsonite-crossite schist, San Onofre breccia, Laguna. Diam. 2 mm. Needles of deep-blue crossite, stumpy prisms of lawsonite, granoblastic quartz, accessory iron oxides, and granular sphene.

B. Lawsonite-chlorite-epidote schist, North Berkeley Hills. Diam. 2 mm. Large porphyroblasts of lawsonite elongated parallel to [100] and tabular parallel to {001}; colorless iron-poor epidote, green chlorite, finely granular sphene.

C. Jadeite-glaucophane-albite schist, Sebastopol Quadrangle, California. Diam. 2 mm. Subradiate clusters of colorless jadeite, prisms of pale glaucophane, and shapeless grains of albite.

Jadeite-Albite Rocks

Rare highly sodic rocks consisting entirely of jadeite and albite occur as blocks enclosed in serpentinite in San Benito County, California. Both constituents are almost pure—chemically analyzed specimens of jadeite approximate Jd_{95}. A rather similar mineralogical oddity, illustrated in Figure 19–9C, consists of jadeite, pure albite, and pale-colored glaucophane.

Rocks of Transitional Facies

At the low end of the blueschist pressure spectrum are schists lacking both glaucophane and sodic pyroxene. These rocks, now designated type members of the lawsonite-albite-chlorite facies, are metagraywackes consisting of quartz, albite, muscovite, chlorite, lawsonite, and aragonite. At

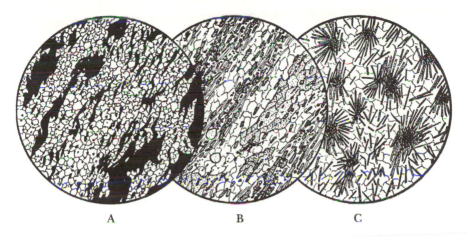

A B C

Figure 19–10. Metacherts of Blueschist Facies

A. Quartz-hematite schist, Colombia, South America. Diam. 3 mm.

B. Piedmontite-quartz-calcite-muscovite schist containing a little manganiferous garnet, Chichibu, Honshu, Japan. Diam. 1 mm.

C. Crossite-aegirine-quartz schist, near contact between serpentine rock and chert, Coast Ranges, California. Diam. 1 mm. Knots of fibrous deep-green aegirine form nuclei from which radiate needles of deep-blue crossite. The granoblastic matrix is of quartz.

the high-pressure end of the blueschist spectrum are coarse-grained garnet and hornblende-bearing glaucophanic metabasalts transitional perhaps into the amphibolite and the eclogite facies.

OVERALL MINERALOGICAL PATTERN IN RELATION TO METAMORPHIC FACIES

By definition, the essence of the concept of metamorphic facies lies in the strong and constant correlation that exists between mineralogical paragenesis and bulk chemistry (i.e., rock type) among rocks of comparable metamorphic grade. This correlation with respect to rocks that we have collectively termed "products of low-grade regional metamorphism" is presented in synoptic form in Table 19–1. This table is not intended to represent a comprehensive or complete picture of low-grade phase assemblages. Only selected mineral assemblages that together display the most characteristic aspects of specific facies are included. Accessory and unessential phases are omitted. So also are assemblages typical

Table 19–1. Low-Grade Mineral Paragenesis in Relation to Facies of Regional Metamorphism (Selected Mineral Assemblages)

Rock type	Zeolite and Pumpellyite Facies	Greenschist Facies	Blueschist Facies
Metapelites	Montmorillonite-illite-quartz–alkali feldspar ± pyrophyllite	Muscovite (phengitic)-chlorite-quartz-albite-epidote ± stilpnomelane or chloritoid Same as above plus biotite ± almandine; stilpnomelane rare	Muscovite (phengitic)-paragonite-lawsonite-chlorite-glaucophane-quartz-albite-sphene
Metagraywackes	Quartz-heulandite ± analcime Quartz-albite-laumontite-prehnite-chlorite ± stilpnomelane Quartz-albite-prehnite-pumpellyite-chlorite ± stilpnomelane	Quartz-albite-epidote-muscovite-chlorite ± stilpnomelane Same as above with biotite ± almandine; stilpnomelane absent	Quartz-jadeite-muscovite-chlorite-lawsonite-glaucophane-sphene Same as above + almandine ± epidote
Metacherts	Quartz + iron oxides	Quartz + iron oxides Quartz-piedmontite-muscovite-spessartine Quartz-magnetite-spessartine-stilpnomelane	Quartz-stilpnomelane-spessartine Quartz-crossite-aegirine ± lawsonite

Calcareous	Calcite ± quartz	Calcite-quartz ± tremolite or talc Calcite-dolomite ± tremolite or talc Calcite-zoisite-grossular (andraditic) Calcite-albite-epidote	Aragonite ± lawsonite ± glaucophane Calcite ± relict aragonite
Metabasalts	Spilitic assemblage: albite-chlorite-epidote or pumpellyite + relict augite	Albite-chlorite-epidote ± stilpnomelane Albite-actinolite-epidote-chlorite ± calcite ± biotite	Albite-lawsonite-pumpellyite-glaucophane-chlorite-stilpnomelane-sphene Albite-epidote-glaucophane-omphacite-chlorite-actinolite Albite-lawsonite-clinozoisite-chlorite ± hornblende ± almandine
Serpentinites and derivative magnesian rocks	Chrysotile and/or lizardite ± brucite	Antigorite-brucite ± tremolite Antigorite-calcite-talc Antigorite-diopside-forsterite Talc-magnesite ± tremolite	Antigorite ± tremolite ± talc

of facies transitions. It should be noted that in rocks of very low grade some of the most typical assemblages include metastable phases and so themselves are not in a state of stable equilibrium. Partly for this reason, and partly because of incomplete documentation, the three facies of lowest grade—zeolite, prehnite-pumpellyite, and pumpellyite-actinolite—are treated collectively.

GENERAL REFERENCE

F. J. Turner, *Metamorphic Petrology*, 2d ed. (Washington, D. C.: Hemisphere, 1981).

REFERENCE NOTES

[1] D. G. Bishop, *Bulletin of the Geological Society of America,* vol. 83 (1972): pp. 3177–3197.

[2] T. W. Bloxam, *American Journal of Science*, vol. 41 (1956): pp. 488–496.

[3] B. Levi, *Contributions to Mineralogy and Petrology*, vol. 24 (1961): pp. 30–49.

[4] On New Caledonia, see especially R. N. Brothers, *Contributions to Mineralogy and Petrology*, vol. 46 (1974): pp. 109–127; P. M. Black, *Tectonophysics*, vol. 43 (1977): pp. 89–107. On the California Coast Ranges, see R. G. Coleman and D. E. Lee, *Journal of Petrology*, vol. 4 (1963): pp. 260–301; W. G. Ernst, *Bulletin of the Geological Society of America*, vol. 76 (1965): pp. 879–914; W. G. Ernst, *Journal of Petrology*, vol. 12 (1971): pp. 413–437.

[5] S. O. Agrell and J. M. Langley, *American Mineralogist*, vol. 50 (1965): pp. 278–279.

20

Products of Regional Metamorphism, II: Rocks of Higher Grades

PLAN OF TREATMENT

Metamorphic grade in rocks of broadly similar chemistry, as in metapelites or in metagraywackes, is essentially a function of total entropy of each mineral assemblage—including, it is assumed, an aqueous phase (or in a carbonate assemblage a CO_2-rich gaseous phase). The rocks in an isochemical series of this kind can be arranged artificially in order of increasing grade; thus, metapelites in the sequence chlorite-mica schist, garnet-muscovite-biotite schist, garnet-staurolite-muscovite-biotite schist, garnet-kyanite (or sillimanite) muscovite-biotite schist. And so arranged, the sequence is one of progressively increasing dehydration. This is, in fact, the identical order of zones of progressive metamorphism that have now been mapped in many regions on the basis of pelitic protolithology.

It might seem tempting therefore to consider any other lithologic sequence of decreasing H_2O content as one of progressive dehydration of initially hydrous parent rocks: for example, greenschist $\rightarrow$ amphibolite $\rightarrow$ pyroxene granulite. But this interpretation is unwarranted where the completely unmetamorphosed parent rock is essentially anhydrous—in the above case, basalt or diabase. Each member of a greenschist $\rightarrow$ amphibolite $\rightarrow$ granulite sequence might have been generated independently, at some point on a continuous temperature gradient, by a degree of hydration consistent with equilibrium—in every case from the same essentially anhydrous parent rock. Indeed it has been demonstrated from field evidence, backed by petrographic observations, that in many Archaean terranes anhydrous pyroxene granulates have become hydrated to amphibolite assemblages on a regional scale. Here the metamorphic sequence is retrograde.

We have chosen therefore to group rocks of medium and high grade in categories based on mineralogical criteria related to grade, without specifying possible paths of metamorphism. It follows that there is a perceptible broad correlation between individual rock groups and metamorphic facies; but the facies concept itself is not the primary basis of presentation. The adopted plan of grouping is as follows:

1. Essentially hydrous rocks
   a. pelitic schists
   b. micaceous quartzo-feldspathic schists and gneisses
   c. amphibolites and hornblende-schists
   d. magnesian schists.
2. Essentially anhydrous rocks
   a. quartzites
   b. carbonate rocks (with calcite and/or dolomite)
   c. granulites
   d. eclogites and rare cofacial rocks.

Group 1 includes rocks of the amphibolite facies, excluding those whose principal components are carbonates (calcite, dolomite). Thus, by implication, we have drawn an arbitrary limit between low and higher grades of metamorphism where albite gives way to a decidedly more calcic plagioclase, $An_{>25}$, in assemblages containing an epidote mineral. By definition, groups 2c (with some exceptions) and 2d fall respectively in the granulite facies (implying high temperatures and variable pressures) and the eclogite facies (implying metamorphism at extreme pressures). Groups 2a and 2b span the combined range of the amphibolite and the granulite facies.

ESSENTIALLY HYDROUS ROCKS

High-Grade Pelitic Schists

Much of what has come to be accepted regarding mineralogic pattern in pelitic rocks affected by regional metamorphism comes from the petrographic descriptions of A. Harker (1932). His petrographic account is based largely on observations made of schists from Paleozoic fold belts in the Grampian Highlands, Scotland—the site of G. Barrow's pioneer work on progressive regional metamorphism. Two distinct pelitic parageneses of medium to high grade have been recognized in this region: one developed in the southern and western sector, the other confined to the northeast (as displayed along the northern coast of Aberdeenshire and Banffshire). They have come to be known as parageneses, respectively, of the Barrovian and of the Buchan type. The former clearly reflects a higher pressure regime (perhaps 5–6 kb) than the latter (possibly 2–3 kb). Similar general parageneses are displayed in other metamorphic regions. Both are represented, for example, in different parts of the Appalachian belt of the northeastern United States. The Barrovian type is familiar in the Caledonides of Norway and in the Lepontine Alps. Large sectors of the Hercynian folds of Europe (e.g., in the Pyrenees) and of late Mesozoic belts in Japan conform approximately to the Buchan pattern.

Since the evidence relating to relative pressure rests largely on differences in pelitic mineralogy, it is appropriate here to treat separately the two contrasting types—remembering always that intermediate types are also known and may be equally abundant.

Pelitic Schists of the Barrovian Type

Even at high grades of metamorphism conducive to development of a coarsely crystalline fabric, pelitic schists of the Barrovian type are strongly foliated rocks, whose ready fissility is produced by abundance of subparallel crystals of mica. Segregation layering and lineation are normal elements in the rock fabric. The regularity of all these parallel structures tends, however, to be disturbed by the presence of large porphyroblasts of garnet, staurolite, kyanite, or alkali feldspar.

Micas are invariably present—usually deep-red-brown biotite accompanied by muscovite. In schists of potassic composition the typical assemblage is biotite-muscovite-orthoclase-plagioclase-quartz. But more commonly a high ratio Al_2O_3/K_2O precludes the presence of K-feldspar, and one or more aluminous minerals (garnet, staurolite, kyanite, sillimanite) are then associated with quartz and mica (Figures 20–1 and 20–2). Of these, pink almandine garnet is most typical. It tends to form sharply bounded porphyroblasts, and may enclose spiral trains of inclusions which indicate that the crystals rotated while they were growing (Figure

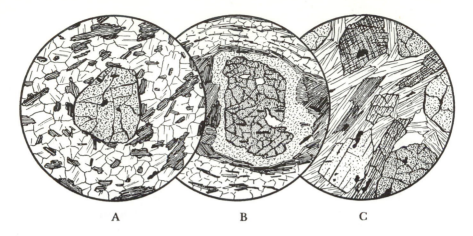

A   B   C

Figure 20–1. High-Grade Pelitic Schists

A. Almandine-biotite-plagioclase schist, sillimanite zone, Scottish Highlands. Diam. 4.5 mm.

B. Staurolite-biotite-muscovite-quartz schist, near Innsbruck, Austria. Diam. 4.5 mm. The central porphyroblast of golden staurolite is marginally altered to finely divided white mica (retrograde metamorphism involving introduction of potassium).

C. Kyanite-staurolite-almandine-muscovite schist with minor biotite and quartz, Gassets, Vermont. Diam. 3 mm. Pale-pink almandine at right and top left margins; golden staurolite, lacking cleavage, at top right and lower right; kyanite prisms have well-developed cleavage (the crystal at lower left is cut parallel to {100} and shows a nearly centered negative bisectrix figure; extinction is at 30° to the cleavage).

16–1A). Staurolite, conspicuous by virtue of its golden-yellow pleochroism, cross-shaped twins, and coarse porphyroblastic habit, is more restricted in its occurrence. It is almost always associated with garnet and micas in rocks with high Fe/Mg.

Kyanite appears mainly in garnet-mica schists and mostly near the top of the temperature range within the amphibolite facies. Kyanite crystals tend to develop strong preferred orientation with {100} in the plane of schistosity. Consequently, in a thin section cut parallel to the schistosity colorless prismatic sections of kyanite are easily distinguished by their orientation normal to α, the presence of two cleavages intersecting at 90°, and inclined extinction $\gamma \wedge z = 30°$); but in sections cut normal to the schistosity kyanite prisms give nearly straight extinction and may be hard to distinguish from orthorhombic pyroxene (which, however, is unknown

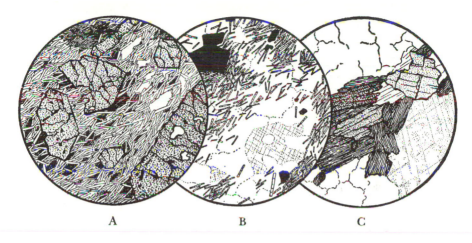

Figure 20-2. High-Grade Pelitic Schists

A. Almandine-staurolite-muscovite schist, Morgan Station, Pennsylvania. Diam. 6 mm. Minor minerals are quartz and brown biotite, the latter locally replaced retrogressively by chlorite.

B. Sillimanite-quartz schist, Bellinzona, Italy. Diam. 2 mm. Contains minor iron oxide and microcline (lower right).

C. Kyanite-biotite-quartz-andesine schist, Bellinzona, Italy. Diam. 3.5 mm. Quartz shows undulatory extinction; andesine is somewhat turbid (at right). Rare flakes of muscovite (near upper crystal of kyanite), and a single zircon enclosed in biotite.

in high-grade metapelites). Sheaves of sillimanite prisms are typical of high-grade pelitic rocks beyond the zone of staurolite. The first appearance of this mineral may simply mark the point where the metamorphic field gradient crosses the kyanite → sillimanite inversion curve (cf. Figure 16–7, curve 3a). Such seems to be the case in the Barrovian zonal sequence in Scotland. Sillimanite schists there and in the sillimanite-orthoclase zone of the northern Appalachians contain sillimanite, K-feldspar, and muscovite in apparent equilibrium. And sillimanite in such rocks may take the form of fine needles ("fibrolite") dispersed within grains of quartz. At maximum grade in other regions sillimanite and K-feldspar develop by gradual elimination (dehydration) of muscovite in the presence of quartz (cf. Figure 16–7, curve 4). A typical assemblage at this stage is quartz-biotite-oligoclase-orthoclase-sillimanite-almandine (± muscovite).

The usual accessories in pelitic schists of medium to high grade are apatite, tourmaline, zircon, and opaque iron-titanium oxides.

Pelitic Schists of the Buchan Type

In the Buchan sector of the Scottish Dalradian terrane medium- and high-grade pelitic rocks are much less deformed than in the Barrovian zone to the south. They have the aspect of hornfelses rather than schists, and clearly retain such textural relics as bedding undisturbed by metamorphic strain. The same may be said of rocks of the same type in the Pyrenees, in the Precambrian of southern Finland (Orijärvi district), and in the late Mesozoic Abukuma belt of Japan. Typical representatives from northern Scotland are illustrated in Figure 20–3. Mineralogically, they are distinguished especially by abundant, irregularly bounded porphyroblasts of andalusite, packed with small inclusions of quartz, micas, and in some cases dusty graphite. Sieved porphyroblasts of either cordierite or staurolite also are commonly present. The enclosing matrix in all cases consists mainly of granular quartz (with or without plagioclase) and abundant small flakes of muscovite and red-brown biotite.

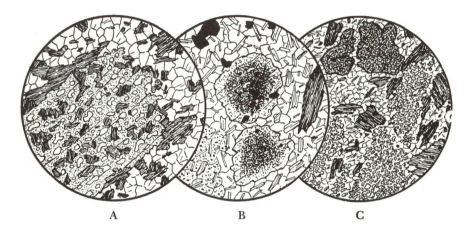

A   B   C

Figure 20–3. Hornfelsic Andalusite Schists of Northeastern Scotland

A. Andalusite-biotite-quartz-plagioclase schist, Coreen Hills, Aberdeenshire. Diam. 2.5 mm. The porphyroblast is of andalusite.

B. Andalusite-muscovite-biotite-cordierite-quartz schist, Coreen Hills, Aberdeenshire. Diam 2.5 mm. The central porphyroblasts are of andalusite, crowded with graphitic inclusions. Cordierite, with dark inclusions surrounded by yellow pleochroic haloes, occurs near the lower left edge.

C. Andalusite-staurolite-biotite schist, west of Banff. Diam. 2.5 mm. Spongy porphyroblasts of staurolite (above) and of andalusite and large flakes of biotite, in a granoblastic matrix of quartz, muscovite, and biotite.

Modifying Features

In particular situations the simple mineralogical and textural patterns of high-grade pelitic schists as just described may be modified in either of two ways: by local alkali metasomatism in high-grade schists adjoining synchronously emplaced granitic plutons and migmatitic complexes, or by irregularly distributed but often widespread minor effects of retrograde metamorphism.

Local development of porphyroblasts of K-feldspar or of oligoclase near granitic plutons is common in regions where plutonism and regional metamorphism were more or less synchronous, for example, in the vicinity of "Older Granite" complexes in the Dalradian. Feldspar porphyroblasts in such rocks appear to have formed at the expense of mica, in places where alkali has been introduced from granite or some other external source. In the field it is sometimes possible to demonstrate complete local transition from pelitic mica schist to granite, with porphyroblastic schist and feldspathic augen gneiss as intermediate stages. Such "granitized" pelitic schists tend to show microscopic intergrowths of vermicular quartz with muscovite, biotite, or oligoclase.

Pelitic mineral assemblages are especially sensitive to change of temperature and pressure. Instances of polymetamorphism, therefore, are by no means unusual in such rocks and leave their record in mixed assemblages of minerals. In polymetamorphic pelitic mica schist, cordierite or andalusite, typical of hornfelses, may be associated with staurolite, kyanite, and garnet formed during a different metamorphic episode. From the degree of deformation and metasomatic alteration exhibited by the cordierite and andalusite, it is possible in most cases to tell whether these minerals belong to an episode of metamorphism that preceded, or to one that followed, the deformational phase of metamorphism responsible for staurolite and garnet. A very common case arises by mineralogical adjustment of high-grade pelitic schists to falling temperature (retrograde metamorphism) under the catalytic influence of percolating aqueous fluids or of superposed local strain. Among the changes thus produced are chloritization of biotite and garnet, conversion of staurolite to chloritoid or chlorite, and partial replacement of kyanite by muscovite. Where chlorite completely replaces biotite, its secondary origin is indicated by enclosed webs of acicular rutile or strings of granular sphene representing titanium set free in the process, or by persistence of pleochroic haloes inherited from the parent biotite.

Quartzo-Feldspathic Schists and Gneisses

High-grade foliated rocks composed mainly of quartz and feldspars are coarser grained and poorer in mica than pelitic schists, and therefore are less fissile and less regularly schistose. Many may be termed gneisses (as defined on p. 448); some indeed are transitional into granulites—rocks typical of, but not confined to, granulite facies.

Quartzo-feldspathic schists are derived mainly from arenaceous sediments and siliceous volcanic rocks. Typically they consist of quartz, plagioclase, and orthoclase (or microcline), with subordinate muscovite and biotite aligned in and defining the foliation. Some contain prominent porphyroblasts of almandine. Many but not all of the gneissic rocks are metamorphosed acid plutonic rocks. Clinozoisitic epidote is typical of rocks in which the plagioclase is relatively sodic (An_{25} to An_{35}). In rocks transitional from the greenschist to the amphibolite facies, just beyond the garnet isograd in the Haast Schist terrane of southwestern New Zealand, two markedly different varieties of plagioclase coexist in apparent equilibrium—albite, An_2–An_7, and oligoclase, An_{25}–An_{30}. The first of these, by reason of its low refractive index, is conspicuous. In contrast, the oligoclase phase was long overlooked since its refractive index is close to that of associated quartz, and its almost monoclinic optic symmetry renders lamellar twinning virtually invisible; but interference figures yielded by sections normal to an optic axis reveal its true identity. Disappearance of albite marks a sharply defined oligoclase isograd; and from this grade upward, quartzo-feldspathic schists are in the amphibolite facies. Scattered grains of almandine are common in rocks lacking K-feldspar. Increase in micas, decrease in feldspar, and in some rocks the presence of minor kyanite mark a gradual transition into pelitic schists. Appearance of hornblende and increase in epidote indicate transition into amphibolites.

In some quartzo-feldpathic schists thin dark micaceous layers parallel to the foliation superficially suggest relict bedding inherited from a sandstone. While the general trend of schistosity imposed during deformational metamorphism is, indeed, likely to correspond with the bedding of the parent rock, it is only rarely that details of cross-bedding and graded bedding will survive high-grade metamorphism involving strong internal deformation. However, structures superficially resembling cross-bedding can develop during metamorphism by shearing out of small-scale folds. Some of the "cross-bedded" fabrics of quartzo-feldspathic schists certainly have such an origin. True bedding is marked, however, by the thin strings of zircon, tourmaline, and opaque oxides (inherited concentrations of heavy minerals) sometimes observed microscopically in quartzo-feldspathic schists. Both true cross bedding, thus defined, and simulated "cross bedding" of deformational origin are well represented among the Moine schists of Scotland.

Granitic and Granodioritic Gneisses

The principal mineral constituents of granitic and granodioritic rocks—quartz, feldspars, micas, hornblende—are stable in the presence of aqueous fluids over the complete temperature-pressure range of the amphibolite facies. Rocks of the same chemical and mineralogical composition, but with a strong metamorphically imprinted gneissic fabric make up the main lithologic element in extensive segments of ancient cratons. There can be little doubt that these rocks are metamorphosed, in some cases repeatedly metamorphosed, plutonic elements of the ancient crust. They are usually termed granitic or granodioritic *gneisses*.

Amphibolites

Amphibolites are foliated metamorphic rocks composed essentially of hornblende and plagioclase. They are diagnostic of the amphibolite facies; and they are among the commonest rocks formed by regional metamorphism of moderate-to-high grade. Segregation layering may or may not be present; and megascopic schistosity is not necessarily conspicuous in types containing no mica. Beneath the microscope, however, there is usually an obvious preferred orientation of hornblende prisms, which defines a schistosity, a lineation, or both.

In amphibolites derived from basic igneous rocks, hornblende and plagioclase tend to be equally abundant. Almandine garnet, epidote, and biotite may all be present. Quartz and biotite are generally minor minerals, though both may be conspicuous in rocks derived from tuffs. Kyanite is found in a restricted group of rocks rather rich in Al_2O_3. Sphene (or rutile), apatite, and opaque oxides are almost always present; some rocks contain accessory sulfides. The hornblende of amphibolites is strongly colored (γ = deep blue-green or, at high grades of metamorphism, greenish-brown). It occurs in prisms bounded by {110}, with ragged terminations, and tending to be crowded with inclusions of quartz or iron oxides. The plagioclase generally ranges from An_{25} to An_{40} and occurs in equant xenoblastic grains, either untwinned or showing a few lamellae of either the albite or the pericline type. The epidote, usually in sharply crystallized prisms, is either colorless or very pale yellowish but nevertheless highly birefringent. It may enclose brownish pleochroic cores of allanite. Sphene may be in the form of coarse wedge-shaped crystals or in clusters of droplike granules. At the highest grades of metamorphism ilmenite tends to take the place of sphene; rutile is much rarer. Typical amphibolites are illustrated in Figure 20–4).

Where amphibolite has formed by metamorphism of gabbro or diabase without strong internal deformation (as in the inner portions of marginally deformed sills), it may be so weakly foliated that it can hardly be distinguished with certainty from strictly igneous diorite, which is also a hornblende-plagioclase rock. Indeed, the term *epidiorite* has been gen-

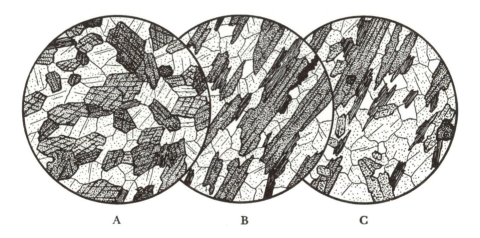

A　　　　　　　B　　　　　　　C

Figure 20–4. Amphibolites Derived from Basic Igneous Rocks

A. Archaean amphibolite, Grand Canyon, Arizona; section cut perpendicular to schistosity and to lineation. Diam. 2.5 mm. Idioblastic deep-green hornblende, xenoblastic oligoclase, brown biotite, accessory sphene.

B. Same rock as A; section cut perpendicular to schistosity but parallel to lineation. Diam. 2.5 mm. Contains a few grains of clear quartz.

C. Epidote-amphibolite, Lake Manapouri, New Zealand. Diam. 2.5 mm. Section cut parallel to schistosity. Main minerals are oligoclase, brown biotite, and deep-green hornblende. Stumpy prisms of colorless, highly birefringent epidote are rather less plentiful (e.g., at lower edge, in feldspar at right, and in biotite at left). Sphene is a constant accessory (right edge). Note alignment of hornblende and biotite parallel to the lineation, which is steeply inclined.

erally used to cover rocks of this category. In such rocks the hornblende may retain structural features, such as parting on {001} or on {100}, inherited from the pyroxene it has replaced; in the plagioclase the originally euhedral form, tabular habit, and complex twinning (e.g., Carlsbad-albite) that are characteristic of igneous plagioclase may still be preserved. Careful examination, moreover, should show several features characteristic of amphibolites but not of diorites: great abundance of hornblende; simple form, ragged terminations, sieve structure, and absence of twinning in this hornblende; relatively sodic composition and partly recrystallized condition of plagioclase; relict ilmenite represented by rings of granular sphene surrounding small cores of opaque oxide; metamorphic constituents such as epidote, garnet, and granular quartz; and some degree of foliation.

Amphibolites derived from mixed calcareous or tuffaceous sediments also contain blue-green hornblende as their most conspicuous mineral. Plagioclase is less abundant, and quartz and biotite usually more so, than in amphibolites of basic igneous origin. Especially characteristic are the presence of abundant greenish diopside (cf. Figure 20–5A) and a general absence of almandine garnet. Diopside may also occur in nongar-

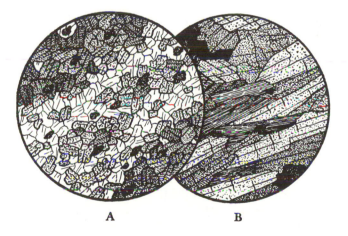

A   B

Figure 20–5. Amphibolite and Amphibole Schist

A. Hornblende-diopside amphibolite, Sierra Nevada, California. Diam. 1 mm. Green hornblende, pale-green diopside (shown in lighter stipple), andesine, and minor sphene.

B. Anthophyllite-hornblende-biotite rock, Snarum, Norway. Diam. 3.5 mm. Colorless anthophyllite is associated with dark-green hornblende (shown closely stippled) and brown biotite. Opaque oxides enclose rutile.

netiferous amphibolites of igneous parentage, but is much less common and never very abundant. Epidote is conspicuous in many metasedimentary amphibolites. Sphene and apatite are normal accessories.

High-Grade Magnesian Schists

A sequence of steps in the progressive dehydration of serpentinites has been traced in magnesian schists of the Lepontine Alps by V. Trommsdorff and B. W. Evans, who correlated the resulting parageneses with zones of progressive metamorphism based on both pelitic and dolomitic lithologies. Chemically expressed, the critical steps in evolution of magnesian schists in this region are represented by these equations:

1. 2 diopside + 5 antigorite → tremolite + 6 forsterite + 9 H_2O
2. 5 antigorite → talc + 6 forsterite + 9 H_2O
3. 9 talc + 4 forsterite → 5 anthophyllite + 4 H_2O
4. anthophyllite + forsterite → 9 enstatite + H_2O
5. tremolite + forsterite → 2 diopside + 5 enstatite + H_2O

All these reactions take place within the amphibolite facies. The first two lead to increase in content of forsterite and elimination of antigorite; the products of reactions 4 and 5 are anhydrous assemblages. The complete reaction sequence is one of progressive dehydration. Critical resulting magnesian parageneses, in order of increasing grade, are:

> Antigorite-forsterite-tremolite; less commonly (in somewhat more siliceous rocks) antigorite-tremolite-talc;
> Forsterite-talc-tremolite; or forsterite-anthophyllite-tremolite;
> Forsterite-enstatite-tremolite (± green spinel);
> Rarely: forsterite-enstatite-diopside (± green spinel).

Among carbonate-bearing magnesian schists in this region the commonest low-grade type consists of antigorite, tremolite, and calcite (in the greenschist facies). In rocks of higher grade the pair antigorite-calcite has been eliminated (by combined dehydration and decarbonation) to yield first magnesite-anthophyllite-tremolite (in the kyanite zone of metapelites) and finally (beyond the sillimanite isograd) magnesite-enstatite-tremolite. Some olivine-rich rocks (*olivinites* of the Italian Alps) possibly are products of extreme dehydration of serpentinite (cf. reactions

4 and 5 above). They commonly consist of olivine with varying amounts of diopside, enstatite, actinolite, and garnet (always with accessory picotite).

In many of the magnesian schists of the Lepontine Alps, newly generated forsterite has a characteristic crystalline habit in the form of long bladed crystals enclosed in a finer-grained matrix of hydrous phases (especially antigorite).

Not all magnesian schists are derivatives of serpentinites or ophicarbonates. Some such schists, typically developed by local Mg-metasomatism of semipelitic and other rocks in the vicinity of granitic plutons that were emplaced synchronously with regional metamorphism, contain aluminous magnesian phases in abundance. Typical of these are coarse, commonly porphyroblastic cordierite-anthophyllite rocks with or without biotite or almandine. They are well represented in the classic low-pressure amphibolite-facies paragenesis of the Orijärvi district in southern Finland. A somewhat comparable anthophyllite-hornblende-biotite schist from Norway is illustrated in Figure 20–5B.

ESSENTIALLY ANHYDROUS ROCKS

Quartzites

Quartzites, as the term is here used, consist almost entirely of recrystallized quartz. Most commonly they are derived from siliceous sediments——quartz arenites or pure cherts. The grade of metamorphism in any specific case can be inferred from the nature of associated rocks of other lithology, reinforced in some cases by the minor mineral components of the quartzites themselves. Relict sedimentary textures are common, especially in rocks of lower grades. Unmetamorphosed quartz-cemented quartz arenites are termed *orthoquartzites*.

High-Grade Marbles and Calc-Granulites

In the amphibolite and granulite facies, as in chemically similar rocks of lower grade, relatively pure limestones and dolomitic sediments have been converted to marbles composed mainly of calcite with or without dolomite. These high-grade marbles differ but slightly from those of lower grade as regards texture. Coarser-grained rocks perhaps are more abundant, and evidence of postcrystalline deformation (strongly elongate grains, dense lamellar twinning) on the whole is less conspicuous in high-grade rocks.

It is in rocks derived from carbonate sediments with significant siliceous and clay impurities that differences in grade become manifest, for the metamorphic silicate components of impure marbles and calc-granulites reveal a sequence of different responses to increasingly high temperature at any given level of pressure. In reactions involving carbonates, dolomite, except under some conditions in the granulite facies, is the principal participant. Dolomite thus is reduced in quantity, even to the point of elimination, in the metamorphic product. Calcite remains stable in the presence of silica except rarely at maximum temperatures and minimum pressures in the granulite facies (cf. Figure 16–6).

Progressive metamorphism of both calcite- and dolomite-rich interbeds among dominantly pelitic schists of the Lepontine Alps has been studied in detail by E. Wenk, V. Trommsdorff, and their associates. They have mapped isograds for both lithologies. The trend of these isograds is more or less parallel to that of staurolite, kyanite, and sillimanite isograds mapped by E. Niggli for pelitic lithology in the same region.[1] In calcite marbles a plagioclase and quartz are common. The composition of plagioclase varies from about An_{28}, close to the kyanite isograde, to values as high as An_{80} or even An_{90} within the sillimanite zone. Dolomitic marbles in the greenschist facies contain only talc as the newly generated silicate phase. But just beyond the kyanite isograd (i.e., within the amphibolite facies as expressed by associated metabasalts) talc is eliminated and tremolite appears in abundance at the expense of dolomite. Resulting calc-schists consist mainly of calcite and tremolite, the former being the more abundant. At higher grades, still within the kyanite zone and continuing beyond the sillimanite isograd, similar schists now consist of calcite, diopside, and forsterite (some with chondrodite), or of calcite, diopside, and tremolite.

The variety of minor phases in marbles and calc-schists the world over is considerable and expresses, in particular, the influence of varying amounts of clay and other impurities in the parent sediments. So along with diopsidic pyroxene and amphibole there appear as accessories many newly generated minerals—phlogopite, plagioclase, or microcline, and combinations of many Ca-Al silicates, particularly grossular-andradite, clinozoisite (or zoisite), and scapolite. These tend to be concentrated in pods or knots scattered through otherwise much purer calcite marbles. Sphene is a widespread accessory; apatite and graphite are characteristic but less common. Here we note just a few of many widely recorded assemblages of the amphibolite facies.

1. Calcite-biotite-quartz-clinozoisite-sphene (Figure 20–6A).
2. Calcite-zoisite-scapolite-quartz—in knots set in calcite marble (Figure 20–6B).
3. Calcite-diopside-scapolite-phlogopite—in a calcite marble from Doubtful Sound, southwest New Zealand. Here parallel alignment of golden phlogopite flakes in a rock composed largely of calcite imparts distinct foliation to the hand specimen. Flaky metallic-looking graphite in a similar rock from the same general region has a similar effect on its macroscopic appearance.
4. Calcite-microcline-zoisite-plagioclase-quartz-hornblende in thin calc-silicate bands—in metasediments of the kyanite and sillimanite zones, Scottish Highlands.

Marbles and calc-granulites are widespread in many regions of Precambrian granulite-facies metamorphism. Typical are those of the central highlands of Sri Lanka (Ceylon) as described particularly by P. G. Cooray

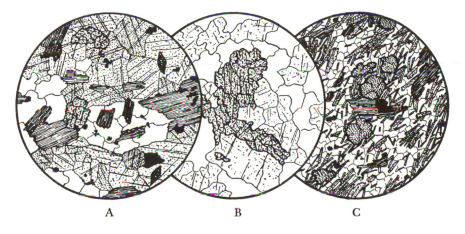

A   B   C

Figure 20–6. High-Grade Calc-Schists

A. Calcite marble with quartz, biotite, clinozoisite, and sphene (lozenge-shaped crystals), Grantown series, Strathspey, Scotland. Diam. 2.5 mm.

B. Zoisite-scapolite-quartz rock, Doubtful Sound, New Zealand. Diam. 2.5 mm. Associated rocks are marbles and high-grade amphibolites.

C. Garnet-hornblende rock, Loch Katrine, Scotland. Diam. 2.5 mm. Pink almandine, deep-green hornblende, red-brown biotite, quartz, and iron oxide.

and by D. J. A. C. Hapuarachchi.[2] Massive beds of dolomitic limestone have been converted to marbles consisting of calcite ± dolomite, with subordinate forsterite, phlogopite, and in some rocks a purple spinel. Impure carbonate bands interbedded with pelitic and quartzo-feldspathic granulites are now represented by calc-granulites consisting solely of silicates. Most consist principally of diopside, scapolite, plagioclase, and in some rocks grossular-andradite. Sphene is almost ubiquitous, and apatite prominent in some rocks. More calcic rocks confined to regions where pressures seem to have been relatively low consist of calcite, scapolite, and wollastonite (± plagioclase or sphene, or both).

Granulites

Granulites include pelitic and quartzo-feldspathic metasediments and rocks of basic igneous parentage most of which bear the mineralogical imprint of the granulite facies. All are essentially anhydrous except for the presence of phlogopitic mica in some metasediments and brown hornblende in some basic rocks (pyroxene granulites). Granulites are almost, but not completely, restricted to the Precambrian continental shields. Where not converted to rocks of the amphibolite facies by regionally superposed retrograde metamorphism, they in fact comprise one of the most conspicuous elements in the ancient continental crust.

Quartzo-Feldspathic Granulites

Quartzo-feldspathic granulites (Figure 20–7B) are characterized by "granulitic fabric" and by the distinctive high-temperature assemblages quartz-orthoclase-plagioclase-kyanite (or sillimanite) and quartz-orthoclase-plagioclase-garnet. Granulitic fabrics have regular planar schistosity determined by alternation of parallel, greatly flattened lenticles of rather coarse quartz, with layers of finely crystalline quartz and feldspar. The garnet of granulites is pink almandine, somewhat higher in MgO than the garnet of the amphibolite facies. Prisms of kyanite or of sillimanite lie in the plane of schistosity. The orthoclase is invariably perthitic. Rutile is an almost constant accessory in contrast with sphene, which is the characteristic titanium mineral of the greenschist facies and common in the amphibolite facies. Where boron has been introduced during metamorphism, tourmaline is present; it is rarely accompanied by the Mg-Al borosilicate prismatine (Figure 20–7A). This is a colorless pris-

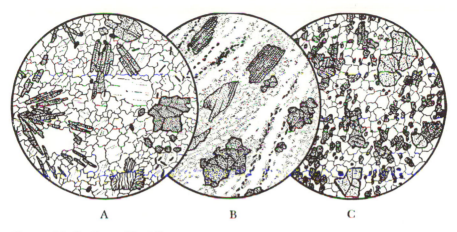

A   B   C

Figure 20–7. Granulites, Saxony

A. Prismatine-tourmaline-quartz granulite, Waldheim. Diam. 7 mm. The main minerals are quartz, plagioclase, and prismatine. In the lower right quadrant are golden tourmaline and coarse sillimanite (lower edge). Nearly opaque needles and prisms of brown rutile occur throughout the section.

B. Almandine-kyanite-quartz granulite, Rohrsdorf. Diam. 2 mm. The porphyroblasts are of kyanite and almandine. The matrix consists of layers alternately rich in quartz (clear) and orthoclase (shown stippled). The texture is coarser in the quartzose than in the feldspathic layers. Biotite flakes occur in fine-grained streaks parallel to the foliation.

C. Pyroxene granulite, Hartmannsdorf. Diam. 2 mm. Porphyroblasts of pink almandine in a granoblastic matrix of hypersthene, diopside, and andesine.

matic mineral, with medium refractive index and birefringence, easily recognized by parallel extinction, negative elongation, small optic axial angle, and negative character.

High-grade metacherts include quartz-rich granulites that completely lack feldspar. Interesting examples are quartz granulites that occur as blocks a few meters wide enclosed in granodioritic hornblende-biotite gneisses (Amîtsoq gneiss)—the most ancient terrestrial rocks yet recorded—in southwestern Greenland. The granulite blocks, themselves presumably in the amphibolite not the granulite facies, display two closely related kinds of mineral assemblage: quartz-diopside (hedenbergitic)-cummingtonite-garnet; and quartz-hedenbergite-ferrohypersthene-garnet. Cummingtonite (monoclinic) carries significant amounts of both Mg and Fe^{2+} and is distinguished from other amphiboles by its positive optic sign. The completely dehydrated state of the second assemblage (one with high Fe/Mg) reflects the relatively low temperature at which ferrous cummingtonite becomes unstable with respect to orthopyroxene. The petrography of these rocks raises an intriguing question. At some very early date in terrestrial history (earlier even than the 3.8 billion year date of Amîtsoq metamorphism) could ankeritic and ferruginous cherts already have been deposited on the floor of a primitive ocean?

The presence of kyanite rather than sillimanite in metasedimentary granulites indicates high pressures—above 7 kb judging from the point of intersection of respective univariant curves for dehydration of muscovite and sillimanite ⇆ kyanite inversion. In contrast, the coexistence of cordierite and garnet in many granulite terranes suggests relatively low pressures—in some cases perhaps only 3–4 kb. Typical cordierite granulites of pelitic parentage from the Grenville province, Ontario, consist of quartz, orthoclase, plagioclase, sillimanite, cordierite, almandine, and biotite. Otherwise similar cordierite granulites from other regions lack any hydrous phase—even biotite.

We have already said something of *charnockites*—so widespread in Precambrian granulite terranes of Madras, Sri Lanka, central Australia, and elsewhere—as a group of quartzo-feldspathic rocks bearing hypersthene and diopside, some of which are thought to be plutonic rocks of the granitic family, strictly igneous in their origin. Other "charnockites" more probably are quartzo-feldspathic granulites (some of them initially acid igneous rocks) whose present mineralogical and textural character is due to metamorphism in the granulite facies. "Charnockites" of this type, composed mainly of quartz, K-feldspar, sodic plagioclase, and garnet, are illustrated in Figure 20–8B by a hornblende-bearing variety from Sweden. The charnockite shown in Figure 20–8C could equally well be of metamorphic or purely igneous origin.

Pyroxene Granulites

The pyroxene granulites (Figure 20–7C) are metamorphic rocks of high grade (granulite facies) composed principally of plagioclase, hypersthene, diopside, and in some cases almandine garnet. Chemically they are equivalent to rocks of the gabbro and basalt family. Mineralogically, also pyroxene granulites that do not contain garnet may be almost identical with norites; for at the high temperatures and pressures inferred for the granulite facies, mineralogical convergence of igneous and metamorphic rocks can be expected.

The plagioclase of pyroxene granulites is generally andesine or labradorite. Like that of amphibolites, it commonly shows either albite or

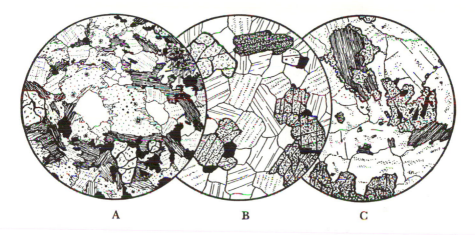

Figure 20–8. Cordierite-Plagioclase Schist and "Charnockites"

A. Cordierite schist, Schwarzwald, Baden. Diam. 3 mm. Cordierite containing pleochroic haloes around inclusions of zircon and sphene, and marginally altered to pale-green chlorite ("pinite"); lamellar-twinned oligoclase; clear quartz; dark-brown biotite locally interleaved with muscovite; almandine garnet (below) and sillimanite needles (upper right); abundant graphite; minor pyrite. The complex mineral composition suggests internal disequilibrium resulting from repeated metamorphism over a considerable range of high temperatures.

B. "Charnockite," Varberg, Sweden. Diam. 3 mm. Garnet and green hornblende (upper part of section), diopside (lower left), and hypersthene (right). Accessory iron oxides and apatite. The colorless minerals are quartz, perthite, and andesine.

C. Charnockite, Stillwell Island, Adelie Land, Antarctica. Diam. 2 mm. Biotite with a rim of garnet (upper left), garnet with inclusions of ilmenite (on right), hypersthene (below). The colorless grains are mainly perthite; a little quartz is also present.

pericline twinning, and lacks the complex twins and euhedral form that are usual in igneous plagioclase. Both types of pyroxene are found by chemical analysis to contain appreciable Al_2O_3. The diopside of granulites has no distinctive optical characteristics, but the hypersthene shows unusually strong pleochroism (from pink to pale green or pale yellow)—attributable not to high Fe/Mg, but to relatively low temperature compared with that of volcanic crystallization. The usual accessories are rutile and magnetite. Some pyroxene granulites contain minor olive-green hornblende, and this cannot in all cases be attributed to retrograde metamorphism (i.e., transition to the amphibolite facies).

Since platy and prismatic minerals are generally lacking, pyroxene granulites tend to be massive rocks with but poorly defined foliation. In some of these rocks, however, there may be rough segregation of feldspathic and pyroxenic bands.

Special types of pyroxene granulites include plagioclase-rich types formed by metamorphism of anorthosite. Others rich in enstatite and carrying olivine are high-grade metaserpentinites or peridotites that have been reworked by metamorphism in the granulite facies. A clue to the metamorphic imprint in some of these rocks (e.g., in the Archaean of Southern India and western Greenland) is the presence of the rare silica-deficient Mg-Al silicate sapphirine. This mineral, blue in hand specimen, can be recognized optically by faint pleochroism in pale blues and yellows, high refractive index, low birefringence, and strong dispersion; $2V_{\alpha}$ is medium to high. Though so rare, sapphirine is now recognized as a highly characteristic component of magnesian silica-poor rocks in granulite terranes.

Rocks of the Amphibolite-Granulite Facies Transition

In many parts of the continental shields, complete zonal transitions have been mapped between rocks of the amphibolite and the granulite facies. Rocks in extensive transitional zones of this kind resemble granulites but contain hydrous phases. Metapelites consist of sillimanite, almandine, cordierite, biotite, K-feldspar, and quartz. The typical basic assemblage is plagioclase-hornblende (typically a brownish variety) -diopside-hypersthene. In these transitional rocks sphene has usually been displaced by rutile. Texturally many such rocks, pelitic or basic, can be called schistose granulites.

Eclogites and Related Rocks

Eclogites

Eclogites are the diagnostic rocks of the eclogite facies and by far the most familiar of its members. In their simplest form they are virtually anhydrous rocks composed essentially of omphacite and magnesian garnet. Many, perhaps most, eclogites now exposed at the earth's surface have been transported tectonically in their present state directly from sites at the base of the continental crust or in the upper mantle—some, judging from the presence of diamond as a minor primary constituent, from depths of 150 km or more. However, in blueschist terranes there repeatedly occur blocks of metamorphic eclogite of either basaltic or diabasic parentage. For example, in the Eoalpine blueschist province of the Alps, direct but still incomplete transitions have been traced from basaltic pillow lavas to eclogites. And in California, heavy veining of certain blueschist blocks of basaltic composition with streaks of omphacite and almandine-spessartine garnet suggests a similar transition in progress. All such rocks must be of crustal origin.

Whatever their source or origin—mantle or deep crust, igneous or metamorphic—eclogites occur in many parts of the world (mostly in Phanerozoic mobile belts) as isolated bodies from a meter or so to several tens of kilometers in diameter. In virtually all cases they are "out of context" in relation to adjacent or enclosing rocks in that they belong to an extreme metamorphic facies of very high pressure. Small masses of eclogitic rock also occur in explosively emplaced igneous breccia pipes, notably those whose igneous components are kimberlite (cf. p. 254) or nephelinic ultrabasic volcanic rocks (cf. p. 200). In both situations eclogite bodies obviously have been mechanically transported to their present sites from much deeper levels.

Eclogites are remarkable for their high density, absence of feldspar, and generally simple mineralogical composition. In bulk chemistry they closely approximate basalts. The two principal components are a pyroxene (omphacite) of the diopside-jadeite series, bright green in hand specimen, and coarsely crystalline red garnet—a magnesian member of the pyrope-almandine series. In thin section the pyroxene is almost colorless with $\gamma \wedge [001]$ 38°–42°, $2V_\gamma = 20°$–80°. Some European eclogites contain, in addition, a pale green amphibole (possibly a retrograde but perhaps a primary constituent). Others carry plentiful kyanite (Figure 20–9A). The associated minor oxide phase invariably is rutile, deep brown and nonpleochroic in thin section.

Their anhydrous composition alone suggests that the environment of their origin itself is water-free; and experimental evidence shows that in the presence of H_2O, even at the very high pressures, the principal phase assemblage of eclogite is unstable in comparison with that of hornblendic amphibolite. This conclusion is clinched by petrographic observations, for eclogite blocks of blueschist belts in California, the Alps, and elsewhere show all stages of retrograde alterations. Garnets show partial or complete replacement by a variety of lower-grade minerals: chlorite in

Table 20–1. High-Grade Mineral Paragenesis in Relation to Facies of Regional Metamorphism (Selected Mineral Assemblages)

Rock Type	Amphibolite Facies	Granulite Facies	Eclogite Facies
Metapelite (micas predominant) and quartzo-feldspathic rocks (quartz and feldspars predominant)	Muscovite-biotite-quartz-plagioclase ± orthoclase[a]-almandine ± staurolite ± kyanite or sillimanite ± chlorite ± epidote Same as above, with cordierite and andalusite as Al_2SiO_5 polymorph[b]	Quartz-K-feldspar–plagioclase-sillimanite (or kyanite)-almandine-phlogopite Same plus cordierite (kyanite excluded)[c]	Quartz-jadeite-phengite-zosite-pyrope-rutile
Granitic	Quartz-plagioclase-orthoclase (or microcline)-biotite ± hornblende or muscovite	Quartz-orthoclase (or microcline)-plagioclase-hypersthene-augite-almandine	
Metacherts	Quartz-diopside (hedenbergitic)-hypersthene-garnet Quartz-diopside-hedenbergite-cummingtonite-garnet	Quartz-hedenbergite-hypersthene-fayalite-magnetite	
Calcareous	Calcite-tremolite-quartz Calcite-diopside-quartz Calcite-diopside-tremolite Calcite-dolomite-forsterite ± clinohumite Calcite-tremolite-forsterite-phlogopite Zoisite-scapolite-quartz Calcite-plagioclase (An$_{>20}$) Diopside-zoisite-plagioclase ± hornblende	Calcite-dolomite-forsterite ± spinel Calcite-diopside-wollastonite[c] Diopside-scapolite-bytownite-grossular-andradite	Garnet (magnesian grossular)-omphacite ± kyanite

Metabasalts and metagabbros	Hornblende-plagioclase ± biotite ± almandine Hornblende-plagioclase-diopside ± almandine Hornblende-plagioclase-epidote ± quartz	Plagioclase-diopside-hypersthene-rutile ± spinel ± sapphirine	Omphacite-pyrope-almandine-rutile ± kyanite ± amphibole
Magnesian schists and granulites	Antigorite-forsterite-tremolite Forsterite-talc-tremolite Forsterite-anthophyllite-tremolite Forsterite-enstatite-tremolite Magnesite-anthophyllite (or enstatite)-tremolite Cordierite-anthophyllite[c]	Forsterite-enstatite-diopside ± spinel	Forsterite-enstatite-diopside-pyrope-spinel[d]

[a] Orthoclase is seldom found (and then only at maximum grades in the amphibolite facies) in assemblages that include kyanite or sillimanite.
[b] Stable only at relatively low pressures.
[c] This is a low-pressure assemblage.
[d] Mineral paragenesis of peridotite enclosing streaks and bands of eclogite (both probably of mantle origin).

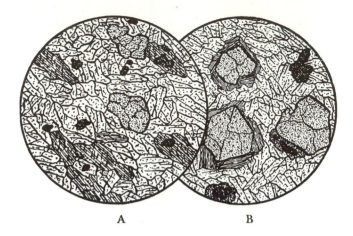

A   B

Figure 20–9. Eclogites

A. Kyanite eclogite, Sulztal, Tyrol. Diam. 3 mm. Pink pyrope, colorless ompha-
    cite, and kyanite, with accessory rutile. Crystals of kyanite (with closely spaced
    cleavage cracks) show strong preferred orientation.

B. Eclogite, closely associated with serpentinite, near Healdsburg, Coast Ranges,
    California. Diam. 3 mm. Idioblastic pink garnets rimmed with chlorite; abun-
    dant colorless omphacite; deep-brown rutile rimmed with granular sphene.
    Sphene and chlorite (and in other sections glaucophane) are products of
    incipient retrograde metamorphism.

many cases, lawsonite, albite, or aragonite in others. Omphacite likewise
becomes replaced by mixtures of glaucophane and epidote; relics of
rutile are enveloped in encroaching granular sphene.

Rare Cofacial Rocks

A rare associate of eclogite fragments in diamond-bearing kimberlite
breccias of Yakutia, Siberia, is *grospydite*, a calcic rock possibly cofacial
with eclogite. It consists of garnet, omphacite, and kyanite; and the garnet proves to be a uniquely magnesian grossular containing about equal
amounts (~17% each) of pyrope and almandine.

   Even more remarkable are granitic rocks closely associated with eclogite in remnants of the hercynian basement that have been preserved in
the Seisia-Lanzo zone of the Alps in the Italian Piedmont. These rocks
still retain typically granitic textures including perthitic microstructures
and now consist of a unique assemblage of minerals; jadeite and zoisite
have completely replaced plagioclase; phengite and magnesian garnet

have replaced biotite. The complete metamorphic assemblage is quartz-jadeite-zoisite-phengite-garnet–K-feldspar. One can only conclude that the initially high-level crustal rocks, not withstanding their low density, have been depressed to upper-mantle levels, and then have returned unchanged during subsequent presumably rapid ascent to the surface.

OVERALL MINERALOGICAL PATTERN IN RELATION TO METAMORPHIC FACIES

We conclude this chapter, like the preceding one, with a synoptic review (Table 20–1) of the overall mineralogical pattern of high-grade region-ally metamorphic rocks in relation to pertinent metamorphic facies. The reader is again reminded that coverage is by no means exhaustive. We include only selected, commonly recorded mineral assemblages that are characteristic of each facies; accessory and unessential phases are omit-ted, as also are minor variations in paragenesis. It is appropriate to end our review of metamorphic petrography on this note; for the key to quantitative interpretation of petrographic data lies in the concept of facies.

GENERAL REFERENCE

A. Harker, *Metamorphism* (London: Methuen, 1932).

REFERENCE NOTES

[1] For references to pertinent literature and for summaries of the findings of these and other Swiss geologists, see F. J. Turner, *Metamorphic Petrology*, 2d ed. (Washington, D.C.: Hemisphere, 1981).

[2] P. G. Cooray, Geological Survey of Ceylon, Memoir no. 2 (1961); P. G. Cooray, *Quarterly Journal of the Geological Society of London*, vol. 118 (1962): pp. 239–273; D. J. A. C. Hapuarachchi, Geological Survey Department of Sri Lanka, Professional Paper no. 4 (1975).

APPENDIX A

Elementary Treatment of Mutual Relationships Between Some Critical Thermodynamic Functions

FUNDAMENTAL SIGNIFICANCE OF MACROSCOPIC MEASUREMENTS

It is emphasized at the outset of this discussion that classic thermodynamics, as developed particularly by J. Willard Gibbs, was based purely on macroscopic measurements and experiments. Relevant data collected and set out in standard sets of thermodynamic tables likewise are based on macroscopic observations made in the laboratory over accessible ranges of temperature T and pressure P. These refer to primary thermodynamic functions of any system—internal energy E, volume V, and entropy S—and some additional functions contrived by combining some of these, along with P and T, with a view to providing the means for solving fundamental problems relating to stability of and equilibrium within systems, and reversibility of postulated phase transformations under prescribed conditions of temperature and pressure. These additional functions include enthalpy H and Gibbs free energy G. Thermodynamic argument[1] concerns variation in each of these functions in specific systems in response to finite changes in P and T.* The requisite

---

*Such reasoning is used chiefly in parts of Chapters 2 and 16.

data for present purposes are laboratory measurements of heat capacity and volume of phases of petrologic interest—minerals and a few fluids (notably H_2O and CO_2) of simple composition at temperatures up to 1000°–1500°C and pressures up to a few kilobars.

HEAT CAPACITY, ENTHALPY, AND ENTROPY
IN RELATION TO TEMPERATURE AT CONSTANT PRESSURE

To illustrate the nature and mutual relationships of certain important thermodynamic functions, discussion designed for the novice in the field will be confined in this section to a single phase, pure albite at atmospheric pressure. An unusually complete set of thermal measurements over the temperature range 15K to 370K (−258.15°C to 96.85°C) is available through the work of R. E. Openshaw and coauthors.[2]

Heat Capacity

Heat capacity of any phase, C_P, measured at constant pressure, is defined as the quantity of heat q required to raise 1 mole (gram-formula weight)—in this case 262.24 g—through a temperature increment of 1K from any initial temperature T. It has dimensions of energy, usually expressed as calories, or joules, per degree Kelvin. Thus, by definition, at any specified temperature and pressure

$$C_P = dq/dT \tag{1}$$

The value of C_P of any phase increases from zero at 0K (−273.15°C) along a sigmoid curve as shown in Figure A-1*. Note that *heat* itself is a form of energy that is measured in a state of flux down a gradient of diminishing temperature.

---

*Figures A-1 to A-4 are based on selected C_P values for low albite (converted from joule deg^{-1} to cal deg^{-1} taken from the data of Openshaw et al. (1976) and set out in Table A-1, column 3.

Enthalpy

Enthalpy, H, is an energy function empirically defined in general terms as

$$H = E + PV \tag{2}$$

It is easily shown that *at constant pressure* this equation reduces to a simple one involving only temperature and a quantity of heat q, that is,

$$\frac{dH}{dT} = \frac{dq}{dT} = C_P \tag{3}$$

or for any finite increment ΔH over an interval of increasing temperature $\Delta T = (T_2 - T_1)$

$$\Delta H = \int_{T_1}^{T_2} C_P \, dT \tag{3A}$$

Graphic solution for the interval 0K to 298.15K (= room temperature, 25°C) is illustrated in Figure A-1, giving a value $\Delta H = (H_{298.15}, - H_0)$ = about 7996 cal mole^{-1}. Only differences in enthalpy, ΔH, not H itself, can be measured calorimetrically. Since the value of H necessarily increases with temperature (at constant pressure) H_0, the value of H at 0K, must be the lower limit of enthalpy of any specific phase at given pressure; but it cannot be assigned any numerical value. So there is no way in which the values of H_0 of any two different phases, say quartz and microcline, can be directly compared. Table A-1 (column 4) gives computed values of $\Delta H = (H_T - H_0)$ for low albite at temperatures T at atmospheric pressure.

Entropy

Definition

The *molar entropy* of a phase at temperature T and some specified pressure is defined empirically by the equation

$$dS = \frac{dq}{T} = \frac{C_P \, dT}{T} = C_P \, d\ln T \tag{4}$$

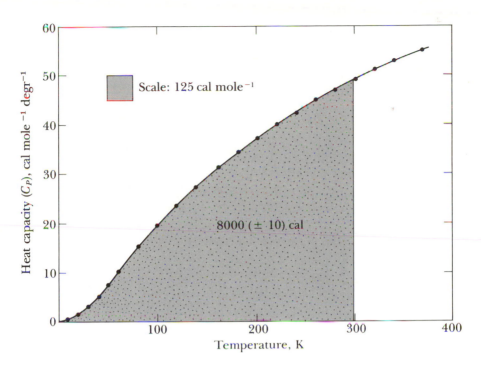

Figure A–1. Heat Capacity, C_P, of Low Albite in Relation to T (in kelvins)

Data source: Openshaw et al. (1976); see Table A–1. Increments in H from 0K to T, $\Delta H = (H_T - H_0)$ in Table A–1 are expressed by areas beneath the curve to left of T (stippled for $T = 298.15°K$): $\Delta H_{298.15} = 7996$ cal mole^{-1}.

whence

$$S_T - S_0 = \int_0^T C_P \, d\ln T \qquad (4A)$$

Graphic solution of this equation for various temperatures in the range 0K–370K is illustrated in Figure A-2 and a set of S values computed from equation 4A is tabulated in column 5 of Table A-1. To retain sensitivity of plotting it is convenient to employ different C_P scales for different intervals of temperature; for example, from 50° to 100°C (Figure A-3A) and from 150°C to 300°C (Figure A-3B). The third law of thermodynamics postulates that the entropy of a perfect crystal at 0K is itself

Table A-1. Thermodynamic Properties of Low Albite at $P = 1$ Bar. (Data, converted to Calories, from Openshaw et al., *U.S. Geological Survey Journal of Research* 4[1976]: 195–204.) All quantities reckoned from zero values at 0K.

Temperature T, (in kelvins)	$\ln T = \log_e T$	Heat Capacity C_P, cal mole^{-1} (Measured Calorimetrically)	Enthalpy $\Delta H = (H_T - H_0)$ cal mole^{-1}	Entropy S, cal mole^{-1} degr^{-1}	Free Energy $\Delta G = (G_T - G_0)$ cal mole^{-1}
10	2.30	0.18	0.4	0.05	−0.1
15	2.71	0.57	1.95		
20	3.00	1.26	6.0	0.40	−2.0
25	3.22	2.14	14.0		
30	3.40	3.14	27.0	1.22	−9.6
40	3.70	5.40	69.2		
50	3.91	7.84	135.0	3.88	−59
60	4.10	10.30	225.6		
70	4.25	12.72	340.2	7.92	−214
80	4.38	15.06	480		
90	4.50	17.34	648	11.04	−346
100	4.61	19.54	825		
110	4.70	21.66	1,030	14.94	−613
120	4.79	23.69	1,258	16.90	−770
140	4.94	27.51	1,770	20.85	−1,149
160	5.08	31.02	2,355	24.76	−1,606
180	5.19	34.27	3,010	28.6	−2,138
200	5.30	37.26	3,724	32.36	−2,748
220	5.39	40.00	4,498	36.04	−3,431
240	5.48	42.57	5,323	39.65	−4,193
260	5.561	45.43	6,198	43.14	−5,018
280	5.635	47.16	7,120	46.56	−5,916
300	5.703	49.20	8,088	49.88	−6,876
320	5.768	51.17	9,085	53.13	−7,917
340	5.829	53.06	10,132	56.29	−9,007
360	5.886	54.66	11,210	59.37	−10,163
370	5.914	55.50	11,762	60.87	−10,760

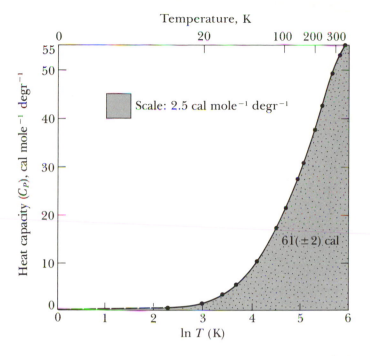

Figure A–2. Heat Capacity C_P of Low Albite in Relation to $\ln T$ (in kelvins)

Data source: Openshaw et al. (1976); see Table A–1. Molar entropy S_T for $T =$ 370K is expressed by the stippled area (60.87 cal mole^{-1} degr^{-1}).

zero.* Assuming that low (perfectly ordered) albite indeed fulfills the requirement of crystalline perfection, S in Table A-1 and in corresponding figures (Figures A-2 and A-3) is extrapolated to zero at 0K. Thus absolute values of entropy at any temperature can be assigned to all perfectly crystalline phases; and these values, unlike those of H, are mutually comparable.

Heat of Reaction in Relation to Entropy

To utilize values of S and H in assessing phase equilibria still requires another type of macroscopic measurement. This is *heat of reaction*, ΔH_r,

---

*But one may ask, What is a perfect crystal? To be more specific, ordered microcline and disordered sanidine cannot both be "perfect" forms of $KAlSi_3O_8$. Some correction must be applied for disordered phases, some of which (like sanidine) remain so—as metastable phases—to temperatures close to 0K. This aspect of entropy, though in some cases important, will be neglected in our elementary treatment.

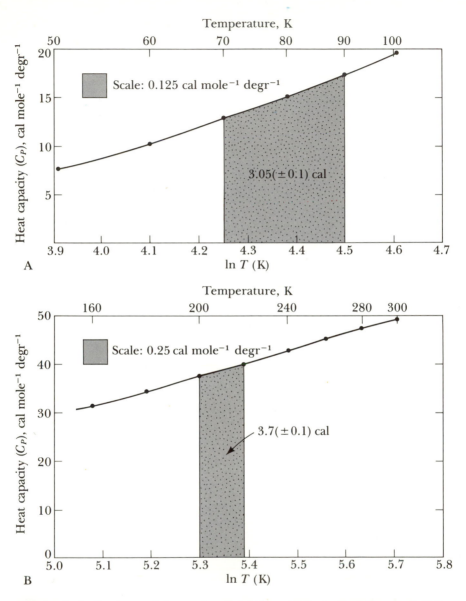

Figure A–3. Sections of Curve for C_P of Low Albite in Relation to $\ln T$ (in kelvins)

Data source: Openshaw et al. (1976); see Table A–1. Stippled areas show ΔS for two temperature intervals

A. 70–90K: 3.12 cal mole^{-1} degr^{-1};

B. 200–220K: 3.68 cal mole^{-1} degr^{-1}.

defined as the quantity of heat, q, absorbed during conversion of one phase or simple phase assemblage A to another B of identical bulk chemical composition—that is, $A + \Delta H_r \rightarrow B$. For endothermic reactions such as the melting of ice to water (ice $+ q \rightarrow$ water) ΔH_r is positive. For exothermic reactions such as hydration of periclase to brucite [MgO + $H_2O \rightarrow Mg(OH)_2 + q$] ΔH_r is negative. Values of ΔH_r for different transformations are directly comparable. This permits indirect simple computation of ΔH_r for a complex reaction by combining values already available for still simpler reactions involving phases in common. For example, at atmospheric temperature (298.15K) and pressure (1.013 bar) the following ΔH_r values are available for formation of the following phases from their component elements:

calcite: $-288,592$ cal mole^{-1}
quartz: $-217,650$ cal mole^{-1}
wollastonite: $-390,640$ cal mole^{-1}
carbon dioxide: $-94,051$ cal mole^{-1}

Remembering that ΔH_r = enthalpy of products B minus enthalpy of reactants A at given P and T, then ΔH_r at the same temperature and pressure, for calcite + quartz $\rightarrow$ wollastonite + CO_2 = $(-390,640 - 94,051 + 288,592 + 217,650)$ = $+21,551$ cal.

Since entropy values of all phases are directly comparable, entropy of reaction, ΔS_r, at any given P and T, can be similarly but more simply computed as entropy of products minus entropy of reactants.

Significance of Entropy

Our discussion inevitably at first raises questions such as these: For what purpose was entropy, an empirical quantity, defined? What role does entropy play in evaluating mineral equilibria? What does it signify regarding some general and more comprehensible property of individual phases?

In the first place wide laboratory experience of thermal changes accompanying spontaneous phase transformations such as freezing of water undercooled to $-1°C$ shows that $(\Delta H_r - T\Delta S_r)$ invariably has a negative sign: that is, $T\Delta S_r > \Delta H_r$. All such reactions are *irreversible*. By contrast, for any truly *reversible* theoretical reaction, such as, ice $\rightleftharpoons$ water at 0°C and atmospheric pressure, ΔH_r and $T\Delta S_r$ have identical values. Here is an unequivocal criterion of equilibrium between two equally stable alternative states of any system at constant pressure and temperature as discussed on p. 464.* The term TS, as will be seen later, is essential to formulation of a function ΔG_r—change in free energy G—that is used to define stability and equilibrium with reference to both temperature and pressure; and at any fixed P and T, ΔG_r is actually $(\Delta H_r - T\Delta S_r)$.

Second, it is soon obvious from general experience that transformations (reversible or irreversible) for which ΔS is positive—melting of ice, vaporization of water, dehydration of brucite, heating of copper, mixing of oxygen and nitrogen, and a host of others—*invariably* lead to an increased state of disorder within the system. This refers of course to relative orientation, situation, and individual motions of the myriad of component atoms, ions, or molecules. Thus, it is clear that entropy of a system, a *macroscopically* defined quantity, in some way is closely connected with the state of internal disorder on the *microscopic* scale. The precise relationship can be quantified only through the science of statistical mechanics, on a level of sophistication far beyond that of our present elementary treatment.

It may be noted in conclusion, however, that in the late part of the last century, Boltzmann proposed a simple relationship expressed by the equation,

$$S = k \ln W \tag{5}$$

The universal constant k that bears Boltzmann's name has dimensions of energy 1.3805×10^{-16} erg degr^{-1} and is directly derived from the gas constant R, but refers to individual particles instead of to tangible molar quantities. In other words, the role of k in microscopic computation is comparable to that of R in macroscopic computation. The number W expresses the *probability* of the system—the multitude of accessible states or complexions of a system—taking into account relative positions and velocities of all component particles consistent with macroscopically prescribed constraints upon the system. W cannot be precisely evaluated except in some very simple systems, and then only through very sophisticated statistical treatment. But the enormous magnitude of W can be appreciated by the very small magnitude of k, itself an inverse function

---

*The basic relations between entropy and heat originally proposed with this end in view are these: for any *reversible* change resulting from a small increment of heat dq to a system at constant P and T, $dq = TdS$ (and for any *irreversible* change $dq < TdS$). From this postulate (the second law of thermodynamics) the definitive equations 4 and 4A logically follow.

of the Avogadro number $N_0 = 6.0226^{23}$ mole^{-1}. Yet even the beginner can appreciate that some added limited element of disorder—such as random distribution of three Si^{4+} and one Al^{3+} among four available sites in the unit cell of $KAlSi_3O_8$—might be evaluated numerically. Indeed, this and similar contributory factors have been computed for specific compounds and applied to correct S values macroscopically measured on one of two polymorphs related by order $\rightleftharpoons$ disorder transitions.

We emphasize, in conclusion, that entropy of any mineral phase is a function that is computed from calorimetric measurement. The important notion of entropy as a precise measure of relative disorder to that extent is secondary. However, just as in atomic physics, the true significance of the macroscopically defined quantity is revealed only through the microscopic approach. "Entropy" and "disorder" are by no means synonymous terms; and attempts to apply the concept of entropy to social, geomorphic, ecological, or similar problems—which have no connection with thermodynamics—are based on utterly false analogy.

GIBBS FREE ENERGY

Definition

The *free energy* of a system, G (cal mole^{-1}), is a thermodynamic function that appears to the beginner to be even more abstract than entropy. It is defined for any given temperature T and pressure P by the empirical equation

$$G = E + PV - TS \tag{6}$$

E, the internal energy of the system, cannot be evaluated absolutely in numerical terms, so the same applies also to G—as to energy functions except S. But relative values of G, measured from some convenient reference datum where G is arbitrarily assigned zero value, can indeed be computed from macroscopic measurements of heat capacity and volume. The resulting numbers, as we shall presently see, are essential to thermodynamic calculations relating to equilibrium and relative stability; and the function G itself was devised with that end in view.

Values of ΔH and ΔG for low albite, as cited in Table A-1, are referred to a common zero point at atmospheric pressure and 0K. The standard state selected in most data tables is atmospheric temperature and pressure—298.15K, 1.013 bar; so to make tabulated values of G and H mutually comparable the functions that are commonly employed are ΔH_f° and ΔG_f° signifying heats and free energies of formation of any specified phase from its component elements in the standard state. How ΔH_f° and ΔG_f° are computed will be discussed shortly.

Free Energy Changes

Any infinitesimal change dG can be evaluated in terms of equation 6 as

$$dG = dE + PdV + VdP - TdS - SdT \tag{7}$$

To get rid of E we may substitute in equation 6 a relationship that follows directly from the first law of thermodynamics, namely, at constant P and T

$$dE = TdS - PdV \tag{8}$$

where $-PdV$ is work done on the system, and TdS is the heat absorbed by it in the course of any small change in E.

Substitution for dE reduces equation 7 to

$$dG = VdP - SdT \tag{9}$$

Changes in G at Constant Pressure

For any system at constant pressure equation 9 reduces further to

$$dG = -SdT \tag{10}$$

or

$$\left(\frac{\delta G}{\delta T}\right)_P = -S \tag{11}$$

By integration of equation 11 over any desired temperature range $\Delta T = (T_2 - T_1)$ we have

$$\Delta G = -\int_{T_1}^{T_2} SdT \tag{12}$$

If T_1 is 0K (here chosen as the zero point for reckoning ΔG values in Table A-1) equation 12 yields

$$G_T = - \int_0^T S dT \tag{13}$$

where G_T is the free energy value at any temperature T.

Graphic integration of equation 13 for low albite at temperatures between 0K and 370K is illustrated in Figure A-4. Representative values (cf. Table A-1) are

$$
\begin{aligned}
G_{298.15°K} &= -6780 \text{ cal mole}^{-1} \\
G_{200°K} &= -2748 \text{ cal mole}^{-1} \\
\Delta G_{200°K \rightarrow 300°K} &= -4128 \text{ cal mole}^{+1}
\end{aligned}
$$

Note that, since S is a positive quantity, equation 10 stipulates that G diminishes with increasing temperature at constant pressure.

Since values of H, S, and G are all derived from the same C_P values, then solution of equation 13 becomes redundant provided H and S values have already been so computed and referred to the same zero temperature datum, 0K. Values of G can then be computed from this simple and important relation (obtained by combining equations 6 and 2):

$$G = H - TS \tag{14}$$

G values so computed for low albite (referred to zero at 0K) are tabulated as ΔG in column 6 of Table A-1. Within limits of error in plotting, these numbers are identical with those obtained by graphic solution of equation 13 in Figure A-4.

Equations analogous to 10 through 14 above similarly express relations between ΔG_r, ΔS_r, and ΔH_r in response to variation in T at constant P and vice versa. These provide the means for extrapolating values of ΔH_f and ΔG_f from temperatures and pressures convenient for experimental determination of ΔH_f to any standard P and T. Values of $\Delta H_f°$ and $\Delta G_f°$ so computed and cited in standard tables are large negative numbers subject, because of difficulties in high-temperature calorimetry, to considerable uncertainty. Thus molar values given for albite at $P = 1.013$ bar and $T = 25°C$ (298.15K) by R. A. Robie[3] are:

$$
\begin{aligned}
\Delta H_f° &= -35,900 \pm 1,500 \text{ cal} \\
\Delta G_f° &= -37,530 \pm 1,800 \text{ cal}
\end{aligned}
$$

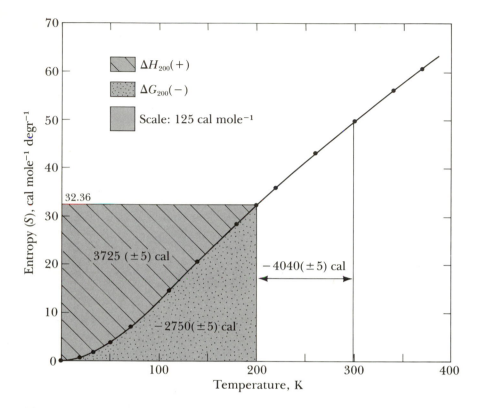

Figure A–4. Molar Entropy S of Low Albite in Relation to T (in kelvins) at $P = 1$ bar

Data source: Openshaw et al. (1976); see Table A–1. Decreases in free energy from 0K to T, tabulated in Table A–1 as $\Delta G = (G_T - G_0)$, are expressed by corresponding areas beneath the curve to the left of T: $\Delta G_{200} = -2750 \pm 5$ cal mole^{-1}; $\Delta G_{298.15} = -6790 \pm 10$ cal mole^{-1}. For significance of hatched area see p. 591.

Changes in G at Constant Temperature

From equation 10 the free energy of one mole of any phase at constant temperature T increases with pressure by an amount ΔG given by

$$\left(\frac{\delta G}{\delta P}\right)_T = V \tag{15}$$

from which

$$\Delta G = \int_{P_1}^{P_2} V dP \tag{16}$$

Since gaseous phases are much more compressible and ΔV values for any prescribed ΔP interval are correspondingly much larger for gaseous than for comparable crystalline phases, we have chosen steam at 500°C and at 800°C to illustrate relationships stipulated in equations 15 and 16 (Figures A-5, A-6). Relevant data have been taken from standard tables

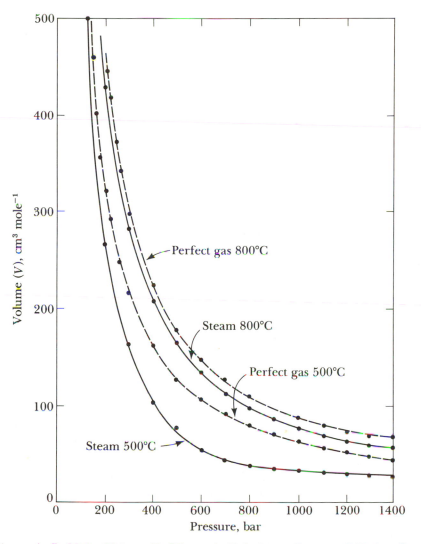

Figure A–5. Molar Volume V of Steam in Relation to Pressure P(Bar) at Constant T = 800°C and at Constant T = 500°C

Data source: Burnham, Holloway, and Davis (1969). Dashed curves are for perfect gas.

compiled by C. W. Burnham, J. R. Holloway, and N. F. Davis.[4] Graphic solutions of equation 16 for steam at 800°C and at 500°C over the interval 500–1000 bar are given in Figure A-6A, B (stippled areas).

The experimentally established equation of state for a *perfect gas* is

$$\frac{PV}{T} = R \tag{17}$$

where R is a universal constant 83.147 bar cm^3 (= 8.3147 joules = 1.987 cal) mole^{-1}. For T 800°C (1073.15K), $V = (83.147 \times 1073.15)/P = 89229/P$. Plots for V against P at 500°C and at 800°C are shown as dashed curves in Figure A-5. Note how at these high temperatures and over a span of low pressure (< 400 bar) the behavior of steam approximates that of a perfect gas. The same is true of gases in general.

Significance of PV and TS terms in ΔG

The terms PV and TS both appear in equations relating, respectively, to internal energy E (e.g., equation 8) and to Gibbs free energy G (e.g., equation 6). In the former case $-PV$ expresses work w done on the system, and TS heat q absorbed by it during some prescribed finite change of state. But in free-energy equations the same terms have no implications regarding either work or heat. Any confusion that may arise in the mind of the reader merely stems from use of alternative units of energy that of themselves have no implications as to specific form of energy— work, heat, or something else. These several energy units, it must be remembered, are completely interconvertible: 1 cal = 4.184 joules = 41.84 bar cm^3. So the particular units that emerge from simple computation (e.g., from equations 12 and 16) do not necessarily indicate the paths by which the corresponding energy increments were reached. The nature of the path is revealed only by the equations by which the numbers in question were derived.

To assess the values of ΔG relating to energy changes that accompany changes within the system from state 1 to state 2, we have already found it convenient to separate isothermal from isobaric changes of state—at constant T and P, respectively. G was defined in the first place to make this possible. Below we recapitulate in synoptic form some of these changes in the G function and compare them with changes in E over the same intervals of P or T. Equations are expressed in terms of molar quantities, and illustrative examples are drawn once more from a system consisting of one mole of a pure phase—steam or low albite.

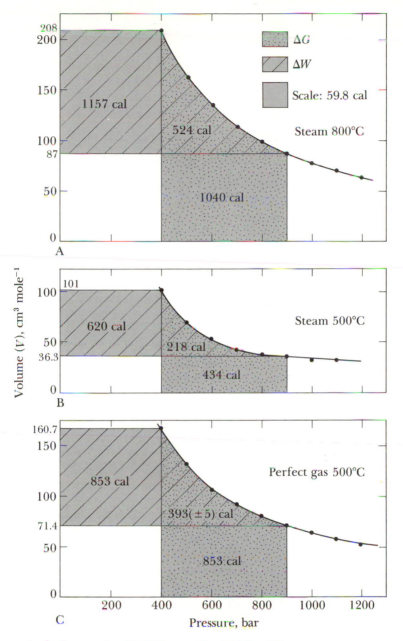

Figure A–6. Segments of *P-V* Curves for 1 Mole of Gas

A. Steam at 800°C

B. Steam at 500°C

C. Corresponding curve for perfect gas at 500°C.
Stippled areas represent increases in free energy over the presure interval 400
bar → 500 bar. Hatched areas represent increments in E due to work done on
the system during accompanying compression from $V_{400\,bar}$ to $V_{900\,bar}$.

1. *Isothermal changes*, with pressure increasing from P_1 to P_2
Work increments in E:

$$dw = -PdV$$

and for an accompanying finite volume change $\Delta V = (V_2 - V_1)$, sign negative, the work increment Δw is positive

$$\Delta w = -\int_{V_1}^{V_2} PdV \tag{18}$$

Increments in G (cf. equations 15 and 16):

$$dG = VdP$$

and for an accompanying finite pressure increment $\Delta P = (P_2 - P_1)$, sign positive

$$\Delta G = \int_{P_1}^{P_2} VdP \tag{19}$$

also a positive quantity.

2. *Isobaric changes*, with temperature increasing from T_2 to T_2
Heat increments in E:

$$dq = dH = TdS$$

and for an accompanying finite entropy increment ΔS, sign positive

$$\Delta H = \int_{T_1}^{T_2} TdS \tag{20}$$

Changes in G (equations 11 and 12):

$$dG = -SdT$$

and for a finite temperature increment $\Delta T = (T_2 - T_1)$, sign positive, the change in G is negative;

$$\Delta G = -\int_{T_1}^{T_2} SdT \tag{21}$$

Data for isothermal changes in V for pure steam at 500°C and 800°C over the pressure span 400–900 bar (from Table A-2) have been plotted in Figure A-6A,B. And graphic solutions of equations 18 and 19 are illustrated by the shaded areas—stippled for ΔG, hatched for Δw (sign positive). Total energy expressed by each of these shaded sectors in both diagrams is given, in calories, by the number enclosed in each. The two diagrams bring out two distinctions of general validity between the PV terms that appear in equations relating respectively to E and to G: Δw and ΔG, each expressed in the first instance as PV (bar cm^3) differ in magnitude, but have the same sign.

at 500°C, for ΔP = $(900 - 400)$ bar
and ΔV = $(36.3 - 101.2)$cm^3
ΔG = 652 cal mole^{-1}
Δw = 838 cal mole^{-1}

Table A-2. Pressure-Volume Relations for Steam and for Perfect Gas at 500°C and 800°C over Pressure Span 200–1400 Bar (Data for steam from Burnham, Holloway, and Davis, *Geological Society of America Special Paper*, no. 132 (1969). Selected ΔG Values Added.

Pressure Bar	*Volume, cm^3 mole^{-1}*			
	Steam		*Perfect Gas*	
	500°C	*800°C*	*500°C*	*800°C*
200	266.1	429	322	446
300	156.5	281	214	298
400	101.2	208	160.7	223
500	70.0	163	129	179
600	53.1	135	107	149
700	44.5	114	92	128
800	39.5	99	81	112
900	36.3	87	71.4	99
1000	34.1	78	65	89
1100	32.4	71	59	81
1200	31.2	65	54	75
1300	30.1	60	50	69
1400	29.3	57	46	64

ΔG 400–900 bar: steam 500°C, 652 cal mole^{-1}
steam 800°C, 1564 cal mole^{-1}

Perfect Gas, 500°C: ΔG = $RT \ln (900/400)$
= $RT \ln (160.7/71.4)$
= 1246 cal mole^{-1}

$$
\begin{aligned}
\text{At } 800°\text{C, for } \Delta P \;&=\; (900 - 400)\ \text{bar} \\
\text{and } \Delta V \;&=\; (87\text{-}208)\ \text{cm}^3 \\
\Delta G \;&=\; 1564\ \text{cal mole}^{-1} \\
\Delta w \;&=\; 1681\ \text{cal mole}^{-1}
\end{aligned}
$$

It is obvious, too, that while the PV term derived from equation 18 represents merely the work increment in the E function, that derived from equation 19 is the total increment ΔG in free energy and is not to be equated with work.

In one respect only there is a unique exception to the above generalizations regarding isothermal changes in energy functions; and this is provided by a perfect gas. From the corresponding equation of state (equation 17), at constant temperature T

$$
P_1 V_1 = P_2 V_2 = RT \text{ (a constant)}
$$

and

$$
\frac{P_1}{P_2} = \frac{V_2}{V_1} \tag{22}
$$

Figure A-6C shows graphic solutions for equations 18 (hatched sector) and 19 (stippled sector) for a perfect gas over the pressure span $400-900$ bar at 500°C. From equation 22 the two simply shaded rectangular sectors are equal in area (both representing 853 cal mole^{-1}). For this unique case, then, Δw and ΔG are of equal magnitude (1246 cal mole^{-1}); and both have the same sign (positive);* ΔG carries no connotation of work done.

In a similar fashion, it is easy to show that in isobaric changes of state the two accompanying changes in energy functions ΔH and ΔG differ

---

*For a perfect gas at constant T,

$$
\begin{aligned}
PV &= RT \text{ (a constant)} \\
d(PV) &= PdV + VdP = d(RT) = 0
\end{aligned}
$$

so

$$
dG = VdP = -PdV = dw
$$

Also, from equation 17

$$
dG = VdP = \frac{RT}{P}dP = RTd\ \ln P
$$

For a pressure increment $\Delta P = (P_2 - P_1)$ and accompanying volume decrease $(V_2 - V_1)$

$$
\begin{aligned}
\Delta G &= RT\ (\ln P_2 - \ln P_1) = RT \ln (P_2/P_1) \\
&= RT \ln (V_1/V_2) = \Delta w
\end{aligned}
$$

both in magnitude and sign as well as in heat implications. In Figure A-4 the stippled and hatched areas, respectively, express graphic solutions of equations 20 and 21 for low albite at atmospheric pressure for the temperature increment $\Delta T = 0\text{K} - 200\text{K}$ and corresponding entropy increment $\Delta S = 32.36$ cal mole^{-1} (both reckoned from the common zero point 0°K). From Table A-1:

$$\Delta H = 3724 \text{ cal mole}^{-1}$$
$$\Delta G = -2748 \text{ cal mole}^{-1}$$

As required by equation 14 the area of the complete shaded rectangle TS equals $\Delta H - \Delta G = 6472$ cal.

Free Energy of Reaction ΔG_r at Constant Temperature and Pressure

Any postulated phase transformation (reaction), $A \rightarrow B$ within a system of given composition at some temperature T and pressure P of interest is accompanied by a finite change in free energy ΔG_r. This quantity $(G_B - G_A)$, by analogy with the definition of free energy at constant P (equation 14), may be written for an isobaric transformation

$$\Delta G_r = \Delta H_r - T\Delta S_r \tag{23}$$

and it may be extrapolated isothermally to any desired higher pressure (at the same temperature) by an equation analogous to equation 16

$$\Delta G_r \text{ increment} = \int_{P_1}^{P_2} \Delta V_r \, dP \tag{24}$$

Similarly, for an isobaric transportation at pressure P by analogy with euqation 12

$$\Delta G_r = -\int_{T_1}^{T_2} S \, dV \tag{25}$$

The Gibbs free-energy function at any given P and T provides the unequivocal criterion of the sense in which any postulated reaction may proceed (spontaneously and irreversibly): ΔG_r must, without exception, be negative.

Thus, for any spontaneous (irreversible) transformation $A \rightarrow B$, from left to right at constant P and T, by analogy with equation 14,

$$\Delta G_r = (\Delta H_r - T\Delta S_r) < 0 \tag{26}$$

Otherwise expressed, neglecting signs of the two terms involved,

$$T\Delta S_r > \Delta H_r \tag{26A}$$

For equilibrium between the two alternative phase assemblages, $A \leftrightarrows B$ at given P and T

$$\Delta G_r = 0 \tag{27}$$

and

$$T\Delta S_r = \Delta H_r \tag{27A}$$

The reaction is reversible, and occurs only if a small quantity of heat enters or leaves the system.

There should be no need to stress further the potentiality of the Gibbs free-energy function in solving problems relating to mineralogical transformations, the direct evidence of which is revealed by the petrographic microscope. Only through an appreciation of the nature of free energy can petrographers interpret the evidence they record for microscopically recognizable phase transformations and also discuss matters relating to equilibrium or relative stability in mineral assemblages.

REFERENCE NOTES

[1]The level of present treatment is extremely elementary. For lucid elaboration of the topics treated here, as they may be applied to largely crystalline phase assemblages such as igneous and metamorphic rocks and products of sedimentary diagenesis, the reader is referred to the outline given by J. Verhoogen in *Igneous and Metamorphic Petrology*, F. J. Turner and J. Verhoogen (New York: McGraw-Hill, 1960), pp. 6–49, and a more comprehensive exposition by M. W. Zemansky and H. C. Van Ness, *Basic Engineering Thermodynamics* (New York: McGraw-Hill, 1966).

[2]R. E. Openshaw et al., *U.S. Geological Survey Journal of Research*, vol. 4 (1976): pp. 195–204.

[3]R. A. Robie, *Geological Society of America Memoir*, vol. 97 (1966): pp. 437–458.

[4]C. W. Burham, J. R. Holloway, and N. F. Davis, *Geological Society of America Special Paper*, no. 132 (1969).

APPENDIX B

Models for Visual Estimation of Percentage Composition of Rocks

Reprinted with permission from Society of Economic Paleontologists and Mineralogists; originally published by R. D. Terry and G. V. Chilingar, "Summary of 'Concerning some additional aids in studying sedimentary formations by M. S. Shvetsov, ' " *Journal of Sedimentary Petrology,* no. 25 (1955): pp. 229–234.

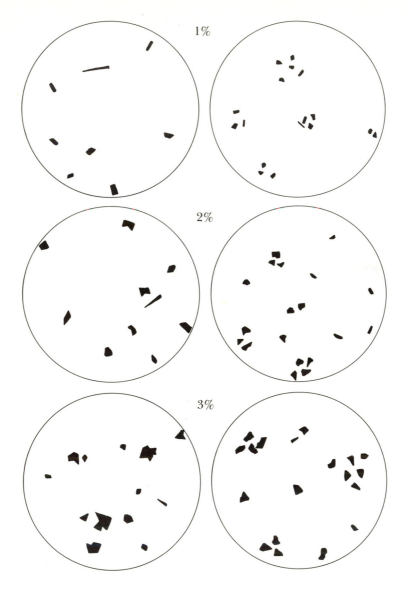

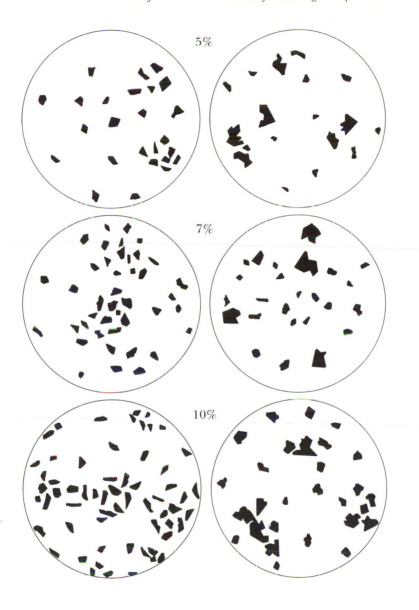

5%

7%

10%

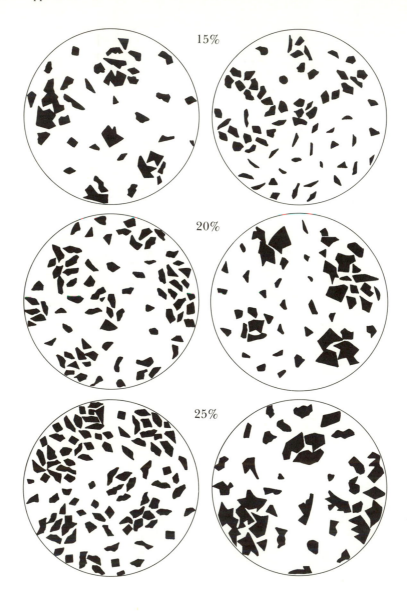

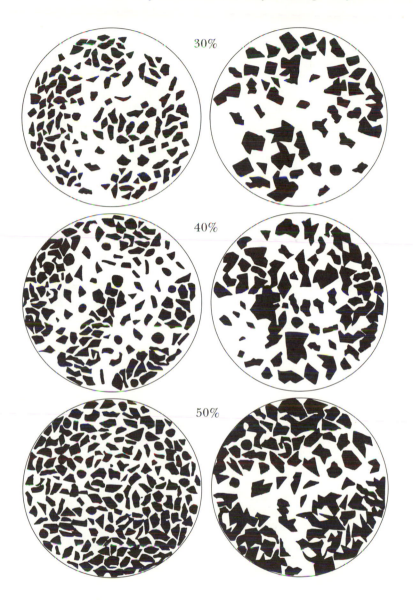

30%

40%

50%

Index of Authors

Index of Localities

References to illustrations in italics.

Index of Subjects

Page references to illustrations in italics.